Lecture Notes in Computer Science 6158

Commenced Publication in 1973
Founding and Former Series Editors:
Gerhard Goos, Juris Hartmanis, and Jan van Leeuwen

Editorial Board

David Hutchison
Lancaster University, UK
Takeo Kanade
Carnegie Mellon University, Pittsburgh, PA, USA
Josef Kittler
University of Surrey, Guildford, UK
Jon M. Kleinberg
Cornell University, Ithaca, NY, USA
Alfred Kobsa
University of California, Irvine, CA, USA
Friedemann Mattern
ETH Zurich, Switzerland
John C. Mitchell
Stanford University, CA, USA
Moni Naor
Weizmann Institute of Science, Rehovot, Israel
Oscar Nierstrasz
University of Bern, Switzerland
C. Pandu Rangan
Indian Institute of Technology, Madras, India
Bernhard Steffen
TU Dortmund University, Germany
Madhu Sudan
Microsoft Research, Cambridge, MA, USA
Demetri Terzopoulos
University of California, Los Angeles, CA, USA
Doug Tygar
University of California, Berkeley, CA, USA
Gerhard Weikum
Max-Planck Institute of Computer Science, Saarbruecken, Germany

Fernando Ferreira   Benedikt Löwe
Elvira Mayordomo   Luís Mendes Gomes (Eds.)

Programs, Proofs, Processes

6th Conference on Computability in Europe, CiE 2010
Ponta Delgada, Azores, Portugal, June 30 – July 4, 2010
Proceedings

 Springer

Volume Editors

Fernando Ferreira
Universidade de Lisboa, Departamento de Matemática
Campo Grande, Edifício C6, 1749-016 Lisboa, Portugal
E-mail: ferferr@cii.fc.ul.pt

Benedikt Löwe
Universiteit van Amsterdam, Institute for Logic, Language and Computation (ILLC)
P.O. Box 94242, 1090 GE Amsterdam, The Netherlands
E-mail: b.loewe@uva.nl

Elvira Mayordomo
Universidad de Zaragoza, Departamento de Informática e Ingeniería de Sistemas
María de Luna 1, 50018, Zaragoza, Spain
E-mail: elvira@unizar.es

Luís Mendes Gomes
Universidade dos Açores, Campus de Ponta Delgada
Apartado 1422, 9501-801 Ponta Delgada, Açores, Portugal
E-mail: luis.mendesgomes@gmail.com

Library of Congress Control Number: 2010929188

CR Subject Classification (1998): F.2, F.1, G.2, F.3, I.1, G.2.2

LNCS Sublibrary: SL 1 – Theoretical Computer Science and General Issues

ISSN   0302-9743
ISBN-10   3-642-13961-2 Springer Berlin Heidelberg New York
ISBN-13   978-3-642-13961-1 Springer Berlin Heidelberg New York

springer.com

© Springer-Verlag Berlin Heidelberg 2010
Printed in Germany

Typesetting: Camera-ready by author, data conversion by Scientific Publishing Services, Chennai, India
Printed on acid-free paper   06/3180

Preface

CiE 2010: Programs, Proofs, Processes
Ponta Delgada, Azores, Portugal, June 30–July 4 2010

The last few years, starting in 2005 with out inaugural conference in Amsterdam, have seen a development from an informal cooperation via an increasingly established conference series to an association, founded in 2008. While the organization form of *Computability in Europe* (CiE) may have changed, the scientific scope is still the same and as interdisciplinary and innovative as it was six year ago when we held the first conference. CiE aims to promote computability-related science in its broadest sense, including mathematics, computer science, applications in various natural and engineering sciences (e.g., physics, biology, computer engineering), and also reaches out to meta-studies such as the history and philosophy of computing. Researchers at CiE conferences wish to advance our theoretical understanding of what can and cannot be computed, by *any* means of computation.

CiE 2010 was the sixth conference of the series, held in a geographically unique and dramatic location, Europe's most westerly outpost, at the University of Azores in Ponta Delgada, Portugal. The theme of CiE 2010 "Programs, Proofs, Processes" points to the usual CiE synergy of computer science, mathematics and logic, with important computability-theoretic connections to science and the real universe. Formal systems, attendant proofs, and the possibility of their computer generation and manipulation (for instance, into programs) have been changing a whole spectrum of disciplines. The conference addressed not only the more established lines of research of computational complexity and the interplay between proofs and computation, but also novel views that rely on physical and biological processes and models to find new ways of tackling computations and improving their efficiency.

The first five meetings of CiE were at the University of Amsterdam in 2005, at the University of Wales Swansea in 2006, at the University of Siena in 2007, at the University of Athens in 2008, and at the University of Heidelberg in 2009. Their proceedings, edited in 2005 by S. Barry Cooper, Benedikt Löwe, and Leen Torenvliet, in 2006 by Arnold Beckmann, Ulrich Berger, Benedikt Löwe, and

John V. Tucker, in 2007 by S. Barry Cooper, Benedikt Löwe, and Andrea Sorbi, in 2008 by Arnold Beckmann, Costas Dimitracopoulos, and Benedikt Löwe, and in 2009 by Klaus Ambos-Spies, Benedikt Löwe, and Wolfgang Merkle were published in Springer's *Lecture Notes in Computer Science*, Volumes 3526, 3988, 4497, 5028, and 5635, respectively.

CiE and its conferences have changed our perceptions of computability and its interface with other areas of knowledge. The large number of mathematicians and computer scientists attending these conferences had their view of computability theory enlarged and transformed: they discovered that its foundations were deeper and more mysterious, its technical development more vigorous, its applications wider and more challenging than they had known. The annual CiE conference has become a major event, and is the largest international meeting focused on computability theoretic issues. Future meetings in Sofia (2011, Bulgaria), and Cambridge (2012, UK) are in planning. The series is coordinated by the CiE Conference Series Steering Committee consisting of Arnold Beckmann (Swansea), Paola Bonizzoni (Milan), S. Barry Cooper (Leeds), Viv Kendon (Leeds), Benedikt Löwe (Amsterdam, Chair), Elvira Mayordomo (Zaragoza), Dag Normann (Oslo), and Peter van Emde Boas (Amsterdam).

The conference series CiE is very much interested in linking the computabilitty-related interdisciplinary research to the studies of philosophy and history of computing. This year, history of computing was represented by the special session "Reasoning and Computation from Leibniz to Boole," as well as several accepted regular papers (one of them in this volume).

The conference was based on invited tutorials and lectures, and a set of special sessions on a range of subjects; there were also many contributed papers and informal presentations. This volume contains 20 of the invited lectures and 31% of the submitted contributed papers, all of which have been refereed. There will be a number of post-conference publications, including special issues of *Annals of Pure and Applied Logic* and *Theory of Computing Systems*.

The Best Student Paper Award was given to Rebecca Steiner for her paper "Computable Fields and Weak Truth-Table Reducibility."

The tutorial speakers were Bruno Codenotti (Pisa) and Jeffrey Bub (Maryland).

The following invited speakers gave talks: Eric Allender (Rutgers), José L. Balcázar (Cantabria), Shafi Goldwasser (Weizmann Institute, MIT), Denis Hirschfeldt (Chicago), Seth Lloyd (MIT), Sara Negri (Helsinki), Toniann Pitassi (Toronto), and Ronald de Wolf (Amsterdam). Toniann Pitassi was the APAL lecturer.

Six special sessions were held:

Biological Computing. *Organizers:* Paola Bonizzoni (Milan) and Shankara Krishna (Mumbai).
Speakers: Nataša Jonoska, Giancarlo Mauri, Yasubumi Sakakibara, Stéphane Vialette.
Computational Complexity. *Organizers:* Luís Antunes (Porto) and Alan Selman (Buffalo NY).

Speakers. Eric Allender, Christian Glaßer, John Hitchcock, Rahul Santhanam.

Computability of the Physical. *Organizers:* Cris Calude (Auckland) and Barry Cooper (Leeds).
Speakers: Giuseppe Longo, Yuri Manin, Cris Moore, David Wolpert.

Proof Theory and Computation; *Organizers:* Fernando Ferreira (Lisbon) and Martin Hyland (Cambridge).
Speakers: Thorsten Altenkirch, Samuel Mimram, Paulo Oliva, Lutz Straßburger.

Reasoning and Computation from Leibniz to Boole: *Organizers:* Benedikt Löwe (Amsterdam) Guglielmo Tamburrini (Naples).
Speakers: Nimrod Bar-Am, Michèle Friend, Olga Pombo, Sara Uckelman.

Web Algorithms and Computation: *Organizers:* Thomas Erlebach (Munich) and Martin Olsen (Aarhus).
Speakers: Hannah Bast, Debora Donato, Alex Hall, Jeannette Janssen.

A special tribute to Professor Marian Pour-El was presented by Ning Zhong (Cincinnati). Marian Pour-El died on June 10, 2009 and was considered a leader in the field of computability and functional analysis, and its applications to physical theory.

The conference CiE 2010 was organized by Luís Antunes (Porto), Fernando Ferreira (Lisbon), Elisabete Freire (Ponta Delgada), Matthias Funk (Ponta Delgada), Hélia Guerra (Ponta Delgada), Luís Mendes Gomes (Ponta Delgada), and João Rasga (Lisbon).

The Program Committee was chaired by Fernando Ferreira and Elvira Mayordomo.

We are delighted to acknowledge and thank the following for their essential financial support: Universidade dos Açores, Direcção Regional da Ciência, Tecnologia e Comunicações, Associação para a Mobilidade Antero de Quental, Fundação para a Ciência e a Tecnologia, Direcção Regional do Turismo, Elsevier, The Elsevier Foundation, SQIG—Instituto de Telecomunicações, Centro de Matemática Aplicada e Tecnologias de Informação, Centro de Matemática e Aplicações Fundamentais.

We were proud to offer the program "Women in Computability" funded by the Elsevier Foundation as part of CiE 2010. The Steering Committee of the conference series CiE-CS is concerned with the representation of female researchers in the field of computability. The series CiE-CS has actively tried to increase female participation at all levels in the past years. Starting in 2008, our efforts have been funded by a grant of the Elsevier Foundation under the title *"Increasing representation of female researchers in the computability community."* As part of this program, we had another workshop, a grant scheme for female researchers, and a mentorship program.

The high scientific quality of the conference was possible through the conscientious work of the Program Committee, the special session organizers and the referees. We are grateful to all members of the Program Committee for their efficient evaluations and extensive debates, which established the final program. We also thank the referees.

We thank Andrej Voronkov for his EasyChair system which facilitated the work of the Program Committee and the editors considerably.

May 2010   Fernando Ferreira
                                               Benedikt Löwe
                                               Elvira Mayordomo
                                               Luís Mendes Gomes

Organization

Program Committee

Klaus Ambos-Spies (Heidelberg)
Luís Antunes (Porto)
Arnold Beckmann (Swansea)
Paola Bonizzoni (Milan)
Alessandra Carbone (Paris)
Steve Cook (Toronto ON)
Barry Cooper (Leeds)
Erzsébet Csuhaj-Varjú (Budapest)
Fernando Ferreira (Lisbon, Chair)
Nicola Galesi (Rome)
Rosalie Iemhoff (Utrecht)
Achim Jung (Birmingham)
Michael Kaminski (Haifa)
Jarkko Kari (Turku)
Viv Kendon (Leeds)

James Ladyman (Bristol)
Kamal Lodaya (Chennai)
Giuseppe Longo (Paris)
Benedikt Löwe (Amsterdam)
Elvira Mayordomo (Zaragoza, Chair)
Luís Mendes Gomes (Ponta Delgada)
Wolfgang Merkle (Heidelberg)
Russell Miller (New York NY)
Dag Normann (Oslo)
Isabel Oitavem (Lisbon)
João Rasga (Lisbon)
Nicole Schweikardt (Frankfurt)
Alan Selman (Buffalo NY)
Peter van Emde Boas (Amsterdam)
Albert Visser (Utrecht)

Referees

Andreas Abel
Michele Abrusci
Pedro Adão
Isolde Adler
Bahareh Afshari
Marco Antoniotti
Toshiyasu Arai
Jeremy Avigad
Serikzhan Badaev
Mohua Banerjee
Nimrod Bar-Am
Georgios Barmpalias
A. Baskar
Andrej Bauer
Julie Baussand
Eli Ben-Sasson
Sivan Bercovici
Ulrich Berger
Luca Bernardinello
Daniela Besozzi

Olaf Beyersdorff
Stephen Binns
Roberto Blanco
Guillaume Bonfante
Paolo Bottoni
Andrey Bovykin
Vasco Brattka
Stanley Burris
Lorenzo Carlucci
Rafi Chen
Ehsan Chiniforooshan
Peter Cholak
Matthias Christandl
Jennifer Chubb
José Manuel Colom
José Félix Costa
Vítor Santos Costa
David Cram
Barbara Csima
Vincent Danos

Valeria de Paiva
Gianluca Della Vedova
David Doty
Stewart Duncan
Matthias Ehrgott
Ali Enayat
Thomas Erlebach
Nissim Francez
Dominik Freydenberger
Pierluigi Frisco
Zsolt Gazdag
Clemens Grabmayer
Phil Grant
Floriana Grasso
Stefano Guerrini
Håkon Robbestad Gylterud
Rupert Hölzl
Katharina Hahn
Magnus M. Halldorsson
André Hernich
Denis Hirschfeldt
Mika Hirvensalo
Clare Horsman
Martin Hyland
Jouni Järvinen
Emil Jeřábek
Nataša Jonoska
Reinhard Kahle
Iskander Kalimullin
Lila Kari
Matty Katz
Akitoshi Kawamura
Thomas Kent
Bakhadyr Khoussainov
Laurence Kirby
Bjørn Kjos-Hanssen
Julia Knight
Pascal Koiran
Michal Kouchy
Thorsten Kräling
Walter Krämer
Laurence Laurence
Massimo Lauria
Hannes Leitgeb
Steffen Lempp

Joop Leo
Alberto Leporati
Peter Leupold
Bruno Loff
Judit X. Madarász
Andreas Malcher
Francisco Martins
Armando Matos
Alexander Melnikov
Russell Miller
Grigori Mints
Andrey Morozov
Anthony Morphett
Philippe Moser
M. Andrew Moshier
André Nies
Paulo Oliva
Martin Olsen
Yasser Omar
Michele Pagani
Paritosh Pandya
Soumya Paul
Volker Peckhaus
Mario Perez Jimene
Sylvain Perifel
Michalis Petropoulos
Giovanni Pighizzini
Alexandre Pinto
Valery Plisko
Carlo Proietti
Heiko Röglin
Charles Rackoff
William Rapaport
Uday Reddy
José Ignacio Requeno
Greg Restall
Sebastian Rudolph
Atri Rudra
Wojciech Rytter
Pavol Safarik
Ivano Salvo
Ute Schmid
Olivier Serre
Jeffrey Shallit
Amir Shpilka

Table of Contents

Avoiding Simplicity Is Complex

Eric Allender

Department of Computer Science, Rutgers University, Piscataway, NJ 08855
allender@cs.rutgers.edu

Abstract. It is a trivial observation that every decidable set has strings of length n with Kolmogorov complexity $\log n + O(1)$ if it has any strings of length n at all. Things become much more interesting when one asks whether a similar property holds when one considers *resource-bounded* Kolmogorov complexity. This is the question considered here: Can a feasible set A avoid accepting strings of low resource-bounded Kolmogorov complexity, while still accepting some (or many) strings of length n?

More specifically, this paper deals with two notions of resource-bounded Kolmogorov complexity: Kt and KNt. The measure Kt was defined by Levin more than three decades ago and has been studied extensively since then. The measure KNt is a nondeterministic analog of Kt. For all strings x, $\mathrm{Kt}(x) \geq \mathrm{KNt}(x)$; the two measures are polynomially related if and only if NEXP $\subseteq$ EXP/poly [5].

Many longstanding open questions in complexity theory boil down to the question of whether there are sets in P that avoid all strings of low Kt complexity. For example, the EXP vs ZPP question is equivalent to (one version of) the question of whether avoiding simple strings is difficult: (EXP = ZPP if and only if there exist $\epsilon > 0$ and a "dense" set in P having no strings x with $\mathrm{Kt}(x) \leq |x|^\epsilon$ [4]).

Surprisingly, we are able to show *unconditionally* that avoiding simple strings (in the sense of KNt complexity) is difficult. Every dense set in NP $\cap$ co-NP contains infinitely many strings x such that $\mathrm{KNt}(x) \leq |x|^\epsilon$ for some ϵ. The proof does not relativize. As an application, we are able to show that if E = NE, then accepting paths for nondeterministic exponential time machines can be found somewhat more quickly than the brute-force upper bound, if there are many accepting paths.

Keywords: Hitting Sets, Kolmogorov Complexity, Complexity Theory.

1 Introduction

It has been observed before that many popular conjectures in complexity theory can be restated equivalently in terms of questions about the resource-bounded Kolmogorov complexity of feasible sets, and that this restatement can serve to highlight some of the tension among these conjectures. For instance, it is common to conjecture that

1. The containment $\mathrm{NTime}(t(n)) \subseteq \mathrm{DTime}(2^{O(t(n))})$ is nearly optimal, and that
2. Cryptographically-secure one-way functions exist.

The first of these two conjectures implies that there are polynomial time-bounded Turing machines that, for infinitely many inputs n, accept some strings in Σ^n, but none

F. Ferreira et al. (Eds.): CiE 2010, LNCS 6158, pp. 1–10, 2010.

having Kt-complexity less than (say) $n/5$ [2], where Kt is a time-bounded version of Kolmogorov complexity defined by Levin [15]. (Definitions will be provided in Section 2.) In contrast, the second conjecture implies that secure pseudorandom generators exist [12], which in turn implies that any polynomial time-bounded machine *must* accept some strings in Σ^n with Kt-complexity much less than $\sqrt{n}$ *if* the machine accepts at least half of the strings in Σ^n [1]. Thus, if the popular conjectures are true, sets in P *can* avoid accepting strings with low Kt-complexity, but *only* if they don't accept many strings of any given length. If a set in P contains a lot of strings of a given input length, then it *cannot* avoid accepting some simple strings, according to the popular belief.

This paper deals with the question of how difficult it is to avoid simple strings (i.e., strings of low resource-bounded Kolmogorov complexity) while still accepting a large number of strings. The main contribution of the paper is to present one setting in which we can replace popular conjecture and vague belief with unconditional theorems, showing that easy-to-compute sets *must* contain simple strings if they contain many strings. We also present an application of this new insight to the question of whether accepting computation paths of nondeterministic exponential time machines are easy to find, assuming "only" E = NE.

Let us introduce some notation, to help us gauge how successfully a set is avoiding simple strings. For any set $A \subseteq \Sigma^*$, define $\mathrm{Kt}_A(n)$ to be $\min\{\mathrm{Kt}(x) : x \in A^{=n}\}$, where $A^{=n} = A \cap \Sigma^n$. (For a definition of Levin's measure $\mathrm{Kt}(x)$, see Section 2.) If $A^{=n} = \emptyset$, then $\mathrm{Kt}_A(n)$ is undefined. Note that the rate of growth of $\mathrm{Kt}_A(n)$ is a measure of how successfully A avoids strings of low Kt complexity. The rate of growth of $\mathrm{Kt}_A(n)$ for sets A in P and P/poly is especially of interest, as can be seen from the following theorem. (We give a more precise definition of "dense" in Section 3, but for now it is sufficient to consider a set to be "dense" if it contains at least $2^n/n$ strings of each length n. An "NE search problem" is the task of mapping an input x to an accepting computation of M on input x, if one exists, where M is an NE machine.)

Theorem 1.   – *There is an NE search problem that is not solvable in time $2^{2^{o(n)}}$ if and only if there is a set $A \in$ P and an $\epsilon > 0$ such that $\mathrm{Kt}_A(n) \neq O(n^\epsilon)$ [2, Theorem 6].*
   – *There is an NE search problem that is not solvable in time $2^{O(n)}$ if and only if there is a set $A \in$ P such that $\mathrm{Kt}_A(n) \neq O(\log n)$ [2, Theorem 6].*
   – *EXP $\not\subseteq$ P/poly if and only if for every dense set $A \in$ P/poly and every $\epsilon > 0$, $\mathrm{Kt}_A(n) = O(n^\epsilon)$ [3, Theorem 12].*
   – *There is a set $B \in$ E and an $\epsilon > 0$ such that, for all large n, there is no circuit of size $2^{\epsilon n}$ accepting $B^{=n}$ if and only if for every dense set $A \in$ P/poly, $\mathrm{Kt}_A(n) = O(\log n)$ [3, Theorem 13].*
   – *EXP $\neq$ ZPP if and only if for every dense set $A \in$ P and every $\epsilon > 0$, $\mathrm{Kt}_A(n) = O(n^\epsilon)$ [4].*

A nondeterministic analog of Levin's Kt measure, denoted KNt, was introduced recently [5]. For any set A, let $\mathrm{KNt}_A(n)$ be $\min\{\mathrm{KNt}(x) : x \in A^{=n}\}$. The rate of growth of $\mathrm{KNt}_A(n)$ is similarly related to open questions in complexity theory:

Theorem 2. *[5, Theorem 44] There is a set $B \in$ NE/lin such that for all large n, there is no nondeterministic circuit of size $2^{\epsilon n}$ accepting $B^{=n}$ if and only if for every dense set A in NP/poly $\cap$ coNP/poly, $\mathrm{KNt}_A(n) = O(\log n)$.*

Theorem 2 presents a condition regarding $\mathrm{KNt}_A(n)$ for dense sets A in a *nonuniform* class, and Theorem 1 contains analogous conditions regarding dense sets in both uniform *and* nonuniform classes. It is natural to wonder if Theorem 2 can be extended, to say something about the corresponding uniform class, and the experience of Theorems 1 and 2 could lead one to expect that such an extension would consist of showing that a statement about the KNt-complexity of dense sets in NP ∩ co-NP is equivalent to some longstanding open question in complexity theory.

Thus it is of interest that our main theorem shows *unconditionally* that $\mathrm{KNt}_A(n)$ grows slowly for all dense sets in NP ∩ co-NP (and even for all of those dense sets lying in (NP ∩ co-NP)/$n^{o(1)}$).

The rest of the paper is organized as follows. Definitions and preliminaries are presented in Section 2. The main results are presented in Section 3. An application to NE search problems is presented in Section 4. And finally, some musings about possible improvements to the main result are presented in Section 5.

2 Preliminaries

We assume that the reader is familiar with complexity classes such as P, ZPP, NP, AM, and PSPACE; for background consult a standard text such as [6]. We use the following notation for deterministic and nondeterministic exponential-time complexity classes: $\mathrm{E} = \mathrm{DTime}(2^{O(n)})$, $\mathrm{NE} = \mathrm{NTime}(2^{O(n)})$, $\mathrm{EXP} = \mathrm{DTime}(2^{n^{O(1)}})$, and $\mathrm{NEXP} = \mathrm{NTime}(2^{n^{O(1)}})$. $\mathrm{P}^{\mathrm{NP}[n]}$ is the class of languages accepted by polynomial-time oracle Turing machines with an oracle from NP, where the oracle machine makes at most n oracle queries on inputs of length n.

For any complexity class $\mathcal{C}$, and function $h(n) : \mathbb{N} \to \mathbb{N}$, let $\mathcal{C}/h(n)$ denote the class of sets B such that, for some "advice function" $a(n) : \mathbb{N} \to \Sigma^{h(n)}$, and some set $A \in \mathcal{C}$, $x \in B$ if and only if $(x, a(|x|)) \in A$. $\mathcal{C}/\mathrm{poly}$ denotes $\bigcup_k \mathcal{C}/n^k + k$; $\mathcal{C}/\mathrm{lin}$ denotes $\bigcup_k \mathcal{C}/kn$. The class P/poly has an equivalent characterization as the class of problems solvable by families of polynomial-size circuits. Note in particular that (NP ∩ co-NP)/poly is quite possibly a proper subset of NP/poly ∩ coNP/poly.

Levin defined $\mathrm{Kt}(x)$ to be $\min\{|d| + \log t : U(d) = x$ in time $t\}$ [15], where U is some fixed universal Turing machine. (It is important to note that Levin's definition is *independent* of any run-time t; the "t" that appears in the definition is a quantity that participates in the minimization expression.) Later, it was observed that $\mathrm{Kt}(x)$ is polynomially related to the oracle circuit size that is required to compute the function that has x as its truth table [5], where the oracle is a complete set for E. In order to obtain a time-bounded notion of Kolmogorov complexity in the spirit of Levin's Kt function that is related to circuit complexity for more general oracles (including the empty oracle), a new measure, called KT, was defined [4]:

Definition 1. *Let U be a universal Turing machine and let B be an oracle. Define the measure $\mathrm{KT}^B(x)$ to be*

$$\mathrm{KT}^B(x) = \min\{\, |d| + t \;:\; U^{B,d} \text{ describes } x \text{ in time } t, \text{ (meaning that}$$
$$\forall b \in \{0, 1, *\} \; \forall i \leq |x| + 1, U^{B,d}(i, b) \text{ accepts in } t \text{ steps iff } x_i = b)\}.$$

4		E. Allender

(The notation "$U^{B,d}(i,b)$" indicates that the machine U has random access (or "oracle access") to both the string d and the oracle B. This allows the running time to be less than $|d|$.) We omit the superscript B if $B = \emptyset$.

It is known that one can pick a complete set C for E such that Levin's definition of $\text{Kt}(x)$ is linearly-related to $\text{KT}^C(x)$ [4].

A nondeterministic analog of Kt called KNt was recently investigated [5], and it was shown that $\text{KNt}(x)$ is linearly related to $\text{KT}^D(x)$ for a set D that is complete for NE. Thus, for this paper, we will let $\text{Kt}(x)$ and $\text{KNt}(x)$ denote $\text{KT}^C(x)$ and $\text{KT}^D(x)$ for this E-complete set C and NE-complete set D, respectively.

For a given set A, and oracle B, we define $\text{KT}_A^B(n)$ to be equal to $\min\{\text{KT}_A^B(x) : x \in A^{=n}\}$. Thus $\text{Kt}_A(n) = \text{KT}_A^C(n)$, and $\text{KNt}_A(n) = \text{KT}_A^D(n)$.

We assume that the reader is familiar with polynomial-time Turing reducibility, denoted $\leq_T^P$. We also need to make use of reductions computed by polynomial-size *circuits*, instead of polynomial-time *machines*. A P/poly-Turing reduction of a set A to a set B is a family of polynomial-size circuits computing A, where the circuits have *oracle gates* for B, in addition to the usual AND and OR gates. (An oracle gate for B outputs 1 if the string y that is presented to it as input is an element of B.) If a P/poly-Turing reduction has the property that there is no path in the circuit from one oracle gate to another, then it is called a P/poly-truth-table reduction, denoted $\leq_{tt}^{P/poly}$.

3   Main Result

The main theorem applies only to languages that have "sufficiently many" strings; we call such sets "dense". The following definition makes precise exactly what sort of "density" is required:

Definition 2. *A set $A \subseteq \{0,1\}^*$ is* dense *if there is a k such that for every n there is some m, $n \leq m \leq n^k + k$ such that $|A \cap \{0,1\}^m| \geq 2^m/m^k$.*

Theorem 3. *Let A be a dense set in $(\text{NP} \cap \text{co-NP})/a(n)$ for some $a(n) = n^{o(1)}$. Then for every $\epsilon > 0$ there are infinitely many $x \in A$ such that $\text{KNt}(x) < |x|^\epsilon$.*

Proof. Most of the work has already been done in an earlier paper, in which it was shown that R_{KNt} (the set of "KNt-random strings", i.e., the set of strings x such that $\text{KNt}(x) \geq |x|$) is not in NP $\cap$ co-NP [5]. It was noticed only shortly after that paper was submitted for publication that the lower bound applied not only to R_{KNt}, but in fact to *every* dense set A such that, for some $\epsilon > 0$, $\text{KNt}_A(n) \neq \Omega(n^\epsilon)$. We need to recall some of the main theorems of earlier work on this topic.

One of the main insights obtained in earlier work is that for "large" complexity classes, dense sets having only strings of high Kolmogorov complexity are hard under P/poly reductions. The following definition captures the property that a "large" class needs to have, in order for the proof to go through:

Definition 3. *A set B is* PSPACE-*robust if* $\text{P}^B = \text{PSPACE}^B$.

The notion of PSPACE-robustness was defined by Babai et al [8], who observed that every set that is complete for EXP under $\leq_T^P$ reductions is PSPACE-robust. Later, it was shown that NEXP also has this property [5].

Theorem 4. *[4, Theorem 31] Let B be any* PSPACE-*robust set. Let A be a set such that for some $\epsilon > 0$ and k, for every n there is some m such that*

- $n \leq m \leq n^k + k$,
- $|A \cap \{0,1\}^m| \geq 2^m/m^k$
- $\mathrm{KT}_A^B(m) \geq m^\epsilon$.

Then B is reducible to A via $\leq_{tt}^{P/\mathrm{poly}}$ reductions.

(This is a slight modification of the statement of the theorem as given in [4]. There, it was assumed that A contains many strings of *every* length, and contains *no* strings of low KT^B complexity. However, the $\leq_{tt}^{P/\mathrm{poly}}$ reduction that is given in [4] has the property that, on inputs of length n, all queries to the oracle A have the same length m, and the reduction works properly as long as, for the given length m, A contains many strings and no strings of low KT^B complexity. Thus, by simply encoding the length m into the nonuniform advice of the $\leq_{tt}^{P/\mathrm{poly}}$ reduction, the proof given in [4] suffices to establish Theorem 4.) We remark also that the proof given in [4] proceeds by showing that A can be used as a test to distinguish truly random strings from pseudorandom strings produced by a pseudorandom generator constructed from B. Thus, the same argument shows that B is reducible to A even if A contains only a *few* strings with low KT^B complexity. Consequently, it is possible to improve the statement of Theorem 3 to say that every dense set in $(\mathrm{NP} \cap \mathrm{co\text{-}NP})/a(n)$ has *many* strings of KNt complexity $\leq n^\epsilon$, for infinitely many n. We do not pursue that generalization here.

Since we are using the definition of KNt as KT^D for some set D that is complete for NE, and since every set that is complete for NE is PSPACE-robust, Theorem 4 immediately yields the following corollary:

Corollary 1. *Let A be any dense set such that $\mathrm{KNt}_A(n) = \Omega(n^\epsilon)$ for some $\epsilon > 0$. Then A is hard for NEXP under $\leq_{tt}^{P/\mathrm{poly}}$ reductions.*

We also need to use the fact that any A that satisfies the hypothesis of Corollary 1 is also hard for PSPACE under probabilistic reductions:

Theorem 5. *[4, Theorem 33 and Lemma 35] Let A be any set of polynomial density, such that $\mathrm{Kt}_A(n) = \Omega(n^\epsilon)$ for some $\epsilon > 0$. Then $\mathrm{PSPACE} \subseteq \mathrm{ZPP}^A$.*

Note that the term "polynomial density" as used in [4] is slightly more restrictive than the term "dense" as defined in this paper, since a set has "polynomial density" if it contains many strings of *every* length.

Corollary 2. *Let A be any dense set in $(\mathrm{NP} \cap \mathrm{co\text{-}NP})/a(n)$ such that $\mathrm{KNt}_A(n) = \Omega(n^\epsilon)$ for some $\epsilon > 0$. Then $\mathrm{PSPACE} \subseteq \bigcup_k (\mathrm{NP} \cap \mathrm{co\text{-}NP})/2a(n^k) + O(\log n)$.*

Proof. Note that $\text{Kt}_A(n) \geq \text{KNt}_A(n) = \Omega(n^\epsilon)$. Thus we would like to modify the proof of Theorem 5 to (nearly) obtain that $\text{PSPACE} \subseteq \text{ZPP}^A$.

The proof of Theorem 5 given in [4] presents a ZPP reduction with the property that, on inputs of length n, there are lengths m_1 and m_2 such that all queries to the oracle have length either m_1 or m_2. (Queries to length m_1 are used to obtain a string of high complexity, which is then used in conjunction with the Impagliazzo-Wigderson construction [14] to derandomize a BPP^A reduction, which only makes queries of length m_2.) The length m_2 can be replaced by any m_2' such that $m_2 \leq m_2' \leq m_2^{O(1)}$, as long as the reduction is given suitable advice, saying which length m_2' has sufficiently many strings, and the length m_1 can be replaced by any m_1' at most polynomially larger than m_2', again if the reduction is given advice, saying which length m_1' is suitable. Thus the ZPP reduction running in time n^k can be simulated by a $(\text{ZPP}^{\text{NP} \cap \text{co-NP}})/2a(n^c) + O(\log n)$ computation, where c depends on k and on the density parameters of A. The corollary follows by observing that $\text{ZPP}^{\text{NP} \cap \text{co-NP}} = \text{NP} \cap \text{co-NP}$. $\qquad\square$

We now proceed with the proof of Theorem 3. The proof is by contradiction: Assume that A is a dense set in $(\text{NP} \cap \text{co-NP})/a(n)$ for some $a(n) = n^{o(1)}$ such that, for all large $x \in A$, we have $\text{KNt}(x) \geq |x|^\epsilon$. That is, $\text{KNt}_A(n) = \Omega(n^\epsilon)$.

By Corollaries 1 and 2 we have $\text{PSPACE} \subseteq (\text{NP} \cap \text{co-NP})/n^{o(1)}$ and $\text{NEXP} \subseteq \text{P}^A/\text{poly} \subseteq \text{P}^{(\text{NP} \cap \text{co-NP})/a(n)}/\text{poly} = (\text{NP} \cap \text{co-NP})/\text{poly}$.

It is known that if $\text{NEXP} \subseteq (\text{NP} \cap \text{co-NP})/\text{poly}$ then $\text{NEXP} = \text{AM}$ [5, Theorem 29]. Thus under our assumptions we have

$$\text{NEXP} = \text{AM} = \text{PSPACE} \subseteq (\text{NP} \cap \text{co-NP})/n^{o(1)}.$$

This is a contradiction, since $(\text{NP} \cap \text{co-NP})/n^{o(1)} \subseteq \text{P}^{\text{NP}[n]}/n$, and it was shown by Buhrman, Fortnow, and Santhanam [10] that NEXP is not contained in $\text{P}^{\text{NP}[n]}/n$. $\qquad\square$

It would be nice to know if a better upper bound on the KNt-complexity of dense sets in $\text{NP} \cap \text{co-NP}$ (or in P) can be proved.

3.1 Does This Relativize?

The proof of Theorem 3 does not relativize, since it relies on Theorem 5 (which in turn relies on the characterization of PSPACE in terms of interactive proofs [16,17]) and also Theorem 28 of [5] (which relies on the characterization of NEXP in terms of interactive proofs [7]). However, we do not know if the statement of Theorem 3 actually fails relative to some oracle.

Spakowski [18] has pointed out that an oracle construction of Buhrman, Fortnow, and Laplante [9] might be relevant. They present a set A such that $\text{CND}^{2^{\sqrt{|x|}}}(x) \geq |x|/4$ for all $x \in A$, where $\text{CND}^{2^{\sqrt{n}}}$ is a notion of "$2^{\sqrt{n}}$-time-bounded nondeterministic distinguishing complexity". It is known that CND-complexity is related to KNt complexity [5], and one can easily show that, for their set $A \in \text{P}^A$, there is some $\epsilon > 0$ such that $\text{KNt}^A(x) \geq |x|^\epsilon$ for all $x \in A$, where $\text{KNt}^A(x)$ is the measure that results when one defines KNt complexity using a universal Turing machine that can access the oracle A. Spakowski suggests that a slight modification of their construction yields a set A satisfying the same conditions, that contains many strings of length n, for infinitely many n. Thus this comes close to being an oracle relative to which Theorem 3 fails.

4   An Application to Search Problems

One of the aspects of the theory of NP-completeness that makes it so widely applicable, is the fact that, for NP-complete problems, *search* is equivalent to *decision*. That is, the problem of *deciding* membership in an NP-complete set is polynomially-equivalent to the problem of *finding* a proof of membership. Hartmanis, Immerman, and Sewelson observed that the proof of equivalence that works for NP-complete problems breaks down for exponential-time computations, and they asked whether search and decision are also equivalent for NE-complete problems [11]. A partial answer was provided by Impagliazzo and Tardos [13], who presented an oracle relative to which E = NE but relative to which there exists a nondeterministic exponential-time machine M such that there is no function computable in exponential time that maps each input x accepted by M to an accepting computation of M on input x. An alternative oracle construction was subsequently given by Buhrman, Fortnow, and Laplante [9].

The trivial brute-force deterministic algorithm for finding accepting computations of NE machines takes doubly exponential time $2^{2^{O(n)}}$. No significantly better upper bound is known, even for the special case of finding accepting computations of probabilistic NE machines, that have *many* accepting computation paths if they have any at all. This has been the case, even under the assumption that E = NE.

As a consequence of the results of Section 3, we can now say something nontrivial about an upper bound on the complexity of finding accepting computations of NE machines if E = NE – at least for certain classes of NE machines. (Actually, it suffices to use the weaker assumption that NEXP $\subseteq$ EXP/poly.) Let ZPE be the exponential-time analog of the complexity class ZPP. That is, B is in ZPE if there are two nondeterministic Turing machines M_0 and M_1 running for time 2^{cn} for some c, where M_0 accepts $\overline{B}$ and M_1 accepts B, with the property that if $x \in B$, then for at least half of the strings r of length 2^{cn}, M_1 accepts x along the computation path given by r, and if $x \notin B$, then for at least half of the strings r M_0 accepts x along the computation path given by r. Thus, for every string x, either half of the strings r of length $2^{c|x|}$ are accepting computations of M_1, or half of the strings r are accepting computations of M_0. A ZPE *search problem* (defined by the machines M_0 and M_1) is the task of taking x as input, and producing a string r as output, that causes either M_0 or M_1 to accept.

Theorem 6. *If* NEXP $\subseteq$ EXP/poly, *then for every* ZPE *search problem, there is a deterministic algorithm* M *solving it with the property that, for every* $\epsilon > 0$, M *runs in time* $2^{2^{|x|^\epsilon}}$ *for infinitely many* x.

Proof. Consider a ZPE search problem defined by machines M_0 and M_1. Let N be an NE machine running in time 2^{cn} that, on input x, guesses a string r of length 2^{cn} and accepts if r causes either M_0 or M_1 to accept on input x. (Note that N accepts *every* string x.)

Let $d : \mathbb{N} \to \{0,1\}^*$ be a standard bijection (e.g., $d(i)$ is the string x such that the binary representation of $i + 1$ is $1x$). Let W_N be the set $\{r : |r| = n^{c+1}$ and (some prefix of) r causes N to accept the string $d(n)\}$. Note that, since $|d(n)| = O(\log n)$, W_N is in P, and A is dense (since it contains at least half of the strings of each length of the form n^{c+1}).

By Theorem 3, for every $\epsilon > 0$ there are infinitely many $r \in W_N$ such that $\mathrm{KNt}(x) < |r|^\epsilon$. Since we are assuming that NEXP $\subseteq$ EXP/poly, it follows that Kt and KNt are polynomially related [5], and thus we have that for every $\epsilon > 0$ there are infinitely many $r \in W_N$ such that $\mathrm{Kt}(r) < |r|^\epsilon$. Let C be the E-complete set such that $\mathrm{Kt}(x) = \mathrm{KT}^C(x)$.

Consider the following algorithm M: On input x, compute n so that $d(n) = x$. For $k = 1$ to n^c, for all descriptions d of length k, see if $U^{C,d}$ describes a string r of length n^c in time k. If so, and if r causes N to accept on input x, then halt and output r.

It is straightforward to verify that the algorithm M has the properties claimed for it in the statement of the theorem.   □

The conclusion of Theorem 6 holds for a great many more NE search problems than merely those in ZPE. It holds for any NE machine N for which the language W_N constructed in the proof of Theorem 6 is dense. (This corresponds to those problems in NE that are accepted by NE machines that have many accepting computation paths for at least one string of every length (or, more generally, at least one string out of every $O(1)$ consecutive lengths).) Rather than creating a new definition to capture this class, we simply state the following corollary:

Corollary 3. *If* NEXP $\subseteq$ EXP/poly, *then for every* NE *search problem defined by an* NE *machine* N *such that the set* W_N *is dense, there is a deterministic algorithm* M *solving it with the property that, for every* $\epsilon > 0$, M *runs in time* $2^{2^{|x|^\epsilon}}$ *for infinitely many* x.

It is natural to wonder if E $=$ NE implies faster algorithms for *all* instances of ZPE, instead of merely for infinitely many inputs x. This is essentially a question of whether polynomial-time computations can accept many strings while avoiding all simple strings for *some* input lengths, but not for others. This topic is discussed at greater length in the next section.

5   Are Σ^n and Σ^m Fundamentally Different, for $n \neq m$?

In this section, we discuss the KNt complexity of dense sets in P. Theorem 3 says that, for every dense set $A \in$ P, there exist infinitely many lengths n such that A contains a string of length n having KNt complexity less than n^ϵ. In this section, we observe that we can essentially swap the quantifiers. There are long segments of consecutive input lengths $S = [i, i + 1, \dots, i^c]$ such that, for every $m \in S$, *every* machine running in time m^k must accept strings in Σ^m with KNt complexity at most m^ϵ if it accepts many strings of length m. There may be long "deserts" of input lengths where, for all we know, polynomial-time machines can behave badly by avoiding all of the simple strings while still accepting many strings. However, we are guaranteed that there are infinitely many large "oases" in the desert, where machines behave as expected (i.e., by accepting some strings with low KNt complexity, if they accept many strings).

Consider the standard universal set for DTime(n^k): $A_k = \{(i, x) : M_i$ accepts x in $|x|^k$ steps$\}$, where we assume an enumeration of machines such that M_0 accepts Σ^* in linear time. Let $a(n)$ be defined to be the index $i \leq n$ of the machine M_i such that, among all of the machines M_j with $j \leq n$ that run in time n^k on inputs of length n and accept at least $2^n/n^k$ strings of length n, KNt$_{L(M_j)}(n)$ is maximized. Note that $a(n)$ is always defined, by our choice of M_0. The set $S_k = \{x : M_{a(|x|)}(x) = 1\}$ is in P$/\log n$ and contains at least $2^n/n^k$ strings of each length n, and has KNt complexity asymptotically as high as any dense set in DTime(n^k).

Define the k-oasis to be the set $\{n : \text{KNt}_{S_k}(n) \leq n^{1/k}\}$. It is immediate that the $(k+1)$-oasis is a subset of the k-oasis, for every k. Also note that, for every $c \geq 1$, and every k, the k-oasis contains infinitely many sequences of consecutive numbers $n, n+1, \ldots, n^c$, since otherwise the set S_k would be a dense set in P$/\log n$ that would be hard for NEXP under $\leq_{tt}^{P/\text{poly}}$ reductions (by Theorem 4) and would also be hard for PSPACE under ZPP$/O(\log n)$ reductions (by Corollary 1), and one would obtain a contradiction exactly as in the proof of Theorem 3.

That is, the k-oases for large k contain superpolynomially-long sequences of consecutive input lengths where all DTime(n^k) machines "behave well", and the k-oases for smaller values of k are even larger.

Since there is not a recursive enumeration of pairs of machines that define sets in NP ∩ co-NP, the strategy that was used in defining "k-oases" for DTime(n^k) must be modified, in order to define an analogous notion of k-oasis for the classes NTime(n^k) ∩ coNTime(n^k). It suffices to make use of a *nonrecursive* enumeration of pairs of machines; details will appear in the full version of this paper.

It seems reasonable to conjecture that each k-oasis is actually ℕ (or at least, that it contains all large natural numbers). Otherwise, for infinitely many lengths m, there are circuits of size m^k that accept a large fraction of the strings of length m, but accept nothing of small KNt complexity, while this is impossible for other lengths m'. This would seem to indicate that Σ^m has some structural property that small circuits are able to exploit, whereas $\Sigma^{m'}$ has no such structure. However, Σ^m seems *devoid* of any useful structure that is not shared by $\Sigma^{m'}$ for $m' \neq m$.

6 Closing Comments

For sufficiently "powerful" forms of resource-bounded Kolmogorov complexity (such as KT^E where E is complete for EXPSPACE), the lexicographically first element of $A^{=n}$ will always have logarithmic complexity, for any $A \in$ P [5]. Conceivably, one could define a version of resource-bounded Kolmogorov complexity related to a low level of the exponential-time hierarchy (with just a few alternations – and therefore conceptually "closer" to KNt than KT^E) where this same technique could be applied. It seems unlikely that KNt is powerful enough to always give the lexicographically least element of $A^{=n}$ logarithmic complexity, for every set A in P, although we know of no unlikely consequences, if that were to be the case.

Acknowledgments

The research of the author is supported in part by NSF Grants DMS-0652582, CCF-0830133, and CCF-0832787. Some of this work was performed while the author was a visiting scholar at the University of Cape Town and at the University of South Africa. We acknowledge helpful discussions with Chris Umans, Holger Spakowski, and Ronen Shaltiel.

References

1. Allender, E.: Some consequences of the existence of pseudorandom generators. Journal of Computer and System Sciences 39, 101–124 (1989)
2. Allender, E.: Applications of time-bounded Kolmogorov complexity in complexity theory. In: Watanabe, O. (ed.) Kolmogorov Complexity and Computational Complexity, pp. 4–22. Springer, Heidelberg (1992)
3. Allender, E.: When worlds collide: Derandomization, lower bounds, and Kolmogorov complexity. In: Hariharan, R., Mukund, M., Vinay, V. (eds.) FSTTCS 2001. LNCS, vol. 2245, pp. 1–15. Springer, Heidelberg (2001)
4. Allender, E., Buhrman, H., Koucký, M., van Melkebeek, D., Ronneburger, D.: Power from random strings. SIAM Journal on Computing 35, 1467–1493 (2006)
5. Allender, E., Koucký, M., Ronneburger, D., Roy, S.: The pervasive reach of resource-bounded Kolmogorov complexity in computational complexity theory. Journal of Computer and System Sciences (to appear)
6. Arora, S., Barak, B.: Computational Complexity, a modern approach. Cambridge University Press, Cambridge (2009)
7. Babai, L., Fortnow, L., Lund, C.: Non-deterministic exponential time has two-prover interactive protocols. Computational Complexity 1, 3–40 (1991)
8. Babai, L., Fortnow, L., Nisan, N., Wigderson, A.: BPP has subexponential time simulations unless EXPTIME has publishable proofs. Computational Complexity 3, 307–318 (1993)
9. Buhrman, H., Fortnow, L., Laplante, S.: Resource-bounded Kolmogorov complexity revisited. SIAM Journal on Computing 31(3), 887–905 (2002)
10. Buhrman, H., Fortnow, L., Santhanam, R.: Unconditional lower bounds against advice. In: Albers, S., et al. (eds.) ICALP 2009, Part I. LNCS, vol. 5555, pp. 195–209. Springer, Heidelberg (2009)
11. Hartmanis, J., Immerman, N., Sewelson, V.: Sparse sets in NP-P: EXPTIME versus NEXPTIME. Information and Control 65(2/3), 158–181 (1985)
12. Håstad, J., Impagliazzo, R., Levin, L., Luby, M.: A pseudorandom generator from any one-way function. SIAM Journal on Computing 28, 1364–1396 (1999)
13. Impagliazzo, R., Tardos, G.: Decision versus search problems in super-polynomial time. In: Proc. IEEE Symp. on Found. of Comp. Sci (FOCS), pp. 222–227 (1989)
14. Impagliazzo, R., Wigderson, A.: $P = BPP$ if E requires exponential circuits: Derandomizing the XOR lemma. In: Proc. ACM Symp. on Theory of Computing (STOC) 1997, pp. 220–229 (1997)
15. Levin, L.A.: Randomness conservation inequalities; information and independence in mathematical theories. Information and Control 61, 15–37 (1984)
16. Lund, C., Fortnow, L., Karloff, H., Nisan, N.: Algebraic methods for interactive proof systems. Journal of the ACM 39, 859–868 (1992)
17. Shamir, A.: IP = PSPACE. Journal of the ACM 39, 869–877 (1992)
18. Spakowski, H.: Personal Communication (2009)

Higher-Order Containers

Thorsten Altenkirch[1,*], Paul Levy[2,**], and Sam Staton[3,***]

[1] University of Nottingham
[2] University of Birmingham
[3] University of Cambridge

Abstract. Containers are a semantic way to talk about strictly positive types. In previous work it was shown that containers are closed under various constructions including products, coproducts, initial algebras and terminal coalgebras. In the present paper we show that, surprisingly, the category of containers is cartesian closed, giving rise to a full cartesian closed subcategory of endofunctors. The result has interesting applications in generic programming and representation of higher order abstract syntax. We also show that the category of containers has finite limits, but it is not locally cartesian closed.

Keywords: Datatypes, Category Theory, Functional Programming.

1 Introduction

Containers are a representation of datatypes, using a set of shapes S and a family of positions P indexed over shapes. The associated datatype is given by a choice of shape and an assignment of data to positions over that shape, clearly this is an endofunctor of **Set**. In previous work [2,1] it was shown that all strictly positive datatypes give rise to containers. To include nested inductive and coinductive definitions it was necessary to introduce n-ary containers, corresponding to n-ary functors. This can be generalized further to indexed containers [5] to modell dependent families.

Containers can be used to analyze generic constructions on datatypes without having to do induction over the syntax of datatypes. E.g. in [4] containers are used to study the notion of a derivative of a datatype.

Other applications of containers are related to container morphisms which are a concrete and complete representations of polymorphic functions, i.e. natural transformations, between datatypes. In [15] this is exploited to derive theorems about polymorphic functions on lists.

The previous results can be stated in terms of properties of the category of containers: it is closed under products, coproducts and exponentiation with a set and the extension functor into sets is full and faithful. Recent work by the 3rd author [17] on using higher order representations in generalized structured

---

* Supported by EPSRC grant EP/G03298X/1.
** Supported by ESPRC Advanced Research Fellowship EP/E056091/1.
*** Supported by EPSRC Postdoctoral Research Fellowship EP/E042414/1.

F. Ferreira et al. (Eds.): CiE 2010, LNCS 6158, pp. 11–20, 2010.

operational semantics raised the question wether the category of containers is cartesian closed. In the present paper we can answer this question positively.

As a simple example consider the functor $\Lambda \in \mathbf{Set} \to \mathbf{Set}$ which assigns to any set of variables the set of untyped lambda terms over this set. This functor can be specified as the inital solution to the following equation in the category of endofunctors

$$\Lambda \simeq \mathrm{Id} + \Lambda^2 + \mathrm{Id} \to \Lambda$$

Here Id is the identity functor, and $\to$ refers to the exponential of endofunctors (which may or may not exist). It turns out that this higher order equation is equivalent to

$$\Lambda \simeq \mathrm{Id} + \Lambda^2 + \Lambda \circ \mathrm{Maybe}$$

where $\mathrm{Maybe}\,X = 1 + X$. Indeed, this leads directly to the well-known representation of λ-terms as a nested datatype in Haskell

```
data Lam a = Var a | App (Lam a) (Lam a) | Lam (Maybe a)
```

which has been studied in [6,7].

The category of containers can be defined wrt. any locally cartesian closed category with coproducts. We are going to use the language of Type Theory to develop our results, which is the internal language of locally cartesian closed categories. Hence the constructions presented here can be intuitively understood as taking place in a naive set theory.

A natural question is whether the category of containers itself is a model of Type Theory, i.e. locally cartesian closed. We are able to construct pullbacks if we assume that the ambient category has quotient types, corresponding to exact coequalizers. However, we show that the right adjoint to pullbacks don't exist in general.

2 Preliminaries

We work in an extensional Type Theory [12] as the internal langauge of locally cartesian closed categories with disjoint coproducts.

Set We use **Set** to denote our universe of small types we identfy families with functions into **Set**.

0, 1 An empty type $0 \in \mathbf{Set}$ and a unit type $1 \in \mathbf{Set}$. Categorically, those correspond to initial and terminal objects. We write $() \in 1$ for the unique inhabitant of 1 and $!_A \in A \to 1$ with $!_A\,a = ()$ for the unique map into 1.

Σ- and Π-types Given $A \in \mathbf{Set}$ and $B \in \mathbf{Set}$ given that $x \in A$ then $\Sigma x \in A.B, \Pi x \in A, B \in \mathbf{Set}$. Elements of Σ-types are pairs, if $a \in A$ and $b \in B[x := a]$ then $(a, b) \in \Sigma x \in A.B$, while elements of Π-types are functions: given $b \in B$ assuming $x \in A$ then $\lambda x.b \in \Pi x \in A.B$.

Equality types Given $a, b \in A \in \mathbf{Set}$ we write $a = b \in \mathbf{Set}$ for the equality type. The constructor for equality is reflexivity $\mathrm{refl}\,a \in a = a$ if $a \in A$.

2 A type of Booleans $0, 1 \in 2 \in \mathbf{Set}$, which is disjoint, i.e. we have that $(0 = 1) \to 0$ is inhabited.

We omit a detailed treatment of eliminators and use functional programming notation as present in Agda [14] and Epigram [13]. All our definitions can be translated into using the standard eliminators at the price of readability. To avoid clutter we adopt the usual type-theoretic device of allowing hidden arguments, if they are inferable from the use. We indicate hidden arguments by subscripting the type, i.e. writing $\Pi_{x \in A} B$ and $\Sigma_{x \in A} B$ instead $\Pi x \in A.B$ and $\Sigma x \in A.B$.

While finite products arise as non-dependent Σ-types, finite coproducts can be represented as

$$A_0 + A_1 = \Sigma b \in 2. \text{if } b \text{ then } A_1 \text{ else } A_0$$

We use the injections $\text{in}_i \in A_i \to A_0 + A_1$ with $\text{in}_i \, a = (i, a)$ for $i \in 2$.

Σ-types can also be used to encode set-comprehension. If the family $B \in \mathbf{Set}$ (given $a \in A$) is propositional, i.e. there is at most one element in B for any $a \in A$, we write $\{a \in A \mid B\}$ for $\Sigma a \in A.B$.

We are going to use type-theoretic representations of categorical concepts. Given a bifunctor $F : \mathbf{Set}^{\text{op}} \to \mathbf{Set} \to \mathbf{Set}$ we define its end as a subset of the polymorphic functions:

$$\int_X F \, X \, X \in \mathbf{Type}$$

$$\int_X F \, X \, X = \{f \in \Pi X \in \mathbf{Set}. F \, X \, X$$

$$\mid \forall A, B \in \mathbf{Set}, g \in A \to B. F \, g \, B \, (f \, B) = F \, A \, g \, (f \, A)\}$$

This just internalizes the categorical definition of an end as a universal wedge. Dually, coends can be defined as quotients of large Σ-types (i.e. abstract datatypes), but we shall not need this here.

Using this notation, the type of natural transformations between two endofunctors $F, G \in \mathbf{Set} \to \mathbf{Set}$ arises as $\int_X F \, X \to G \, X$. The Yoneda lemma becomes:

$$F \, X \simeq \int_Y (X \to Y) \to F \, Y \qquad \text{Yoneda}$$

As it is well known the category of endofunctors has products which can be calculated pointwise:

$$(F \times G) \, X = F \, X \times G \, X$$

If we assume that the exponential of two endofunctors $F \to G$ exists then it must have a certain form reminiscent of the Kripke interpretation of implication:

$$\begin{aligned}
(F \to G) \, X \\
\simeq \int_Y (X \to Y) \to (F \to G) \, Y \quad &\text{Yoneda} \\
\simeq \int_Y (X \to Y) \times F \, Y \to G \, Y \quad &\text{adjoint} \\
\simeq \int_Y (X \to Y) \to F \, Y \to G \, Y \quad &\text{curry}
\end{aligned} \tag{1}$$

However, for size reasons $F \to G$ doesn't always exist. E.g. in the category of classical sets, which after all is a model of Type Theory, we have that the collection of $\int_X \mathcal{P} X \to \mathcal{P}(\mathcal{P}X)$, where $\mathcal{P}$ is the covariant powerset functor, is not a set. Indeed, there is a natural transformation $\alpha_\kappa \in \int_X \mathcal{P} X \to \mathcal{P}(\mathcal{P}X)$ for every cardinal κ, where $\alpha_\kappa X S = \{T \subseteq S \mid \operatorname{card} T < \kappa\}$ for every set X and $S \subseteq X$.

3   Containers

A container is given by a set of shapes $S \in \mathbf{Set}$ and a family of positions indexed over shapes $P \in S \to \mathbf{Set}$ - we write $S \lhd P$. We shall also use $\lhd$ as a binder, writing $s \in S \lhd P$ for $S \lhd \lambda s.P$. A container represents an endofunctor

$$\llbracket S \lhd P \rrbracket \in \mathbf{Set} \to \mathbf{Set}$$
$$\llbracket S \lhd P \rrbracket X = \Sigma s \in S.P\, s \to X$$

Given containers $S \lhd P$ and $T \lhd Q$ a morphism $f \lhd r$ is given by

$$f \in S \to T$$
$$r \in \Pi_{s \in S} Q\,(f\,s) \to P\,s$$

This constitutes the category of containers $\mathbf{Cont}$ with the obvious definitions of identity and composition. $\llbracket - \rrbracket$ extends to a functor $\llbracket - \rrbracket \in \mathbf{Cont} \to (\mathbf{Set} \to \mathbf{Set})$ assigning natural transformations to container morphisms:

$$\llbracket f \lhd r \rrbracket \in \int_X \llbracket S \lhd P \rrbracket X \to \llbracket T \lhd Q \rrbracket X$$
$$\llbracket f \lhd r \rrbracket X\,(s,p) = (f\,s, p \circ r)$$

Indeed $\mathbf{Cont}$ gives rise to a full subcategory of the category of endofunctors as shown in [2]:

Proposition 1. $\llbracket - \rrbracket \in \mathbf{Cont} \to (\mathbf{Set} \to \mathbf{Set})$ *is full and faithful.*

Containers also give rise to two modalities which operate on families: given $B \in A \to \mathbf{Set}$ we have

$$\Box_{S \lhd P}\, B, \Diamond_{S \lhd P}\, B \in \llbracket S \lhd P \rrbracket A \to \mathbf{Set}$$
$$\Box_{S \lhd P}\, B\,(s, h) = \Pi p \in P\, s.B\,(h\,p)$$
$$\Diamond_{S \lhd P}\, B\,(s, h) = \Sigma p \in P\, s.B\,(h\,p)$$

$\Box$ can be defined for any functor because it corresponds to applying the functor to the representation of the family as an arrow.

The identity functor is given by $\mathrm{Id} = 1 \lhd 1$ and given $S \lhd P$ and $T \lhd Q$ we can construct their composite:

$$(S \lhd P) \circ (T \lhd Q) = \llbracket S \lhd P \rrbracket T \lhd \Diamond_{S \lhd P} Q$$

Composition is functorial but we shall not explore the 2-categorical structure of **Cont** any further.

In [2] it is shown that containers are closed under finite products and coproducts. Indeed they are closed under arbitrary products and coproducts. Given a family of containers $F \in I \to$ **Cont** this family can be isomorphically presented as

$$S \in I \to \mathbf{Set}$$

$$P \in \Pi_{i \in I} S\, i \to \mathbf{Set}$$

with $F\, i = S\, i \triangleleft P_i$. We write $S \triangleleft P$ for this family. We now define the coproduct and the product of $S \triangleleft P$:

$$\Sigma(S \triangleleft P) = (i, s) \in \Sigma i \in I.S\, i \triangleleft P\, i\, s$$

$$\Pi(S \triangleleft P) = f \in \Pi i \in I.S\, i \triangleleft \Sigma i \in I.P\, i\, (f\, i)$$

We summarize the operations on containers:

Proposition 2. *Containers contain and are closed under:*

identity

$$[\![\mathrm{Id}]\!]\, A \simeq A$$

composition

$$[\![(S \triangleleft P) \circ (T \triangleleft Q)]\!] \simeq [\![S \triangleleft P]\!] \circ [\![T \triangleleft Q]\!]$$

coproducts

$$[\![\Sigma\,(S \triangleleft P)]\!]\, A \simeq \Sigma i \in I.[\![S\, i \triangleleft P_i]\!]$$

products

$$[\![\Pi\,(S \triangleleft P)]\!]\, A \simeq \Pi i \in I.[\![S\, i \triangleleft P_i]\!]$$

It is important to realize that the infinite coproducts and products are internal wrt. to the ambient category. The case of constant exponentiation in [2] arises as a constant product.

4   Containers Are Cartesian Closed

Our central observation is that exponentiation with a container which has only one shape $1 \triangleleft P$, i.e. a container representing an internal hom functor $[\![1 \triangleleft P]\!]\, X = P \to X$, is straightforward.

$$([\![1 \triangleleft P]\!] \to F)\, X$$

$$\simeq \int_Y (X \to Y) \to [\![1 \triangleleft P]\!]\, Y \to F\, Y \qquad\qquad \text{using (1)}$$

$$= \int_Y (X \to Y) \to (P \to Y) \to F\, Y$$

$$\simeq \int_Y (X \to Y) \times (P \to Y) \to F\, Y \qquad\qquad \text{uncurry}$$

$$\simeq \int_Y (X + P \to Y) \to F\, Y \qquad\qquad \text{adjunction}$$

$$\simeq F(X + P)$$

To summarize we have that

$$[\![1 \lhd P]\!] \to F \simeq F \circ (+P) \tag{2}$$

where $(+P)\, X = X + P$. Extensionally, every container is the a coproduct of hom containers:

$$[\![S \lhd P]\!]\, X \tag{3}$$
$$\simeq \Sigma s \in S.P\, s \to X$$
$$\simeq \Sigma s \in S.[\![1 \lhd P\, s]\!]\, X$$

Because of proposition 1 this is also true in the category of containers;

$$S \lhd P \simeq \Sigma s \in S.1 \lhd P\, s \tag{4}$$

Combining these observations we can see that exponentiation by a container is always possible and can be constructed using products and composition:

$$[\![S \lhd P]\!] \to F$$
$$\simeq [\![\Sigma s \in S.1 \lhd P\, s]\!] \to F \qquad \text{using (4)}$$
$$\simeq \Pi s \in S.[\![1 \lhd P\, s]\!] \to F \qquad \text{adjunction}$$
$$\simeq \Pi s \in S.F \circ (+P\, s) \qquad \text{using (2)}$$

Proposition 3. *Given a container $S \lhd P$ and a functor $F \in \mathbf{Set} \to \mathbf{Set}$ we have:*

$$[\![S \lhd P]\!] \to F \simeq \Pi s \in S.F \circ (+P\, s)$$

Using proposition 2 we know that if F is a container then $\Pi s \in S.F \circ (+P\, s)$ is a container. Since containers are a full subcategory of $\mathbf{Set} \to \mathbf{Set}$ (prop. 1) this implies our main result:

Corollary 1. *The category of containers is cartesian closed, and the embedding $[\![-]\!] \in \mathbf{Cont} \to (\mathbf{Set} \to \mathbf{Set})$ preserves the cartesian closed structure.*

We can spell out the construction of the exponential by expanding the definitions of the operations involved. Note that $+P\, s$ is given by $l : 1 + P\, s \lhd l = \mathrm{in}_0\, ()$:

$$S \lhd P \to T \lhd Q$$
$$\simeq \Pi s \in S.(T \lhd Q) \circ (+P\, s)$$
$$\simeq \Pi s \in S.(T \lhd Q) \circ (l : 1 + P\, s \lhd l = \mathrm{in}_0\, ())$$
$$\simeq \Pi s \in S.[\![T \lhd Q]\!](1 + P\, s) \lhd \lozenge_{T \lhd Q}(\lambda l.l = \mathrm{in}_0\, ())$$
$$\simeq f \in \Pi s \in S.[\![T \lhd Q]\!](1 + P\, s) \lhd \Sigma s \in S.\lozenge_{T \lhd Q}(\lambda l.l = \mathrm{in}_0\, ())\,(f\, s)$$
$$\simeq f \in \Pi s \in S.\Sigma t \in T.Q\, t \to 1 + P\, s$$
$$\qquad \lhd \Sigma s \in S.\Sigma q \in Q\, s.(f\, s).2\, q = \mathrm{in}_0\, ()$$

We can also obtain a less direct proof of Corollary 1 by piecing together some known facts. Recall that an endofunctor on **Set** is a container if and only if it preserves wide pullbacks (see e.g. [8]). Thus the containers are precisely those functors (**Set** → **Set**) that are orthogonal to certain cones (see [11, Ch. 6]). In fact, they form an orthogonality class in the category of *small* functors (**Set** → **Set**), which is cartesian closed (see e.g. [16, Prop. 1]). As an orthogonality class that is closed under products, containers form an exponential ideal in the category of small functors (see [9, Sec. 11.2] and also [10]).

5   Local Cartesian Closure?

The previous section shows that we can interpret the simply typed λ-calculus within the category of containers. Can we go further, i.e. can we interpret dependent types in **Cont**?

Dependent types correspond to locally cartesian closed categories, LCCCs. A category is locally cartesian closed, if it has a terminal object and pullbacks (i.e. all finite limits) and a (fibred) right adjoint to the pullback functor. We will show that pullbacks can indeed be constructed, if we have quotient types (i.e. exact coequalizers) in the underlying Type Theory. However, we show that the right adjoints do not exist in general and hence while the category of containers has finite limits, it is not locally cartesian closed.

We know from the previous section that **Cont** has a terminal object $1 \lhd 0$ because this is just the nullary case of a finite product. Pullbacks correspond to products in the slice category, i.e. given

$$f_i \lhd a_i \in \mathbf{Cont}\,(S_i \lhd P_i)\,(T \lhd Q) \qquad\qquad i \in 2$$

we need to construct a new container $U \lhd R = (f_0 \lhd a_0) \times_{T \lhd Q} (f_1 \lhd a_1)$ together with projections:

$$g_i \lhd h_i \in \mathbf{Cont}\,(U \lhd R)\,(S_i \lhd P_i) \qquad\qquad i \in 2$$

We define

$$U \in \mathbf{Set}$$
$$U = \{(s_0, s_1) \in S_0 \times S_1 \mid f_0\, s_0 = f_1\, s_1\}$$
$$R \in U \to \mathbf{Set}$$
$$R\,(s_0, s_1) = (P_0\, s_0 + P_1\, s_1)/\sim$$

where $\sim$ is an equivalence relation on $P_0\, s_0 + P_1\, s_1$ generated by

$$\mathrm{in}_0\,(a_0\, q) \sim \mathrm{in}_1\,(a_1\, q)$$

for $q \in Q\,(f_0\, s_0)$ which due to the assumption $f_0\, s_0 = f_1\, s_1$ is equivalent to $q \in Q\,(f_1\, s_1)$. The definition of $g_i \lhd h_i$ of the projections is straightforward : $g_i \lhd h_i = \pi_i \lhd \mathrm{in}_i$.

Proposition 4. $U \lhd R, g_i \lhd h_i$ *is a pullback of* $f_i \lhd a_i$ *in* **Cont**.

We omit the laborious verification of the equational conditions. As a consequence we have:

Corollary 2. *The category of containers has all finite limits, and the embedding* $[\![-]\!] \in \textbf{Cont} \to (\textbf{Set} \to \textbf{Set})$ *preserves finite limits.*

The restriction to finite limits isn't essential, it is not hard to generalize the construction to arbitrary limits, again in the appropriate internal sense.

Pullbacks are products in the slice category, i.e. for a given container A the slice category $\textbf{Cont}/A$ has as objects morphisms with codomain A and as morphisms commuting triangles. Given arrows $\alpha, \beta \in \textbf{Cont}/A$ their pullback $\alpha \times_A \beta$ is the product in $\textbf{Cont}/A$. For the category to be locally cartesian closed we need a right adjoint to $\alpha \times_A -$: assume $\gamma, \delta \in \textbf{Cont}/A$, there is a $\gamma \to_A \delta \in \textbf{Cont}/A$ such that for all $\alpha \in \textbf{Cont}/A$,

$$\textbf{Cont}/A\,(\alpha \times_A \gamma)\,\delta \simeq \textbf{Cont}/A\,\alpha\,(\gamma \to_A \delta)$$

There is the additional condition that the local exponentials are fibred. However, this doesn't matter here because we are going to construct a counterexample, showing that already the isomorphism above doesn't exist in general. We let $\text{Id}^2 = 1 \lhd 2$ and we will show that the slice $\textbf{Cont}/\text{Id}^2$ is not cartesian closed. We set

$$\gamma = !_1 \lhd !_2 \in \textbf{Cont}\,\text{Id}\,\text{Id}^2 \qquad\qquad \text{(there is only one)}$$
$$\delta = !_1 \lhd \lambda x.\,0 \in \textbf{Cont}\,\text{Id}^2\,\text{Id}^2$$
$$\alpha = !_1 \lhd \text{id} \in \textbf{Cont}\,\text{Id}^2\,\text{Id}^2$$
$$\beta = !_1 \lhd (0 \mapsto 0, 1 \mapsto 2) \in \textbf{Cont}\,\text{Id}^3\,\text{Id}^2 \qquad \text{(where } \text{Id}^3 = 1 \lhd 3)$$

Suppose that the exponential $\gamma \to_{\text{Id}^2} \delta$ exists. Let

$$\gamma \to_{\text{Id}^2} \delta = !_S \lhd f \in \textbf{Cont}\,(S \lhd P)\,\text{Id}^2 \qquad \text{with} \qquad f \in \Pi s \in S.2 \to P\,s.$$

We will derive a contradiction. Let us investigate the structure of this supposed exponential.

The pullback $\alpha \times_{\text{Id}^2} \gamma$ is again γ. There is exactly one morphism $\textbf{Cont}/\text{Id}^2\,\gamma\,\delta$ since there is only one morphism in $\textbf{Cont}\,\text{Id}\,\text{Id}^2$. Thus there is exactly one morphism in $\textbf{Cont}/\text{Id}^2\,\alpha\,(\gamma \to_{\text{Id}^2} \delta)$.

To give a morphism in $\textbf{Cont}/\text{Id}^2\,\alpha\,(\gamma \to_{\text{Id}^2} \delta)$ is to give a shape $s \in S$ together with a function $g \in P\,s \to 2$ such that $g \circ (f\,s) = \text{id}_2$. Because there is a unique morphism $\textbf{Cont}/\text{Id}^2\,\alpha\,(\gamma \to_{\text{Id}^2} \delta)$ there must be a shape t such that

1. The function $f\,t \in 2 \to P\,t$ is a bijection, with inverse $g \in P\,t \to 2$.
2. For all $s \neq t$, $f\,s$ is not injective, i.e. $f\,s\,0 = f\,s\,1$.

To give a morphism in $\textbf{Cont}/\text{Id}^2\,\beta\,(\gamma \to_{\text{Id}^2} \delta)$ is to give a shape $s \in S$ together with a function $h \in P\,s \to 3$ such that $h \circ (f\,s) = (0 \mapsto 0, 1 \mapsto 2)$. In this situation, $f\,s$ must be injective, so, by the two conditions above, we must have $s = t$. Thus there is at most one morphism in $\mathbf{Cont}/\mathrm{Id}^2\,\beta\,(\gamma \to_{\mathrm{Id}^2} \delta)$.

On the other hand, the pullback $\beta \times_{\mathrm{Id}^2} \gamma$ is δ. There are two morphisms $\mathbf{Cont}/\mathrm{Id}^2\,\delta\,\delta$. So, if $\gamma \to_{\mathrm{Id}^2} \delta$ was an exponential, there would be two morphisms in $\mathbf{Cont}/\mathrm{Id}^2\,\beta\,(\gamma \to_{\mathrm{Id}^2} \delta)$. Hence it is not an exponential and $\mathbf{Cont}$ is not locally cartesian closed.

6   Conclusions and Further Work

The category of containers is a full subcategory of the category of endofunctors with a number of interesting closure properties. The initial motivation was to find a subcategory which is closed under taking initial algebras and terminal coalgebras and has the necessary infrastructure to define datatypes, i.e. products and coproducts. The fact that this category is also cartesian closed is an additional benefit and shows that we can interpret higher order constructions. Finite limits are also an interesting feature which may help in modelling dependent types directly, however, the failure of local cartesian closure might indicate that we should look for a different category. Quotient containers [3] might be an interesting alternative but an initial analysis indicates that they are not locally cartesian closed either.

It is usual to work in an ambient category in which initial algebras of containers exist (W-types). However, the container Λ for λ-terms, in the introduction, is an initial algebra of a container, not in the category of sets, but in the category of containers. An investigation into the nature of W-types in the category of containers is left for future work.

References

1. Abbott, M.: Categories of Containers. PhD thesis, University of Leicester (2003)
2. Abbott, M., Altenkirch, T., Ghani, N.: Containers - constructing strictly positive types. Theoretical Computer Science 342, 3–27 (2005); Applied Semantics: Selected Topics
3. Abbott, M., Altenkirch, T., Ghani, N., McBride, C.: Constructing polymorphic programs with quotient types. In: Kozen, D. (ed.) MPC 2004. LNCS, vol. 3125, pp. 2–15. Springer, Heidelberg (2004)
4. Abbott, M., Altenkirch, T., Ghani, N., McBride, C.: ∂ for data: derivatives of data structures. Fundamenta Informaticae 65(1,2), 1–128 (2005)
5. Altenkirch, T., Morris, P.: Indexed containers. In: Twenty-Fourth IEEE Symposium in Logic in Computer Science, LICS 2009 (2009)
6. Altenkirch, T., Reus, B.: Monadic presentations of lambda terms using generalized inductive types. In: Computer Science Logic (1999)
7. Bird, R., Paterson, R.: Generalised folds for nested datatypes. Formal Aspects of Computing 11(3) (1999)
8. Carboni, A., Johnstone, P.: Connected limits, familial representability and artin glueing. Math. Structures Comput. Sci. 5, 441–459 (1995)
9. Fiore, M.P.: Enrichment and representation theorems for categories of domains and continuous functions (March 1996)

10. Fiore, M.P., Menni, M.: Reflective Kleisli subcategories of the category of Eilenberg-Moore algebras for factorization monads. Theory Appl. of Categ. 15 (2005)
11. Kelly, G.M.: Basic concepts of enriched category theory. Lecture Notes in Mathematics, vol. 64. Cambridge University Press, Cambridge (1982); Republished in Reprints in Theory and Applications of Categories 10, 1-136 (2005)
12. Martin-Löf, P.: An intuitionistic theory of types: Predicative part. In: Rose, H.E., Shepherdson, J.C. (eds.) Proceedings of the Logic Colloquium, pp. 73–118. North-Holland, Amsterdam (1974)
13. McBride, C., McKinna, J.: The view from the left. Journal of Functional Programming 14(1) (2004)
14. Norell, U.: Towards a Practical Programming Language based on Dependent Type Theory. PhD thesis, Chalmers University of Technology (2007)
15. Prince, R., Ghani, N., McBride, C.: Proving Properties of Lists using Containers. In: FLOPS (2008)
16. Rosický, J.: Cartesian closed exact completions. J. Pure Appl. Algebra 142(3), 261–270 (1999)
17. Staton, S.: General structural operational semantics through categorical logic. In: Symposium on Logic in Computer Science, pp. 166–177 (2008)

On the Completeness
of Quantum Computation Models

Pablo Arrighi[1] and Gilles Dowek[2]

[1] École normale supérieure de Lyon
LIP, 46 allée d'Italie, 69008 Lyon, France
and Université de Grenoble
LIG, 220 rue de la chimie, 38400 Saint Martin d'Hères, France
pablo.arrighi@imag.fr
[2] École polytechnique and INRIA,
LIX, École polytechnique, 91128 Palaiseau Cedex, France
gilles.dowek@polytechnique.edu

Abstract. The notion of computability is stable (*i.e.* independent of the choice of an indexing) over infinite-dimensional vector spaces provided they have a finite "tensorial dimension". Such vector spaces with a finite tensorial dimension permit to define an absolute notion of completeness for quantum computation models and give a precise meaning to the Church-Turing thesis in the framework of quantum theory.

1 Introduction

In classical computing, algorithms are sometimes expressed by boolean circuits. *E.g.* the algorithm mapping the booleans x and y to $and(x, not(y))$ can be expressed by a circuit formed with an *and* gate and a *not* gate. A typical result about such circuit models is that a given set of gates is complete, *i.e.* that all functions from $\{0,1\}^n$ to $\{0,1\}^p$ can be expressed with a circuit with n inputs and p outputs. Yet, circuit models are limited because an algorithm must take as input and return as output a fixed number of boolean values, and not, for instance, an arbitrary natural number or sequence of boolean values of arbitrary length. Thus, other computation models, such as Turing machines, λ-calculus, cellular automata, ... have been designed. Such models cannot express all functions from $\mathbb{N}$ to $\mathbb{N}$, but a typical completeness result is that they can express all *computable* functions. Thus, before stating and proving such a result, we need to define a subset of the functions from $\mathbb{N}$ to $\mathbb{N}$: the *computable* functions.

In quantum computing, both types of models exist. Most algorithms are expressed using the quantum circuit model, but more evolved models, allowing to express algorithms working with unbounded size data exist also, *e.g.* quantum Turing machines [8,6,12], quantum λ-calculi [20,16,17,1,2], quantum cellular automata [15,5,4]. To define a notion of completeness for these quantum models of computation, we must first define a set of computable quantum functions.

Quantum functions are linear functions from a vector space to itself, and when this vector space is finite or countable (*i.e.* when it is a vector space of finite or

F. Ferreira et al. (Eds.): CiE 2010, LNCS 6158, pp. 21–30, 2010.
© Springer-Verlag Berlin Heidelberg 2010

countable dimension over a finite or countable field), the notion of computability can be transported from the natural numbers to this vector space, with the use of an *indexing*.

The robustness of the notion of computability over the natural numbers has been pointed out for long, and is embodied by the Church-Turing thesis. In contrast, it is well-known that the notion of computability over an arbitrary countable set depends on the choice of an indexing: the composition of an indexing with a non computable function yields another indexing that defines a different set of computable functions [9,7]. Thus, the robustness of the computability over the natural numbers does not directly carry through to countable vector spaces and the naive notion of completeness for such a quantum computational model is relative to the choice of an indexing.

On the other hand, a vector space is not just a set, but a set equipped with an addition and a multiplication by a scalar. More generally, sets are rarely used in isolation, but come in algebraic structures, equipped with operations. As their name suggests, these operations should be computable. Thus, in algebraic structures, we can restrict to the so called *admissible* indexings that make these operations computable [13]. In many cases, restricting to such admissible indexings makes the set of computable functions independent of the choice of the indexing. In such a case, computability is said to be *stable* over the considered algebraic structures [18].

Unfortunately, we show in this paper that, although computability is stable over finite-dimensional vector spaces, it is not stable over infinite-dimensional ones, such as those we use to define quantum functions on unbounded size data. Indeed, any non-computable permutation of the vectors of a basis changes the set of computable functions without affecting the computability of the operations. Thus, making the operations of the vector space computable is not enough to have a stable notion of computability. This result may be considered as worrying because infinite-dimensional vector spaces are rather common in physics: they are the usual sort of space in which one defines some dynamics, whether quantum or classical. As one may wish to study the ability of a physical machine or a dynamics, to compute, one would like to have a stable notion of computability on such a space, and not a notion relative to the choice of an indexing. The fact that, in quantum theory, one does not directly observe the state vector but rather the induced probability distribution over measurement outcomes does not circumvent this problem: as was pointed out [10], uncomputable amplitudes can still be read out probabilistically.

Fortunately, we show that if we add a tensor product as a primitive operation of the algebraic structure, and restrict to indexings that make also this operation computable, then computability may be stable despite the infinite dimension of the space. This theorem is obtained from a novel general result on extensions of algebraic structures which preserve stable computability. Most proofs are not included in this short version but are available in [3].

2   Stable Computability

We assume familiarity with the notion of a computable function over the natural numbers, as presented for instance in [14,11]. As usual, this notion is transported to finite trees labeled in a finite set through the use of an indexing. To define this indexing, we first associate an index $\ulcorner f \urcorner$ to each element f of the set of labels and we define the indexing of the tree $f(t_1, \ldots, t_l)$ as $\ulcorner f(t_1, \ldots, t_l) \urcorner = \ulcorner f \urcorner; \ulcorner t_1 \urcorner; \ldots; \ulcorner t_l \urcorner; 0$ where ; is a computable one-to-one mapping from $\mathbb{N}^2$ to $\mathbb{N} \setminus \{0\}$, e.g. $n; p = (n+p)(n+p+1)/2 + n + 1$. The *indexing* of this set of trees is the partial function i mapping the index $\ulcorner t \urcorner$ to the term t. Notice that this indexing is independent of the choice of the indexing of the labels, that is if i_1 and i_2 are two indexings of the same set of trees built on different indexings of the labels, there exists a computable function h such that $i_2 = i_1 \circ h$.

When A is a set of natural numbers, we say that A is *effectively enumerable* if it is either empty or the image of a total computable function from $\mathbb{N}$ to $\mathbb{N}$. This is equivalent to the fact that A is the image of a partial computable function, or again to the fact that it is the domain of partial computable function.

Definition 1 (Indexing). *An indexing of a set E is a partial function i from $\mathbb{N}$ to E, such that*

- *dom(i) is effectively enumerable,*
- *i is surjective,*
- *there exists a computable function eq from dom$(i) \times$ dom(i) to $\{0,1\}$ such that eq$(x,y) = 1$ if $i(x) = i(y)$ and eq$(x,y) = 0$ otherwise.*

An indexing of a family of sets $E_1, \ldots, E_n$ is a family of indexings $i_1, \ldots, i_n$ of the sets $E_1, \ldots, E_n$.

Notice that this definition is slightly more general than the usual one (see for instance [18, Definition 2.2.2.], Definition 2.2.2) as we only require $dom(i)$ to be effectively enumerable, instead of requiring it to be decidable. Requiring full decidability of $dom(i)$ is not needed in this paper and this weaker requirement simplifies the study of extensions of finite degrees of the field of rationals.

Definition 2 (Computable function over a set equipped with an indexing). *Let $E_1, \ldots, E_m, E_{m+1}$ be a family of sets and $i_1, \ldots, i_m, i_{m+1}$ an indexing of this family. A function from $E_1 \times \ldots \times E_m$ to E_{m+1} is said to be computable relatively to this indexing if there exists a computable function $\hat{f}$ from dom$(i_1) \times \ldots \times$ dom(i_m) to dom(i_{m+1}) such that for all x_1 in dom(i_1), $\ldots, x_m$ in dom(i_m), $\hat{f}(x_1, \ldots, x_m)$ is in dom(i_{m+1}) and $i_{m+1}(\hat{f}(x_1, \ldots, x_m)) = f(i_1(x_1), \ldots, i_m(x_m))$.*

Definition 3 (Admissible indexing of an algebraic structure [13]). *An indexing of an algebraic structure $\langle E_1, \ldots, E_n, op_1, \ldots, op_p \rangle$ is an indexing of the family $E_1, \ldots, E_n$. Such an indexing is said to be admissible if the operations $op_1, \ldots, op_p$ are computable relatively to the indexings of their domains and co-domain.*

Definition 4 (Stable computability [18]). *Computability is said to be stable over a structure $\langle E_1, \ldots, E_n, op_1, \ldots, op_p \rangle$ if there exists an admissible indexing $i_1, \ldots, i_n$ of this structure and for all admissible indexings $j_1, \ldots, j_n$ and $j_1', \ldots, j_n'$ of this structure, there exist computable functions h_1 from $dom(j_1')$ to $dom(j_1)$, ..., h_n from $dom(j_n')$ to $dom(j_n)$ such that $j_1' = j_1 \circ h_1$, ..., $j_n' = j_n \circ h_n$.*

It can be proved that a structure has stable computability if and only if the set of computable function over a structure is independent of the choice of an admissible indexing. See the long version of the paper for the details.

3   Relative Finite Generation

We now want to prove that computability is stable over finitely generated structures. Intuitively, a structure is finitely generated if all its elements can be constructed with the operations of the structure from a finite number of elements $a_0, \ldots, a_{d-1}$. For instance, the structure $\langle \mathbb{N}, S \rangle$ is finitely generated because all natural numbers can be generated with the successor function from the number 0. This property can also be stated as the fact that for each element b of the structure there exists a term t, *i.e.* a finite tree labeled with the elements $a_0, \ldots, a_{d-1}$ and the operations of the structure, such that the denotation $[\![t]\!]$ of t is b. Yet, in this definition, we must take into account two other elements.

First, in order to prove that computability is stable, we must add the condition that equality of denotations is decidable on terms. This leads us to sometimes consider not the full set of terms $\mathcal{T}$ generated by the operations of the structure, but just an effectively enumerable subset T of this set, which is chosen large enough to contain a term denoting each element of the structure. This is in order to avoid situations where the equality of denotations may be undecidable, or more at least more difficult to prove decidable, on the full set of terms.

Secondly, when a structure has several domains, such as a vector space that has a domain for scalars and a domain for vectors, we want, in some cases, to assume first that some substructure, *e.g.* the field of the scalars, has stable computability, and still be able to state something alike finite generation for the other domains, *e.g.* the domain of the vectors. Thus, we shall consider structures $\langle E_1, \ldots, E_m, E_{m+1}, \ldots, E_{m+n}, op_1, \ldots, op_p, op_{p+1}, \ldots, op_{p+q} \rangle$ with two kinds of domains, assume that the structure $\langle E_1, \ldots, E_m, op_1, \ldots, op_p \rangle$ has stable computability, and consider terms that generate the elements of $E_{m+1}, \ldots, E_{m+n}$ with the operations $op_{p+1}, \ldots, op_{p+q}$. In these terms, the elements of $E_1, \ldots, E_m$ are expressed by their index which is itself expressed using the symbols 0 and S, in order to keep the language finite. The case $m = 0$ is the usual definition of finite generation, whereas in the case $m > 0$ we say that the algebraic structure $\langle E_{m+1}, \ldots, E_{m+n}, op_{p+1}, \ldots, op_{p+q} \rangle$ is finitely generated relatively to $\langle E_1, \ldots, E_m, op_1, \ldots, op_p \rangle$, which in turn may or may not be finitely generated.

Definition 5 (Terms, denotation). *Let $E_1, \ldots, E_m$ be sets equipped with indexings $i_1, \ldots, i_m$, $E_{m+1}, \ldots, E_{m+n}$ be sets, $A = \{a_0, \ldots, a_{d-1}\}$ be a finite set* *of elements of* $E_{m+1}, \ldots, E_{m+n}$, *and* $op_{p+1}, \ldots, op_{p+q}$ *be operations whose arguments are in* $E_1, \ldots, E_m, E_{m+1}, \ldots, E_{m+n}$ *but whose values are in* $E_{m+1}, \ldots, E_{m+n}$. *The set* $\mathcal{T}_k$ *of terms of sort* E_k *is inductively defined as follows*

- *if the natural number* x *is an element of* $dom(i_k)$ *(for* k *in* $1, \ldots, m$*), then* $S^x(0)$ *is a term of sort* E_k,
- *if* a *is an element of* A *and* E_k, *then* a *is a term of sort* E_k,
- *if* t_1 *is a term of sort* E_{k_1}, *...,* t_l *is an term of sort* E_{k_l} *and* op *is one of the functions from* $op_{p+1}, \ldots, op_{p+q}$ *from* $E_{k_1} \times \ldots \times E_{k_l}$ *to* $E_{k_{l+1}}$ *then* $op(t_1, \ldots, t_l)$ *is a term of sort* $E_{k_{l+1}}$.

The denotation of a term of sort E_k *is the element of* E_k *defined as follows*

- $[\![S^x(0)]\!]_{E_k} = i_k(x)$,
- $[\![a]\!]_{E_k} = a$,
- $[\![op(t_1, \ldots, t_l)]\!]_{E_{k_{l+1}}} = op([\![t_1]\!]_{E_{k_1}}, \ldots, [\![t_l]\!]_{E_{k_l}})$.

As a term is a tree labeled in the finite set $a_0, \ldots, a_{d-1}, op_{p+1}, \ldots, op_{p+q}, 0, S$, *we can associate an index* $\ulcorner t \urcorner$ *to each term* t *in a canonical way.*

Definition 6 (Finite generative set). *Let* $E_1, \ldots, E_m$ *be a family of sets equipped with indexings* $i_1, \ldots, i_m$ *and let* $\langle E_{m+1}, \ldots, E_{m+n}, op_{p+1}, \ldots, op_{p+q} \rangle$ *be a structure. A finite set* A *of elements of the sets* $E_{m+1}, \ldots, E_{m+n}$ *is said to be a* finite generative set *of the structure* $\langle E_{m+1}, \ldots, E_{m+n}, op_{p+1}, \ldots, op_{p+q} \rangle$ *relatively to* $E_1, \ldots, E_m, i_1, \ldots, i_m$, *if there exist effectively enumerable subsets* $T_{m+1}, \ldots, T_{m+n}$ *of the sets* $\mathcal{T}_{m+1}, \ldots, \mathcal{T}_{m+n}$ *of terms of sort* $E_{m+1}, \ldots, E_{m+n}$, *such that for each element* b *of a set* E_k *there exists a term* t *in* T_k *such that* $b = [\![t]\!]_{E_k}$ *and, for each* k, *there exists a computable function* eq_k *such that for all* t *and* u *in* T_k, $eq_k(\ulcorner t \urcorner, \ulcorner u \urcorner) = 1$ *if* $[\![t]\!] = [\![u]\!]$ *and* $eq_k(\ulcorner t \urcorner, \ulcorner u \urcorner) = 0$ *otherwise.*

Definition 7 (Finitely generated). *Let* $E_1, \ldots, E_m$ *be a family of sets equipped with indexings* $i_1, \ldots, i_m$. *The structure* $\langle E_{m+1}, \ldots, E_{m+n}, op_{p+1}, \ldots, op_{p+q} \rangle$ *is said to be finitely generated relatively to* $E_1, \ldots, E_m, i_1, \ldots, i_m$, *if it has a finite generative set relatively to* $E_1, \ldots, E_m, i_1, \ldots, i_m$.

Remark 1. This notion of finite generation generalizes to arbitrary structures the notion of finite-dimensional vector space. More generally, we can define the *dimension* of a structure as the minimal cardinal of a finite generative set of this structure.

Theorem 1. *Let* $\langle E_1, \ldots, E_m, op_1, \ldots, op_p \rangle$ *be a structure with stable computability and* $s_1, \ldots, s_m$ *be an admissible indexing of this structure. Then, if the structure* $\langle E_{m+1}, \ldots, E_{m+n}, op_{p+1}, \ldots, op_{p+q} \rangle$ *is finitely generated relatively to* $E_1, \ldots, E_m, s_1, \ldots, s_m$, *then computability is stable over the structure* $\langle E_1, \ldots, E_m, E_{m+1}, \ldots, E_{m+n}, op_1, \ldots, op_p, op_{p+1}, \ldots, op_{p+q} \rangle$.

Proof. We build an indexing such that for k in $m+1, \ldots, m+n$, i_k maps the index $\ulcorner t \urcorner$ of the term t to $[\![t]\!]_{E_k}$ and we prove that all indexings are equivalent to this one. See the long version of the paper for the details.

As corollaries, we get stable computability for well-known cases.

Proposition 1 (Natural numbers). *Computability is stable over the structures $\langle \mathbb{N}, S \rangle$ and $\langle \mathbb{N}, + \rangle$.*

Proposition 2 (Rational numbers). *Computability is stable over the structure $\langle \mathbb{Q}, +, -, \times, / \rangle$.*

Proposition 3. *Computability is stable over the structure $\langle \mathbb{Q}, +, \times \rangle$.*

Proof. We prove that the indexings admissible for $\langle \mathbb{Q}, +, \times \rangle$ and $\langle \mathbb{Q}, +, -, \times, / \rangle$ are the same. If the indexing i is admissible for $\langle \mathbb{Q}, +, -, \times, / \rangle$ then it is obviously admissible for $\langle \mathbb{Q}, +, \times \rangle$. Conversely, if i is admissible for $\langle \mathbb{Q}, +, \times \rangle$, then there exist computable functions $\hat{+}$ and $\hat{\times}$, from $\mathbb{N}^2$ to $\mathbb{N}$, such that $i(x \hat{+} y) = i(x) + i(y)$ and $i(x \hat{\times} y) = i(x) \times i(y)$. The set $dom(i)$ is non empty and recursively enumerable, thus it is the image of a computable function g. Let n and p be two natural numbers, we define $n \hat{-} p = g(z)$ where z is the least natural number such that $p \hat{+} g(z) = n$. We have $i(x \hat{-} y) = i(x) - i(y)$. We build the function $\hat{/}$ in a similar way. Thus, i is admissible for $\langle \mathbb{Q}, +, -, \times, / \rangle$.

Remark 2. Computability is stable over the structure $\langle \mathbb{Q}, +, \times \rangle$, but this structure is not finitely generated. Indeed, consider a finite number of rational numbers $a_0 = p_0/q_0, \ldots, a_{d-1} = p_{d-1}/q_{d-1}$ and call q a common multiple of the denominators $q_0, \ldots, q_{d-1}$. All numbers that are the denotation of a term in the language $a_0, \ldots, a_{d-1}, +, \times$ have the form p/q^k for some p and k. The numbers $q + 1$ and q are relatively prime, thus $q + 1$ is not a divisor of q^k for any k, the number $1/(q + 1)$ does not have the form p/q^k, and it cannot be expressed by a term. Thus, finite generation is a sufficient condition for stable computability, but it is not a necessary one.

The stability of computability over the structure $\langle \mathbb{Q}, +, \times \rangle$ can be explained by the fact that, although subtraction and division are not operations of the structure, they can be effectively defined from these operations. Such a structure is said to be *effectively generated*.

There are several ways to define this notion. For instance, [19] proposes a definition based on a simple imperative programming language with the operations of the algebraic structure as primitive. A more abstract definition is the existence of a language, that needs not be the language of terms built with operations on the structure, but may, for instance, be a programming language, and a denotation function associating an element of the algebraic structure to each expression of the language, verifying the following properties:

- the set of well-formed expressions is effectively enumerable,
- each element is the denotation of some expression,
- equality of denotations is decidable,
- an expression denoting $op(a_1, \ldots, a_n)$ can be built from ones denoting $a_1, \ldots, a_n$,
- for any admissible indexing, an index of a can be computed from an expression denoting a.

Replacing expressions by their indices, we get this way exactly the definition of stable computability. Thus, stability of computability and effective generation are trivially equivalent in this approach.

4 Vector Spaces

As a corollary of Theorem 1, we also get the following.

Proposition 4 (Vector spaces). *If computability is stable over the field $\langle K, +, \times \rangle$ and the vector space $\langle K, E, +, \times, +, . \rangle$ is finite-dimensional, then computability is stable over $\langle K, E, +, \times, +, . \rangle$.*

We can also prove (but the proof is more complex) that computability is stable over the finite extensions of the field of rationals, including for instance $\mathbb{Q}[\sqrt{2}]$ and $\mathbb{Q}[i, \sqrt{2}]$. See the long version of the paper for the details.

Proposition 5. *Computability is not stable over the structure $\langle \mathbb{N} \rangle$.*

Proof. Let f be a non computable one-to-one function from $\mathbb{N}$ to $\mathbb{N}$, for instance the function mapping $2n$ to $2n+1$ and $2n+1$ to $2n$ if $n \in U$ and $2n$ to $2n$ and $2n+1$ to $2n+1$ otherwise, where U is any undecidable set.

The function f and the identity are both admissible indexings of the structure $\langle \mathbb{N} \rangle$. If computability were stable over this structure, then there would exist a computable function h such that $f = id \circ h$. Thus, $f = h$ would be computable, which is contradictory.

Proposition 6. *Computability is stable over the vector space $\langle K, E, +, \times, +, . \rangle$ if and only if it is stable over the field $\langle K, +, \times \rangle$ and $\langle K, E, +, \times, +, . \rangle$ is a finite-dimensional vector space.*

Proof. If the dimension of E is not countable, then the set E itself is not countable, and hence computability is not stable over this structure.

If the dimension of E is finite or countable and computability is not stable over the field $\langle K, +, \times \rangle$, then it is not stable over the vector space $\langle K, E, +, \times, +, . \rangle$ either. Indeed, if s and s' are two non equivalent indexings of $\langle K, +, \times \rangle$, and $e_0, e_1, \ldots$ is a finite or countable basis of E, then, we let i be the function mapping $\ulcorner (S^{\lambda_0}(0).e_0 + (S^{\lambda_1}(0).e_1 + (\ldots S^{\lambda_n}(0).e_n))) \urcorner$ where $\lambda_0, \lambda_1, \ldots, \lambda_n$ are in $dom(s)$ to $s(\lambda_0).e_0 + s(\lambda_1).e_1 + \ldots + s(\lambda_n).e_n$ and i' be the function mapping $\ulcorner (S^{\lambda_0}(0).e_0 + (S^{\lambda_1}(0).e_1 + (\ldots S^{\lambda_n}(0).e_n))) \urcorner$ where $\lambda_0, \lambda_1, \ldots, \lambda_n$ are in $dom(s')$ to $s'(\lambda_0).e_0 + s'(\lambda_1).e_1 + \ldots + s'(\lambda_n).e_n$, and s, i and s', i' are two non equivalent indexings of $\langle K, E, +, \times, +, . \rangle$.

We have proved in Proposition 4 that if computability is stable over the field $\langle K, +, \times \rangle$ and the vector space $\langle K, E, +, \times, +, . \rangle$ is finite-dimensional, then computability is stable over this space. Thus, all that remains to be proved is that if computability is stable over the field $\langle K, +, \times \rangle$ and the vector space $\langle K, E, +, \times, +, . \rangle$ has a countably infinite dimension, then computability is not stable over this vector space.

Let s be an indexing of $\langle K, +, \times \rangle$, $e_0, e_1, \ldots$ be a basis of E, and i be the function mapping $\ulcorner (S^{\lambda_0}(0).e_0 + (S^{\lambda_1}(0).e_1 + (\ldots S^{\lambda_n}(0).e_n))) \urcorner$ where $\lambda_0, \lambda_1, \ldots \lambda_n$ are in $dom(s)$ to $s(\lambda_0).e_0 + s(\lambda_1).e_1 + \ldots + s(\lambda_n).e_n$. As the function s is surjective and $e_0, e_1, \ldots$ is a basis, the function i is surjective. As there exist computable functions $\hat{+}$ and $\hat{\times}$ and eq on scalars, we can build computable functions $\hat{+}$ and $\hat{.}$ and eq on such terms. Thus, the function i is an admissible indexing of E.

Then, let f be a non computable one-to-one function from $\mathbb{N}$ to $\mathbb{N}$. Let ϕ be the one-to-one linear function from E to E mapping the basis vector e_p to $e_{f(p)}$ for all p. As ϕ is a one-to-one mapping between E and itself, $\phi \circ i$ is a surjection from $dom(i)$ to E. As ϕ is linear, $\phi(i(u \hat{+} v)) = \phi(i(u) + i(v)) = \phi(i(u)) + \phi(i(v))$ and $\phi(i(\lambda \hat{.} u)) = \phi(s(\lambda).i(u)) = s(\lambda).\phi(i(u))$. Thus, $\phi \circ i$ is an admissible indexing of E. And if computability were stable over $\langle K, E, +, \times, +, . \rangle$, there would exists a computable function g from $dom(i)$ to $dom(i)$ such that $\phi \circ i = i \circ g$.

Let z be any number such that $s(z)$ is the scalar 0 and u be any number such that $s(u)$ is the scalar 1. Let B be the computable function, from $\mathbb{N}$ to $\mathbb{N}$, mapping p to $\ulcorner S^z(0).e_0 + \ldots + S^z(0).e_{p-1} + S^u(0).e_p \urcorner$. We have $i(B(p)) = e_p$. Let C be the partial computable function from $\mathbb{N}$ to $\mathbb{N}$ mapping the index $\ulcorner(S^{\lambda_0}(0).e_0 + (S^{\lambda_1}(0).e_1 + (\ldots S^{\lambda_n}(0).e_n)))\urcorner$ to the least p such that $eq(\lambda_p, u) = 1$. If $i(x) = e_p$ then $C(x) = p$. Let h be the computable function $C \circ g \circ B$. Let p be an arbitrary natural number. We have $i(g(B(p))) = \phi(i(B(p))) = \phi(e_p) = e_{f(p)}$. Thus, $h(p) = C(g(B(p))) = f(p)$. Thus, $f = h$ would be computable, which is contradictory.

5   Tensor Spaces

In Proposition 6, we have shown that infinite-dimensional vector spaces do not have a stable notion of computability. Intuitively, the lack stable computability for infinite-dimensional vector spaces happens for the same reason as it does for $\langle \mathbb{N} \rangle$, as can be seen from the proofs of Proposition 5 and 6. This is because neither algebraic structures are effectively generated, *i.e.* there is an infinity of elements (the natural numbers in one case, the vectors of a basis in the other) which are unrelated from one another.

In the case of the natural numbers this can be fixed by requiring that the successor be computable, *i.e.* by considering the structure $\langle \mathbb{N}, S \rangle$. On the set of finite sequences of elements taken in a finite set, the problem would be fixed in a similar way by adding the *cons* operation, that adds an element at the head of the list. On an infinite set of trees, we would add the operation that builds the tree $f(t_1, \ldots, t_n)$ from f and $t_1, \ldots, t_n, \ldots$ These operations express that, although these data types are infinite, they are finitely generated.

In the same way, in quantum computing, an infinite data type comes with some structure, *i.e.* the vector space used to describe these data has a basis that is finitely generated with the tensor product. Typically, the basis vectors have the form $b_1 \otimes \ldots \otimes b_n$ where each b_i is either $|0\rangle$ or $|1\rangle$.

Definition 8 (Tensor space). *A tensor space $\langle K, E, +, \times, +, ., \otimes \rangle$ is a vector space with an extra operation $\otimes$ that is a bilinear function from $E \times E$ to E.*

Theorem 2. *Let $\langle K, E, +, \times, +, ., \otimes \rangle$ be a tensor space such that computability is stable over the field $\langle K, +, \times \rangle$ and there exists a finite subset A of E such that the set of vectors of the form $b_1 \otimes (b_2 \otimes \ldots \otimes (b_{n-1} \otimes b_n) \ldots)$, for $b_1, \ldots, b_n$ in A, is a basis of E. Then, computability is stable over the structure $\langle K, E, +, \times, +, ., \otimes \rangle$.*

Proof. Let s be an indexing of the field $\langle K, +, \times \rangle$. Consider the terms of the form $(S^{\lambda_0}(0).(b_1^0 \otimes \ldots \otimes b_{n_0}^0) + (S^{\lambda_1}(0).(b_1^1 \otimes \ldots \otimes b_{n_1}^1) + \ldots$ where $\lambda_0, \lambda_1, \ldots$ are in $dom(s)$ and $b_1^0, b_2^0, \ldots$ are in A. The denotation of such a term is the vector $s(\lambda_0).(b_1^0 \otimes \ldots \otimes b_{n_0}^0) + s(\lambda_1).(b_1^1 \otimes \ldots \otimes b_{n_1}^1) + \ldots$ As s is surjective and the vectors $b_1^0 \otimes \ldots \otimes b_{n_0}^0, \ldots$ form a basis, every element of E is the denotation of a term. As equality of denotations can be decided on scalars, it can be decided on terms. Thus, the structure $\langle E, +, ., \otimes \rangle$ is finitely generated relatively to the field $\langle K, +, \times \rangle$. As computability is stable over this field, it is stable over $\langle K, E, +, \times, +, ., \otimes \rangle$.

6   Conclusion

The robustness of the notion of computability over the natural numbers has been pointed out for long. But it has also been pointed out that this robustness does not extend to other countable domains, where computability is relative to the choice of an indexing. We have shown that this robustness is, in fact, shared by many algebraic structures: all the extensions of finite degree of the field of rationals (*e.g.* $\mathbb{Q}[\sqrt{2}]$ and $\mathbb{Q}[i, \sqrt{2}]$), all the finite-dimensional vector spaces over such a field, and all tensor spaces over such a field that are finite-dimensional (as tensor spaces) even if they are infinite-dimensional (as vector spaces).

For the vector spaces used in quantum computing, it is not the dimension as a vector space that matters, but the dimension as a tensor space. Indeed, the vector space operations handle the superposition principle, but the finite generation of the data types is handled by the tensor product. Finite-dimensional tensor space over extensions of finite degree of the field of rationals are probably sufficient to express all quantum algorithms.

Whether such spaces are sufficient to express all quantum physics is related to the possibility to decompose a system that has an infinite number of base states into finite but *a priori* unbounded number of subsystems, each having a finite number of base states.

More generally, when a structure E has stable computability, and some dynamics of a physical system is described as a function f E to E, we can then consider its associated function $\hat{f}$, and ask ourselves whether this function is computable, universal, or uncomputable. Hence, this provides a formal sense in which some dynamics may respect, or break the Church-Turing thesis [10].

References

1. Altenkirch, T., Grattage, J., Vizzotto, J.K., Sabry, A.: An algebra of pure quantum programming. In: Third International Workshop on Quantum Programming Languages. ENTCS, vol. 170C, pp. 23–47 (2007)
2. Arrighi, P., Dowek, G.: Linear-algebraic lambda-calculus: higher-order, encodings and confluence. In: Voronkov, A. (ed.) RTA 2008. LNCS, vol. 5117, pp. 17–31. Springer, Heidelberg (2008)
3. Arrighi, P., Dowek, G.: On the completeness of quantum computation models. Long version of this paper, available as an ArXiv preprint (2010)

4. Arrighi, P., Grattage, J.: Intrinsically universal n-dimensional quantum cellular automata. ArXiv preprint: arXiv:0907.3827 (2009)
5. Arrighi, P., Nesme, V., Werner, R.: Unitarity plus causality implies localizability. QIP 2010, ArXiv preprint: arXiv:0711.3975 (2007)
6. Bernstein, E., Vazirani, U.: Quantum complexity theory. In: Proceedings of the Twenty-Fifth Annual ACM Symposium on Theory of Computing, pp. 11–20. ACM, New York (1993)
7. Boker, U., Dershowitz, N.: The church-turing thesis over arbitrary domains. In: Avron, A., Dershowitz, N., Rabinovich, A. (eds.) Pillars of Computer Science. LNCS, vol. 4800, pp. 199–229. Springer, Heidelberg (2008)
8. Deutsch, D.: Quantum theory, the Church-Turing principle and the universal quantum computer. Proceedings of the Royal Society of London. Series A, Mathematical and Physical Sciences (1934-1990) 400(1818), 97–117 (1985)
9. Montague, R.: Towards a general theory of computability. Synthese 12(4), 429–438 (1960)
10. Nielsen, M.A.: Computable functions, quantum measurements, and quantum dynamics. Phys. Rev. Lett. 79(15), 2915–2918 (1997)
11. Odifreddi, P.: Classical Recursion Theory. North-Holland, Amsterdam (1988)
12. Perdrix, S.: Partial observation of quantum Turing machine and weaker wellformedness condition. In: Proceedings of Joint Quantum Physics and Logic & Development of Computational Models (Joint 5th QPL and 4th DCM) (2008)
13. Rabin, M.O.: Computable algebra, general theory and theory of computable fields. Transactions of the American Mathematical Society 95(2), 341–360 (1960)
14. Rogers, H.: Theory of Recursive Functions and Effective Computability. MIT Press, Cambridge (1967)
15. Schumacher, B., Werner, R.: Reversible quantum cellular automata. ArXiv preprint quant-ph/0405174 (2004)
16. Selinger, P.: Towards a quantum programming language. Mathematical Structures in Computer Science 14(4), 527–586 (2004)
17. Selinger, P., Valiron, B.: A lambda calculus for quantum computation with classical control. Mathematical Structures in Computer Science 16(3), 527–552 (2006)
18. Stoltenberg-Hansen, V., Tucker, J.V.: Effective algebras, pp. 357–526. Oxford University Press, Oxford (1995)
19. Tucker, J.V., Zucker, J.I.: Abstract versus concrete computability: The case of countable algebras. In: Stoltenberg-Hansen, V., Väänänen, J. (eds.) Logic Colloquium 2003. Lecture Notes in Logic, vol. 24, pp. 377–408 (2006)
20. Van Tonder, A.: A lambda calculus for quantum computation, arXiv:quant-ph/0307150 (2003)

The Ordinal of Skolem + Tetration Is τ_0

Mathias Barra and Philipp Gerhardy*

Department of Mathematics, University of Oslo, P.B. 1053,
Blindern, 0316 Oslo, Norway
{georgba,philipge}@math.uio.no

Abstract. In [1], we proved that a certain family of number theoretic functions S_* is *well-ordered* by the *majorisation relation* '$\preceq$'. Furthermore, we proved that a *lower bound* on the ordinal $O(S_*, \preceq)$ of this well-order is *the least critical epsilon number* τ_0. In this paper we prove that τ_0 is also an *upper bound* for its ordinal, whence our sought-after result,
$$O(S_*, \preceq) = \tau_0,$$
is an immediate consequence.

1 Introduction and Previous Results

In [5] Thoralf Skolem defined an ordered class S of number-theoretic functions corresponding to the ordinal ε_0. The motivation for this was to find a more mathematical representation than the usual set-representation one. Once an ordered class of functions corresponding to an interesting ordinal has been found, a natural question is: *given some other class of functions and an order-relation, is this a well-order too, and if so, what is its ordinal?* The question which we began to answer in [1] and complete here is of this kind.

Before we state our new result, we recall some definitions and results from [1], and we adopt all notation from [1]. In particular, given a class of functions F and an order-relation $\leq_F$ on F such that the structure $(F, \leq_F)$ is a well-order, we denote by $O(F, \leq_F)$ the order-type (ordinal) of the order. For $f \in F$ we write $O(f)$ for the ordinal of the initial segment of $(F, \leq_F)$ determined by f. Latin letters f, g etc. range over functions, Greek letters α, β etc. (except φ) range over ordinals, and **ON** is the class of ordinals.

Definition 1 (Majorisation). *Define the* majorisation relation '$\preceq$' *on* $\mathbb{N}^{\mathbb{N}}$ *by*

$$f \preceq g \overset{\text{def}}{\Leftrightarrow} \exists_{N \in \mathbb{N}} \forall_{x \geq N} (f(x) \leq g(x)).$$

We say that g majorises *f when $f \preceq g$, and as usual $f \prec g \overset{\text{def}}{\Leftrightarrow} f \preceq g \wedge g \not\preceq f$. We say that f and g are* comparable *if $f \prec g$, $f = g$ or $g \prec f$.*

* The second author is supported by a grant from the Norwegian Research Council. We would like to thank the anonymous referees for their careful reading and helpful suggestions.

Hence g majorises f if g *is almost everywhere (a.e.) greater than or equal to* f.

We assume the reader is familiar with ordinals and their arithmetic (see e.g. [4]), in particular basic knowledge of the epsilon numbers; the fixed points $\varepsilon_0, \varepsilon_1, \ldots, \varepsilon_\alpha, \ldots$ of the map $\alpha \mapsto \omega^\alpha$. We denote by τ_0 the *least critical epsilon number*, viz. the smallest fixed point to the map $\alpha \mapsto \varepsilon_\alpha$.

In [5] Skolem describes a class of functions S, that is well-ordered by the majorisation relation and corresponds to ε_0. At the end of [5] Skolem considers two extensions of his class S and asks whether they are well-ordered by $\preceq$ and what their ordinals might be. For the second of these extensions these questions were addressed and partially solved in [1]. For the accurate formulation of Skolem's second problem we define the number-theoretic function $t \colon \mathbb{N}^2 \to \mathbb{N}$ of *tetration* by[1]

$$t(x, y) \overset{\text{def}}{=} x_y \overset{\text{def}}{=} \left\{ \begin{array}{l} 1 \qquad\quad\ , y = 0 \\ x^{(x_{y-1})} \ , y > 0 \end{array} \right\} = x^{\cdot^{\cdot^{\cdot^{x}}}} \Big\}_y.$$

Let us now restate the second problem here:

Problem 2 (Skolem [5]). Define S_* by $0, 1 \in S_*$, and, if $f, g \in S_*$ then $f + g, f \cdot g, x^f, x_f \in S_*$. Is $(S_*, \preceq)$ a well-order? If so, what is $\mathrm{O}\,(S_*, \preceq)$?

The key to establishing the well-orderedness of S_* by $\preceq$ and the corresponding ordinal is to define suitable unique normal-forms for terms[2] in S_*. The well-order and ordinal bounds may then be investigated using nice properties of these normal forms.

The first step to finding unique normal-forms was to observe that – since addition and multiplication are commutative and distributive – every S_*-term can be written as a sum of products, a so-called $\Sigma\Pi$-pre-normal-form:

Definition 2 (Pre-normal-form). *For* $f \in S$, *we call the unique normal-form for* f *from [5] the Skolem normal-form (SNF) of* f.

Let s, t *range over* S. *We say that a function* $f \in S_*$ *is on* $(\Sigma\Pi\text{-})$ *pre-normal-form* $((\Sigma\Pi\text{-})$ *PNF) if* $f \in S$ *and* f *is on SNF, or if* $f \in \overline{S_*}$ *and* f *is of the form* $f = \sum_{i=1}^n \prod_{j=1}^{n_i} f_{ij}$ *where either*

$f_{i,j} \equiv x_g$ *where* g *is in* $\Sigma\Pi\text{-}PNF$;
$f_{i,j} \equiv x^g$ *where* g *is in* PNF, $g \notin S$, $g \equiv (\prod_{i=1}^{n_g} g_i) \not\equiv x_h$, *and* $g_{n_g} \not\equiv (s+t)$;
$f_{i,j} \equiv s$ *where* s *is in* SNF, *and* $j = n_i$.

An f_{ij} *on one of the above three forms is an* S_*-*factor, a product* $\prod f_i$ *of* S_*-*factors is an* S_*-*product, also called a* Π-*PNF. We say that* x_h *is a tetration-factor, that* $x^{\prod g_j}$ *is an exponent-factor with exponent-product* $\prod g_j$, *and that* $s \in S$ *is a Skolem-factor.*

---

[1] What 0^0 should be is irrelevant, it may be set to any natural number without affecting any results.

[2] As in [1], we blur the distinction between functions and the terms that represent them. We feel particularly justified to do so as we established the existence of unique normal- forms.

Adding further constraints, such as reducing x^{x_h} to x_{h+1} or reducing x^{f+g} to $x^f x^g$, we arrived at the following normal form:

Definition 3 (Normal-form). *Let $f \in S_*$. We say that the pre-normal-form $\Sigma\Pi f_{ij}$ is a* normal-form (NF) *for f if*

(NF1) $f_{ij} \succeq f_{i(j+1)}$, *and* $f_{ij} \succ f_{i(j+1)} \Rightarrow \forall_{\ell\in\mathbb{N}} \left(f_{ij} \succ \left(f_{i(j+1)} \right)^{\ell} \right)$;

(NF2) $\forall_{s\in S} \left(\prod_j f_{ij} \succ \left(\prod_j f_{(i+1)j} \right) \cdot s \right)$;

(NF3) *If f_{ij} is of the form x_h or x^h, then h is in NF.*

A $\Sigma\Pi$-normal-form with one summand only is called a Π-normal-form.

Terms in $\Sigma\Pi$-normal-form can be compared subterm by subterm. In fact, two functions $f = \Sigma\Pi f_{ij}$ and $g = \Sigma\Pi g_{ij}$, satisfy $f \prec g$ if and only if $f_{ij} \prec g_{ij}$ for the (lexicographically) first indices i, j where they differ. This fact, and also that all functions in S_* admit a unique normal-form, was proved in [1].

Next, by defining a countable sequence of functions $\{F_k\}$ in S_*, such that the corresponding sequence of ordinals $\mathrm{O}(F_k)$ was cofinal in τ_0, we obtained the lower bound.

Theorem 1 (Barra and Gerhardy [1])

1. $(S_*, \preceq)$ *is a well-order;*
2. $\tau_0 \leq \mathrm{O}(S_*, \preceq)$.

For the details of these two results, see [1].

Hence, to fully answer Skolem's second problem from [5], we need to establish that $\mathrm{O}(S_*, \preceq) \leq \tau_0$, which is done throughout the next section.

2   Main Results

Our strategy will be to describe an order-preserving map $\varphi : S_* \to \tau_0$. Since we have unique normal-forms for the functions in S_*, it suffices to define φ for terms in normal-form only. The definition of φ may then be given in a straightforward way, namely by translating the arithmetical operations on the functions into ordinal counterparts. Intuitively this translation maps every term from S_* in normal-form to an ordinal term by the simple substitution $x := \omega$. However, to obtain injectivity and at the same time stay below τ_0, care must be given to the definition of 'ordinal tetration', so that φ avoids the ε-numbers.

To elaborate on this issue a little bit, consider the definition of $t(x, y) = x_y$ above. Clearly e.g. $x_{x+1} = x^{x_x}$. Now, an ordinal generalisation of tetration as iterated ordinal exponentiation with canonical limit-clause $t(\alpha, \gamma) \stackrel{\mathrm{def}}{=} \sup_{\beta<\gamma} t(\alpha, \beta)$ does not yield a satisfactory normal[3] function on the ordinals;

---

[3] A normal function on the ordinals is continuous and increasing. The standard arithmetical operations, addition, multiplication and exponentiation, on the ordinals are all normal in the second argument.

e.g. one easily verifies $t(\omega, \omega + 1) = \omega^{t(\omega,\omega)} = \omega^{\varepsilon_0} = \varepsilon_0 = t(\omega, \omega)$. Basically, $t(\alpha, \beta)$ becomes stationary at the next ε-number after α for $\beta \geq \omega$.

Thus, assume the function x_x gets mapped to ε_0 (we know that in the actual order $(S_*, \preceq)$, ε_0 is the ordinal corresponding to x_x). Then a naive definition of φ 'should' map x_{x+1} to ω^{ε_0} – but that mapping is not order-preserving as, of course, $\omega^{\varepsilon_0} = \varepsilon_0$.

For our purposes, at the sucessor steps, we can define our version of ordinal tetration in the obvious way: by increasing the number of exponentiations by one. In the limit step though, we have to make sure we avoid the ε-numbers. The definition below achieves these objectives.

Definition 4. *Define the following version of ordinal (ω-)tetration $\alpha \mapsto \omega_\alpha$ by:*

$$\omega_0 = 1 \;,\; \omega_{\alpha+1} = \omega^{\omega_\alpha} \;,\; \omega_\gamma = (\sup_{\beta < \gamma} \omega_\beta)\omega.$$

Of course, this function is not normal either, $\sup_{n < \omega} \omega_n = \varepsilon_0 < \varepsilon_0 \omega = \omega_{\sup_{n < \omega}}$, i.e. it is not continuous. Therefore it may not represent the ideal choice for ordinal tetration. However, it is monotone increasing, a paramount property for the purpose it is meant to serve here: a key ingredient in the construction of φ.

Before we move on to the definition of φ, we establish some important properties of our definition of ordinal tetration. But first, we recall the following, useful description of the ε-numbers:

- $\varepsilon_0 = \sup\{\omega, \omega^\omega, \omega^{\omega^\omega}, \ldots\}$.
- $\varepsilon_{\alpha+1} = \sup\{\varepsilon_\alpha + 1, \omega^{\varepsilon_\alpha+1}, \omega^{\omega^{\varepsilon_\alpha+1}}, \ldots\}$.
- $\varepsilon_\gamma = \sup_{\beta < \gamma}\{\varepsilon_\beta\}$ for γ a limit ordinal.

Proposition 1. *1. For all $\alpha, \beta \in \mathbf{ON}$ we have that $\omega_\alpha \neq \varepsilon_\beta$, that is, no ω_α is an ε-number. 2. If $\omega_\alpha = \beta$, then $\omega_{\alpha+\omega} = \varepsilon' \cdot \omega$, where ε' is the smallest ε-number strictly greater than β.*

Proof. $\omega_0 = 1$ is not an ε-number, and $\omega_{\alpha+1}$ is an ε-number only if ω_α is an ε-number. In the limit case, observing that $\alpha\omega$ is never an ε-number, proves 1. For 2. Since $\beta = \omega_\alpha$ is not an ε-number by 1., the result follows directly from basic properties of the ε-numbers. ☐

So $\omega_\omega = \varepsilon_0 \cdot \omega$, $\omega_{\omega \cdot 2} = \varepsilon_1 \cdot \omega$, and $\omega_{\omega^2} = \varepsilon_\omega \cdot \omega$.

Proposition 2. *Let $\alpha < \beta$. Then $\omega_\alpha < \omega_\beta$.*

Proof. By induction on β.

$\beta = 1$: $\omega_0 = 1 < \omega = \omega_1$.

$\beta = \beta' + 1$: Then $\alpha \leq \beta'$, and, since, by Proposition 1, ω_α cannot be an ε-number, $\omega_\alpha < \omega^{\omega_\alpha} \leq \omega^{\omega_{\beta'}} \overset{\text{def}}{=} \omega_\beta$.

β is a limit ordinal: Since $\alpha < \beta$, we also have $\omega_\alpha \leq \sup_{\gamma < \beta}\{\omega_\gamma\}$, and then $\omega_\alpha < (\sup_{\gamma < \beta}\{\omega_\gamma\})\omega \overset{\text{def}}{=} \omega_\beta$. ☐

Remark 1. The above two propositions would already hold, if we defined $w_\gamma = (\sup_{\beta<\gamma} w_\beta) + 1$ instead of $w_\gamma = (\sup_{\beta<\gamma} w_\beta)w$. However, Proposition 3 would *not* hold, and that proposition is crucial for showing that the map φ will be order-preserving (hence injective).

Proposition 3. *Let* $\alpha < \beta$ *and* $k \in \mathbb{N}$. *Then* $(w_\alpha)^k < w_\beta$.

Proof. Obviously, $\alpha + 1 \leq \beta$, and in principle we only prove that $(w_\alpha)^k < w_{\alpha+1}$. The full result then follows by Proposition 2. The proof is by induction on α.

$\alpha = 0$: $(w_0)^k = 1 < w \leq w_\beta$.

$\alpha = \alpha' + 1$: $(w_\alpha)^k = (w^{w_{\alpha'}})^k = w^{w_{\alpha'}k} < w^{w^{w_{\alpha'}}} \leq w_\beta$, where the strict inequality is justified by Proposition 1.

α is a limit ordinal:

$$(w_\alpha)^k \overset{\text{def}}{=} ((\sup_{\gamma<\alpha}\{w_\gamma\})w)^k < (\sup_{\gamma<\alpha}\{w_\gamma\})^w \overset{\dagger}{\leq} w^{(\sup_{\gamma<\alpha}\{w_\gamma\})w} \overset{\text{def}}{=} w^{w_\alpha} \leq w_\beta$$

where the $\overset{\dagger}{\leq}$ follows from the simple observation that $\beta \leq w^\beta$ for all β , where in this case $\beta = \sup_{\gamma<\alpha}\{w_\gamma\}$. $\square$

Proposition 4. *Let* $\alpha \geq w^w$ *be an additive prime[4]. Then* $w_\alpha = \varepsilon_\alpha \cdot w$.

Proof. Assume first that $w_\alpha = \varepsilon_\alpha w$ for some α. This means we have induction start for a proof of the assertion that

$$w_{\alpha+w\cdot\beta} = \varepsilon_{\alpha+\beta} \cdot w.$$

In the successor case we first have $w_{\alpha+w\cdot(\beta+1)} = w_{\alpha+w\cdot\beta+w}$. By Proposition 1 and the IH we obtain $w_{\alpha+w\cdot\beta+w} = \varepsilon_{\alpha+\beta+1} \cdot w$. The limit case remains. Then

$$w_{\alpha+w\cdot\gamma} = \left(\sup_{\beta<\alpha+w\cdot\gamma} w_\beta\right) \cdot w = \left(\sup_{\beta<\gamma} w_{\alpha+w\cdot\beta}\right) \cdot w = \left(\sup_{\beta<\gamma}(\varepsilon_{\alpha+\beta} \cdot w)\right) \cdot w = \varepsilon_{\alpha+\gamma} \cdot w ,$$

which proves the assertion above. Above, the third equality is the IH, and the second is justified by the fact that $\beta \mapsto \alpha + w \cdot \beta$ is continuous, so that the suprema coincide.

Straightforward computations, e.g. using Proposition 1, now show that indeed $w_{w^w} = \varepsilon_{w^w} \cdot w$. Hence, for ordinals of the form $w^\beta + \rho$ such that both $w^w + w^\beta = w^\beta$ and $w \cdot (w^\beta + \rho) = w^\beta + \rho$, the lemma follows immediately from the above, since then also

$$w^w + w \cdot (w^\beta + \rho) = w^w + w^\beta + \rho = w^\beta + \rho$$

Since all additive primes above w^w have the form w^β with $\beta \geq w$ and thus do meet these conditions, we are done. $\square$

---

[4] By *additive prime* we mean an ordinal α which cannot be written as a (nontrivial) sum of two ordinals. Such ordinals thus have Cantor normal form w^β for $\beta \leq \alpha$ (where the inequality is strict exactly when α is not an ε-number).

Actually, we have proved that if $\alpha = \omega^{\alpha_k} + \cdots + \omega^{\alpha_0}$ is in Cantor normal form, and if $\alpha_0 \geq \omega$, then $\omega_\alpha = \varepsilon_\alpha \cdot \omega$.

We now define the mapping φ, and next argue that it is indeed order-preserving when restricted to the normal-forms of S_*, and that it maps each function to an ordinal below τ_0.

Definition 5. *We define the map* $\varphi : S_* \to \mathbf{ON}$ *as follows:*

$$\varphi(0) = 0, \ \varphi(1) = 1, \ \varphi(x) = \omega, \ \varphi(f+g) = \varphi(f) + \varphi(g) \ ,$$
$$\varphi(fg) = \varphi(f)\varphi(g), \ \varphi(x^f) = \omega^{\varphi(f)}, \ \varphi(x_f) = \omega_{\varphi(f)}.$$

Note that φ is not well-defined if we allow it to be applied to terms not in normal-form. For example,

$$\varphi(1 + x) = 1 + \omega = \omega \neq \omega + 1 = \varphi(x + 1) \ ,$$

though $1 + x$ and $x + 1$ merely are different terms for the same function, where $x + 1$ is the normal form.

By the definitions of φ, Proposition 4, and basic properties of ε-numbers, we immediately obtain

Lemma 1. *Let* $2 \leq k \in \mathbb{N}$, *let* F_k *denote the map* $x \mapsto x^{\cdot^{\cdot^{x}}} \Big\}_k$ *(i.e.* F_k *has tetration depth* $k - 1$*), and define a sequence of ordinals* δ_k *by* $\delta_2 \overset{def}{=} \varepsilon_0 \cdot \omega$; $\delta_{k+1} \overset{def}{=} \varepsilon_{\delta_k} \cdot \omega$. *Then* $\varphi(F_k) = \delta_k$. *Furtermore,* $\delta_k < \varepsilon^{\cdot^{\cdot^{\cdot}}}_{\ \ \varepsilon_0} \Big\}_{k+1} < \tau_0$. $\square$

Proof. Clearly $\omega^\omega + \omega \cdot \delta_k = \delta_k$, and $\varphi(F_2) = \delta_2$ by definition. Consequently, by induction on $k \in \mathbb{N}$, we obtain:

$$\varphi(F_{k+1}) \overset{def}{=} \omega_{\varphi(F_k)} \overset{IH+P.4}{=} \varepsilon_{\delta_k} \cdot \omega \overset{def}{=} \delta_{k+1}.$$

That $\delta_k < \tau_0$ for all $k \in \mathbb{N}$ is obvious; indeed $\delta_k < \varepsilon^{\cdot^{\cdot^{\cdot}}}_{\ \ \varepsilon_0} \Big\}_{k+1}$. $\square$

In order to establish the full injectivity of the mapping, we must show that when mapping terms in $\Sigma\Pi$-NF a crucial property of these terms is preserved: *the finite power property (FPP).* Let two $\Sigma\Pi$-normal-forms $\Sigma\Pi f_{ij}$ and $\Sigma\Pi g_{ij}$ be given, and let ij be the lexicographically least indices where f_{ij} and g_{ij} differ. The FPP then guarantees that $f_{ij} \succ g_{ij}$ is sufficient for $f \succ g$ to hold. We need this property to hold also for the ordinal terms into which φ maps $\Sigma\Pi f_{ij}$ and $\Sigma\Pi g_{ij}$. In particular, this property must hold of $\varphi(f_{ij})$ and $\varphi(g_{ij})$.

Lemma 2. *Let* $f, g \in S_*$ *be given in their* $\Sigma\Pi$-*NFs. If* $x_x \preceq f \prec g$ *then:*

1. $\varphi(f) < \varphi(g)$;
2. $\varphi((x_f)^k) < \varphi(x_g)$ *for any* $k \in \mathbb{N}$;
3. $\varphi((x^f)^k) < \varphi(x^g)$ *for any* $k \in \mathbb{N}$ *if, in addition,* x^f, x^g *are in normal form.*

The conditions in item 3. ensure that x^f (resp. x^g) cannot be reduced to e.g. x_{h+1} or e.g. $x^{f'} x^{f'}$ for $f = f' + f'$.

Note that when $f, g \preceq x_x$, i.e. f and g belong in Skolem's class S, then φ simply computes the actual ordinal of the initial segment determined by f in the well-order $(S, \preceq)$ and we really need only show injectivity above x_x.

Proof. We prove the lemma by induction on the tetration depth (i.e. the number of nested tetrations), with a sub-induction on the tower height.

The item 1. for tetration depth 0 – i.e. terms in Skolem's class S – is folklore.

In the induction step, the item 1. follows from the IH on items 2. and 3.; just as the analogous results for the $\preceq$-order on S_* as proved in Lemmata 1 and 3 in [1].

The item 2. follows directly from Proposition 3. For 3. recall that we already have that $(x^f)^k \prec x^g$ on the functional side. Letting $f = \prod f_i$ and $g = \prod g_i$, we also have that $(x^f)^k \prec x^g \Leftrightarrow (\prod f_i) \cdot k \prec \prod g_i$. But by the IH (on the tower height), we already have $\varphi((\prod f_i) \cdot k) < \varphi(\prod g_i)$, and the result follows.   □

Corollary 1. $\mathrm{O}\,(S_*, \preceq) \leq \tau_0$.

Proof. φ is order-preserving (hence injective) by Lemma 2. Because every $f \in S_*$ is majorised by F_k for some $k \in \mathbb{N}$ (for F_k as in Lemma 1), we have, by Lemma 1, that $\varphi(f) < \tau_0$, for all $f \in S_*$. Hence $\mathrm{O}\,(S_*, \preceq) \leq \tau_0$ as required.   □

We immediately obtain – by combining Corollary 1 and Theorem 1 – our desired main theorem:

Main Theorem 1. $\mathrm{O}\,(S_*, \preceq) = \tau_0$.   □

Note that a similar result has been obtained by R. McBeth in [3] for a related but slightly simpler class of functions than S_*. McBeth considers a class EP^*, which is our class S^* without multiplication, and obtains the ordinal τ_0 for the well-order $(EP^*, \preceq)$.

The class EP^* arises as follows: EP is the class of functions S as Skolem defined it, and EP^* is the closure under tetration. The subtle issue is this: In [5] Skolem defined S without explicitly adding multiplication. This is possible because for the class of functions defined by addition and exponentiation, one can define multiplication in the two other operations. When one adds tetration this is no longer the case! Thus e.g. the function $x_x x$ is definable in our class S_*, but not in McBeth's class EP^*. It is interesting to observe that the two well-orders have the same order type.

There are three immediate avenues of future work.

1. Combined with the results in [1], we have now established the exact ordinal of the well-order $(S_*, \preceq)$. We can even give very coarse upper and lower bounds on the ordinal of the initial segment defined by a given S_*-function f:

$$\varepsilon^{\cdot^{\cdot^{\cdot}}}_{\varepsilon_0} \Big\}_k \leq \mathrm{O}\,(f) < \varphi(f) \leq \varepsilon^{\cdot^{\cdot^{\cdot}}}_{\varepsilon_0} \Big\}_{k+1} \ ,$$

where k is the tetration depth of f. What we cannot do is compute the exact order type of initial segments associated with even very simple functions in S_*.

For example, we know the ordinal of x_x is ε_0. Based on computations, we are fairly sure that the ordinal of x_{x+1} is ε_0^ω. We have no clear intuition about what the ordinal of x_{x+2} is, let alone of more complicated functions. Continuing the study of $(S_*, \preceq)$ may reveal more about the nature of τ_0, an ordinal spacious enough to absorbe and accomodate the added complexity arising when extending EP^* to S_*.

2. In [2], H. Levitz extends Skolem's function class S by base-n exponentiation for all $n \in \mathbb{N}$ (i.e. if f is a function, then so is n^f). He proceeds to show that this class is well-ordered too and that the order type, as for Skolem's class, is ε_0. Levitz establishes the order type by defining a clever mapping of functions into ordinals. It would be interesting to see whether Levitz's ideas can be transferred to the tetration setting (i.e. allow for the formation of n_f as well).

3. Finally, the question remains how to extend these results to higher levels, i.e. 'pentation' (iterated tetration), 'hexation' (iterated pentation), etc. We are confident that it is possible to define suitable normal forms and establish the well-orderedness of the various function classes by $\preceq$, in a manner similar to the results in [1]. Developing suitable order-preserving maps into the ordinals to establish the order type may turn out more complicated than for the class S_*. In particular, the suitable definitions of 'ordinal pentation', 'ordinal hexation', etc. will require some care to ensure that functions are related to ordinals in a meaningful way.

References

1. Barra, M., Gerhardy, P.: Skolem + Tetration is Well-Ordered. In: Ambos-Spies, K., Löwe, B., Merkle, W. (eds.) Mathematical Theory and Computational Practice. LNCS, vol. 5635, pp. 11–20. Springer, Heidelberg (2009)
2. Lebitz, H.: An ordered set of Arithmetic Functions Representing the Least ε-number. Zeitschr. f. math. Logik und Grundlagen d. Math. 21, 115–120 (1975)
3. McBeth, R.: Exponential Polynomials of Linear Height. Zeitschr. f. math. Logik und Grundlagen d. Math. 26, 399–404 (1980)
4. Sierpiński, W.: Cardinal and Ordinal Numbers. PWN-Polish Scientific Publishers, Warszawa (1965)
5. Skolem, T.: An ordered set of arithmetic functions representing the least ϵ-number. Det Kongelige Norske Videnskabers selskabs Forhandlinger 29(12), 54–59 (1956)

Proofs, Programs, Processes

Ulrich Berger and Monika Seisenberger

Swansea University, Swansea, SA2 8PP, Wales, UK
{u.berger,m.seisenberger}@swansea.ac.uk

Abstract. We study a realisability interpretation for inductive and coinductive definitions and discuss its application to program extraction from proofs. A speciality of this interpretation is that realisers are given by terms that correspond directly to programs in a lazy functional programming language such as Haskell. Programs extracted from proofs using coinduction can be understood as perpetual processes producing infinite streams of data. Typical applications of such processes are computations in exact real arithmetic. As an example we show how to extract a program computing the average of two real numbers w.r.t. to the binary signed digit representation.

Keywords: program extraction, realisability, coinduction, exact real number computation.

1 Introduction

The purpose of this paper is to provide a theoretical foundation for the extraction of programs from proofs involving inductive and coinductive definitions. Its title is motivated by the fact that programs extracted from proofs using coinduction can often be understood as perpetual processes producing infinite streams of data. Typical applications of such processes are computations in exact real arithmetic. An informal introduction into this subject focusing on intuitive explanations and illustrating examples, but refraining from a stringent formal development, was given in [2]. In the present paper we provide this formal development. We give a realisability interpretation of our theory of inductive and coinductive definitions with a *Soundness Theorem* stating that from a proof of a formula one can extract a functional program provably realising it.

The programming language where the realisers are taken from is formally represented by an ML-style polymorphically typed λ-calculus with full recursion. The fact that terms are typed and full recursion is available has two major advantages:

1. The realisation of induction and coinduction is simple and elegant.
2. Realisers can be directly understood as programs in a typed functional programming language with a lazy semantics such as, for example, Haskell.

Another aspect of our system which is of great practical importance is the fact that quantifiers are treated uniformly in the realisability interpretation. This

F. Ferreira et al. (Eds.): CiE 2010, LNCS 6158, pp. 39–48, 2010.
© Springer-Verlag Berlin Heidelberg 2010

entails that realisers never depend on variables of the logical system (which represents a piece of formalised mathematics) and do not produce output in that language. Therefore, abstract mathematical objects do not need to be "constructivised", and it is possible to directly extract programs from abstract mathematical proofs. The uniform treatment of first-order quantifiers can be seen as a special case of the interpretations studied by Schwichtenberg [17], Hernest and Oliva [10] and Ratiu and Trifonov [16], which allow for a fine control of the amount of computational information extracted from proofs.

We illustrate our interpretation by extracting the average function on the real interval $[-1, 1]$ from a proof. The average function (and other arithmetic operations) were implemented and verified before in [6,15]. We would like to stress that while in these papers the programs where "guessed" and verified afterwards, we are able to extract these programs from proofs. This saves ourselves the work to implement not only the algorithm, but also the underlying data structure (streams, in this case) as these are generated automatically through the extraction mechanism. Finally we get the correctness proof for free. So far, the extraction has been carried out "by hand". The implementation in the interactive proof system Minlog [1,12] of the extraction method extended to coinductive definitions is ongoing work.

Related interpretations of systems with coinductive definitions were studied by Tatsuta [18], Miranda-Perea [13] and by the first author [3]. The latter paper defines realisability with respect to an untyped λ-calculus and gives an operational and denotational semantics of the calculus. It also discusses the main differences to the former two works. In the present paper we omit the semantics as it would be very similar to [3]. Instead, we concentrate on realisability and soundness where we significantly differ from and improve over the above cited work. To our knowledge, the application to the extraction of exact real number algorithms is new.

Due to lack of space we only give the main definitions and state the results. We intend to give a more complete account of our system with full proofs and more substantial applications in a future publication.

2   Inductive and Coinductive Definitions

We fix a first-order language $\mathcal{L}$. $\mathcal{L}$-terms, $r, s, t \ldots$, are built from constants, first-order variables and function symbols as usual. Formulas, $A, B, C \ldots$, are $s = t$, $\mathcal{P}(t)$ where $\mathcal{P}$ is a predicate (see below), $A \wedge B$, $A \vee B$, $A \rightarrow B$, $\forall x\, A$, $\exists x\, A$. A predicate is either a predicate constant P, or a predicate variable X, or a comprehension term also written $\{x \mid A\}$, or an inductive predicate $\mu X.\mathcal{P}$, or a coinductive predicate $\nu X.\mathcal{P}$ where $\mathcal{P}$ is a predicate of the same arity as the predicate variable X and which is strictly positive (s.p.) in X, i.e. X does not occur free in any premise of a subformula of $\mathcal{P}$ which is an implication. The application, $\mathcal{P}(t)$, of a predicate $\mathcal{P}$ to a list of terms t is a primitive syntactic construct, except when $\mathcal{P}$ is a comprehension term, $\mathcal{P} = \{x \mid A\}$, in which case $\mathcal{P}(t)$ stands for $A[t/x]$. It will sometimes write $x \in \mathcal{P}$ instead of $\mathcal{P}(x)$, $\mathcal{P} \subseteq \mathcal{Q}$

for $\forall \boldsymbol{x}\,(\mathcal{P}(\boldsymbol{x}) \rightarrow \mathcal{Q}(\boldsymbol{x}))$ and $\mathcal{P} \cap \mathcal{Q}$ for $\{\boldsymbol{x} \mid \mathcal{P}(\boldsymbol{x}) \wedge \mathcal{Q}(\boldsymbol{x})\}$ etc. We also write $\{t \mid A\}$ as an abbreviation for $\{x \mid \exists \boldsymbol{y}\,(x = t \wedge A)\}$ where x is a fresh variable and $\boldsymbol{y} = \mathrm{FV}(t) \cap \mathrm{FV}(A)$, as well as $f(\mathcal{P})$ for $\{f(x) \mid x \in \mathcal{P}\}$. Furthermore, we introduce *operators* $\varPhi := \lambda \boldsymbol{X}.\mathcal{P}$, and write $\varPhi(\mathcal{Q})$ for the predicate $\mathcal{P}[\mathcal{Q}/\boldsymbol{X}]$ where the latter is the usual substitution of the predicates $\mathcal{Q}$ for the predicate variables $\boldsymbol{X}$. $\varPhi$ is called a *s.p. operator* if $\mathcal{P}$ is s.p. in X. In this case we also write $\mu\varPhi$ and $\nu\varPhi$ for $\mu X.\mathcal{P}$ and $\nu X.\mathcal{P}$. A formula, predicate, or operator is called *non-computational*, if it contains neither free predicate variables nor the propositional connective $\vee$. Otherwise it is called *computational*.

The *proof rules* are the usual ones of intuitionistic predicate calculus with equality augmented by rules expressing that $\mu\varPhi$ and $\nu\varPhi$ are the least and greatest fixed points of the operator $\varPhi$. As is well-known, the fixed point property can be replaced by appropriate inclusions. Hence we stipulate the axioms

Closure	$\varPhi(\mu\varPhi) \subseteq \mu\varPhi$	Induction	$\varPhi(\mathcal{Q}) \subseteq \mathcal{Q} \rightarrow \mu\varPhi \subseteq \mathcal{Q}$
Coclosure	$\nu\varPhi \subseteq \varPhi(\nu\varPhi)$	Coinduction	$\mathcal{Q} \subseteq \varPhi(\mathcal{Q}) \rightarrow \mathcal{Q} \subseteq \nu\varPhi$

for all s.p. operators $\varPhi$ and predicates $\mathcal{Q}$. In addition we allow any axioms expressible by non-computational formulas that hold in the intended model. We write $\varGamma \vdash A$ if A is derivable from assumptions in $\varGamma$ in this system. If A is derivable without assumptions we write $\vdash A$, or even just A. We define falsity as $\bot := \mu X.X$ where X is a propositional variable (i.e. a 0-ary predicate variable). From the induction axiom for $\bot$ follows directly $\bot \rightarrow A$ for every formula A. As a running example we use the first-order language of the ordered real numbers. As axioms we adopt any non-computational formulas that are true in the structure of real numbers, e.g. the axioms of a real closed field where anti-symmetry and linearity of the order are expressed non-computationally, e.g. by $\forall x, y\,((y \not< x \wedge x \not< y) \leftrightarrow x = y)$. All sets we define in the following are subsets of the set of real numbers. We define the set $\mathbb{N}$ of natural numbers as usual inductively by

$$\mathbb{N} := \mu X.\{0\} \cup \{x + 1 \mid X(x)\}$$

Note that, since $\forall x, y \in \mathbb{N}\,(x < y \vee x = y \vee y < x)$ is (easily) provable, equality of natural number is decidable. Next we define coinductively a constructive analogue of the closed interval $\mathbb{I} := [-1, 1] \subseteq \mathbb{R}$. Let $\mathrm{SD} := \{0, 1, -1\}$ be the set of signed binary digits. We define coinductively

$$C_0 := \nu X.\{(i + x)/2 \in \mathbb{I} \mid \mathrm{SD}(i) \wedge X(x)\}$$

It is easy to see that, classically, C_0 coincides with $\mathbb{I}$. The point is that from a constructive proof of $C_0(x)$ we can extract a program computing an infinite signed digit representation of x.

3   Programs

We now introduce an extended λ-calculus which we will use in Sect. 4 as our programming language, i.e. language of realisers, as well as a typing discipline which will facilitate the definition of map operators, iterators and coiterators as realisers of monotonicity, induction and coinduction. Program terms (which we will simply call "terms" in the following) are given by the grammar

$$M, N, K, L ::= x \mid \lambda x.M \mid M\,N \mid \text{rec}\,x\,.\,M \mid C(M_1, \ldots, M_n) \mid$$
$$\text{case}\,M\,\text{of}\{C_1(\boldsymbol{x_1}) \to R_1\,;\,\ldots;\,C_n(\boldsymbol{x_n}) \to R_n\}$$

where x ranges over a set of variables, C ranges over a set of constructors (each with a fixed arity) and in case $M\,\text{of}\{C_1(\boldsymbol{x_1}) \to R_1\,;\,\ldots;\,C_n(\boldsymbol{x_n}) \to R_n\}$ all constructors C_i are distinct and each $\boldsymbol{x_i}$ is a vector of distinct variables. We axiomatise this calculus by the equations

$$(\lambda x.M)N = M[N/x] \qquad \text{rec}\,x\,.\,M = M[\text{rec}\,x\,.\,M/x]$$
$$\text{case}\,C_i(\boldsymbol{K})\,\text{of}\{\ldots;C_i(\boldsymbol{x_1}) \to R_i\,;\,\ldots\} = R_i[\boldsymbol{K}/\boldsymbol{x_i}]$$

We write $\vdash M = N$ if the equation $M = N$ can be derived from these axioms by the usual rules of equational logic.

The typing we introduce now serves two purposes. First, types are used as indices for terms realising monotonicity, induction and coinduction. Second, we will show that extracted programs are typeable and hence are valid programs in typed functional programming languages such as Haskell or ML. Types are constructed from type variables $\alpha, \beta, \ldots \in \text{TVar}$ according to the grammar

$$\text{Type} \ni \rho, \sigma, \tau ::= \alpha \mid \rho \to \sigma \mid \mathbf{1} \mid \rho \times \sigma \mid \rho + \sigma \mid \text{fix}\,\alpha.\rho$$

We consider the instance of our term language determined by the constructors Nil (nullary), Left, Right (unary), Pair (binary), and $\text{In}_{\text{fix}\,\alpha.\rho}$ (unary) for every fixed point type $\text{fix}\,\alpha.\rho$. We inductively define the relation $\Gamma \vdash M : \rho$ (term M is of type ρ in environment Γ.)

Variable, recursion and λ-calculus rules.

$$\Gamma, x : \rho \vdash x : \rho \qquad \frac{\Gamma, x : \tau \vdash M : \tau}{\Gamma \vdash \text{rec}\,x\,.\,M : \tau}$$

$$\frac{\Gamma, x : \rho \vdash M : \sigma}{\Gamma \vdash \lambda x.M : \rho \to \sigma} \qquad \frac{\Gamma \vdash M : \rho \to \sigma \qquad \Gamma \vdash N : \rho}{\Gamma \vdash M\,N : \sigma}$$

Constructor rules.

$$\Gamma \vdash \text{Nil} : \mathbf{1} \qquad \frac{\Gamma \vdash M : \rho \qquad \Gamma \vdash N : \sigma}{\Gamma \vdash \text{Pair}(M, N) : \rho \times \sigma}$$

$$\frac{\Gamma \vdash M : \rho}{\Gamma \vdash \text{Left}(M) : \rho + \sigma} \qquad \frac{\Gamma \vdash M : \sigma}{\Gamma \vdash \text{Right}(M) : \rho + \sigma}$$

$$\frac{\Gamma \vdash M : \rho_0(\rho(\boldsymbol{\sigma}), \boldsymbol{\sigma})}{\Gamma \vdash \text{In}_\rho(M) : \rho(\boldsymbol{\sigma})} \quad \text{where } \rho = \rho(\boldsymbol{\alpha}) = \text{fix}\,\alpha.\rho_0(\alpha, \boldsymbol{\alpha}).$$

Elimination rules for constructors. The constructor rules for each type constructor determine one elimination rule. For example:

$$\frac{\Gamma \vdash M : \rho + \sigma \qquad \Gamma, x_1 : \rho \vdash L : \tau \qquad \Gamma, x_2 : \sigma \vdash R : \tau}{\Gamma \vdash \text{case } M \text{ of}\{\text{Left}(x_1) \to L \,; \text{Right}(x_2) \to R\} : \tau}$$

$$\frac{\Gamma \vdash M : \rho(\boldsymbol{\sigma}) \qquad \Gamma, x : \rho_0(\rho(\boldsymbol{\sigma}), \boldsymbol{\sigma}) \vdash N : \tau}{\Gamma \vdash \text{case } M \text{ of}\{\text{In}_\rho(x) \to N\} : \tau} \quad (\rho = \rho(\boldsymbol{\alpha}) = \text{fix } \alpha.\rho_0(\alpha, \boldsymbol{\alpha}))$$

As a preparation for the realisability interpretation in Sect. 4 we define map-terms, iterators and coiterators that will be used as realisers for the monotonicity of strictly positive predicate transformers, induction and coinduction. The following definition refers to a fixed one-to-one assignment of variables f_α to type variables α. For every list of type variables $\boldsymbol{\alpha}$ and every type ρ which is s.p. in $\boldsymbol{\alpha}$ we define the term $\mathbf{map}_{\boldsymbol{\alpha};\rho} := \lambda f_{\alpha_1}, \ldots, f_{\alpha_n} . \mathbf{Map}_{\boldsymbol{\alpha};\rho}$ where

$$\mathbf{Map}_{\boldsymbol{\alpha};\alpha_i} = f_i, \qquad \mathbf{Map}_{\boldsymbol{\alpha};\rho} = \lambda x.\, x, \text{ if no } \alpha_i \text{ occurs in } \rho$$

$$\mathbf{Map}_{\boldsymbol{\alpha};\rho+\sigma} = \lambda x.\text{case } x \text{ of}\{\text{Left}(y) \to \text{Left}(\mathbf{Map}_{\boldsymbol{\alpha};\rho}y) \,;$$
$$\text{Right}(z) \to \text{Right}(\mathbf{Map}_{\boldsymbol{\alpha};\sigma}z)\}$$

$$\mathbf{Map}_{\boldsymbol{\alpha};\rho\times\sigma} = \lambda x.\, \text{case } x \text{ of}\{\text{Pair}(y, z) \to \text{Pair}(\mathbf{Map}_{\boldsymbol{\alpha};\rho}y, \mathbf{Map}_{\boldsymbol{\alpha};\sigma}z)\}$$

$$\mathbf{Map}_{\boldsymbol{\alpha};\rho\to\sigma} = \lambda x.\, \lambda y.\, \mathbf{Map}_{\boldsymbol{\alpha};\sigma}(x\,y)$$

$$\mathbf{Map}_{\boldsymbol{\alpha};\text{fix } \alpha.\rho} = \text{rec } f_\alpha.\, \lambda x.\, \text{case } x \text{ of}\{\text{In}_{\text{fix } \alpha.\rho}(y) \to \text{In}_{\text{fix } \alpha.\rho}(\mathbf{Map}_{\boldsymbol{\alpha},\alpha;\rho}y)\}$$

A type is called *regular* if in its construction the clause fix $\alpha.\rho$ is applied only if ρ is s.p. in α. In the following all mentioned types are assumed to be regular.

Lemma 1 (Typing of map). *Let* $\rho = \rho(\boldsymbol{\alpha})$ *be s.p. in* $\boldsymbol{\alpha}$. *Then*

$$\vdash \mathbf{map}_{\boldsymbol{\alpha};\rho} : (\boldsymbol{\sigma} \to \boldsymbol{\tau}) \to \rho(\boldsymbol{\sigma}) \to \rho(\boldsymbol{\tau})$$

For every (regular) type fix $\alpha.\rho$ we define iterator and coiterator by

$$\mathbf{It}_{\text{fix } \alpha.\rho} := \lambda s.\, \text{rec } f.\, \lambda x.\, \text{case } x \text{ of}\{\text{In}_{\text{fix } \alpha.\rho}(y) \to s(\mathbf{map}_{\boldsymbol{\alpha};\rho}fy)\}$$
$$\mathbf{Coit}_{\text{fix } \alpha.\rho} := \lambda s.\text{rec } f.\, \lambda x.\text{In}_{\text{fix } \alpha.\rho}(\mathbf{map}_{\boldsymbol{\alpha};\rho}f(s\,x))$$

Lemma 2 (Typing of iterator and coiterator). *For all types* $\sigma, \boldsymbol{\sigma}$

$$\vdash \mathbf{It}_{\text{fix } \alpha.\rho} : (\rho(\sigma, \boldsymbol{\sigma}) \to \sigma) \to \text{fix } \alpha.\rho(\boldsymbol{\sigma}) \to \sigma$$
$$\vdash \mathbf{Coit}_{\text{fix } \alpha.\rho} : (\sigma \to \rho(\sigma, \boldsymbol{\sigma})) \to \sigma \to \text{fix } \alpha.\rho(\boldsymbol{\sigma})$$

In the following let $\rho = \text{fix } \alpha.\rho_0(\alpha)$. We set $\mathbf{in}_\rho := \lambda y.\, \text{In}_\rho(y)$ and $\mathbf{out}_\rho := \lambda x.\, \text{case } x \text{ of}\{\text{In}_\rho(y) \to y\}$. Note that $\vdash \mathbf{in}_\rho : \rho_0(\rho) \to \rho$ and $\vdash \mathbf{out}_\rho : \rho \to \rho_0(\rho)$, i.e. $\mathbf{in}_\rho$ resp. $\mathbf{out}_\rho$ is an algebra resp. coalgebra for the functor $\alpha \mapsto \rho_0(\alpha)$. The next lemma states that the iterator resp. coiterator witnesses the initiality resp. finality of $\mathbf{in}_\rho$ resp. $\mathbf{out}_\rho$. ($M \circ N := \lambda z.M(Nz)$, z fresh.)

Lemma 3 (Initiality and finality)

(a) $\vdash \mathbf{It}_\rho s \circ \mathbf{in}_\rho = s \circ \mathbf{map}_{\boldsymbol{\alpha};\rho_0(\alpha)}(\mathbf{It}_\rho s)$
(b) $\vdash \mathbf{out}_\rho \circ \mathbf{Coit}_\rho s = \mathbf{map}_{\boldsymbol{\alpha};\rho_0(\alpha)}(\mathbf{Coit}_\rho s) \circ s$

4 Realisability

We now introduce a formalised realisability interpretation of the theory of inductive and coinductive definitions of Sect. 2. To this end we need a system that can talk about mathematical objects *and* realisers. Therefore we extend our first-order language $\mathcal{L}$ to a language $\mathbf{r}(\mathcal{L})$ by adding a new sort for program terms. All logical operations including inductive and coinductive definitions, as well as axioms and rules for $\mathcal{L}$ including closure, induction, coclosure and coinduction and the rules for equality, are extended mutatis mutandis for $\mathbf{r}(\mathcal{L})$. In addition, we have as extra axioms the equations given in Sect. 3.

We assign to every $\mathcal{L}$-formula A a unary $\mathbf{r}(\mathcal{L})$-predicate $\mathbf{r}(A)$ on program terms. Intuitively, $\mathbf{r}(A)(a)$, sometimes also written $a\,\mathbf{r}\,A$, states that a "realises" A. The predicate $\mathbf{r}(A)$ is defined relative to a fixed one-to-one mapping from $\mathcal{L}$-predicate variables X to $\mathbf{r}(\mathcal{L})$-predicate variables $\tilde{X}$ with one extra argument place for program terms. The definition of $\mathbf{r}(A)$ is such that if the formula A has the free predicate variables $X_1, \ldots, X_n$, then the predicate $\mathbf{r}(A)$ has the free predicate variables $\tilde{X}_1, \ldots, \tilde{X}_n$. Simultaneously with $\mathbf{r}(A)$ we define a predicate $\mathbf{r}(\mathcal{P})$ for every predicate $\mathcal{P}$, where $\mathbf{r}(\mathcal{P})$ has one extra argument place for domain elements. We also define regular types $\tau(A)$ and $\tau(\mathcal{P})$ relative to a fixed assignment of a type variable α_X to each predicate variable X.

If A and $\mathcal{P}$ are non-computational:

$$\mathbf{r}(A) \quad = \{\mathrm{Nil} \mid A\} \qquad\qquad\qquad \tau(A) \quad = \mathbf{1}$$
$$\mathbf{r}(\mathcal{P}) \quad = \{(\mathrm{Nil}, \boldsymbol{x}) \mid \mathcal{P}(\boldsymbol{x})\} \qquad \tau(\mathcal{P}) \quad = \mathbf{1}$$

If A is non-computational but B is:

$$\mathbf{r}(A \wedge B) = \mathbf{r}(B \wedge A) = \{x \mid A \wedge \mathbf{r}(B)(x)\} \quad \tau(A \wedge B) = \tau(B \wedge A) = \tau(B)$$
$$\mathbf{r}(A \to B) = \{x \mid A \to \mathbf{r}(B)(x)\} \qquad\qquad \tau(A \to B) = \tau(B)$$

In all other cases:

$$\mathbf{r}(\mathcal{P}(\boldsymbol{t})) \quad = \{x \mid \mathbf{r}(\mathcal{P})(x, \boldsymbol{t})\} \qquad\qquad \tau(\mathcal{P}(\boldsymbol{t})) \quad = \tau(\mathcal{P})$$
$$\mathbf{r}(A \wedge B) \quad = \mathrm{Pair}(\mathbf{r}(A), \mathbf{r}(B)) \qquad\qquad \tau(A \wedge B) \quad = \tau(A) \times \tau(B)$$
$$\mathbf{r}(A \vee B) \quad = \mathrm{Left}(\mathbf{r}(A)) \cup \mathrm{Right}(\mathbf{r}(B)) \qquad \tau(A \vee B) \quad = \tau(A) + \tau(B)$$
$$\mathbf{r}(A \to B) = \{f \mid f(\mathbf{r}(A)) \subseteq \mathbf{r}(B)\} \qquad\quad \tau(A \to B) = \tau(A) \to \tau(B)$$
$$\mathbf{r}(\forall y\, A) \quad = \{x \mid \forall y\, (\mathbf{r}(A)(x))\} \qquad\qquad \tau(\forall y\, A) \quad = \tau(A)$$
$$\mathbf{r}(\exists y\, A) \quad = \{x \mid \exists y\, (\mathbf{r}(A)(x))\} \qquad\qquad \tau(\exists y\, A) \quad = \tau(A)$$
$$\mathbf{r}(\{\boldsymbol{x} \mid A\}) = \{(y, \boldsymbol{x}) \mid \mathbf{r}(A)(y)\} \qquad\quad \tau(\{\boldsymbol{x} \mid A\}) = \tau(A)$$
$$\mathbf{r}(X) \quad = \tilde{X} \qquad\qquad\qquad\qquad\qquad \tau(X) \quad = \alpha_X$$
$$\mathbf{r}(\mu X.\mathcal{P}) = \mu \tilde{X}.\{(\mathrm{In}(y), x) \mid \mathbf{r}(\mathcal{P})(y, x)\} \quad \tau(\mu X.\mathcal{P}) = \mathrm{fix}\,\alpha_X.\tau(\mathcal{P})$$
$$\mathbf{r}(\nu X.\mathcal{P}) = \nu \tilde{X}.\{(\mathrm{In}(y), x) \mid \mathbf{r}(\mathcal{P})(y, x)\} \quad \tau(\nu X.\mathcal{P}) = \mathrm{fix}\,\alpha_X.\tau(\mathcal{P})$$

where in the last two equations $\mathrm{In} := \mathrm{In}_{\mathrm{fix}\,\alpha_X.\tau(\mathcal{P})}$.

For example, $\tau(\mathbb{N}) = \operatorname{fix}\alpha.\mathbf{1} + \alpha$, the usual recursive definition of the data type of unary natural numbers. Its canonical inhabitants are the numerals $\underline{k} := \operatorname{inr}^k(\operatorname{inl}(\operatorname{Nil}))$ ($k \in \mathbb{N}$) where $\operatorname{inl}(x) := \operatorname{In}(\operatorname{Left}(x))$ and $\operatorname{inr}(x) := \operatorname{In}(\operatorname{Right}(x))$. Realisability for $\mathbb{N}$, $\mathbf{r}(\mathbb{N})$, is the least relation such that

$$\mathbf{r}(\mathbb{N}) = \{(\operatorname{inl}(\operatorname{Nil}), 0)\} \cup \{(\operatorname{inr}(n), x + 1) \mid \mathbf{r}(\mathbb{N})(n, x)\}$$

Hence, we have for a term d and $k \in \mathbb{R}$ that $d\,\mathbf{r}\,\mathbb{N}(k)$ holds iff k is a natural number and $d = \underline{k}$, i.e. d is a unary representation of k.

Regarding C_0, if we identify the set SD with the type $\mathbf{1} + \mathbf{1} + \mathbf{1}$ and every $i \in$ SD with its corresponding program term, then $\tau(C_0) = \operatorname{fix}\alpha.\operatorname{SD} \times \alpha$, the type of infinite streams of signed digits, and

$$\mathbf{r}(C_0) = \nu\tilde{X}.\{(\operatorname{Pair}(i, a), (i + x)/2) \mid i \in \operatorname{SD} \wedge |(i + x)/2| \leq 1 \wedge \tilde{X}(a, x)\}$$

It is easy to see that $\mathbf{r}(C_0)(a, x)$ means that the signed digit stream $a = a_0, a_1, \ldots$ represents x i.e. $x = \Sigma_{i=0}^{\infty} 2^{-(i+1)} * a_i$.

Lemma 4 (Map). *Let $\Phi = \lambda X.\mathcal{P}$ be a (strictly positive) operator in the language $\mathcal{L}$, $\alpha := \alpha_X$, and $\rho := \tau(\mathcal{P})$. Then $\operatorname{map}_{\alpha;\rho}$ realises the monotonicity of Φ, that is*

$$\operatorname{map}_{\alpha;\rho} \mathbf{r} \left(\mathcal{P} \subseteq \mathcal{Q} \to \Phi(\mathcal{P}) \subseteq \Phi(\mathcal{Q})\right)$$

for all $\mathcal{L}$-predicates $\mathcal{P}$ and $\mathcal{Q}$.

The previous lemmas are the essential facts needed to prove:

Theorem 1 (Soundness). *From a closed derivation of a formula A one can extract a program term M such that $\mathbf{r}(A)(M)$ and $M\!:\!\tau(A)$ are derivable.*

From the Soundness Theorem one can easily conclude that the program extracted from the proof of a *data formula*, i.e. a formula without free predicate variables and such that every subformula of the form $A \to B$ or $\nu\Phi(t)$ is non-computational, "evaluates" to a data term, i.e. a closed term built from constructors only, which realises A. For details about evaluating terms w.r.t. a call-by-name semantics see [3].

5   Example: Extraction of the Average Function

As an example we extract from a proof that the predicate C_0 is closed under averages a program computing the average of two real numbers in the interval $\mathbb{I}$ w.r.t. the signed digit representation.

Lemma 5. *If $x, y \in C_0$, then $\frac{x+y}{2} \in C_0$.*

Proof. Set $X := \{\frac{x+y}{2} \mid x, y \in C_0\}$. In order to show $X \subseteq C_0$, it suffices to find a set $Y \subseteq \mathbb{R}$ such that $X \subseteq Y$ and $Y \subseteq \Phi(Y)$ (hence $Y \subseteq C_0$, by coinduction). Setting $\operatorname{SD}_2 := \{-2, -1, 0, 1, 2\}$ we define

$$Y := \{\frac{x + y + i}{4} \mid x, y \in C_0, i \in \operatorname{SD}_2\}$$

Proof of "$X \subseteq Y$": Let $x, y \in C_0$. We have to show that $z := \frac{x+y}{2} \in Y$. By the coclosure axiom, $x = \frac{x'+d}{2}$, $y = \frac{y'+e}{2}$, for some $d, e \in SD$ and $x', y' \in C_0$. Hence $z = \frac{x'+y'+d+e}{4} \in Y$.

Proof of "$Y \subseteq \Phi(Y)$": Let $x, y \in C_0$ and $i \in SD_2$. We have to show that $z := \frac{x+y+i}{4} \in \Phi(Y)$. By the coclosure axiom, $x = \frac{x'+d'}{2}$ and $y = \frac{y'+e'}{2}$, for some $d', e' \in SD$ and $x', y' \in C_0$. Hence $z = \frac{x'+y'+d'+e'+2i}{8}$. Since $z \in \mathbb{I}$, it suffices to find $d \in SD$ and $\tilde{z} \in Y$ such that $z = \frac{\tilde{z}+d}{2}$. By the definition of Y this means that we must find $\tilde{x}, \tilde{y} \in C_0$ and $j \in SD_2$ such that $\tilde{z} = \frac{\tilde{x}+\tilde{y}+j}{4}$, i.e. $z = \frac{\tilde{x}+\tilde{y}+j+4d}{8}$. Choosing $\tilde{x} := x'$ and $\tilde{y} := y'$ we are left with the equation $d' + e' + 2i = j + 4d$ which is clearly solvable with suitable $d \in SD$ and $j \in SD_2$.

Applying our method of program extraction to this proof one obtains the following Haskell program `average`. It takes two infinite streams and first reads the first digits d and e on both of them; this corresponds to the proof of "$X \subseteq Y$". The functional `aux` recursively calls itself; this corresponds to the use of the coinduction principle with coiteration as realiser. The remainder of the program links to the computational content of the proof of "$Y \subseteq \Phi(Y)$" in an obvious way.

```
type SD  = Int   -- -1, 0, 1 only
type SDS = [SD] -- infinite streams only
type SD2 = Int   -- -2, -1, 0, 1, 2 only

average :: SDS -> SDS -> SDS
average (d:ds) (e:es) = aux (d+e) ds es   where

    aux :: SD2 -> SDS -> SDS -> SDS
    aux i (d':ds) (e':es) = d : aux j ds es   where

      k = d'+e'+2*i

      d | abs k <= 2  = 0
        | k > 2       = 1
        | otherwise   = -1

      j = k-4*d
```

As a demo we run the extracted program with inputs $[1,0,1,0,0,0,\ldots] = \frac{5}{8}$ and $[1,1,0,0,0,\ldots] = \frac{3}{4}$. Looking at the first 10 digits of the computed stream

```
Main> take 10 (average ([1,0,1]++[0,0..]) ([1,1]++[0,0..]))
[1,1,0,-1,0,0,0,0,0,0]
```

one sees that the result is correct, as $(\frac{5}{8} + \frac{3}{4})/2 = \frac{11}{16} = \frac{1}{2} + \frac{1}{4} - \frac{1}{16}$.

6   Conclusion and Further Work

In our opinion, one of the main advantages of program extraction over the traditional specify-implement-verify method is that it is possible to carry out proofs in a very simple formal system. Neither complicated data types (lists, streams, trees, function types, etc.) nor programming constructs (recursion, lambda-abstraction) need to be formalised by the user; these are all generated by the realisability interpretation automatically.

On the basis of the results of this paper one can now begin to formalise parts of constructive analysis and other branches of mathematics where inductive and coinductive definitions are used (or can be used), with the aim of extracting nontrivial certified programs. Currently, we are investigating a generalisation of the predicate $C_0 \subseteq \mathbb{R}$ (one of our running examples) to predicates $C_n \subseteq \mathbb{R}^{\mathbb{I}^n}$ characterising the (constructively) uniformly continuous function from $\mathbb{I}^n$ to $\mathbb{I}$ [2]. For $n = 1$ the definition is

$$C_1 := \nu F.\mu G.\{f \in \mathbb{I}^{\mathbb{I}} \mid \exists i \in \mathrm{SD}\, \exists f'\, (f = \mathrm{av}_i \circ f' \wedge F(f')) \vee \bigwedge_{i \in \mathrm{SD}} G(f \circ \mathrm{av}_i)\}$$

where F and G range over subsets of $\mathbb{R}^{\mathbb{I}}$ and $\mathrm{av}_i(x) := (i + x)/2$. To see the analogy with C_0 it is useful to rewrite the definition of the latter equivalently as

$$C_0 := \nu X.\{x \in \mathbb{I} \mid \exists i \in \mathrm{SD}\, \exists x'\, (x = \mathrm{av}_i(x') \wedge X(x'))\}$$

The predicate C_0 characterises real numbers in $\mathbb{I}$ as objects perpetually emitting digits. A continuous function $f : \mathbb{I} \to \mathbb{I}$, which can be viewed as a real number in $\mathbb{I}$ that depends on an input in $\mathbb{I}$, perpetually emits digits as well, but before an emission can take place f may have to gain information about the input by absorbing finitely many digits from it in order to decide which digit to emit. The absorption part is formalised in C_1 by the inner "$\mu G \ldots G(f \circ \mathrm{av}_i)$". The data type associated with C_1 is

$$\tau(C_1) = \nu\alpha.\mu\beta.\mathrm{SD} \times \alpha + \beta^3$$

which is the type of non-wellfounded trees with two kinds of nodes: one kind labelled by a signed digit and one child (emitting a digit), the other kind without label and three children (absorbing a digit). The fact that β is quantified by μ means that only those trees are legal members of $\tau(C_1)$ that have on each path infinitely many emitting nodes. A similar type of trees has been studied independently in [9], however, not in the context of analysis and realisability. The definition of C_1 is motivated by earlier works on the development and verification of exact real number algorithms based on the signed digit representation of real numbers [11,8,7] some of which make use of coinductive methods [6,5,4,14].

Based on the characterisation of uniformly continuous functions by the predicates C_n implementations of elementary arithmetic functions have been extracted [2]. Further work in progress is the automatization of this form of program extraction in the Minlog proof system [1], and the study of integration and analytic functions based on this approach.

References

1. Benl, H., Berger, U., Schwichtenberg, H., Seisenberger, M., Zuber, W.: Proof theory at work: Program development in the Minlog system. In: Bibel, W., Schmitt, P.H. (eds.) Automated Deduction – A Basis for Applications. Applied Logic Series, vol. II, pp. 41–71. Kluwer, Dordrecht (1998)
2. Berger, U.: From coinductive proofs to exact real arithmetic. In: Grädel, E., Kahle, R. (eds.) CSL 2009. LNCS, vol. 5771, pp. 132–146. Springer, Heidelberg (2009)
3. Berger, U.: Realisability and adequacy for (co)induction. In: Bauer, A., Hertling, P., Ko, K.-I. (eds.) 6th Int'l Conf. on Computability and Complexity in Analysis (Ljubljana), Dagstuhl, Germany. Schloss Dagstuhl - Leibniz-Zentrum fuer Informatik, Germany (2009)
4. Berger, U., Hou, T.: Coinduction for exact real number computation. Theory of Computing Systems 43, 394–409 (2008)
5. Bertot, Y.: Affine functions and series with co-inductive real numbers. Math. Struct. Comput. Sci. 17, 37–63 (2007)
6. Ciaffaglione, A., Di Gianantonio, P.: A certified, corecursive implementation of exact real numbers. Theor. Comput. Sci. 351, 39–51 (2006)
7. Edalat, A., Heckmann, R.: Computing with real numbers: I. The LFT approach to real number computation; II. A domain framework for computational geometry. In: Barthe, G., Dybjer, P., Pinto, L., Saraiva, J. (eds.) Applied Semantics - Lecture Notes from the International Summer School, Caminha, Portugal, pp. 193–267. Springer, Heidelberg (2002)
8. Geuvers, H., Niqui, M., Spitters, B., Wiedijk, F.: Constructive analysis, types and exact real numbers. Math. Struct. Comput. Sci. 17(1), 3–36 (2007)
9. Ghani, N., Hancock, P., Pattinson, D.: Continuous functions on final coalgebras. Electr. Notes in Theoret. Comput. Sci. 164 (2006)
10. Hernest, M.D., Oliva, P.: Hybrid functional interpretations. In: Beckmann, A., Dimitracopoulos, C., Löwe, B. (eds.) CiE 2008. LNCS, vol. 5028, pp. 251–260. Springer, Heidelberg (2008)
11. Marcial-Romero, J.R., Escardo, M.H.: Semantics of a sequential language for exact real-number computation. Theor. Comput. Sci. 379(1-2), 120–141 (2007)
12. The Minlog System, http://www.mathematik.uni-muenchen.de/~minlog/
13. Miranda-Perea, F.: Realizability for monotone clausular (co)inductive definitions. Electr. Notes in Theoret. Comput. Sci. 123, 179–193 (2005)
14. Niqui, M.: Coinductive formal reasoning in exact real arithmetic. Logical Methods in Computer Science 4(3-6), 1–40 (2008)
15. Plume, D.: A Calculator for Exact Real Number Computation. PhD thesis, University of Edinburgh (1998)
16. Ratiu, D., Trifonov, T.: Exploring the computational content of the infinite pigeon-hole principle. Journal of Logic and Computation (to appear 2010)
17. Schwichtenberg, H.: Realizability interpretation of proofs in constructive analysis. Theory of Computing Systems 43(3-4), 583–602 (2008)
18. Tatsuta, M.: Realizability of monotone coinductive definitions and its application to program synthesis. In: Jeuring, J. (ed.) MPC 1998. LNCS (LNM), vol. 1422, pp. 338–364. Springer, Heidelberg (1998)

Ergodic-Type Characterizations
of Algorithmic Randomness

Laurent Bienvenu[1], Adam Day[2], Ilya Mezhirov[3], and Alexander Shen[4,*]

[1] LIAFA, CNRS & Université de Paris 7, France
laurent.bienvenu@liafa.jussieu.fr
[2] Victoria University of Wellington, New Zealand
adam.day@msor.vuw.ac.nz
[3] Technical University of Kaiserslautern
mezhirov@gmail.com
[4] LIF, CNRS & Université d'Aix-Marseille 1, France
alexander.shen@lif.univ-mrs.fr

Abstract. A theorem of Kučera states that given a Martin-Löf random infinite binary sequence ω and an effectively open set A of measure less than 1, some tail of ω is not in A. We show that this result can be seen as an effective version of Birkhoff's ergodic theorem (in a special case). We prove several results in the same spirit and generalize them via an effective ergodic theorem for bijective ergodic maps.

1 Introduction

The classical setting for the ergodic theorem is as follows. Let X be a space with a probability measure μ on it, and let $T\colon X \to X$ be a measure-preserving transformation. Let f be a real-valued integrable function on X. Birkhoff's ergodic theorem (see for example [7]) says that the time-average

$$\frac{f(x) + f(T(x)) + f(T(T(x))) + \ldots + f(T^{(n-1)}(x))}{n}$$

has a limit (as $n \to \infty$) for all x except for some null set, and this limit (the "time average") equals the space average $\int f(x)\,d\mu(x)$ if the transformation T is ergodic (i.e., has no non-trivial invariant subsets).

The classical example is the left shift on Cantor space Ω (the set of infinite binary sequences, denoted also by $2^{\mathbb{N}}$ or 2^ω): $\sigma(\omega_0\omega_1\omega_2\ldots) = \omega_1\omega_2\ldots$ It preserves Lebesgue measure (a.k.a. uniform measure) μ on Ω and is ergodic. Therefore, the time and space averages coincide for almost every starting point x. For a special case where f is an indicator function of some (measurable) set A, we conclude that almost surely (for all x outside some null set) the fraction of terms in the sequence $x, \sigma(x), \sigma(\sigma(x)), \ldots$ that are inside A, converges to the measure of A. Assuming that A has positive measure, we conclude that almost surely at least

* Supported by ANR Sycomore, NAFIT ANR-08-EMER-008-01, RFBR 09-01-00709-a grants and Shapiro visitors program at Penn State University.

one element of this sequence belongs to A. Switching to complements: if A has measure less than 1, then (almost surely) some elements of this sequence are outside A. Kučera [5] proved an effective version of this statement:

Theorem 1. *If A is an effectively open set of measure less than 1, then for every Martin-Löf random sequence ω at least one of ω, $\sigma(\omega)$, $\sigma(\sigma(\omega))$,...does not belong to A.*

Recalling the definition of Martin-Löf randomness (a sequence is random if it is outside any effectively null set) we can reformulate Kučera's theorem as follows:

> Let A be an effectively open set of measure less than 1. Consider the set A^* of all sequences ω such that every tail $\sigma^{(n)}(\omega)$ belongs to A. Then A^* is an effectively null set.

Before presenting the proof, let us mention an interpretation of this result. Recall that the universal Martin-Löf test is a computable sequence $U_1, U_2, \ldots$ of effectively open sets such that $\mu(U_i) \leq 1/2^i$ and the intersection $\cap_i U_i$ is the maximal effectively null set, i.e., the set of all non-random sequences. Kučera's theorem shows that randomness can be (in a paradoxical way) characterized by U_1 alone: a sequence is non-random if and only if all its tails belong to U_1. (In one direction it is Kučera's theorem, in the other direction we need to note that a tail of a non-random sequence is non-random.)

Proof (of Kučera's theorem). We start with the following observation: it is enough to show that for every interval I, we can uniformly construct an effectively open set $J \subset I$ that contains $I \cap A^*$ and such that $\mu(J) \leq r\mu(I)$ for some fixed $r < 1$ (here we call an *interval* any set of type $x\Omega$, where x is some finite string, which is the set of infinite binary sequences that start with x). Then we represent the effectively open set A of measure $r < 1$ as a union of disjoint intervals $I_1, I_2, \ldots$, construct the sets J_i for every I_i and note that the union A_1 of all J_i is an effectively open set that contains A^* and has measure r^2 or less. Splitting A_1 into disjoint intervals and repeating this argument, we get a set A_2 of measure at most r^3, etc. In this way we get a effectively open cover for A^* of arbitrarily small measure, so A^* is an effectively null set.

It remains to show how to find J given I. The interval I consists of all sequences that start with some fixed prefix x, i.e., $I = x\Omega$. Since sequences in A^* have all their tails in A, the intersection $I \cap A^*$ is contained in xA, and the latter set has measure $r\mu(I)$ (where $r = \mu(A)$).   □

Note that this proof also shows the following: suppose A is an effectively open set of measure less than 1, and A can be written as a disjoint union of intervals $A = x_1\Omega \cup x_2\Omega \cup \ldots$. Let ω be an infinite sequence that can be written as $\omega = w_1 w_2 w_3 \ldots$ where for all i, $w_i = x_j$ for some j. Then ω is not random. (If A contains all non-random sequences, the reverse implication is also true, and we get yet another criterion of randomness.)

Effective versions of the ergodic theorem in a general setting have been studied in several papers (see for example [9,10,2,3]). In this paper, we present characterizations of randomness that resemble Kučera's and which (to the best of our knowledge) cannot be directly derived from previous papers.

2   Effective Kolmogorov 0-1-Law

Trying to find characterizations of randomness similar to Kučera's theorem, one may look at Kolmogorov's 0-1-law. It says that any measurable subset A of the Cantor space that is stable under finite changes of bits (i.e. if $\omega \in A$ and ω' is equal to ω up to a finite change of bits, then $\omega' \in A$) has measure 0 or 1. It can be reformulated as follows: let A be a (measurable) set of measure less than 1. Consider the set A^* defined as follows: $\omega \in A^*$ if and only if all sequences that are obtained from ω by changing finitely many terms, belong to A. Then A^* has measure zero (indeed, A^* is stable and cannot have measure 1). Note also that we may assume without loss of generality that A is open (replacing it by an open cover of measure less than 1).

A natural effective version of Kolmogorov's 0-1-law can then be formulated as follows.

Theorem 2. *Let A be an effectively open set of measure $r < 1$. Consider the set A^* of all sequences that belong to A and remain in A after changing finitely many terms. Then A^* is an effectively null set.*

(As we have seen, the last two sentences can be replaced by the following claim: *any Martin-Löf random sequence can be moved outside A by changing finitely many terms.*)

Proof. To prove this effective version of the 0-1-law, consider any interval I. As before, we want to find an effectively open set $U \subset I$ that contains $A^* \cap I$ and has measure at most $r\mu(I)$. Let x be a prefix that defines I, i.e., $I = x\Omega$. For every string y of the same length as x, consider the set $A_y = \{\omega \mid y\omega \in A\}$. It is easy to see that the average measure of A_y (over all y of a given length) equals $\mu(A) = r$. Therefore, the set $B = \bigcap_y A_y$ (which is effectively open as an intersection of an effectively defined finite family of open sets) has measure at most r. Now take $U = xB$. Let us show that U is as wanted. First U is an effectively open set, contained in I, and of measure $r\mu(I)$. Also, it contains every element of $A^* \cap I$. Indeed, if $\alpha \in A^* \cap I$, x is a prefix of α, so one can write $\alpha = x\beta$. Since $\alpha \in A^*$, any finite variation of α is in A, so for all y of the same length as x, $y\beta \in A$. Therefore, β is in all A_y, and therefore is in B. Since $\alpha = x\beta$, it follows that α is in $xB = U$. □

3   Adding Prefixes

We have considered left shifts (deletion of prefixes) and finite changes. Another natural question is about *adding* finite prefixes. It turns out that a similar result can be proven in this case (although the proof becomes a bit more difficult).

Theorem 3. *Let A be an effectively open set of measure $r < 1$. Let A^* be the set of all sequences ω such that $x\omega \in A$ for every binary string x. Then A^* is an effectively null set.*

(Reformulation: *for every Martin-Löf random sequence ω there exists a string x such that $x\omega \notin A$.*)

Proof. To prove this statement, consider again some interval $I = x\Omega$. We want to cover $A^* \cap I$ by an effectively open set of measure $r\mu(I)$. (In fact, we get a cover of measure $s\mu(I)$ for some constant $s \in (r, 1)$, but this is enough.) Consider some string z. We know that the density of A^* in I does not exceed the density of A in $zI = zx\Omega$. Indeed, $x\omega \in A^*$ implies $zx\omega \in A$ by definition of A^*.

Moreover, for any finite number of strings $z_1, \ldots, z_k$ the set A^* is contained in the intersection of sets $\{\omega \mid z_i\omega \in A\}$, and the density of A^* in I is bounded by the minimal (over i) density of A in $z_iI = z_ix\Omega$.

Now let us choose $z_1, \ldots, z_k$ in such a way that the intervals $z_ix\Omega$ are disjoint and cover Ω except for a set of small measure. This is possible for the same reason as in a classic argument that explains why the Cantor set in $[0, 1]$ has zero measure. We start, say, with $z_1 = \Lambda$ and get the first interval $x\Omega$. The rest of Ω can be represented as a union of disjoint intervals, and inside each interval $u\Omega$ we select a subinterval $ux\Omega$ thus multiplying the size of the remaining set by $(1 - 2^{-|x|})$. Since this procedure can be iterated indefinitely, we can make the rest as small as needed.

Then we note that the density of A in the union of disjoint intervals (and this density is close to r if the union covers Ω almost entirely) is greater than or equal to the density of A in one of the intervals, so the intersection (an effectively open set) has density at most s for some constant $s \in (r, 1)$, as we have claimed. (We need to use the intersection and not only one of the sets since our construction should be effective even when we do not know for which interval the density is minimal.)   □

4   Bidirectional Sequences and Shifts

Recall the initial discussion in terms of ergodic theory. In this setting it is more natural to consider bi-infinite binary sequences, i.e., mappings of type $\mathbb{Z} \to \mathbb{B} = \{0, 1\}$; the uniform Bernoulli measure μ can be naturally defined on this space, too. On this space the transformation T corresponding to the shift to the left is reversible: any sequence can be shifted left or right.

The result of Theorem 1 remains true in this setting.

Theorem 4. *Let A be an effectively open set of measure $r < 1$. The set A^* of all sequences that remain in A after any arbitrary shift (any distance in any direction) is an effectively null set.*

To prove this statement, consider any $s \in (r, 1)$. As usual, it is enough to find (effectively) for every interval I_x an effectively open subset of I_x that contains $A^* \cap I_x$ and has measure at most $s\mu(I_x)$. Here x is a finite partial function from $\mathbb{Z}$ to $\mathbb{B}$ and I_x is the set of all its extensions. (One may assume that x is contiguous, since every other interval is a finite union of disjoint contiguous intervals, but this is not important for us.) Then we may iterate this construction, replacing each interval of an effectively open set by an open set inside this interval, and so on until the total measure (s^k, where k is the number of iterations) becomes smaller than any given $\varepsilon > 0$.

Assume that some I_x is given. Note that A^* is covered by every shift of A, so any intersection of I_x with a finite collection of shifted versions of A (i.e. sets of type $T^n(A)$ for $n \in \mathbb{Z}$) is a cover for $I_x \cap A^*$. It remains to show that the intersection of properly chosen shifts of A has density at most s inside I_x. To estimate the measure of the intersection, it is enough to consider the minimum of measures, and the minimum can be estimated by estimating the average measure.

More formally, we first note that by reversibility of the shift and the invariance of the measure, we have

$$\mu\big(I_x \cap T^{-n}(A)\big) = \mu\big(A \cap T^n(I_x)\big)$$

for all n. Then we prove the following lemma:

Lemma 1. *Let $J_1, \ldots, J_k$ be independent intervals of the same measure d corresponding to disjoint functions $x_1, \ldots, x_k$ of the same length. Then the average of the numbers*

$$\mu(A \cap J_1), \ldots, \mu(A \cap J_k)$$

does not exceed sd if k is large enough. Moreover such a k can be found effectively.

Proof (of Lemma 1). The average equals

$$\frac{1}{k} \sum_i \mathsf{E}(\chi_A \cdot \chi_i)$$

where χ_A is the indicator function of A and χ_i is the indicator function of J_i. Rewrite this as

$$\mathsf{E}\left(\chi_A \cdot \frac{1}{k} \sum_i \chi_i\right),$$

and note that

$$\frac{1}{k} \sum_i \chi_i$$

is the frequency of successes in k independent trials with individual probability d. (Since the functions x_i are disjoint, the corresponding intervals J_i are independent events.) This frequency (as a function on the bi-infinite Cantor space) is close to d everywhere except for a set of small measure (by the central limit theorem; in fact Chebyshev's inequality is enough). The discrepancy and the measure of this exceptional set can be made as small as needed using a large k, and the difference is then covered by the gap between r and s. This ends the proof of the lemma.

Now, given an interval I_x, we cover $I_x \cap A^*$ as follows. First, we take a integer N larger than the size of the interval I_x. The intervals

$$T^N(I_x), T^{2N}(I_x), T^{3N}(I_x), \ldots$$

are independent and have the same measure as I_x, so we can apply the above lemma and effectively find a k such that the average of

$$\mu(A \cap T^N(I_x)), \ldots, \mu(A \cap T^{kN}(I_x))$$

does not exceed $s\mu(I_x)$. This means that for some $i \leq k$ one has

$$\mu(I_x \cap T^{-iN}(A)) = \mu(A \cap T^{iN}(I_x)) \leq s\mu(I_x)$$

Therefore, $I_x \cap \bigcap_{i \leq k} T^{-iN}(A)$ is an effectively open cover of A^* of measure at most $s\mu(I_x)$. □

The statement can be strengthened: we can replace all shifts by any infinite enumerable family of shifts.

Theorem 5. *Let A be an effectively open set (of bi-infinite sequences) of measure $\alpha < 1$. Let S be an computably enumerable infinite set of integers. Then the set*

$$A^* = \{\omega \mid \omega \text{ remains in } A \text{ after shift by } s, \text{ for every } s \in S\}$$

is an effectively null set.

(Reformulation: *let A be an effectively open set of measure less than 1; let S be an infinite computably enumerable set of integers; let α be a Martin-Löf random bi-infinite sequences. Then there exists $s \in S$ such that the s-shift of ω is not in A.*)

Proof. The proof remains the same: indeed, having infinitely many shifts, we can choose as many disjoint shifts of a given interval as we want. □

Our last argument is more complicated than the previous ones (that do not refer to the central limit theorem): previously we were able to use disjoint intervals instead of independent ones. In fact the results about shifts in unidirectional sequences (both) are corollaries of the last statement. Indeed, let A be an effectively open set of right-infinite sequences of measure less than 1. Let ω be a right-infinite Martin-Löf random sequence. Then it is a part of a bi-infinite random sequence $\bar\omega$ (one may use, e.g., van Lambalgen's theorem [8] on the random pairs, see last section for a precise statement). So there is a right shift that moves $\bar\omega$ outside $\bar A$, and also a left shift with the same property (here by $\bar A$ we denote the set of bi-infinite sequences whose right halves belong to A).

5   A Generalization: Bijective Ergodic Transformations

The reversibility of the shift in the space $\mathbb{Z} \to \mathbb{B}$ was of crucial use in the above proof. In fact, our proof extends to *all* reversible ergodic maps.

Theorem 6. *Let μ be a computable measure on Ω, and $T : \Omega \to \Omega$ a reversible (=bijective) computable, ergodic, and measure-preserving transformation. Let A be an effectively open subset of Ω of measure less than 1. Let A^* be the set of points $x \in X$ such that $T^n(x) \in A$ for all $n \geq 0$. Then, A^* is an effectively null set.*

Proof. First, notice that since T is a computable bijective ergodic function, then so is T^{-1}. Let r be a real number such that $\mu(A) < r < 1$. As before, given an interval I, we want to effectively find an n such that $I \cap \bigcap_{i \leq n} T^{-i}(A)$ has measure at most $r\mu(I)$.

This is done as follows. For all n, $T^n(I)$ is an effectively open set (by computability of T^{-1}), which, since T is measure preserving, has measure $\mu(I)$, which is computable. Therefore, for any n and $\varepsilon > 0$, one can uniformly approximate $T^n(I)$ by a subset which is a finite union U of intervals such that $\mu(T^n(I) \setminus U) < \varepsilon$. This means that the value

$$\mathsf{E}\left\{ \left| \frac{\chi_1 + \ldots + \chi_n}{n} - \mu(I) \right| \right\}$$

(where χ_i is the indicator function of $T^i(I)$, i.e., $\chi_i = \chi_I \circ T^{-i}$) is computable uniformly in n. By Birkhoff's theorem we know that this quantity tends to 0 so for any $\varepsilon' > 0$ one can effectively find an N such that

$$\mathsf{E}\left\{ \left| \frac{\chi_1 + \ldots + \chi_N}{N} - \mu(I) \right| \right\} < \varepsilon'$$

Given such an N, one has

$$\mathsf{E}\left(\frac{\chi_1 + \ldots + \chi_N}{N} \cdot \chi_A \right) < \varepsilon' + \mu(A)\mu(I)$$

and thus for ε' small enough, we can find an N such that

$$\mathsf{E}\left(\frac{\chi_1 + \ldots + \chi_N}{N} \cdot \chi_A \right) < r\mu(I)$$

But this implies that for some k we have $\mathsf{E}(\chi_k \cdot \chi_A) < r\mu(I)$, i.e. $\mu(T^k(I) \cap A) < r\mu(I)$. By measure preservation and bijectivity, $\mu(T^k(I) \cap A) = \mu(I \cap T^{-k}(A))$, so $I \cap \bigcap_{i=1}^{N} T^{-i}(A)$ has measure less than $r\mu(I)$, which is what we wanted. □

Remark 1. It is possible to further extend Theorem 6 to the general setting of computable probability spaces (see [1] and [4]), but we do not discuss this here due to space restrictions.

Now we get the previous theorems as corollaries: the effective ergodic theorem for the bidirectional shift (Theorem 4) immediately follows as the bidirectional shift is clearly computable, bijective, measure-preserving and ergodic (technically we only proved Theorem 6 for the Cantor space Ω, but observe that the space of functions $\mathbb{Z} \to \mathbb{B}$ on which the bidirectional shift is defined is computably isomorphic to Ω). Moreover, we have already seen that from this theorem one can derive both Theorem 1 (Kučera's theorem for deletion of finite prefixes) and Theorem 3 (addition of finite prefixes).

It turns out that even Theorem 2 (finite change of bits) can also be proven in this setting. Indeed, let us consider the map F defined on Ω by:

$$F(1^n 0\omega) = 0^n 1\omega \text{ for all } n, \text{ and } F(11111\ldots) = 00000\ldots$$

(F adds 1 to the sequence in the dyadic sense). It is clear that F is computable, bijective and measure-preserving. That it is ergodic comes from Kolmogorov's 0-1 law, together with the observation that any two binary sequences ω, ω' that agree on all but finitely many bits are in the same orbit: $\omega' = F^n(\omega)$ for some $n \in \mathbb{Z}$. The reverse is also true except for the case when sequences have finitely many zeros or finitely many ones. This cannot happen for a random sequence, so this exceptional case does not prevent us to derive Theorem 2 from Theorem 5.

Remark 2. Theorem 5 asserts that given a random ω, and a c.e. open set U, there exists an n such that $T^n(\omega) \notin U$ (T being the bidirectional shift), and that moreover n can be taken in a computable enumerable set fixed in advance. This of course still holds for the unidirectional shift on Ω (by the above discussion), but this does not hold for all ergodic maps. Indeed, this fact follows from the so-called *strong mixing property* of the shift, which not all ergodic maps have (e.g. a rotation of the circle by an irrational angle is an ergodic map but does not have this property).

6   An Application

The celebrated van Lambalgen theorem [8] asserts that in the probability space Ω^2 (pairs of binary sequences with independent uniformly distributed components) a pair (ω_0, ω_1) is random if and only if ω_0 is random and ω_1 is ω_0-random (random relative to the oracle ω_0). This can be easily generalized to k-tuples: an element $(\omega_0, \omega_1, \ldots, \omega_{k-1})$ of Ω^k is random if and only if ω_0 is random and ω_i is $(\omega_0, \ldots, \omega_{i-1})$-random for all $i = 1, 2 \ldots, k - 1$. Can we generalize this statement to infinite sequences? Not completely: there exists an infinite sequence $(\omega_i)_{i \in \mathbb{N}}$ such that ω_0 is random and ω_i is $(\omega_0, \ldots, \omega_{i-1})$-random for all $i \geq 1$ and nevertheless $(\omega_i)_{i \in \mathbb{N}}$ is non-random as an element of $\Omega^{\mathbb{N}}$. To construct such an example, take a random sequence in $\Omega^{\mathbb{N}}$ and then replace the first i bits of ω_i by zeros.

Informally, in this example all ω_i are random, but their "randomness deficiency" increases with i, so the entire sequence (ω_i) is not random (in $\Omega^{\mathbb{N}}$). K. Miyabe [6] has shown recently that one can overcome this difficulty allowing finitely many bit changes in each ω_i (number of changed bits may depend on i):

Theorem 7 (Miyabe). *Let $(\omega_i)_{i \in \mathbb{N}}$ be a sequence of elements of Ω such that ω_0 is random and ω_i is $(\omega_0, \ldots, \omega_{i-1})$-random for all $i \geq 1$. Then there exists a sequence $(\omega'_i)_{i \in \mathbb{N}}$ such that*

- *For every i the sequence ω'_i is equal to ω_i except for a finite number of places.*
- *The sequence $(\omega'_i)_{i \in \mathbb{N}}$ is a random element of $\Omega^{\mathbb{N}}$.*

Informally, this result can be explained as follows: as we have seen (Theorem 2), a change in finitely many places can decrease the randomness deficiency (starting from any non-random sequence, we get a sequence that is not covered by a first set of a Martin-Löf test) and therefore can prevent "accumulation" of randomness deficiency.

This informal explanation can be formalized and works not only for finite changes but also for adding/removing prefixes. In fact, the results of this paper allow us to get a simple proof of the following generalization of Miyabe's result (Miyabe's original proof used a different approach, namely martingale characterizations of randomness). We restrict ourselves to the uniform measure, but the same argument works for arbitrary computable measures.

Theorem 8. *Let $(\omega_i)_{i \in \mathbb{N}}$ be a sequence of elements of Ω such that ω_0 is random and ω_i is $(\omega_0, \ldots, \omega_{i-1})$-random for all $i \geq 1$. Let $T : \Omega \to \Omega$ be a computable bijective ergodic map. Then, there exists a sequence $(\omega_i')_{i \in \mathbb{N}}$ such that*

- *For every i, the sequence ω_i' is an element of the orbit of ω_i (i.e. $\omega_i' = T^{n_i}(\omega_i)$ for some integer n_i).*
- *The sequence $(\omega_i')_{i \in \mathbb{N}}$ is a random element of $\Omega^{\mathbb{N}}$.*

Proof. Let U be the first level of a universal Martin-Löf test on $\Omega^{\mathbb{N}}$, with $\mu(U) \leq 1/2$. We will ensure that the sequence $(\omega_i')_{i \in \mathbb{N}}$ is outside U, and this guarantees its randomness.

Consider the set V_0 consisting of those $\alpha_0 \in \Omega$ such that the section

$$U_{\alpha_0} = \{(\alpha_1, \alpha_2, \ldots) \mid (\alpha_0, \alpha_1, \alpha_2, \ldots) \in U\}$$

has measure greater than $2/3$. The measure of V_0 is less than 1, otherwise we would have $\mu(U) > 1/2$. It is easy to see that V_0 is an effectively open subset of Ω. Since ω_0 is random, by Theorem 6 there exists an integer n_0 such that $\omega_0' = T^{n_0}(\omega_0)$ is outside V_0. This ω_0' will be the first element of the sequence we are looking for.

Now we repeat the same procedure for $U_{\omega_0'}$ instead of U. Note that it is an open set of measure at most $2/3$, and, moreover, an effectively open set with respect to oracle ω_0'. Since ω_0 and ω_0' differ by a computable transformation, the set $U_{\omega_0'}$ is effectively open with oracle ω_0. We repeat the same argument (where $1/2$ and $2/3$ are replaced by $2/3$ and $3/4$ respectively) and conclude that there exists an integer n_1 such that the sequence $\omega_1' = T^{n_1}(\omega_1)$ has the following property: the set

$$U_{\omega_0'\omega_1'} = \{(\alpha_2, \alpha_3, \ldots) \mid (\omega_0', \omega_1', \alpha_2, \alpha_3, \ldots) \in U\}$$

has measure at most $3/4$. (Note that we need to use ω_0-randomness of ω_1, since we apply Theorem 6 to an ω_0-effectively open set.)

At the next step we get n_2 and $\omega_2' = T^{(n_2)}\omega_2$ such that

$$U_{\omega_0'\omega_1'\omega_2'} = \{(\alpha_3, \alpha_4, \ldots) \mid (\omega_0', \omega_1', \omega_2', \alpha_3, \alpha_4, \ldots) \in U\}$$

has measure at most $4/5$, etc.

Is it possible that the resulting sequence $(\omega_0', \omega_1', \omega_2', \ldots)$ is covered by U? Since U is open, it would be then covered by some interval in U. This interval may refer only to finitely many coordinates, so for some m all sequences

$$(\omega_0', \omega_1', \ldots, \omega_{m-1}', \alpha_m, \alpha_{m+1}, \ldots)$$

would belong to U (for every $\alpha_m, \alpha_{m+1}, \ldots$). However, this is impossible, because our construction ensures that the measure of the set of all $(\alpha_m, \alpha_{m+1}, \ldots)$ with this property is less than 1. □

Acknowledgements. We are thankful to Mathieu Hoyrup and three anonymous referees for their very helpful comments and suggestions.

References

1. Gács, P.: Lecture notes on descriptional complexity and randomness (manuscript), http://www.cs.bu.edu/fac/gacs/recent-publ.html
2. Gács, P., Hoyrup, M., Rojas, C.: Randomness on computable probability spaces - a dynamical point of view. In: Symposium on Theoretical Aspects of Computer Science (STACS 2009). Dagstuhl Seminar Proceedings. Schloss Dagstuhl - Leibniz-Zentrum fuer Informatik, Germany Internationales Begegnungs- und Forschungszentrum fuer Informatik (IBFI), Schloss Dagstuhl, vol. 9001, pp. 469–480 (2009)
3. Hoyrup, M., Rojas, C.: Applications of effective probability theory to Martin-löf randomness. In: Albers, S., et al. (eds.) ICALP 2009. LNCS, vol. 5555, pp. 549–561. Springer, Heidelberg (2009)
4. Hoyrup, M., Rojas, C.: Computability of probability measures and Martin-Löf randomness over metric spaces. Information and Computation 207(7), 2207–2222 (2009)
5. Kučera, A.: Measure, Π_1^0 classes, and complete extensions of PA. In: Ebbinghaus, H.-D., Müller, G.H., Sacks, G.E. (eds.) Recursion Theory Week. Proceedings of a Conference held in Oberwolfach, West Germany, April 15-21. LNM, vol. 1141, pp. 245–259 (1985)
6. Miyabe, K.: An extension of van Lambalgen's theorem to infinitely many relative 1-random reals. The Notre Dame Journal of Formal Logic (to appear)
7. Shiryaev, A.: Probability, 2nd edn. Springer, Heidelberg (1996)
8. van Lambalgen, M.: Random sequences. PhD dissertation, University of Amsterdam, Amsterdam (1987)
9. V'yugin, V.: Effective convergence in probability and an ergodic theorem for individual random sequences. SIAM Theory of Probability and Its Applications 42(1), 39–50 (1997)
10. V'yugin, V.: Ergodic theorems for individual random sequences. Theoretical Computer Science 207(2), 343–361 (1998)

How Powerful Are Integer-Valued Martingales?

Laurent Bienvenu[1], Frank Stephan[2], and Jason Teutsch[3]

[1] LIAFA, CNRS & Université de Paris 7, France
laurent.bienvenu@liafa.jussieu.fr
[2] National University of Singapore
fstephan@comp.nus.edu.sg
[3] Center for Communications Research–La Jolla, USA
jrteuts@ccrwest.org

Abstract. In the theory of algorithmic randomness, one of the central notions is that of computable randomness. An infinite binary sequence X is computably random if no recursive martingale (strategy) can win an infinite amount of money by betting on the values of the bits of X. In the classical model, the martingales considered are real-valued, that is, the bets made by the martingale can be arbitrary real numbers. In this paper, we investigate a more restricted model, where only integer-valued martingales are considered, and we study the class of random sequences induced by this model.

1 Gambling with or without Coins

One of the main approaches to define the notion of random sequence is the so-called "unpredictability paradigm". We say that an infinite binary sequence is "random" if there is no effective way to win arbitrarily large amounts of money by betting on the values of its bits. The main notion arising from this paradigm is computable randomness, but other central notions such as Martin-Löf randomness, Schnorr randomness, and Kurtz randomness, can be formulated in this setting. For all of these notions, we consider models of games where the player can, at each turn, bet *any* amount of money between 0 and his current capital. In "practice" however, one cannot go into a casino and bet arbitrarily small sums of money: there is always a unit value, and any bet made has to be a multiple of this value. Some casinos (and games) also impose upper limits on the amount of capital the one can gamble in each round of play. In the following exposition, we examine the consequences of restricting betting amounts to integers and finite sets.

To formalize the unpredictability paradigm, we need the central notion of *martingale*. A martingale is a betting strategy for a fair game and is formally represented by a function that corresponds to the gambler's fortune at each moment in time. Let $\{0,1\}^*$ denote the set of all finite binary sequences, and $\{0,1\}^\omega$ is the set of all countably infinite binary sequences (a.k.a *reals*). Any function $M : \{0,1\}^* \to \mathbb{R}^+$ which satisfies the *fairness condition*

$$M(\sigma) = \frac{M(\sigma 0) + M(\sigma 1)}{2} \tag{1.1}$$

F. Ferreira et al. (Eds.): CiE 2010, LNCS 6158, pp. 59–68, 2010.

for all $\sigma \in \{0,1\}^*$ is called a *martingale*. $M(\sigma)$ corresponds to the gambler's capital after having already bet on the finite sequence σ. The fairness condition (1.1) says that the amount of money gained from an outcome of "0" is the same that would be lost from an outcome of "1". It is important to note that our definition of martingale is a very restricted version of what is usually referred to as "martingale" in probability theory, where it is defined to be a sequence $X_0, X_1, \ldots$ of real-valued random variables (possibly taking negative values) such that for all n

$$\mathbb{E}[X_{n+1}|X_0, X_1, \ldots, X_n] = X_n.$$

To make the distinction, we call such a sequence a *martingale process*. A martingale is called *recursive* if M is a recursive function. Throughout this exposition, "martingale" and "recursive martingale" will be used synonymously.

For any $A \in \{0,1\}^\omega$, $A \upharpoonright n$ is the finite binary sequence, or *initial segment*, consisting of the first n digits of A. We also identify sets with their characteristic sequences. A martingale M *succeeds* on $A \in \{0,1\}^\omega$ if M achieves arbitrary sums of money over A, that is, $\limsup_n M(A \upharpoonright n) = \infty$. Otherwise A *defeats* M. M *Schnorr-succeeds* on a set A if M succeeds on A and there exists a recursive, non-decreasing, unbounded function f such that $f(n) < M(A \upharpoonright n)$ for infinitely many n. M *Kurtz-succeeds* on a set A if M succeeds on A and there exists a recursive, non-decreasing, unbounded function f such that $f(n) < M(A \upharpoonright n)$ for all n. We can now define the main classical notions of randomness in terms of martingales.

Definition 1. *A sequence $A \in \{0,1\}^\omega$ is called* computably random *if A defeats every martingale. If no martingale Schnorr-succeeds on A, then A is* Schnorr random. *If no martingale Kurtz-succeeds on A, then A is* Kurtz random *(equivalently, A is Kurtz random if and only if A does not belong to any Π_1^0 subset of $\{0,1\}^\omega$ of measure 0).*

In this paper, we shall consider games where the player can only make bets of integer value. For M a martingale and $\sigma \in \{0,1\}^*$, $|M(\sigma 0) - M(\sigma)|$ is called the *wager at σ*. Now, given a set V of non-negative integers, we say that a martingale is *V-valued* if for all σ the wager of M at σ belongs to V, unless M does not have enough capital in which case the wager at σ is 0. Formally, M is V-valued if for all $\sigma \in \{0,1\}^*$ and $a \in \{0,1\}$, $M(\sigma) < \min(V) \Rightarrow M(\sigma a) = M(\sigma)$ and $M(\sigma) \geq \min(V) \Rightarrow |M(\sigma a) - M(\sigma)| \in V$. A martingale whose wagers are integers is called an *integer-valued* martingale. In case V is finite we say that M is *finitely-valued* and if V is a singleton, that M is *single-valued*.

Definition 2. *A real X is* V-valued random *if no V-valued martingale succeeds on X. A real X is a* finitely-valued / integer-valued / single-valued random *if no finitely-valued / integer-valued / single-valued martingale succeeds on X.*

The rest of the paper studies how these new notions of randomness interact with the classical ones. We will prove the implications of the following diagram:

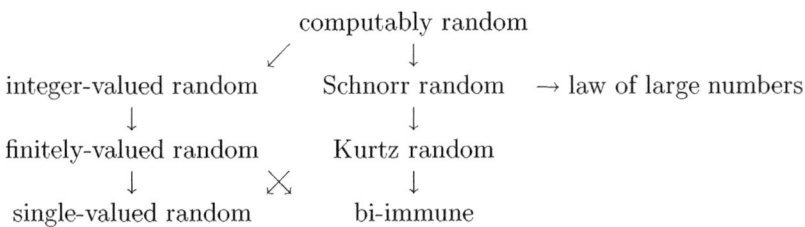

and we shall further see that no other implication than those indicated (and their transitive closure) holds.

If we were to ask someone what the absolute minimum one could expect from a set called "random," you might receive one of the following two responses:

1. The set obeys the law of large numbers.
2. The set is bi-immune.

The person who says "1" believes that a set which does not follow the law of large numbers exhibits a probabilistic bias in its distribution of 0's and 1's. The person who says "2" believes that a set with an infinite recursive subset of 0's or 1's yields algorithmic bias. There exists, however, a third possibility:

3. The set is single-valued random.

"3" closely matches our intuition in the sense that one should not be able to predict successive outcomes resulting from a "random" process. From a practical point-of-view, single-valued randomness also makes sense. If you have to sit out 2^{1000} rounds of roulette before placing a sure bet, as might occur when gambling on a non-bi-immune set, then with probability 1 the casino has already closed while you were waiting for this opportunity. In Section 3, we shall prove that notion "3" indeed differs from notions "1" and "2."

The separation of Kurtz randomness and Schnorr randomness is folklore (we will see in a moment how it can be proven). A somewhat more difficult result is the separation of computable randomness and Schnorr randomness. The separation of these two notions was proven by Wang who constructed a Schnorr random sequence X together with a martingale M that succeeds on X. It turns out that in Wang's construction, the martingale M is already $\{0,1\}$-valued, hence it immediately follows that Schnorr randomness (a fortiori Kurtz randomness) does not imply finitely-valued randomness (and a fortiori integer-valued randomness).

Theorem 1 (Wang [15]). *There exists a Schnorr random $X \in \{0,1\}^\omega$ and a $\{0,1\}$-valued martingale M such that M succeeds on X.*

In Section 2 we shall see that conversely, integer-valued randomness does not imply Schnorr randomness, and a fortiori computable randomness.

2 Integer-Valued Martingales and Genericity

There is an essential difference between rational-valued and integer-valued martingales. The latter can always be permanently defeated while in general the former cannot be. Consider the example of a player starting with an initial capital of 1 who at each turn bets half of its capital on the value 1 (that is, the corresponding martingale M satisfies $M(\sigma 0) = M(\sigma)/2$ and $M(\sigma 1) = 3M(\sigma)/2$ for all $\sigma \in \{0,1\}^*$). This is a rational-valued martingale with the following property. Pick a stage s of the game; no matter how unlucky the player has been before that stage, she always has a chance to recover. More precisely, for any finite sequence of outcomes $\sigma \in \{0,1\}^*$, no matter how small $M(\sigma)$ is, the player can still win the game if the remaining of the outcomes contains a lot of 0's (for example the player wins against the sequence $\sigma 0000\ldots$). This phenomenon no longer holds for integer-valued martingales, and in fact the opposite is true, that is, no matter how lucky the player has been up to stage s, there is always a risk for her to see her strategy permanently defeated at some stage $s' > s$. This is expressed by the following lemma.

Lemma 1. *Let M be an integer-valued martingale. For any $\sigma \in \{0,1\}^*$, there exists an extension $\tau(\sigma, M) \in \{0,1\}^*$ of σ such that $M(\tau') = M(\tau(\sigma, M))$ for all extensions τ' of $\tau(\sigma, M)$ (in particular the strategy M does not succeed on any $X \in \{0,1\}^\omega$ extending $\tau(\sigma, M)$).*

From a topological perspective, the above result shows that any integer-valued martingale M is defeated on a dense open set. Indeed, for any σ, M is defeated by every sequence $X \in [\tau(\sigma, M)]$ hence M is defeated by any sequence in the dense open set

$$\mathcal{U}_M = \bigcup_{\sigma \in \{0,1\}^*} [\tau(\sigma, M)]$$

(it is dense as for any σ, $[\tau(\sigma, M)] \subseteq [\sigma]$ by construction). Therefore, the set of integer-valued random sequences contains the intersection over all integer-valued martingales $\bigcap \mathcal{U}_M$. This is a countable intersection of dense open sets, hence the following corollary.

Corollary 1. *The set of integer-valued random sequences is co-meager.*

This shows that as a notion of randomness, integer-valued randomness is quite weak. Indeed, one of the most basic properties that we can expect from a random sequence X is that it satisfies the *law of large numbers*, that is, the number of 0's in $X \upharpoonright n$ is $n/2 + o(n)$. It is a routine exercise to show that the set of sequences X satisfying the law of large numbers is a meager set (contained in a countable union of closed set with empty interior). Therefore, in the sense of Baire category, most sequences are integer-valued random but do not satisfy the law of large numbers. On the other hand, it is well-known that any Schnorr random sequence must satisfy the law of large numbers, which yields a further corollary.

Corollary 2. *There exists a sequence $X \in \{0,1\}^\omega$ which is integer-valued random but not Schnorr random.*

If we now want to compare integer-valued randomness and Kurtz randomness, the above results are insufficient, as the set of Kurtz random sequences is also a co-meager set. We will prove that Kurtz randomness does not imply integer-valued randomness by looking at the classical counterpart of Baire category, namely genericity. Recall that a set $W \subseteq \{0,1\}^*$ is *dense* if the open set $\bigcup_{\sigma \in W} [\sigma]$ is dense or equivalently if for any string σ there exists a string in W extending σ. We say that $X \in \{0,1\}^\omega$ is *weakly n-generic* if X has a prefix in every dense Σ_n^0 set. We further say that X is *n-generic* if for any (not necessarily dense) Σ_n^0 set of strings W, either X has a prefix in W or there exists a prefix of X which has no extension in W. For all $n \geq 0$ it holds that

$$\text{weakly (n+1)-generic} \quad \Rightarrow \quad \text{n-generic} \quad \Rightarrow \quad \text{weakly n-generic}.$$

Kurtz showed that weakly 1-genericity is enough to ensure Kurtz randomness.

Proposition 1 (Kurtz [7]). *Any weakly 1-generic sequence $X \in \{0,1\}^\omega$ is Kurtz random.*

The next two theorems show that more genericity is needed to ensure integer-valued randomness. That is, weak 2-genericity is sufficient, but 1-genericity is not.

Theorem 2. *Let $X \in \{0,1\}^\omega$ be any weakly 2-generic sequence. Then X is integer-valued random.*

Proof. We have shown in Lemma 1 that for any martingale $M \in \mathfrak{D}$, the set of strings

$$W_M = \{\sigma \ : \ M(\sigma') = M(\sigma) \text{ for all extensions } \sigma' \text{ of } \sigma\}$$

is dense. It is also easy to see that this set is recursive in $\mathbf{0}'$, in particular W_M is Σ_2^0. By definition, a weak-2-generic sequence X must have a prefix in W_M for all integer-valued martingales M, and it is clear that if X has a prefix in W_M, M does not succeed on X.   □

Theorem 3. *There exists a 1-generic sequence $X \in \{0,1\}^\omega$ and a $\{0,1\}$-valued martingale M such that M succeeds on X.*

Corollary 3. *There exists a sequence $X \in \{0,1\}^\omega$ which is Kurtz random but not integer-valued random.*

The converse of this result is also true, that is there exists a sequence X which is integer-valued random but not Kurtz random. To prove this, we will need a different approach, via measure-theoretic arguments, which we will outline in Section 4.

Strictly speaking, integer-valued martingales not only impose a lower limit on betting amounts but also require that all wagers be a multiple of the minimum bet. We are therefore left with a question regarding the robustness of integer-valued randomness: if we remove the requirement that wagers must be a multiple of the minimum bet, do we still obtain the same notion of randomness?

Open question 1. *Let V be the set of all computable reals greater than or equal to 1 unioned with $\{0\}$. Is V-valued random the same as integer-valued random?*

3   Finitely-Valued Martingales

We now consider the effects of imposing betting limits on martingale strategies. First we separate integer-valued randomness from finitely-valued randomness.

Theorem 4. *There exists an integer-valued martingale which succeeds on a finitely-valued random.*

Schnorr showed that for any set A, a real-valued martingale succeeds on A if and only if a rational-valued martingale succeeds on A (see [12], or [10] p.270). His proof, however, does not carry over to the finitely-valued case.

Open question 2. *If we allow finitely-valued martingales to bet real values instead of rationals, do we get the same class of finitely-valued randoms?*

3.1   On Single-Valued Randoms

For the following discussion, it is useful to keep in mind that a real is single-valued random if and only if it is $\{1\}$-valued random; the particular dollar amount which is bet each round is immaterial. For comparison with Kurtz randomness, we appeal directly to a theorem of Doob ([4] p.324). The following version for "non-negative" martingales appears in Ross's book ([11], p.316).

Theorem 5 (Doob's Martingale Convergence Theorem). *For every martingale M, the set of reals on which M succeeds has measure zero. Furthermore, the capital of M converges to some finite value with probability 1.*

Proposition 2. *Every Kurtz random is single-valued random.*

Proof. Suppose that some single-valued martingale M succeeds on a real X. Let F denote the set of reals on which M converges to some finite value. Then X does not belong to F, and F has measure one by Doob's Martingale Convergence Theorem. Hence X belongs to the measure zero set $\overline{F}$.

By definition of single-valued, M is required to bet at every position of the input real. Hence the only way for M to converge to a finite value is to reach the value 0 and become constant. Therefore F is, in fact, the set of reals on which M eventually goes broke. Thus $\overline{F}$ is the set of all infinite paths through the tree $\{\sigma : M(\sigma) > 0\}$. It follows that $\overline{F}$ is a recursive Π_1^0 class. In summary, X belongs to a recursive Π_1^0 class of measure zero and therefore is not Kurtz random. □

The above argument shows even more: every Kurtz random is V-valued random for any positive set of integers V.

We now separate the incomparable notions of bi-immunity, single-valued random and law of large numbers. A set is called *bi-immune* if neither the set nor its complement contains an infinite r.e. subset.

Proposition 3. *There exists a single-valued random X such that neither X nor $\overline{X}$ contains an infinite r.e. set. Hence X is not bi-immune.*

Even stronger than bi-immune, a set $A = \{a_0 < a_1 < a_2 < \cdots\}$ is called *bi-hyperimmune* if there is no recursive function g such that $g(n) > a_n$ for all n and likewise for $\overline{A}$.

Theorem 6. *There exists a bi-hyperimmune set which is recursive in the halting set but not single-valued random.*

Proof. Let $\varphi_0, \varphi_1, \varphi_2, \ldots$ be a list of the partial recursive functions. Define a function f recursive in the halting set satisfying

$$(\forall e < n)\, [\varphi_e(f(n))\!\downarrow \implies f(n+1) > \varphi_e(f(n)) + 2]$$

and let

$$A = \{x : (\exists n)\, [f(2n) \le x < f(2n+1)]\}.$$

Now $A = \{a_0, a_1, a_2, \ldots\}$ is bi-hyperimmune because for any recursive φ_k,

$$\varphi_k[f(2k+1)] < f(2k+2) - 2 \le a_{f(2k+1)} - 1,$$

and a similar inequality holds for the complement of A. On the other hand, the single-valued martingale strategy which bets on $A(n+1)$ what the gambler saw at $A(n)$ will succeed on A. This strategy indeed succeeds because each time $A(n+1)$ disagrees with $A(n)$, we have that $A(n+2)$ and $A(n+3)$ agree with $A(n+1)$. So over each three consecutive rounds of betting, the gambler increases his capital by at least \$1. $\qquad\square$

Proposition 2 and Theorem 6 together give:

Corollary 4. *There exists a bi-immune set which is not Kurtz random.*

Corollary 1 can be used to show incomparability between the law of large numbers and the other two weak randomness notions. Finally, we note that it is possible to separate single-valued randomness from finite-valued randomness using an argument along the lines of Proposition 3.

Proposition 4. *There exists a $\{1, 2\}$-valued martingale which succeeds on a single-valued random.*

3.2   On $\{0,1\}$-Valued Randoms

We can also separate single-valued randomness from finite-valued randomness.

Proposition 5. *Let V be any set containing 0 and at least one other number n. Then any V-valued random is bi-immune.*

Proof. Let A be a set which is not bi-immune; without loss of generality assume that A contains an infinite recursive set B. Then a V-valued martingale strategy which bets n dollars on members of B and 0 on $A - B$ will succeed on A. $\qquad\square$

The following corollary is a consequence of the definition of finitely-valued random and Proposition 5.

Corollary 5. *finitely-valued random* $\implies$ $\{0,1\}$-*valued random* $\implies$ *bi-immune.*

Since single-valued random does not imply bi-immune (Theorem 3), we obtain from Corollary 5:

Corollary 6. *There exists a* $\{0,1\}$-*valued random which is not single-valued random.*

Although we were able to separate single-valued randomness from $\{1,2\}$-valued randomness (Proposition 4), the comparison between $\{0,1\}$-valued randoms and $\{0,1,2\}$-valued randoms seems less clear. We leave the reader with the following interesting question.

Open question 3. *Is* $\{0,1\}$-*valued random the same as finitely-valued random?*

4   Integer-Valued Martingales and Bernoulli Measures

In this last section, we present a proof of the fact that integer-valued randomness does not imply Kurtz randomness. We will get a counter example by choosing a sequence X at random with respect to some carefully-chosen probability measure.

Intuitively speaking, the Lebesgue measure λ on the space $\{0,1\}^\omega$ corresponds to the random trial where all bits are obtained by independent tosses of a balanced 0/1-coin. An interesting generalization of Lebesgue measure is the class of *Bernoulli measures*, where for a given parameter $\delta \in [-1/2, 1/2]$ we construct a sequence X by independent tosses of a coin with bias δ (that is, the coin gives 1 with probability $1/2 + \delta$ and 0 with probability $1/2 - \delta$. This can be further generalized by considering an infinite sequence of independent coin tosses where the nth coin tossed has bias δ_n. This leads to the notion of *generalized Bernoulli measures*. Formally, on the space $\{0,1\}^\omega$, given a sequence $(\delta_n)_{n\in\mathbb{N}}$ of numbers in $[-1/2, 1/2]$, a *generalized Bernoulli measure* of parameter $(\delta_n)_{n\in\mathbb{N}}$ is the unique measure μ such that for all $\sigma \in \{0,1\}^*$:

$$\mu([\sigma]) = \prod_{n\,:\,\sigma(n)=0} (1-p_n) \prod_{n\,:\,\sigma(n)=1} p_n$$

where $p_n = 1/2 + \delta_n$. One can expect that if the δ_n are very small (that is, δ_n tends to 0 quickly), then the generalized Bernoulli measure of parameter $(\delta_n)_{n\in\mathbb{N}}$ will not differ much from Lebesgue measure. This was made precise by Kakutani.

Theorem 7 (Kakutani [6]). *Let μ be the generalized Bernoulli measure of parameter $(\delta_n)_{n\in\mathbb{N}}$. If the condition*

$$\sum_{n\in\mathbb{N}} \delta_n^2 < \infty \tag{4.1}$$

holds, then μ is equivalent to Lebesgue measure λ, that is, for any subset $\mathcal{X}$ of $\{0,1\}^\omega$, $\mu(\mathcal{X}) = 0$ if and only if $\lambda(\mathcal{X}) = 0$. If condition (4.1) does not hold, then μ and λ are inconsistent, that is, there exists some $\mathcal{Y}$ such that $\mu(\mathcal{Y}) = 0$ while $\lambda(\mathcal{Y}) = 1$.

If we want to work in a computability setting, we need to consider *computable* generalized Bernoulli measures, that is, those for which the parameter $(\delta_n)_{n \in \mathbb{N}}$ is a recursive sequence of reals. Vovk [14] showed a constructive analogue of Kakutani's theorem for computable generalized Bernoulli measures in relation with Martin-Löf randomness (perhaps the most famous effective notion of randomness, but we do not need it in this paper). The Kakutani-Vovk result has been used many times in the literature [1,8,9,13]. In particular, Bienvenu and Merkle proved the following.

Theorem 8 (Bienvenu and Merkle [1]). *Let μ be a computable generalized Bernoulli measure of parameter $(\delta_n)_{n \in \mathbb{N}}$. If $\sum_n \delta_n^2 = +\infty$, then the class of Kurtz random sequences has μ-measure 0.*

To prove that integer-valued randomness does not imply Kurtz randomness, we will construct a computable generalized Bernoulli measure μ whose parameter $(\delta_n)_{n \in \mathbb{N}}$ converges to 0 sufficiently slowly to have $\sum_n \delta_n^2 = +\infty$ (hence by the above μ-almost all sequences X are not Kurtz random, which we will make even more precise) but sufficiently quickly to make μ close to Lebesgue measure and ensure that μ-almost all sequences are integer-valued random.

Theorem 9. *There exists a sequence $X \in \{0,1\}^\omega$ which is integer-valued random but not Kurtz random.*

Proof (Sketch). We obtain X by choosing a random sequence with respect to the generalized Bernoulli measure of parameter (δ_n) with

$$\delta_n = \frac{1}{\sqrt{n}\,\ln n}$$

By Theorem 8 we obtain almost surely a sequence X that is not Kurtz random. The difficulty is to show that X is almost surely integer-valued random. The argument goes as follows. When a bit has probability $1/2 + \delta_n$ to be 0, the player's best move (when there is no restriction on how much can be bet) is to bet a *fraction* $2\delta_n$ of her capital on the value 0 (e.g. if $\delta = 0.1$, the player should bet 20% of her capital). Using this strategy, the player can roughly expect to have a capital $O(\sum_{i=0}^n \delta_n^2)$ after stage n. However, if the player plays too risky, i.e. bets at each turn a fraction ρ_n of her capital such that $\rho_n/\delta_n \to \infty$, then almost surely she will lose the game.

With the value of the δ_n we have chosen, the optimal strategy will yield a gain of $O(\sum_{i=0}^n 1/(n \ln n))$, i.e. $O(\ln \ln n)$. But then, if the player is forced to make bets of integer value, even a bet of \$1 will represent a fraction ρ_n of her capital of at least $1/\ln \ln n$, which is much bigger than δ_n, hence the player would lose the game almost surely if making such bets. But the only alternative is to bet 0, which also causes the player to lose the game.

Open question 4. *Do there exist other characterizations for integer-valued, finite-valued, or single-valued randoms in terms of Kolmogorov complexity or Martin-Löf statistical tests?*

NB: A complete version of this paper, containing the omitted proofs and additional results, can be found at `http://arxiv.org/abs/1004.0838`.

References

1. Bienvenu, L., Merkle, W.: Constructive equivalence relations for computable probability measures. Annals of Pure and Applied Logic 160, 238–254 (2009)
2. Billingsley, P.: Probability and measure, 3rd edn. Wiley Series in Probability and Mathematical Statistics. John Wiley & Sons Inc., New York (1995); A Wiley-Interscience Publication
3. Downey, R., Griffiths, E., Reid, S.: On Kurtz randomness. Theoretical Computer Science 321(2-3), 249–270
4. Doob, J.L.: Stochastic Processes. John Wiley & Sons Inc., New York (1953)
5. Hoeffding, W.: Probability inequalities for sums of bounded random variables. Journal of the American Statistical Association 58, 13–30 (1963)
6. Kakutani, S.: On equivalence of infinite product measures. Annals of Mathematics 49(214-224) (1948)
7. Kurtz, S.: Randomness and genericity in the degrees of unsolvability. PhD dissertation, University of Illinois at Urbana (1981)
8. Merkle, W., Miller, J.S., Nies, A., Reimann, J., Stephan, F.: Kolmogorov-Loveland randomness and stochasticity. Annals of Pure and Applied Logic 138(1-3), 183–210 (2006)
9. Muchnik, A.A., Semenov, A., Uspensky, V.: Mathematical metaphysics of randomness. Theoretical Computer Science 207(2), 263–317 (1998)
10. Nies, A.: Computability and Randomness. Oxford Logic Guides, vol. 51. Oxford University Press, Oxford (2009)
11. Ross, S.M.: Stochastic Processes, 2nd edn. Wiley Series in Probability and Statistics: Probability and Statistics. John Wiley & Sons Inc., New York (1996)
12. Schnorr, C.-P.: Zufälligkeit und Wahrscheinlichkeit. Eine algorithmische Begründung der Wahrscheinlichkeitstheorie. LNM, vol. 218. Springer, Berlin (1971)
13. Shen, A.: On relations between different algorithmic definitions of randomness. Soviet Mathematics Doklady 38, 316–319 (1989)
14. Vovk, V.: On a criterion for randomness. Soviet Mathematics Doklady 294(6), 1298–1302 (1987)
15. Wang, Y.: A separation of two randomness concepts. Information Processing Letters 69(3), 115–118 (1999)

A Faster Algorithm for Finding
Minimum Tucker Submatrices

Guillaume Blin[1], Romeo Rizzi[2], and Stéphane Vialette[1]

[1] Université Paris-Est, LIGM - UMR CNRS 8049, France
{gblin,vialette}@univ-mlv.fr
[2] Dipartimento di Matematica ed Informatica (DIMI)
Universit degli Studi di Udine, Italy
rrizzi@dimi.uniud.it

Abstract. A binary matrix has the *Consecutive Ones Property* (C1P) if its columns can be ordered in such a way that all 1s on each row are consecutive. Algorithmic issues of the C1P are central in computational molecular biology, in particular for physical mapping and ancestral genome reconstruction. In 1972, Tucker gave a characterization of matrices that have the C1P by a set of forbidden submatrices, and a substantial amount of research has been devoted to the problem of efficiently finding such a minimum size forbidden submatrix. This paper presents a new $O(\Delta^3 m^2(m\Delta + n^3))$ time algorithm for this particular task for a $m \times n$ binary matrix with at most Δ 1-entries per row, thereby improving the $O(\Delta^3 m^2(mn + n^3))$ time algorithm of Dom *et al.* [17].

1 Introduction

A binary matrix has the *Consecutive Ones Property* (C1P) if its columns can be ordered in such a way that all 1s on each rows are consecutive. Both deciding if a given binary matrix has the C1P and finding the corresponding columns permutation can be done in linear time [9,18,19,23,24,25,28,31]. The C1P of matrices has a long history and it plays an important role in combinatorial optimization, including application fields such as scheduling [6,21,22,36], information retrieval [26], and railway optimization [29,30,33] (see [16] for a recent survey). Furthermore, algorithmic aspects of the C1P turn out to be of particular importance for physical mapping [2,13,27] and ancestral genome reconstruction [1,12]. (see also [10,3,4,5,14,32] for other applications in computational molecular biology). Actually, our main motivation for studying algorithmic aspects of the C1P comes from *minimal conflicting sets* in binary matrices in the context of ancestral genome reconstruction [11]. A minimal conflicting set of rows in a binary matrix is a set of rows R that does not have the C1P but such that any proper subset of R has the C1P (a similar definition applies for columns). The aim of this paper is to lay the foundations for efficiently computing minimal conflicting sets by presenting a new efficient algorithm for finding such a minimum size forbidden Tucker submatrix [8].

F. Ferreira et al. (Eds.): CiE 2010, LNCS 6158, pp. 69–77, 2010.
© Springer-Verlag Berlin Heidelberg 2010

Let us turn the C1P into an optimization problem. Recently, Dom *et al.* [17] investigated natural problems arising when a matrix M does not have the C1P property (the C1P is indeed a desirable property than often leads to efficient algorithms):

- Min-COS-C (*"Consecutive Ones Submatrix by Column Deletion"*) – find a minimum-cardinality set of columns to delete such that the resulting matrix has the C1P.
- Min-COS-R (*"Consecutive Ones Submatrix by Row Deletion"*) – find a minimum-cardinality set of rows to delete such that the resulting matrix has the C1P.
- Min-CO-1E (*"Consecutive Ones by Flipping 1-Entries"*) – find a minimum-cardinality set of 1-entries in the matrix that shall be flipped (that is, re-placed by 0-entries) such that the resulting matrix has the C1P.

All these problems are **NP**-hard even for simple instances [20,34], and hence Dom *et al.* have focussed on approximation and parameterized complexity issues. To this end, they have provided a technical solution based on efficiently detecting forbidden Tucker submatrices [35]. For the sake of presentation, let us introduce these forbidden submatrices by graphs.

Let M be a $m \times n$ binary matrix. Its corresponding vertex-colored bipartite graph $G(M) = (V_M, E_M)$ is defined as follows: for every row (resp. column) of M there is a black (resp. white) vertex in V_M, and there is an edge between a black vertex v_i and a white vertex v_j, *i.e.*, an edge between the vertices that correspond to the ith row and the jth column of M, if and only of $M[i,j] = 1$. Equivalently, M is the reduced adjacency matrix of $G(M)$. See Figure 1 for an illustration. In the sequel, we shall speak indistinctly about binary matrices and their corresponding vertex-colored bipartite graphs. Recall now that an *asteroidal triple*, is an independent set of three vertices such that each pair is joined by a path that avoids the neighborhood of the third. Most of the interest in this definition stems from the following theorem.

Theorem 1 ([35], Theorem 6). *A binary matrix has the C1P if and only if its corresponding vertex-colored bipartite graph does not contain a white asteroidal triple.*

Moreover, Tucker has characterized the binary matrices that have the C1P by a set of *forbidden submatrices*.

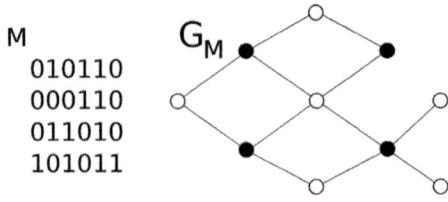

M
010110
000110
011010
101011

Fig. 1. A binary matrix and its corresponding vertex-colored bipartite graph

Theorem 2 ([35], Theorem 9). *A binary matrix has the C1P if and only if it contains none of the matrices M_{I_k}, M_{II_k}, M_{III_k} $(k \geq 1)$, M_{IV} and M_V depicted Figure 2.*

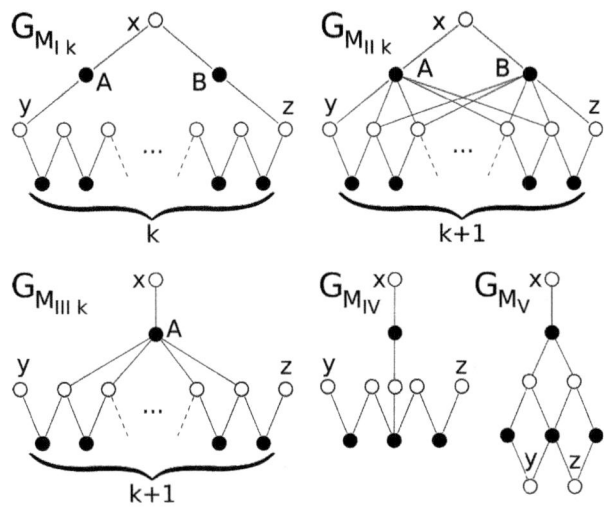

Fig. 2. Forbidden Tucker submatrices represented as vertex-colored bipartite graphs [35]. Black and white vertices correspond to rows and columns, respectively.

In [17], Dom *et al.* provided an algorithm for finding a forbidden Tucker submatrix (*i.e.*, one of $T = \{M_{I_k}, M_{II_k}, M_{III_k}, M_{IV}, M_V\}$) in a given binary matrix. The general algorithm is as follows. For each white asteroidal triple u, v, w of $G(M)$, compute the sum of the lengths of three shortest paths connecting two by two u, v and w (each path has to avoid the closed neighborhood of the third vertex). Select an asteroidal triple u, v, w of $G(M)$ with minimum sum and return the rows and columns of M that correspond to the vertices that occur along the three shortest paths. The authors proved that the returned submatrix does contain a forbidden Tucker submatrix of T but which is not necessarily of minimum size (for M_{III_k}, M_{IV} and M_V). Indeed, since the three shortest paths may share some vertices, the sum of the lengths of the three paths is not necessarily the number of vertices in the union of the three paths. However, Dom *et al.* showed that the returned submatrix contains at most three extra columns (resp. five extra rows) compared with a forbidden Tucker submatrix with minimum number of columns (resp. rows). To overcome this problem, they provided another algorithm devoted to M_{III_k}, M_{IV} and M_V submatrices. More precisely, they used the similarity between M_{III_k} and M_{I_k} to reduce the problem to a minimum-size hole search. For M_{IV} and M_V, they provided an exhaustive search. On the whole, Dom *et al.* provided an algorithm for finding a forbidden Tucker submatrix in a given matrix M (assuming M does not have the C1P) in $O(\Delta^3 m^2 n(m + n^2))$ time, where m is the number of rows of M, n is the number of columns of n, and Δ is the maximum number of 1-entries in a row. More precisely, the authors provided a $O(\Delta mn^2 + n^3)$ time algorithm for finding a M_{I_k} or M_{II_k} submatrix, a $O(\Delta^3 m^3 n + \Delta^2 m^2 n^2)$ time algorithm for finding a M_{III_k} submatrix, a $O(\Delta^3 m^2 n^3)$ time algorithm for finding a M_{IV} submatrix, and a $O(\Delta^4 m^2 n)$ time algorithm for finding a M_V submatrix.

	Dom et al.
M_{I_k} and M_{II_k}	$O(\Delta mn^2 + n^3)$
M_{III_k}	$O(\Delta^3 m^3 n + \Delta^2 m^2 n^2)$
M_{IV}	$O(\Delta^3 m^2 n^3)$
M_V	$O(\Delta^4 m^2 n)$
Total	$O(\Delta^3 m^2 n(m + n^2))$

The main contribution of this paper is a simple $O(\Delta^3 m^2 (m\Delta + n^3))$ time algorithm for finding a minimum size forbidden Tucker submatrix. Our algorithm is based on shortest paths and two graph pruning techniques: *clean* and *anticlean* (to be defined in the next section). Graph pruning techniques were introduced by Conforti *et al.* [15]. One has to note that graph pruning technique not always succeed in the detection of induced configurations. Indeed, in [7], Bienstock gave negative results among which one can find an **NP**-completeness proof for the problem of deciding whether a graph contains an odd hole containing a given vertex. This negative result, which in attacking the perfect graph conjecture was useful in posing limits in what could have been a reasonable approach, also demonstrates that not everything can be done with the detection of induced configurations.

2 Fast Detection of Minimum Size Forbidden Tucker Submatrices

Let us introduce the `clean` and `anticlean` cleaning operations. Let M be a binary matrix and $G(M) = (V_M, E_M)$ be the corresponding vertex-colored bipartite graph. For any node v of $G(M)$, `clean`(v) results in the graph where any neighbor of v has been deleted, *i.e.*, $G(M)[V_M \setminus N(v)]$. For any node v of $G(M)$, `anticlean`(v) results in the graph where any node that does not belong to the same partition nor the neighborhood of v has been deleted, *i.e.*, $G(M)[V_M \setminus \{u : u \notin N(v) \text{ and } \text{color}(u) \neq \text{color}(v)\}]$.

We now focus on the bipartite graphs that represent Tucker configurations (see Figure 2). Define the `guess`$(V \subseteq \{x, y, z, A, B\})$ operation as follows: given a Tucker configuration $T \in \mathcal{T}$, identifies by a brute-force algorithm all the vertices of V among the vertices of $G(M)$. In other words, the `guess` operation tries all possible matching between vertices labeled by x, y, z, A or B in T and vertices of $G(M)$. Of particular importance, guessed vertices will never be affected (*i.e.*, deleted) by the `clean` and `anticlean` operations.

Lemma 1. *Let M be $m \times n$ binary matrix with at most Δ 1-entries per row. One can find the smallest submatrix $G(M_{I_k})$ in $G(M)$ in $O(m^2 \Delta^3 (n + \Delta m))$ time (if such a submatrix exists).*

Algorithm 1. Find $G(M_{I_k})$ in $G(M)$

Proof.   1: guess($\{x, y, z, A, B\}$) and add them to S
  2: clean(x, A, B)
  3: find a shortest path p in the pruned graph between y and z after having removed
      A and B
  4: **if** p exists **then**
  5:     Add the vertices of p to S
  6:     **return** the induced subgraph $G(M)[S]$
  7: **end if**

We apply Algorithm 1 to $G(M)$. Let us first prove that if $G(M_{I_k})$ occurs in $G(M)$, then Algorithm 1 finds it. Suppose $G' = G(M_{I_k})$ occurs in $G(M)$. Then among all the guessed 5-uplets x, y, z, A, B (Line 1), there should be at least one guess such that x, y, z, A, B are part of the vertices of G'. By definition, G' is a hole, and hence does not have a chord. Therefore, clean(x, A, B) preserves G' since, in G', (1) x is only connected to vertices A and B, (2) A is only connected to vertices x and y, and (3) B is only connected to x and z. Moreover, looking for a shortest path p in the pruned graph between y and z after having removed A and B ensures the minimality of the returned graph which is indeed an hole.

The guessing can be done in $O(m^2 \Delta^3)$ time. Indeed, once A has been identified, one can select x and y among the at most Δ neighbors of A and then identify B and one of its at most Δ neighbors as z such that $x \in N(B)$ and $z \notin \{x, y\}$. For each such guessing, the cleaning of x, A, B can be done in $O(\Delta + m)$ time. Finally, one can find a shortest path between y and z by a breadth-first search in the pruned graph after having removed A and B which has at most $m + n$ vertices and Δm edges in $O(n + \Delta m)$ time. On the whole, Algorithm 1 is $O(m^2 \Delta^3 (n + \Delta m))$ time.   □

Lemma 2. *Let M be a $m \times n$ binary matrix with at most Δ 1-entries per row. One can find the smallest submatrix $G(M_{II_k})$ in $G(M)$ in $O(m^2 \Delta^3 (n + \Delta m))$ time (if such a submatrix exists).*

We apply Algorithm 2 to $G(M)$. Let us first prove that if $G(M_{II_k})$ occurs in $G(M)$, then Algorithm 2 finds it. Suppose $G' = G(M_{II_k})$ occurs in $G(M)$. Then

---

Algorithm 2. Find $G(M_{II_k})$ in $G(M)$

Proof.   1: guess($\{x, y, z, A, B\}$) and add them to S
  2: anticlean(A, B)
  3: clean(x)
  4: find a shortest path p in the pruned graph between y and z after having removed
      A and B
  5: **if** p exists **then**
  6:     Add the vertices of p to S
  7:     **return** the induced subgraph $G(M)[S]$
  8: **end if**

Algorithm 3. Find $G(M_{III_k})$ in $G(M)$

Proof. 1: `guess`($\{x, y, z, A\}$) and add them to S
  2: `anticlean`(A)
  3: `clean`(x)
  4: find a shortest path p in the pruned graph between y and z after having removed
     A
  5: **if** p exists **then**
  6:    Add all the nodes of p to S
  7:    **return** the induced subgraph $G(M)[S]$
  8: **end if**

---

among all the guessed 5-uplets x, y, z, A, B in Line 1, there must be at least one guess such that x, y, z, A, B are indeed part of the vertices of G'. By definition, in G', any unguessed white node is in the neighborhood of both A and B. Thus, `anticlean`(A, B) preserves G' since, in G', (1) y which is the only white node not in the neighborhood of B has been guessed and (2) z which is the only white node not in the neighborhood of A has been guessed. Moreover, in G', x should be only connected to A and B. Thus, `clean`(x) preserves G'. Finally, looking for a shortest path p in the pruned graph between y and z after having removed A and B ensures the minimality of the returned graph which is indeed $G(M_{II_k})$.

The guessing can be done in $O(m^2 \Delta^3)$ time. For each such guessing, the cleaning/anticleaning of x, A, B can be done in $O(n + m)$ time. Finally, one can find a shortest path between y and z by a breadth-first search in the pruned graph after having removed A and B which has at most $\Delta + n$ vertices and Δm edges in $O(n + \Delta m)$ time. On the whole, Algorithm 2 is $O(m^2 \Delta^3 (n + \Delta m))$ time.   □

If we compare Algorithm 1 and Algorithm 2, in both cases we are looking for a $y - z$ shortest path in the pruned graph after having removed A and B. Moreover, if we refer to Figure 2, the final structural topology of the $y - z$ path is similar in the M_{I_k} and M_{II_k} matrices. Therefore, one may reasonably think that the total number of path vertices should be equal in both cases. This is not true due to different pruning techniques: cleaning in Algorithm 1 *versus* cleaning/anticleaning in Algorithm 2.

Lemma 3. *Let M be a $m \times n$ binary matrix with at most Δ 1-entries in each row. One can find the smallest $G(M_{III_k})$ in $G(M)$ in $O(m \Delta n^2 (n + \Delta m))$ time (if such a submatrix exists).*

We apply Algorithm 3 to $G(M)$. Let us first prove that if $G(M_{III_k})$ occurs in $G(M)$, then Algorithm 3 finds it. Suppose $G' = G(M_{III_k})$ occurs in $G(M)$. Then among all the guessed 4-uplets x, y, z, A in Line 1, there must be at least one guess such that x, y, z, A are indeed part of the vertices of G'. By definition, in G', any unguessed white node is in the neighborhood of A. Thus, `anticlean`(A) preserves G' since, in G', y and z which are the only white nodes not in the neighborhood of A have been guessed. Moreover, in G', x is only connected to A. Thus, `clean`(x) preserves G'. Finally, looking for a shortest path p in the

Table 1. Comparing our results with Dom *et al.* [17]

	Dom et al.	Our contribution
M_{I_k} and M_{II_k}	$O(\Delta mn^2 + n^3)$	$O(m^2\Delta^3(n + \Delta m))$
M_{III_k}	$O(\Delta^3 m^3 n + \Delta^2 m^2 n^2)$	$O(m\Delta n^2(n + \Delta m))$
M_{IV}	$O(\Delta^3 m^2 n^3)$	
M_V	$O(\Delta^4 m^2 n)$	
Overall	$O(\Delta^3 m^2(mn + n^3))$	$O(\Delta^3 m^2(m\Delta + n^3))$

pruned graph between y and z after having removed A ensures the minimality of the returned graph which is indeed $G(M_{III_k})$.

The guessing can be done in $O(m\Delta n^2)$ time. Indeed, once A has been identified, one can select x among the at most Δ neighbors of A and then identify y and z among the n white nodes such that $x \neq y \neq z$. For each such guessing, the cleaning/anticleaning of x, A can be done in $O(n + m)$ time. Finally, one can find a shortest path between y and z by a breadth-first search in the pruned graph after having removed A which has at most $\Delta + n$ vertices and Δm edges in $O(n + \Delta m)$ time. On the whole, Algorithm 3 is $O(m\Delta n^2(n + \Delta m))$ time.   □

Considering $G(M_{IV})$ and $G(M_V)$, a simple brute-force search yield the following

Lemma 4 ([17], Proposition 5.3). *Let M be a $m \times n$ binary matrix with at most Δ 1-entries per row. One can find the smallest $G(M_{IV})$ (resp. $G(M_V)$) in $G(M)$ in $O(\Delta^3 m^2 n^3)$ (resp. $O(\Delta^4 m^2 n)$ time) if it exists.*

We are now ready to state the main result of this paper (Table 1 compares our results with Dom *et al.* [17].).

Theorem 3. *Let M be a $m \times n$ binary matrix with at most Δ 1-entries per row that does not have the C1P. A minimum size forbidden Tucker submatrix that occurs in M can be found in $O(\Delta^3 m^2(m\Delta + n^3))$ time.*

3   Matrices with Unbounded Δ

As mentioned in [17], a natural question would be to investigate the complexity of the problem when the number of 1s per row is unbounded. One can thus distinguish two subcases: the maximum number of 1s per column is bounded (say by C) or not. Due to space constraint, the two following results are given without proof.

Theorem 4. *Let M be a $m \times n$ binary matrix with at most C 1-entries per column. A minimum size forbidden Tucker submatrix that occurs in M can be found in $O(C^2 n^3(m + C^2 n))$ time.*

Theorem 5. *Let M be $m \times n$ binary matrix. A minimum size forbidden Tucker submatrix that occurs in M can be found in $O(n^4 m^4)$ time.*

References

1. Adam, Z., Turmel, M., Lemieux, C., Sankoff, D.: Common intervals and symmetric difference in a model-free phylogenomics, with an application to streptophyte evolution. J. Comput. Biol. 14, 436–445 (2007)
2. Alizadeh, F., Karp, R., Weisser, D., Zweig, G.: Physical mapping of chromosomes using unique probes. J. Comput. Biol. 2, 159–184 (1995)
3. Althaus, E., Canzar, S., Emmett, M.R., Karrenbauer, A., Marshall, A.G., Meyer-Baese, A., Zhang, H.: Computing h/d-exchange speeds of single residues from data of peptic fragments. In: ACM Press (ed.) SAC 2008, pp. 1273–1277 (2008)
4. Atkins, J.E., Boman, E.G., Hendrickson, B.: A spectral algorithm for seriation and the consecutive ones problem. SIAM J. Comput. 28(1), 297–310 (1998)
5. Atkins, J.E., Middendorf, M.: On physical mapping and the consecutive ones property for sparse matrices. Discrete Appl. Math. 71(13), 23–40 (1996)
6. Bartholdi, J.J., Orlin, J.B., Ratliff, H.D.: Cyclic scheduling via integer programs with circular ones. Oper. Res. 28(5), 1074–1085 (1980)
7. Bienstock, D.: On the complexity of testing for odd holes and induced odd paths. Discrete Math. 90(1), 85–92 (1991)
8. Blin, G., Rizzi, R., Vialette, S.: General framework for minimal conflicting set. Technical report, Université Paris Est, I.G.M (January 2010)
9. Booth, K.S., Lueker, G.S.: Testing for the consecutive ones property, interval graphs, and graph planarity using pq-tree algorithms. J. Comput. System Sci. 13, 335–379 (1976)
10. Chauve, C., Manŭch, J., Patterson, M.: On the gapped consecutive ones property. In: Proc. 5th European Conference on Combinatorics, Graph Theory and Applications (EuroComb), Bordeaux, France. Electronic Notes on Discrete Mathematics, vol. 34, pp. 121–125 (2009)
11. Chauve, C., Stephen, T., Haus, U.-U., You, V.: Minimal conflicting sets for the consecutive ones property in ancestral genome reconstruction. In: Ciccarelli, F.D., Miklós, I. (eds.) RECOMB-CG 2009. LNCS, vol. 5817, pp. 48–58. Springer, Heidelberg (2009)
12. Chauve, C., Tannier, É.: A methodological framework for the reconstruction of contiguous regions of ancestral genomes and its application to mammalian genome. PLoS Comput. Biol. 4, paper e1000234 (2008)
13. Christof, T., Jünger, M., Kececioglu, J., Mutzel, P., Reinelt, G.: A branch-and-cut approach to physical mapping of chromosome by unique end-probes. J. Comput. Biol. 4, 433–447 (1997)
14. Christof, T., Oswald, M., Reinelt, G.: Consecutive ones and a betweenness problem in computational biology. In: Bixby, R.E., Boyd, E.A., Ríos-Mercado, R.Z. (eds.) IPCO 1998. LNCS, vol. 1412, pp. 213–228. Springer, Heidelberg (1998)
15. Conforti, M., Rao, M.R.: Structural properties and decomposition of linear balanced matrices. Mathematical Programming 55, 129–168 (1992)
16. Dom, M.: Algorithmic aspects of the consecutive-ones property. Bull. Eur. Assoc. Theor. Comput. Sci. EATCS 98, 27–59 (2009)
17. Dom, M., Guo, J., Niedermeier, R.: Approximation and fixed-parameter algorithms for consecutive ones submatrix problems. Journal of Computer and System Sciences, Press, Corrected Proof (2009)
18. Fulkerson, D.R., Gross, O.A.: Incidence matrices and interval graphs. Pacific J. Math. 15(3), 835–855 (1965)

19. Habib, M., McConnell, R.M., Paul, C., Viennot, L.: Lex-bfs and partition refinement, with applications to transitive orientation, interval graph recognition and consecutive ones testing. Theoret. Comput. Sci. 234(12), 59–84 (2000)
20. Hajiaghayi, M., Ganjali, Y.: A note on the consecutive ones submatrix problem. Information Processing Letters 83(3), 163–166 (2002)
21. Hassin, R., Megiddo, N.: Approximation algorithms for hitting objects with straight lines. Discrete Applied Mathematics 30(1), 29–42 (1991)
22. Hochbaum, D.S., Levin, A.: Cyclical scheduling and multi-shift scheduling: Complexity and approximation algorithms. Discrete Optimization 3(4), 327–340 (2006)
23. Hsu, W.-L.: A simple test for the consecutive ones property. J. Algorithms 43(1), 1–16 (2002)
24. Hsu, W.-L., McConnell, R.M.: Pc trees and circular-ones arrangements. Theoret. Comput. Sci. 296(1), 99–116 (2003)
25. Korte, N., Mhring, R.H.: An incremental linear-time algorithm for recognizing interval graphs. SIAM J. Comput. 18(1), 68–81 (1989)
26. Kou, L.T.: Polynomial complete consecutive information retrieval problems. SIAM J. Comput. 6(1), 67–75 (1977)
27. Lu, W.-F., Hsu, W.-L.: A test for the consecutive ones property on noisy data – application to physical mapping and sequence assembly. J. Comput. Biol. 10, 709–735 (2003)
28. McConnell, R.M.: A certifying algorithm for the consecutive-ones property. In: ACM Press (ed.) 15th Annual ACMSIAM Symposium on Discrete Algorithms SODA 2004, pp. 768–777 (2004)
29. Mecke, S., Schbel, A., Wagner, D.: Station location complexity and approximation. In: 5th Workshop on Algorithmic Methods and Models for Optimization of Railways ATMOS 2005, Dagstuhl, Germany (2005)
30. Mecke, S., Wagner, D.: Solving geometric covering problems by data reduction. In: Albers, S., Radzik, T. (eds.) ESA 2004. LNCS, vol. 3221, pp. 760–771. Springer, Heidelberg (2004)
31. Meidanis, J., Porto, O., Telles, G.P.: On the consecutive ones property. Discrete Appl. Math. 88, 325–354 (1998)
32. Oswald, M., Reinelt, G.: The simultaneous consecutive ones problem. Theoret. Comput. Sci. 410(2123), 1986–1992 (2009)
33. Ruf, N., Schbel, A.: Set covering with almost consecutive ones property. Discrete Optimization 1(2), 215–228 (2004)
34. Tan, J., Zhang, L.: The consecutive ones submatrix problem for sparse matrices. Algorithmica 48(3), 287–299 (2007)
35. Tucker, A.C.: A structure theorem for the consecutive 1s property. Journal of Combinatorial Theory. Series B 12, 153–162 (1972)
36. Veinott, A.F., Wagner, H.M.: Optimal capacity scheduling. Oper. Res. 10, 518–547 (1962)

Processes in Space

Luca Cardelli[1] and Philippa Gardner[2]

[1] Microsoft Research Cambridge
luca@microsoft.com
[2] Imperial College London
pg@doc.ic.ac.uk

Abstract. We introduce a geometric process algebra based on affine geometry, with the aim of describing the concurrent evolution of geometric structures in 3D space. We prove a relativity theorem stating that algebraic equations are invariant under rigid body transformations.

Keywords: Process Algebra, Affine Geometry.

1 Introduction

Context and Aims. Process Algebra provides a fundamental study of interacting computing systems. From initial work based on simple handshake interaction, the field has expanded to model systems with dynamic interaction patterns [10], static spatial distribution [6], and nested, topologically-changing distribution[2]. In applications to biological systems, interaction and distribution are sufficient to characterize well-mixed chemical systems subdivided into nested compartments, as are commonly found in cellular biology. There are many situations, however, both in computer systems (in robotics and sensor networks) and in biological systems (in growth and development) where a geometrical distance is also necessary beyond purely topological organization.

One of the key motivating examples of the π-calculus [10], for example, is the handover protocol of mobile phones: a mobile phone is connected to a fixed tower, and receives a new frequency to connect to a different tower. In actuality, the handover is based on the relative distance (signal power) between the mobile device and the fixed towers, and hence the protocol depends on geometry. The motivating examples of the Dpi calculus [6] and the Ambient Calculus [2] involve movement through space, but lack a notion of distance. More challenging examples can be found in developmental biology, which deals with the dynamic spatial arrangements of cells, and with forces and interactions between them. Many computational approaches have been developed for modeling geometric systems, including Cellular Automata, extended L-systems [12], and graph models, but few cover complex geometry, dynamic interaction, and dynamic organization together. In particular, the richness of interaction present in Process Algebra, is not found in other approaches.

We therefore start from Process Algebra and we extend it towards geometrical modeling, taking inspiration from a well-developed body of formal work in developmental biology. Concretely, we develop a calculus of processes located in 3-

F. Ferreira et al. (Eds.): CiE 2010, LNCS 6158, pp. 78–87, 2010.

dimensional geometric space. While it may seem in principle logical to 'add a position (and possibly a velocity) to each process', naive attempts result in awkward formal systems with too many features: coordinates, position, velocity, identity, extent, force, collision detection, communication range, and so on. In addition, application areas such as developmental biology are challenging in that the coordinate space is not fixed: it effectively expands, moves, and warps as an organism is developing, making approaches based on fixed coordinate systems or grids awkward. Our aim is thus to begin incorporating flexible and general geometric capabilities in Process Algebra, and among those we must certainly count the geometric transformations of space.

Other important concepts are present in this paper in embryonic form only. Time flow is needed in addition to position to define speed, acceleration and force. Although there is causality in 3π, there is no time, and the forces we can express are therefore not quantitative. This can be fixed by adapting any standard addition of time flow from the pi-calculus to 3π, e.g. synchronous or stochastic. With this approach, the time dimension remains with the evolution of processes, and not with a geometric notion of space-time. Geometric objects, and the spatial extent of an object, are not represented: each process is a point, and its spatial extent can be encoded only by convention through the patterns of interactions with other processes. Spatial extent is an important consideration, but it would entail potentially unbounded complexity, which again is likely to be application specific. Membrane-based abstractions have been investigated in Process Algebra as extensions of π-calculus and could be usefully integrated with geometry, to express notions such as volume-dependent splitting of compartments. The most compelling applications will come from situations where communication, transformations, and forces/extents are connected, such as in developmental biology.

Contributions. In this paper, we introduce a geometric process algebra, called 3π, that combines the interaction primitives of the π-calculus with geometric transformations. In particular we introduce a single new geometric construct, *frame shift*, which applies a 3-dimensional affine transformation to an evolving process. This calculus is sufficient to express many dynamic geometric behaviors, thanks to the combined power of Affine Geometry and Process Algebra. It remains a relatively simple π-calculus, technically formulated in a familiar way, with a large but standard and fairly orthogonal geometric subsystem. From a Process Algebra point of view, we add powerful geometric data structures and transformations; standard notions of process equivalence give rise to geometric invariants. From an Affine Geometry point of view, we add a notion of interacting agents performing geometric transformations.

Introducing 3π. During biological development, tissues expand, split and twist, and there is no fixed coordinate system that one can coherently apply. To capture examples such as these, it is natural to turn to *affine geometry*, which is the geometry of properties that are invariant under linear transformations and translations. Affine geometry already comprises a well-studied set of fundamental geometric primitives. Our challenge is to choose how the geometry relates to the processes living within it, by working out how to combine naturally these affine primitives with the primitives of the π-calculus. How should the position of a process be represented? How should a process move from one position to another? How should processes at different positions interact?

In 3π, processes have access to the standard affine basis consisting of the origin +
and the orthogonal unit vectors $\uparrow_x, \uparrow_y, \uparrow_z$; each process 'believes' this basis to be the
true coordinate system. However, geometric data is interpreted relative to a global
frame A, which is an affine map. In particular, what a process believes to be the
origin, +, is actually $A(+)$, and this is seen as the *actual location* of the process in the
global frame. The true size and orientation of the basis vectors is also determined by
A, as they are interpreted as $A(\uparrow_x), A(\uparrow_y), A(\uparrow_z)$. The global frame A is inaccessible to
processes. Although processes can carry out *observations* that may reveal some
information about A, such as using dot product to compute the absolute size of $\uparrow_x$,
they have no way to obtain other information, such as the value of $A(+)$.

Processes are positioned via a *frame shift* operation M[P], which is our main
addition to the π-calculus. If process M[P] is in a global frame A, and M evaluates to
an affine map B, then P is interpreted in the shifted global frame $A{\circ}B$. The process
M[P] | N[Q] therefore indicates that processes P and Q are in different frames, with P
shifted by M and Q by N. Conversely, the process M[P] | M[Q] = M[P | Q] indicates
that P and Q are in the same frame. Frame shift operations can also be nested, with the
process $M[N_1[P] | N_2[Q]]$ indicating that P is in the frame shifted first by N_1 and then
M, whereas Q is shifted by N_2 then M. Since M denotes a general affine map, frame
shift is more than just a change of location: it generalizes the Dpi [6] notion of
multiple discrete process locations to multiple process frames in continuous space.

Processes interact by exchanging data messages consisting of channel names or
geometric data; such interactions are not restricted by the distance between processes.
Geometric data is evaluated in its current frame and transmitted 'by value' to the
receiver. Consider an output !x(+) to input ?x(z) interaction on channel x:

$$P = M[!x(+).Q] \mid N[?x(z).R] \ _A{\rightarrow} \ M[Q] \mid N[R\{z{\backslash}\varepsilon\}]$$

where M evaluates to B in the global frame A and + evaluates to $\varepsilon = A{\circ}B(+)$.
Technically, this interaction across frame shifts is achieved via the equality:

$$P = !x(M[+]).M[Q] \mid ?x(z).N[R]$$

which distributes the frame shifts throughout the process, thus exposing the output
and input for interaction. In addition to communication, processes can compare data
values. If R is z=+.R' in our above example, then after interaction this process
computes whether $A{\circ}B(+) = A{\circ}C(+)$, where C is the evaluation of N in A, and evolves
to R' only if the original output and input processes are at the same position.

Example: Distance between processes. Let us assume that the global frame is just
the identity map. Process P below is located at -1 on the x axis, because X applies a
translation $T(-\uparrow_x)$ to it. Similarly, process Q is located at +1 on the x axis by Y.

$$X = T(-\uparrow_x)[P] \qquad \text{where} \quad P = !m(+)$$
$$Y = T(\uparrow_x)[Q] \qquad \text{where} \quad Q = ?m(x). \|x\text{-}+\|=2. \ R$$

When P outputs its origin, the actual value being communicated is thus the point
$\langle -1,0,0 \rangle$: this is a *computed value* that is not subject to any further transformation.
Process Q receives that value as x, and computes the size of the vector x-+ obtained
by a point difference. In the frame of Q that computation amounts to the size of the

vector $\langle$-1,0,0$\rangle$ - $\langle$1,0,0$\rangle$, which is 2. Therefore, the comparison $\|x$-+$\|$=2 succeeds, and process R is activated, having verified that the distance between P and Q is 2.

Example: Force fields. A force field is a process that repeatedly receives the location of an 'object' process (and, if appropriate, a representation of its mass or charge), and tells it how to move by a discrete step. The latter is done by replying to the object with a transformation that the object applies to itself. This transformation can depend on the distance between the object and the force field, and can easily represent inverse square and linear (spring) attractions and repulsions. By nondeterministic interaction with multiple force fields, an object can be influenced by several of them. Here P* is process replication (P* $\equiv$ P | P*) for repeated interaction with the field channel f:

Force = (?f(x,p). !x(M{p}))*   Object = (νx) !f(x,+). ?x(Y). Y[Object]

A uniform field ('wind'): $M\{p\} = T(\uparrow_x)$
A linear attractive field at q ('spring'): $M\{p\} = T(\frac{1}{2}(q\text{-}p))$
An inverse-square repulsive field at q ('charge'): $M\{p\} = T((p\text{-}q)/\|p\text{-}q\|^3)$

The ability to express force fields is important for modeling constraints in physical systems. For example, by multiple force fields, one can set up an arbitrary and time-varying network of elastic forces between neighboring cells in a cellular tissue.

Example: Orthogonal bifurcation in lung development. Lung development in mice is based on three splitting processes [9], which demonstrate a relatively simple example of a developmental process. We show how to represent the third process (orthogonal bifurcation, Orth), which is a proper 3D process of recursive tree growth where bifurcations alternate between orthogonal planes.

Orth = !c(+). (M90(π/6)[Orth] | M90(-π/6)[Orth])
M90(θ) = R(M(θ)[$\uparrow_y$],π/2)$\circ$M(θ)
M(θ) = Sc($\frac{1}{2}$)$\circ$R($\uparrow_z$,θ)$\circ$T($\uparrow_y$)

The output of the origin + to the c channel at each iteration provides a trace of the growing process that can be plotted. The transformation M(θ) applies a translation T($\uparrow_y$) by $\uparrow_y$, a rotation R($\uparrow_z$,θ) by θ around $\uparrow_z$, and a uniform scaling Sc($\frac{1}{2}$) by $\frac{1}{2}$. The transformation M90(θ) first applies an M(θ) transformation in the XY plane, and then applies a further 90° rotation around the 'current' direction of growth, which is M(θ)[$\uparrow_y$], therefore rotating out of the XY plane for the next iteration. Opposite 30° rotations applied recursively to Orth generate the branching structure; note that because of parallel composition ('|') the tree grows nondeterministically.

2 Processes

We introduce a process algebra, 3π, where 3-dimensional geometric data (points, vectors, and affine maps, as well as channel names) can be exchanged between processes, and where processes can be executed in different frames. This is a proper extension of π-calculus with by-value communication of geometric data Δ, data comparisons $\Delta=\Delta'.P$, and frame shifting M[P]. By-value communication over named channels is achieved via an evaluation relation $\Delta_{A}\rightarrow \varepsilon$, which evaluates a *data term* Δ to a *data value* ε relative to a global frame A. The data comparison process $\Delta=\Delta'.P$ evaluates to P if Δ and Δ' evaluate to the same value. Frame shifting is the characteristic construct of 3π: the frame shift process M[P] means running the process P in the global frame A shifted by the affine map obtained by evaluating M.

The syntax of processes depends on the syntax of data Δ, given in Section 3. For now, it is enough to know that each data term Δ has a *data sort* σ, where the channel variables $x_c \in Var_c$ have sort c, and the sort of M[Δ] is the sort of Δ.

Definition: Syntax of Processes

$\Delta ::= x_c \mid ... \mid M[\Delta]$	Data terms
$\pi ::= ?_\sigma x(x') \mid !_\sigma x(\Delta) \mid \Delta=_\sigma\Delta'$	Action terms
$P ::= 0 \mid \pi.P \mid P+P' \mid P\vert P' \mid (vx)P \mid P^* \mid M[P]$	Process terms

An action term π can be an *input* $?_\sigma x(x')$, an *output* $!_\sigma x(\Delta)$, or a *data comparison* $\Delta=_\sigma\Delta'$. The input and output actions are analogous to π-calculus actions, where the input receives a data value of sort σ along channel x which it binds to x', and the output sends the value of Δ with sort σ along x. Process interaction only occurs between inputs $?_\sigma$, and outputs $!_\sigma$ of the same sort σ. A comparison of two data terms of sort σ blocks the computation if the terms do not match when evaluated using $_A\rightarrow$. The syntax of actions is restricted by sorting constraints: the x in $?_\sigma x(x')$ and $!_\sigma x(\Delta)$ must have a channel sort c; the x' in $?_\sigma x(x')$ must have sort σ; the Δ in $!_\sigma x(\Delta)$ must have sort σ; and the Δ,Δ' in $\Delta=_\sigma\Delta'$ must have sort σ. We often omit sorting subscripts, and we assume that variables of distinct sorts are distinct.

Process terms look like standard π-calculus terms. We have the standard empty process 0, the action process $\pi.P$ for action π (when $\pi=?x(x')$, the x' binds any free x' in P), choice P+P', parallel composition P $\vert$ P', channel restriction (vx)P where x has sort c (the x binds any free x in P), and replication P*. In addition, we have the non-standard process frame shifting M[P], which represents a shifted frame given by M. We shall see in Section 3 that channel variables do not occur in M; hence, in (vx)M[P], there is no possibility that any variable in M is bound by x.

We now give a *reduction* relation on process terms, written $_A\rightarrow$, which relates two processes relative to the global frame A. Reduction depends on an evaluation relation $\Delta_{A}\rightarrow \varepsilon$ from data Δ to values ε in a global frame A, discussed in Section 3. The reduction rules for process terms are simply the rules of a by-value π-calculus with data terms Δ. Data evaluation is used in the (Red Comm) and (Red Cmp) rules. Data comparison $\Delta=_\sigma\Delta'.P$ requires the data evaluation $\Delta_{A}\rightarrow\leftarrow \Delta'$, meaning there is an ε such that $\Delta_{A}\rightarrow \varepsilon$ and $\Delta'_{A}\rightarrow \varepsilon$.

There is nothing specific in these rules about the use of the global frame A: this is simply handed off to the data evaluation relation. There is also no rule for process frame shifting, M[P], which is handled next in the structural congruence relation.

Definition: Reduction

(Red Comm) $\Delta_{A}\rightarrow \varepsilon \Rightarrow !_\sigma x(\Delta).P + P' \mid ?_\sigma x(y).Q + Q'_{A}\rightarrow P \mid Q\{y\backslash\varepsilon\}$

(Red Cmp) $\Delta_{A}\rightarrow\leftarrow \Delta' \Rightarrow \Delta=_\sigma\Delta'.P_{A}\rightarrow P$

(Red Par)	$P_A \rightarrow Q \Rightarrow P \mid R_A \rightarrow Q \mid R$
(Red Res)	$P_A \rightarrow Q \Rightarrow (vx)P_A \rightarrow (vx)Q$
(Red $\equiv$)	$P' \equiv P,\ P_A \rightarrow Q,\ Q \equiv Q' \Rightarrow P'_A \rightarrow Q'$

In the standard 'chemical' formulation [1] of π-calculus, the *structural congruence* relation has the role of bringing actions 'close together' so that the communication rule (Red Comm) can operate on them. We extend this idea to bringing actions together even when they are initially separated by frame shifts, so that the standard (Red Comm) rule can still operate on them. Therefore, structural congruence, $\equiv$, consists of the normal π-calculus rules plus additional rules for frame shifting: the ($\equiv$ Map...) rules. These map rules essentially enable us to erase frame shifts from the process syntax and to push them to the data. In this sense, process frame shift M[P] is an illusion, or syntactic sugar, for a π-calculus with frame shift only on the data. However, frame shift is important for modularity because, without it, we would have to modify the process code to apply the frame to all the data items individually.

Definition: Structural Congruence (non-standard $\equiv$Map rules)

($\equiv$ Map)	$P \equiv P' \Rightarrow M[P] \equiv M[P']$	($\equiv$ Map Sum)	$M[P+Q] \equiv M[P]+M[Q]$
($\equiv$ Map Cmp)	$M[\Delta=_\sigma\Delta'.P] \equiv M[\Delta]=_\sigma M[\Delta'].M[P]$	($\equiv$ Map Par)	$M[P \mid Q] \equiv M[P] \mid M[Q]$
($\equiv$ Map Out)	$M[!_\sigma x(\Delta).P] \equiv !_\sigma x(M[\Delta]).M[P]$	($\equiv$ Map Res)	$M[(vx)P] \equiv (vx)M[P]$
($\equiv$ Map In)	$M[?_\sigma x(y).P] \equiv ?_\sigma x(y).M[P]\ (y \notin fv_\sigma(M))$	($\equiv$ Map Comp)	$M[N[P]] \equiv (M{\circ}M[N])[P]$

Many other rules can be derived, e.g., for communication across frames shifts at different depths, and for data comparison inside a local frame. In summary, the application of the structural congruence rules allows us to 'flatten' the local frames so that the rules of reduction can be applied directly. There still remains the issue of correctness, or plausibility, of the new structural congruence rules. This issue can be explored by analyzing the expected derived rules, as we briefly mentioned above, and by establishing general properties of the whole system, as done in Section 4.

We have not discussed recursion, which was used in the introductory examples. However, recursive definitions in π-calculus can be encoded, and this extends in 3π to recursive definitions under frame shift by the ability to communicate transformations.

3 Geometric Data

Geometric data consists of points, vectors and transformations, with the operations of affine geometry [5]. We are interested in three main groups of transformations over $\mathbf{R}^3$. The *General Affine Group GA(3)* is the group of *affine maps* over $\mathbf{R}^3$, which include rotation, translation, reflection, and stretching of space, and are indicated by script letters A, B, C. Affine maps are presented as pairs $\langle A,p \rangle$ where A is 3x3 invertible matrix representing a linear transformation, and p is a point in $\mathbf{R}^3$. The *Euclidean Group E(3)* is the subgroup of $GA(3)$ where $A^T = A^{-1}$: namely, it is the group of isometries of $\mathbf{R}^3$ consisting of rotations, translations and reflections. The *Special Euclidean Group SE(3)* is the subgroup of $E(3)$ where *det A = 1*: namely, the *direct isometries* consisting of rotations and translations, but not reflections. Elements of $SE(3)$ are known as the *rigid body transformations*, preserving distances, angles, and handedness. An affine map A has a canonical associated affine frame, namely the frame+ $A(+),A(\uparrow_x),A(\uparrow_y),A(\uparrow_z)$; we therefore refer to A itself as a frame.

We next introduce data terms Δ and data values ε, and show how to compute data values relative to a global affine frame A. Each data term and value has a sort $\sigma \in \Sigma = \{c,a,p,v,m\}$, denoting channels, scalars, points, vectors, and maps respectively.

Definition: Data Values

The set of data values $\varepsilon \in Val$ is the union of the following five sets: $x_c \in Val_c = Var_c$ are the channels; $a \in Val_a = \mathbf{R}$ are the scalars; $p \in Val_p = \mathbf{R}^3$ are the points, which we write $\langle x,y,z \rangle$; $v \in Val_v$ are the vectors, a set isomorphic to Val_p with a bijection $\uparrow : Val_p \rightarrow Val_v$ with inverse $\downarrow = \uparrow^{-1}$ (elements of Val_v are written $\uparrow\langle x,y,z \rangle$); $A \in Val_m = \{\langle A,p \rangle \in \mathbf{R}^{3\times3} \times \mathbf{R}^3 \mid A^{-1} \text{ exists}\}$ are the affine maps.

Definition: Data Terms

$\Delta ::= x_c \mid a \mid p \mid v \mid M \mid M[\Delta]$		Data
$a ::= r \mid f(a_i) \mid v \bullet v' \mid x_a \mid a$	$(i \in 1..arity(f))$	Scalars
$p ::= + \mid v+p \mid x_p \mid p$		Points
$v ::= \uparrow_x \mid \uparrow_y \mid \uparrow_z \mid p\text{-}p' \mid a\cdot v \mid v+v' \mid v\times v' \mid x_v \mid v$		Vectors
$M ::= \langle a_{ij},a_k \rangle \mid M \circ M' \mid M^{-1} \mid x_m \mid A$	$(i,j,k \in 1..3)$	Maps

Data terms consist of *pure data terms* in roman style, which form the 'user syntax', *data values* in italic style, which are inserted during by-value substitutions resulting from process interaction, and variables $x_\sigma \in Var_\sigma$ of each sort $\sigma \in \Sigma$. Channels are regarded both as pure terms and values. Each term Δ has the appropriate sort σ; the sort of a data frame shift $M[\Delta]$ is the sort of Δ. The substitution $\Delta\{x\backslash\varepsilon\}$ distributes over the structure of Δ, with base cases $x\{x\backslash\varepsilon\} = \varepsilon$, $y\{x\backslash\varepsilon\} = y$ for $y\neq x$, $\varepsilon'\{x\backslash\varepsilon\} = \varepsilon'$.

The scalar terms include the real number literals r, the dot product of vectors $v\bullet v'$, giving the ability to measure distances and angles, and basic functions $f(a_i)$, $i \in 1..arity(f)$, for real arithmetic and trigonometry. The point terms include the origin (+) and the addition of a vector to a point. The vector terms include the unit vectors ($\uparrow_x$, $\uparrow_y$, $\uparrow_z$), point subtraction, the vector space operations ($\cdot,+$), and cross product $v\times v'$, which gives the ability to measure areas and volumes and to detect handedness.

The map terms include the base map terms $\langle a_{ij},a_k \rangle$, composition, and inverse. In the term $\langle a_{ij},a_k \rangle$ for $i,j,k \in 1..3$ the first 9 elements represent a 3x3 square matrix, and the last 3 elements represent a translation vector. We require the 3x3 matrix to be invertible, which is verified by a run-time check of the determinant.

The data term $M[\Delta]$ describes a *data frame shift*. Note that $M[\Delta] = \Delta$ is not always true even on scalars; e.g., $M[v\bullet v']$ is not the same as $v\bullet v'$ when M does not preserve distances. Hence, $M[\Delta]$ does not mean apply M to the data value produced by Δ; it means *shift frame* and evaluate the term Δ in the frame obtained from M and composed with the global frame.

The evaluation relation $\Delta_A \rightarrow \varepsilon$, describes the *computation* of a closed data term Δ to value ε, relative to global frame A. The relation is a partial function defined by induction on the structure of terms. Most cases simply follow the structure of terms;

the key rules are the evaluation of the origin in a frame: $+ \, _A\!\!\rightarrow A(\langle 0,0,0 \rangle)$ (and similarly for the unit vectors like $\uparrow_x \, _A\!\!\rightarrow A(\uparrow\langle 1,0,0 \rangle)$), and the evaluation of frame shift: the value of M[Δ] in frame A is uniquely determined as the value of Δ in frame $A \circ B$, provided that the value of M in frame A is B.

4 Process Observation and Equivalence

We establish the invariance of process behavior under certain transformations of the global frame. We base our results on *barbed congruence*, which is one of the most general notions of process equivalence in process algebra [4,7,10] and gives rise to a definition of algebraic process equations. For 3π, we relativize equations to affine frames, and investigate how the validity of the equality changes when shifting frames.

Barbed congruence is defined using barbs and observation contexts. Barbs identify what can be observed by the process environment; in our case, barbs are outputs on channels. Observation contexts define the process environment: different strengths of observation can be characterized by different classes of contexts. We choose to observe processes only by interaction on channels and by restricting the interaction channels. Therefore, we do not allow observation contexts that have the flavor of manipulating a whole process, like injecting a process into the observer's code, or injecting a process into a frame.

Definition (Barbed Congruence)
- An *Observation Context* Γ is given by: $\Gamma ::= []\ |\ P|\Gamma\ |\ \Gamma|P\ |\ (vx)\Gamma$ where $[]$ only occurs once in Γ. The process $\Gamma[Q]$ is obtained by replacing the unique $[]$ in Γ with Q.
- *Strong Barb on x*: $P\!\downarrow_x\ =\ P \equiv (vy_1)..(vy_n)\ (!x(\Delta).P'\ |\ P'')$ with $x \neq y_1..y_n$.
- $_A$*Barb on x*: $P_A\!\Downarrow_x\ =\ \exists P'.\ P\,_A\!\!\rightarrow^* P' \wedge P'\!\downarrow_x$.
- $_A$*Candidate Relation*: $\boldsymbol{R}$ is an $_A$*candidate relation* iff for all $P\boldsymbol{R}Q$: (1) if $P\!\downarrow_x$ then $Q_A\!\Downarrow_x$; conversely if $Q\!\downarrow_x$ then $P_A\!\Downarrow_x$; (2) if $P\,_A\!\!\rightarrow P'$ then there is Q' such that $Q\,_A\!\!\rightarrow^*$ Q' and $P'\boldsymbol{R}Q'$; if $Q\,_A\!\!\rightarrow Q'$ then there is P' such that $P\,_A\!\!\rightarrow^* P'$ and $P'\boldsymbol{R}Q'$; (3) for all observation contexts Γ, we have $\Gamma[P]\ \boldsymbol{R}\ \Gamma[Q]$.
- $_A$*Barbed Congruence*: $_A\!\approx$ is the union of all $_A$candidate relations, which is itself an $_A$candidate relation.

In order to state our theorems, we need *compatibility* relations, $A\propto\Delta$ and $A\propto P$, constraining the frame A by a simple analysis of the vector operators used in data Δ and process P. A closed data term is *affine* if it does not contain $v\bullet v'$ and $v\times v'$ subterms, *Euclidean* if it does not contain $v\times v'$ subterms, and *rigid* otherwise.

Definition: Frame and Group Compatibility
For $A\in GA(3)$ and closed data term Δ, write $A\propto\Delta$ (A compatible with Δ) iff: if Δ contains $\bullet$ then $A\in E(3)$; if Δ contains $\times$ then $A\in SE(3)$; otherwise, no restriction on A. For group $G\subseteq GA(3)$ and closed data term Δ, write $G\propto\Delta$ iff $\forall A\in G.\ A\propto\Delta$. Write $A\propto P$ and $G\propto P$ if A and G are compatible with all the data terms in P.

Hence we have: $GA(3)\propto\Delta$ implies Δ is affine; $E(3)\propto\Delta$ implies Δ is Euclidean; $SE(3)\propto\Delta$ implies Δ is rigid (i.e., $SE(3)\propto\Delta$ always).

We are normally interested only in equations between processes without computed values; we now restrict our attention to such process terms, which we call *pure* terms.

Definition: Pure Terms

A data term Δ and process term P is *pure* if it does not contain a value subterm ϵ of sort $\sigma \in \{a,p,v,m\}$. We use Δ^{π} and P^{π} to denote such pure terms.

The invariance of equations between pure terms under certain maps is described by a relativity theorem. The key property is that G-equations are G-invariant, meaning that, for a group G, the validity or invalidity of equations that are syntactically compatible with G is not changed by G transformations.

Definition: Equations and Laws

An *equation* is a pair of pure process terms P^{π},Q^{π}, written $P^{\pi} = Q^{\pi}$. It is: -(i) a G-*equation* for $G \subseteq GA(3)$ iff $G \propto P^{\pi}$ and $G \propto Q^{\pi}$; -(ii) a *law in A* for $A \in GA(3)$ iff $P^{\pi}_{\ A} \approx Q^{\pi}$; -(iii) a *law in G* for $G \subseteq GA(3)$ iff $\forall A \in G$ it is a law in A; -(iv) *B-invariant* for $B \in GA(3)$ iff $\forall A \in GA(3)$ it is a law in A iff it is a law in BoA; -(v) *G-invariant* for $G \subseteq GA(3)$ iff $\forall B \in G$ it is *B-invariant*; -(vi) *invariant across G* for $G \subseteq GA(3)$ iff $\forall A,B \in G$ it is a law in B if it is a law in A.

Theorem: Relativity

G-equations are G-invariant, and hence invariant across G.

For the three main transformation groups of interest, our theorem has the following corollaries: (1) $GA(3)$-equations (those not using $\bullet$ or $\times$) are $GA(3)$-invariant: that is, **affine equations are invariant under all maps**; (2) $E(3)$-equations (those not using $\times$) are $E(3)$-invariant: that is, **Euclidean equations are invariant under isometries**; (3) $SE(3)$-equations (*all* equations, since $SE(3)$ imposes no syntactic restrictions) are $SE(3)$-invariant: that is, **all equations are invariant under rigid-body maps**. Further, 'G-equations are invariant across G' can be read as 'G laws are the same in all G frames', obtaining: (1) **affine laws are the same in all frames**; (2) **Euclidean laws are same in all Euclidean frames**; (3) **all laws are the same in all rigid body frames**.

For example, the Euclidean equation $(\uparrow_x \bullet \uparrow_x = 1. \ P^{\pi}) = P^{\pi}$ is a law in the *id* frame, and hence is a law in all Euclidean frames. Moreover, this equation may be valid or not in some initial frame (possibly a non-Euclidean one like a scaling $S(2 \uparrow_y)$), but its validity does not change under any further Euclidean transformation. Note also that this equation can be read from left to right as saying that $\uparrow_x \bullet \uparrow_x = 1. P^{\pi}$ computes to P^{π}. Hence, equational invariance implies also computational invariance (but this only for computations from pure terms to pure terms, where any value introduced by communication is subsequently eliminated by data comparison).

As a second example, for any three points p^{π},q^{π},r^{π}, the affine equation $((q^{\pi}-p^{\pi}) + (r^{\pi}-q^{\pi}) = (r^{\pi}-p^{\pi}). P^{\pi}) = P^{\pi}$ is a law in the *id* frame, and so is a law in all frames; in fact, it is the head-to-tail axiom of affine space. As a third example, for any point p^{π}, the equation $(p^{\pi}=+. \ P^{\pi}) = P^{\pi}$ is invariant under all translations (because all equations are invariant under rigid-body maps); hence, the comparison $p^{\pi}=+$ gives the same result under all translations, and cannot be used to test the true value of the origin no matter how p^{π} is expressed, as long as it is a pure term.

Conclusions. We have introduced 3π, an extension of the π-calculus based on affine geometry, to describe the concurrent evolution of geometric structures in 3D space. We have proven a relativity theorem stating that all algebraic equations are invariant under all rigid body transformations (rotations and translations, not reflections), implying that no pure process can observe the location of the origin, nor the orientation of the basis vectors in the global frame. Moreover, processes that do not perform absolute measurements (via • and ×) are invariant under all affine transformations, meaning that they are also unable to observe the size of the basis vectors and the angles between them. Finally, processes that use • but not × are invariant under all the isometries, meaning that they cannot observe whether they have been reflected. Therefore, these results describe the extent to which a process can observe its geometric frame, and describe the behavior of a process in different geometric frames.

References

1. Berry, G., Boudol, G.: The Chemical Abstract Machine. In: Proc. POPL 1989, pp. 81–94 (1989)
2. Cardelli, L., Gordon, A.D.: Mobile Ambients. In: Le Métayer, D. (ed.) Theoretical Computer Science, Special Issue on Coordination, vol. 240(1), pp. 177–213 (June 2000)
3. Coxeter, H.S.M.: Introduction to geometry. Wiley, Chichester (1961)
4. Fournet, C., Gonthier, G.: A Hierarchy of Equivalences for Asynchronous Calculi. In: Larsen, K.G., Skyum, S., Winskel, G. (eds.) ICALP 1998. LNCS, vol. 1443, pp. 844–855. Springer, Heidelberg (1998)
5. Gallier, J.: Geometric Methods and Applications for Computer Science and Engineering. Springer, Heidelberg (2001)
6. Hennessy, M.: A Distributed Pi-Calculus. Cambridge University Press, Cambridge (2007)
7. Honda, K., Yoshida, N.: On Reduction-Based Process Semantics. Theoretical Computer Science 152(2), 437–486 (1995)
8. John, M., Ewald, R., Uhrmacher, A.M.: A Spatial Extension to the π Calculus. Electronic Notes in Theoretical Computer Science 194(3), 133–148 (2008)
9. Metzger, R.J., Klein, O.D., Martin, G.R., Krasnow, M.A.: The branching programme of mouse lung development. Nature 453(5) (June 2008)
10. Milner, R.: Communicating and Mobile Systems: The π-Calculus. Cambridge University Press, Cambridge (1999)
11. Milner, R., Sangiorgi, D.: Barbed Bisimulation. In: Kuich, W. (ed.) ICALP 1992. LNCS, vol. 623, pp. 685–695. Springer, Heidelberg (1992)
12. Prusinkiewicz, P., Lindenmayer, A.: The Algorithmic Beauty of Plants. Springer, Heidelberg (1991)

Computability of Countable Subshifts*

Douglas Cenzer, Ali Dashti, Ferit Toska, and Sebastian Wyman

Department of Mathematics, University of Florida,
P.O. Box 118105, Gainesville, Florida 32611
Phone: 352-392-0281; Fax: 352-392-8357
cenzer@math.ufl.edu

Abstract. The computability of countable subshifts and their members is examined. Results include the following. Subshifts of Cantor-Bendixson rank one contain only eventually periodic elements. Any rank one subshift, in which every limit point is periodic, is decidable. Subshifts of rank two may contain members of arbitrary Turing degree. In contrast, effectively closed (Π_1^0) subshifts of rank two contain only computable elements, but Π_1^0 subshifts of rank three may contain members of arbitrary c. e. degree. There is no subshift of rank ω.

Keywords: Computability, Symbolic Dynamics, Π_1^0 Classes.

1 Introduction

There is a long history of interaction between computability and dynamical systems. A Turing machine may be viewed as a dynamical system which produces a sequence of configurations or words before possibly halting. The reverse notion of using an arbitrary dynamical system for general computation has generated much interesting work. See for example [1,7]. In this paper we will consider computable aspects of certain dynamical systems over the Cantor space $2^{\mathbb{N}}$.

The study of computable dynamical systems is part of the Nerode program to study the effective content of theorems and constructions in analysis. Weihrauch [16] has provided a comprehensive foundation for computability theory on various spaces, including the space of compact sets and the space of continuous real functions.

Computable analysis is related as well to the so-called reverse mathematics of Friedman and Simpson [13], where one studies the proof-theoretic content of various mathematical results. The study of reverse mathematics leads in turn to the concept of *degrees of difficulty* [9,15] Here we say that $P \leq_M Q$ if there is a Turing computable functional F which maps Q into P; thus the problem of finding an element of P can be uniformly reduced to that of finding an element of Q, so that P is "less difficult" than Q. The degrees of difficulty of effectively closed sets (also known as Π_1^0 classes) have been intensively investigated in several recent papers [5,12].

---

* This research was partially supported by NSF grants DMS 0532644 and 0554841 and 652372.

F. Ferreira et al. (Eds.): CiE 2010, LNCS 6158, pp. 88–97, 2010.

The computability of Julia sets in the reals has been studied by Cenzer [3] and Ko [8]; the computability of complex dynamical systems has been investigated by Rettinger and Weihrauch [11] and by Braverman and Yampolski [2].

The connection between dynamical systems and subshifts is the following. Certain dynamical systems may be given by a continuous function F on a *symbolic space* $\mathcal{X}$ (one with a basis of clopen sets). For each $X \in \mathcal{X}$, the sequence $(X, F(X), F(F(X)), \dots)$ is the *trajectory* of X. Given a fixed partition $U_0, \dots, U_{k-1}$ of $\mathcal{X}$ into clopen sets, the *itinerary* $It(X)$ of a point X is the sequence $(a_0, a_1, \dots) \in k^{\mathbb{N}}$ where $a_n = i$ iff $F^n(X) \in U_i$. Let $It[F] = \{It(X) : X \in \mathcal{X}\}$. Note that $It[F]$ will be a closed set. We observe that, for each point X with itinerary $(a_0, a_1, \dots)$, the point $F(X)$ has itinerary $(a_1, a_2, \dots)$. Now the shift operator σ on $k^{\mathbb{N}}$ is defined by $\sigma(a_0, a_1, \dots) = (a_1, a_2, \dots)$. It follows that $It[F]$ is closed under the shift operator, that is, $It[F]$ is a *subshift*.

Computable subshifts and the connection with effective symbolic dynamics were investigated by Cenzer, Dashti and King [4] in a recent paper. A total, Turing computable functional $F : 2^{\mathbb{N}} \to 2^{\mathbb{N}}$ is always continuous and thus will be termed *computably continuous* or just computable. Effectively closed sets (also known as Π_1^0 classes) are a central topic in computability theory; see [6] and Section 2 below. It was shown for any computably continuous function $F : 2^{\mathbb{N}} \to 2^{\mathbb{N}}$, $It[F]$ is a decidable Π_1^0 class and, conversely, any decidable Π_1^0 subshift P is $It[F]$ for some computable map F. In this paper, Π_1^0 subshifts are constructed in $2^{\mathbb{N}}$ and in $2^{\mathbb{Z}}$ which have no computable elements and are not decidable. Thus there is a Π_1^0 subshift with non-trivial Medvedev degree. J. Miller [10] showed that *every* Π_1^0 Medvedev degree contains a Π_1^0 subshift. Simpson [14] studied Π_1^0 subshifts in two dimensions and obtained a similar result there.

Now every nonempty countable Π_1^0 class contains a computable element, so that all countable Π_1^0 classes have Medvedev degree $\mathbf{0}$, and many uncountable classes also have degree $\mathbf{0}$. In the present paper, we will consider more closely the structure of countable subshifts, using the Cantor-Bendixson (CB) derivative. We will compare and contrast countable subshifts of finite CB rank with Π_1^0 subshifts of finite CB rank as well as with arbitrary Π_1^0 classes of finite rank.

The outline of this paper is as follows. Section 2 contains definitions and preliminaries. Section 3 focuses on subshifts of rank one and has some general results about periodic and eventually periodic members of subshifts. We show that every member of a subshift of rank one is eventually periodic (and therefore computable) but there are rank one subshifts of arbitrary Turing degree and rank one Π_1^0 subshifts of arbitrary c. e. degree, so that rank one undecidable Π_1^0 subshifts exist. We give conditions under which a rank one subshift must be decidable. We show that there is no subshift of rank ω. In section 4, we study subshifts of rank two and three. We show that subshifts of rank two may contain members of arbitrary Turing degree. In contrast, we show that Π_1^0 subshifts of rank two contain only computable elements, but Π_1^0 subshifts of rank three may contain members of arbitrary c. e. degree.

2 Preliminaries

We begin with some basic definitions. Let $\mathbb{N} = \{0, 1, 2, ...\}$ denote the set of natural numbers. For any set Σ and any $i \in \mathbb{N}$, Σ^i denotes the strings of length i from Σ, Σ^* denotes the set of all finite strings from Σ, and $\Sigma^{\mathbb{N}}$ denotes the set of countably infinite sequences from Σ. We write $\Sigma^{<n}$ for $\bigcup_{i<n} \Sigma^i$. For any set A, we let $card(A)$ denote the cardinality of the set A.

For a string $w = (w(0), w(1), \ldots, w(n-1))$, $|w|$ denotes the length n of w; the shift function on strings is defined by $\sigma(w) = (w(1), \ldots, w(n-1))$. The empty string has length 0 and will be denoted by λ. A length n string of k's will be denoted k^n. For $m < |w|$, $w \lceil m$ is the string $(w(0), \ldots, w(m-1))$. We say w is an *initial segment* of v (written $w \preceq v$) if $w = v \lceil m$ for some m. Given two strings v and w, the *concatenation* $v^\frown w$ is defined by

$$v^\frown w = (v(0), v(1), \ldots, v(m-1), w(0), w(1), \ldots, w(n-1)),$$

where $|v| = m$ and $|w| = n$. For $a \in \Sigma$, we write $w^\frown a$ (or just wa) for $w^\frown(a)$ and we write $a^\frown w$ (or just aw) for $(a)^\frown w$. For any $X \in \Sigma^{\mathbb{N}}$ and any finite n, the *initial segment* $X \lceil n$ of X is $(X(0), \ldots, X(n-1))$. For a string $w \in \Sigma^*$ and any $X \in \Sigma^{\mathbb{N}}$, we write $w \prec X$ if $w = X \lceil n$ for some n. For any $w \in \Sigma^n$ and any $X \in \Sigma^{\mathbb{N}}$, we let $w^\frown X = (w(0), \ldots, w(n-1), X(0), X(1), \ldots)$. We denote by v^ω the infinite concatenation $v^\frown v^\frown \cdots$.

The topology on $2^{\mathbb{N}}$ has a basis of *intervals*, which are clopen sets of the form

$$J[w] = \{X : w \prec X\}.$$

A subset of $2^{\mathbb{N}}$ is clopen if and only if it is a finite union of basic intervals.

A *tree* T over Σ^* is a set of finite strings from Σ^* which contains the empty string λ and which is closed under initial segments. We say that $w \in T$ is an *immediate successor* of $v \in T$ if $w = va$ for some $a \in \Sigma$. We will assume that $\Sigma \subseteq \mathbb{N}$, so that $T \subseteq \mathbb{N}^*$.

For any tree T, an *infinite path* through T is a sequence $(X(0), X(1), \ldots)$ such that $X \lceil n \in T$ for all n. We let $[T]$ denote the set of infinite paths through T. It is well-known that a subset Q of $2^{\mathbb{N}}$ is closed if and only if $Q = [T]$ for some tree T. A subset P of $\mathbb{N}^{\mathbb{N}}$ is a Π_1^0 *class* (or *effectively closed set*) if $P = [T]$ for some computable tree T. A node $w \in T$ is *extendible* if there exists $X \in [T]$ such that $w \prec X$. The set of extendible nodes forms a tree T_P which is a co-c.e. subset of Σ^* but is not in general computable. P is said to be *decidable* (or *computable*) if T_P is a computable set.

A tree $T \subseteq 2^{\mathbb{N}}$ said to be *subsimilar* if for every v and w, $vw \in T$ implies $w \in T$. The closed set P is subsimilar (or a *subshift*) if T_P is subsimilar.

An element X of a set $P \subseteq \mathbb{N}^{\mathbb{N}}$ is said to be *isolated* in P if there exists n such that $P \cap J[X \lceil n] = \{X\}$. For any compact $P \subset \mathbb{N}^{\mathbb{N}}$, define the the *Cantor-Bendixson derivative* $D(P)$, to be the set of non-isolated elements of P. This derivative can be applied iteratively to define $D^\alpha(P)$ for any ordinal α.

1. $D^0(P) = P$
2. $D^{\alpha+1}(P) = D(D^\alpha(P))$
3. $D^\lambda(P) = \bigcap_{\alpha < \lambda} D^\alpha(P)$ for limit ordinals λ.

For any ordinal α and any compact P, $D^\alpha(P)$ is also compact. For any countable compact set P, $D^\alpha(P) = \emptyset$ for some countable ordinal α. Then we define *the Cantor-Bendixson rank* of a countable compact P as the least ordinal α such that $D^{\alpha+1}(P) = \emptyset$. Also the Cantor-Bendixson rank of an element X in any class P is defined as the least ordinal α such that $X \notin D^{\alpha+1}(P)$. For more background on computability and on the Cantor-Bendixson derivative, see [6].

An element X of $2^{\mathbb{N}}$ is said to be *periodic* if $X = v^\omega$ for some finite string v; the period of X is the minimal length of v such that $X = v^\omega$. X is said to be *eventually periodic* if for some strings u and v, $X = u^\frown v^\omega$.

We will need the following simple connection between periodicity and the shift.

Lemma 1. *(a) X is periodic if and only if $\sigma^n X = X$ for some n.*
(b) X is eventually periodic if and only if $\sigma^{m+n} X = \sigma^m X$ for some m and n.

3   Countable Subshifts

Lemma 2. *If Q is a finite subshift, then Q contains a periodic element and every element of Q is eventually periodic.*

Proof. Let $Y \in Q$, where Q is a finite subshift. Then for each i, $\sigma^i(Y) \in Q$. Since Q is finite, there must exist m and n such that $\sigma^{m+n}(Y) = \sigma^n(Y)$, so that Y is eventually periodic. Then $X = \sigma^m Y$ is a periodic member of Q. $\square$

Lemma 3. *Let $P \subseteq 2^{\mathbb{N}}$ be any subshift. Then $D^\alpha \sigma(P) = \sigma(D^\alpha(P))$ for any ordinal α.*

Proposition 1. *For any countable subshift Q and any ordinal α, $D^\alpha(Q)$ is a subshift.*

Proof. For any $X \in D^\alpha(Q)$, it follows from Lemma 3 that $\sigma(X) \in D^\alpha(\sigma Q)$ and hence $\sigma(X) \in D^\alpha(Q)$. $\square$

Proposition 2. *For any subshift Q of rank α, $D^\alpha(Q)$ has a periodic element and any element of Q having rank α is eventually periodic.*

Proof. Let Q have rank α, so that $D^{\alpha+1}(Q) = \emptyset$ and $D^\alpha(Q)$ is finite. Then $D^\alpha(Q)$ is a subshift by Proposition 1 and the result follows by Lemma 2. $\square$

Proposition 3. *Given any natural number m there is an at most countable decidable subshift P such that its rank is equal to m and all of its elements are eventually periodic.*

Proof. Define P as follows:

$P = \bigcup_{1 \le n \le m} P_n$, where P_n is defined as follows:

$$P_n = \begin{cases} \{0^{k_{n-1}}1^{k_{n-2}}...0^{k_1}1^{\mathbb{N}} : k_j \ge 0, 1 \le j \le n-1\} \ n \text{ is even,} \\ \{0^{k_{n-1}}1^{k_{n-2}}...1^{k_1}0^{\mathbb{N}} : k_j \ge 0, 1 \le j \le n-1\} \ n \text{ is odd,} \end{cases}$$

P is decidable since $v \in T_P$ if and only if v has at most $n-1$ alternations between 1 and 0.

$P_0 = \{0^\omega, 1^\omega\}$ and has rank 0. P_1 has limit elements 0^ω and 1^ω and isolated paths $0^{n+1}1^\omega$ and $1^{n+1}0^\omega$ for all n. Ths P_1 has rank 1. For each n, $D(P_{n+1}) = P_n$ so that P_{n+1} has rank $n+1$. $\qquad\square$

Note that for the sequence of sets P_n defined above, $\bigcup_n P_n$ is not closed and in fact is dense in $2^{\mathbb{N}}$.

Lemma 4. *Suppose $X \in 2^{\mathbb{N}}$ is not eventually periodic . Then for any $k \in \mathbb{N}$, there are at least $k+1$ distinct factors of length k that occur infinitely often in X.*

Proof. The proof proceeds by induction on k. Clearly the lemma is true for $k = 1$, since both 0 and 1 will occur infinitely often in a not eventually periodic sequence. Now suppose that the lemma holds for k, so that X has at least $k+1$ different factors of length k which occur infinitely often in X. If there are more than $k+1$ such factors $u_1, ..., u_\ell$ where $\ell \ge k+2$, then clearly X is going to have at least $k+2$ factors of length $k+1$, since either $u_i \frown 0$ or $u_i \frown 1$ would occur infinitely often as a factor of X for all $i \le \ell$. Thus we may suppose there are exactly $k+1$ distinct factors of length k that occur infinitely often in X; denote them by $u_0, ..., u_k$.

It suffices to show that for at least one i, both $u_i \frown 0$ and $u_i \frown 1$ occur infinitely often in X. Suppose, by way of contradiction, that this is not the case. Then for each i there exists $e_i \in \{0, 1\}$ and there exists n such that

1. For every $m > n$, the factor $X(m), \dots, X(m+k-1)$ is in the set $\{u_0, \dots, u_k\}$.
2. For every occurence of u_i which occurs past $X \restriction n$ u_i is followed in X by e_i.

Now we can show that X must be eventually periodic, which will provide the necessary contradiction. Now consider the sequence of factors $x_i = (X(n+i), x(n+i+1), \dots, X(n+i+k-1))$ and let $j \le k$ be the least such that $x_j = x_0$. Without loss of generality the sequence has the form $u_0, u_1, \dots, u_j, u_0$ for some $j \le k$, so that $x_i = u_i$ for $i \le j$ and $x_{j+1} = u_0$. Since each u_i must be followed by e_i, it follows that for $i > j$, $u_{i+i} = \sigma u_i \frown e_i$ and that $u_0 = \sigma u_j \frown e_j$. Thus $x_{j+2} = u_1$, $x_{j+3} = u_2$ and so on, so that

$$X = X \restriction (n+k)(e_0 e_1 \dots e_j)^\omega.$$

This contradiction shows that at least one of the factors u_i of length k must have two possible extensions of length $k+1$ which occur infinitely often in X. $\square$

Theorem 1. *For any subshift Q of rank one, every member of Q is eventually periodic.*

Proof. Suppose Q is a rank one subshift, that $X \in Q$ and that X is not eventually periodic . Let k be arbitrary and let $w_0, \ldots, w_k$ be distinct factors of X which occur infinitely often in X. Then for each $i < k$, there are infinitely many n such that $\sigma^n X \in J[w_i] \cap Q$. Since X is not eventually periodic , $m \neq n$ implies that $\sigma^m X \neq \sigma^n X$. Thus $J[w_i] \cap Q$ has a limit path for each $i < k$. It follows that Q has at least $k + 1$ limit paths. Since k was arbitrary, P has infinitely many limit paths and thus $rk(Q) > 1$.   □

We note here that there are Π_1^0 classes of rank one with noncomputable elements [6]. Hence we have the following.

Corollary 1. *There is a Π_1^0 class $Q \subseteq 2^{\mathbb{N}}$ of rank one such that, for any Π_1^0 subshift P of rank one, there is no degree-preserving homeomorphism between P and Q.*

Next we consider the decidability of rank one subshifts.

Theorem 2. *(a) For any Turing degree $\mathbf{d}$, there is a subshift Q of rank one such that T_Q has degree $\mathbf{d}$*
(b) For any c. e. degree $\mathbf{d}$, there is a Π_1^0 subshift Q of rank one such that T_Q has degree $\mathbf{d}$.

Proof. Let A be any set of natural numbers of degree $\mathbf{d}$ and let Q contain limit points 0^ω and $1^\frown 0^\omega$, along with isolated points $0^n 10^\omega$, for $n > 0$ and $1^\frown 0^n 10^\omega$ for $n \in A$. Then Q is a rank one subshift and $T_Q \equiv_T A$.
For (b), just take a c. e. set B of degree $\mathbf{d}$ and let $A = \mathbb{N} - B$.   □

We will next consider a special case in which rank one subshifts are decidable. The following lemma is needed.

Lemma 5. *Let Q be a subshift, let $X = v^\omega$ be a periodic element of Q with period k and, for each $i < k$ and each n, let $Q_{i,n} = \{Z : v^n \frown (v \upharpoonright i) \frown (1 - v(i)) \frown Z \in Q\}$. Then $Q_{i,n+1} \subseteq Q_{i,n}$ for each n.*

Proof. If $Z \in Q_{i,n+1}$, then $v^{n+1}(v \upharpoonright i)(1 - v(i))Z \in Q$, so that, since Q is a subshift, $v^n(v \upharpoonright i)(1 - v(i))Z \in Q$ and therefore $Z \in Q_{i,n}$.   □

Theorem 3. *Let Q be a subshift of rank one such that every element of $D(Q)$ is periodic. Then Q is decidable and every element of Q is computable.*

Proof. Every element of $D(Q)$ is eventually periodic, hence computable by Theorem 1. Let $X = v^\omega$ be a periodic element of $D(Q)$ with period k and let $v = X \upharpoonright k$. Let $Q_{i,n}$ be defined as in Lemma 5. Since X has rank one, there exists, for each $i < k$, some n such that $Q_{i,n}$ is finite for all $m \geq n$; let n_i be the least such n. Since $Q_{i,n+1} \subseteq Q_{i,n}$ for all i and n, it follows that the sequence $\{Q_{i,n} : n \in \mathbb{N}\}$ converges to some fixed finite subset P_i of $2^{\mathbb{N}}$. Let D_i be the decidable set T_{P_i}. Now let $S(v) = \{(i, n) : i < k \ \& \ Q_{i,n} \text{ is finite}\}$ and let $A(v)$ be the set of strings of the form $v^n(v \upharpoonright i)w$ for some $(i, n) \in S(v)$. Then $S(v)$ is computable since it is a cofinite set. It follows that $A(v)$ is computable, since it the union of finitely many computable sets together with $\{v^n(v \restriction i)(1 - v(i))w : (i,n) \in S(v) \ \& \ w \in P_i\}$.

For each of the finitely many limit paths $v^\omega \in Q$, we may similarly define the set $A(v)$ of strings in T_Q which branch off from v^ω where the appropriately defined set $Q_{i,n}$ is finite. We claim that T_Q is the union of the finitely many computable sets $A(v)$ together with the words of the form $v^n(v \restriction i)$ for some i and n and is therefore decidable. Certainly each such string is in T_Q. Now suppose that u is some string in T_Q which is not an initial segment of any of the limit paths. Choose v so that u has the longest agreement with v^ω of the limit paths and choose i and n so that $v^n(v \restriction i)(1 - v(i)) \sqsubseteq u$. Then $v^n(v \restriction i)(1 - v(i))$ disagrees with every limit path so that $Q_{i,n}$ is finite and hence $u \in A(v)$.   □

Corollary 2. *For any subshift Q of rank one, there is some finite n such that $\sigma^n(Q)$ is decidable.*

Proof. By Theorem 1, $D(Q)$ is a finite set of eventually periodic points. For each $X \in D(Q)$, $\sigma^m(X)$ is periodic for some m; just let n be the maximumum over $X \in D(Q)$. Then by Lemma 3, $D(\sigma^n(Q)) = \sigma^n(D(Q))$ and thus contains only periodic points, so that Theorem 3 applies.   □

There is another interesting consequence of Lemma 5

Theorem 4. *There is no subshift of rank ω.*

Proof. Suppose by way of contradiction that Q has rank ω. Then $D^{\omega+1}(Q) = \emptyset$ and $D^\omega(Q)$ is finite. Then there is a periodic element X of rank ω by Proposition 2. Let X have period k and let the sets $Q_{i,n}$ be defined as in Lemma 4. Since X has rank ω, there is some n such that for all i and all $m \geq n$, $Q_{i,m}$ has rank $< \omega$. Suppose that $Q_{i,n}$ has rank $r_i < \omega$ and let $r = max\{r_i : i > k\}$. Then by Lemma 5, $rk(Q_{i,m}) \leq r$ for all $m > n$. But this implies that $rk(X) \leq r + 1$, which is the desired contradiction.   □

It follows from the proof that there is no subshift of rank λ, for any limit ordinal λ.

In the next section, we will show that a rank two subshift can have members which are not eventually periodic and indeed not even computable.

4   Subshifts of Rank Two and Three

Proposition 4. *For any increasing sequence $n_0 < n_1 < \ldots$, there is a subshift Q of rank two which contains the element $0^{n_0}10^{n_1}1\ldots$*

Proof. The subshift Q will have the following elements:

(0) For each $k > 0$ and each $n \leq n_k$, the isolated element $0^n10^{n_k+1}10^{n_k+2}1\ldots$
(1) For every n, the element 0^n10^ω which will have rank one in Q.
(2) The unique element 0^ω of rank 2 in Q.   □

Theorem 5. *For any Turing degree* **d**, *there is a rank two subshift* Q *which contains a member of Turing degree* **d** *and such that* T_Q *has Turing degree* **d**.

Proof. Let $A = \{a_0, a_1, \dots\}$ be any infinite set of degree **d** and let $n_i = a_0 + a_1 + \cdots + a_i$ for each i. Now apply Proposition 4.   □

This result can be improved as follows. For $Q \subseteq 2^{\mathbb{N}}$, let $\mathcal{D}(Q)$ be the set of Turing degrees of members of Q.

Theorem 6. *For any countable set* **D** *of Turing degrees containing* **0**, *there is a rank two subshift* Q *such that* $\mathcal{D}(Q) = \mathbf{D}$.

For effectively closed subshifts, the result is quite different.

Theorem 7. *If* Q *is a* Π_1^0 *subshift of rank two, then all of its members are computable.*

Proof. $\mathcal{D}(Q)$ is a subshift of rank one and hence all of its members are eventually periodic and therefore computable. The remaining members of Q are isolated and therefore computable by Theorem 3.12 of [6].   □

Finally, we consider Π_1^0 subshifts of rank three with noncomputable elements.

Theorem 8. *For any c. e. set* A, *there exists* $B \equiv_T A$ *and a* Π_1^0 *subshift* Q *of rank three with* $B \in Q$.

Proof. Let $A = \bigcup_s A_s$ where A_0 is empty and for each s, $A_{s+1} - A_s$ contains at most one number. Let $B \equiv_T A$ have the form

$$0^{n_0} 1 0^{n_1} 1 \cdots$$

where $n_0 < n_1 < \cdots$ represent a modulus of convergence for A. That is,

1. $n_0 = 0$ if $0 \notin A$ and otherwise n_0 is the least s such that $0 \in A_s$.
2. For each k, $n_{k+1} = n_k + 1$ if $k+1 \notin A$ and otherwise n_{k+1} is the least $s > n_k$ such that $k \in A_s$.

For any stage s and $k < s$, we have other sequences $m_0, \dots, m_k$ which are candidates to be the actual modulus sequence. That is, $m_0, \dots, m_k$ is a *candidate for the modulus* at stage s if the following holds.

1. m_0 is the least $t \leq s$ such that $0 \in A_t$ if such s exists and otherwise $m_0 = 0$.
2. For each k, m_{k+1} is the least $t \leq s$ such that $k \in A_s$ and $t > m_k$, if such t exists. Otherwise $m_{k+1} = m_k + 1$.

Note that the set of candidate sequences at stage s is uniformly computable. Since the sequence A_s converges to A, it follows that if $(m_0, \dots, m_k)$ is a candidate sequence for all large enough stages s, then $m_i = n_i$ for all $i \leq k$.

Define the tree T to contain all finite strings of length s of the following form,

$$0^{a_0} {}^\frown 1 {}^\frown 0^{a_1} {}^\frown \cdots {}^\frown 0^{a_{k-1}} {}^\frown 1 {}^\frown 0^{a_k},$$

such that there exists $j < s$ and $m_0 < m_1 < \cdots < m_{k-1}$, a candidate sequence at stage s, such that $a_0 \leq m_j$ and for all $i < k - 1$, $a_{j+i+1} = m_{j+1}$. Here we only require $a_0 \leq m_j$ since we are trying to construct a subshift. Observe that there is no restriction on a_k. This is in case $j + k \notin A$, so that any $m > m_{k-1}$ is a candidate to be m_{j+k}. T is a computable tree so that $Q = [T]$ is a Π_1^0 class.

It can be verified that Q is a subshift and has the following elements.

1. 0^ω has rank 3 in Q.
2. For each $n > 0$, $0^n 1^\omega$ has rank 2 in Q.
3. For all j and all $n \leq n_j$, $X = 0^n \frown 1 \frown 0^{n_{j+1}} \frown 0^{n_{j+2}} \frown 1 \frown \cdots$ has rank 1 in Q. Each of these elements is a shift of B and is Turing equivalent to A. These are the elements of Q with infinitely many 1's.
4. For each j and k and each $n \leq n_j$, $X = 0^n \frown 1 \frown 0^{n_{j+1}} \frown 1 \cdots \frown 0^{n_{j+k}} \frown 1 \frown 0^\omega$ is an isolated element of Q. These are the elements of Q with more than one but only finitely many 1's. $\square$

This result can probably be improved to obtain elements of arbitrary Δ_2^0 degree in a rank three Π_1^0 subshift. We can also modify the proof as in the proof of Theorem 6 to obtain elements of different degrees in the same Π_1^0 subshift.

Proposition 5. *For any Π_1^0 subshift Q of rank three, every element of Q is Δ_3^0.*

Proof. Let Q be a Π_1^0 subshift of rank three. Then $D^2(Q)$ has rank one, so that its members are all eventually periodic . Thus any element of rank two or three in Q is computable. The isolated members of Q are also computable. Finally, suppose that X has rank one in Q. Then X is Δ_3^0 by Theorem 3.15 of [6]. $\square$

On the other hand, an arbitrary Π_1^0 class of rank three may contain members which are not Σ_6^0 and even a Π_1^0 class of rank two may contain members which are not Σ_4^0.

5   Conclusions and Future Research

In this paper, we have investigated subshifts of finite Cantor-Bendixson rank and compared them with Π_1^0 subshifts and with arbitrary Π_1^0 classes of finite rank. We showed that every member of a subshift of rank one is eventually periodic (and therefore computable) but there are rank one subshifts of arbitrary Turing degree and rank one Π_1^0 subshifts of arbitrary c. e. degree, so that rank one undecidable Π_1^0 subshifts exist. We gave conditions under which a rank one subshift must be decidable. We showed that there is no subshift of rank ω. We showed that subshifts of rank two may contain members of arbitrary Turing degree, that Π_1^0 subshifts of rank two contain only computable elements, but Π_1^0 subshifts of rank three may contain members of arbitrary c. e. degree.

Future investigation includes the possible improvement of Theorem 8 to show that Π_1^0 subshifts of rank three may contain members of arbitrary Δ_2^0 degree and possibly elements which are Δ_3^0 and not Δ_2^0. The observed complexity of the members of a rank three Π_1^0 subshift corresponds to the complexity of the members of a Π_1^0 class of rank one. We want to develop this connection further. One conjecture would be that for any Π_1^0 class P of rank one, there is a Π_1^0 subshift Q of rank three such that $\mathcal{D}(P) = \mathcal{D}(Q)$.

Another area for investigation is the nature of the functions F for which the subshift $It[F]$ of itineraries is countable.

References

1. Bournez, O., Cosnard, M.: On the computational power of dynamical systems and hybrid systems. Theoretical Computer Science 168, 417–459 (1996)
2. Braverman, M., Yampolski, M.: Non-computable Julia sets. J. Amer. Math. Soc. 19, 551–578 (2006)
3. Cenzer, D.: Effective dynamics. In: Crossley, J., Remmel, J., Shore, R., Sweedler, M. (eds.) Logical Methods in honor of Anil Nerode's Sixtieth Birthday, pp. 162–177. Birkhauser, Basel (1993)
4. Cenzer, D., Dashti, A., King, J.L.F.: Computable Symbolic Dynamics. Math. Logic Quarterly 54, 524–533 (2008)
5. Cenzer, D., Hinman, P.G.: Degrees of difficulty of generalized r. e. separating classes. Arch. for Math. Logic 45, 629–647 (2008)
6. Cenzer, D., Remmel, J.B.: Π_1^0 classes, in Handbook of Recursive Mathematics, Vol. 2: Recursive Algebra, Analysis and Combinatorics. In: Ersov, Y., Goncharov, S., Marek, V., Nerode, A., Remmel, J. (eds.). Elsevier Studies in Logic and the Foundations of Mathematics, vol. 139, pp. 623–821 (1998)
7. Delvenne, J.-C., Kurka, P., Blondel, V.D.: Decidability and Universality in Symbolic Dynamical Systems. Fund. Informaticae (2005)
8. Ko, K.: On the computability of fractal dimensions and Julia sets. Ann. Pure Appl. Logic 93, 195–216 (1998)
9. Medvedev, Y.: Degrees of difficulty of the mass problem. Dokl. Akad. Nauk SSSR 104, 501–504 (1955)
10. Miller, J.: Two notes on subshifts (preprint)
11. Rettinger, R., Weihrauch, K.: The computational complexity of some Julia sets. In: Goemans, M.X. (ed.) Proc. 35th ACM Symposium on Theory of Computing, San Diego, June 2003, pp. 177–185. ACM Press, New York (2003)
12. Simpson, S.G.: Mass problems and randomness. Bull. Symbolic Logic 11, 1–27 (2005)
13. Simpson, S.G.: Subsystems of Second Order Arithmetic, 2nd edn. Cambridge U. Press, Cambridge (2009)
14. Simpson, S.G.: Medvedev degrees of two-dimensional subshifts of finite type. Ergodic Theory and Dynamical Systems (to appear)
15. Sorbi, A.: The Medvedev lattice of degrees of difficulty. In: Cooper, S.B., et al. (eds.) Computability, Enumerability, Unsolvability: Directions in Recursion Theory. London Mathematical Society Lecture Notes, vol. 224, pp. 289–312. Cambridge University Press, Cambridge (1996)
16. Weihrauch, K.: Computable Analysis. Springer, Heidelberg (2000)

The Limits of Tractability in Resolution-Based Propositional Proof Systems

Stefan Dantchev and Barnaby Martin*

School of Engineering and Computing Sciences, Durham University,
Science Labs, South Road, Durham DH1 3LE, U.K.

Abstract. We study classes of propositional contradictions based on the Least Number Principle (LNP) in the refutation system of Resolution and its generalisations with bounded conjunction, Res(k). We prove that any first-order sentence with no finite models that admits a Σ_1 interpretation of the LNP, relativised to a set that is quantifier-free definable, generates a sequence of propositional contradictions that have polynomially-sized refutations in the system Res(k), for some k. When one considers the LNP with total order we demonstrate that a Π_1 interpretation of this is sufficient to generate such a propositional sequence with polynomially-sized refutations in the system Res(k). On the other hand, we prove that a very simple first-order sentence that admits a Π_1 interpretation of the LNP (with partial and not total order) requires exponentially-sized refutations in Resolution.

1 Introduction

Many of the outstanding examples of propositional tautologies (contradictions) used in proving lower bounds for propositional proof (refutation) systems are derived in a uniform fashion from first-order (fo) principles. For refutation systems, one takes an fo sentence ϕ without finite models and derives a sequence of propositional contradictions the nth of which asserts that ϕ has a model of size n. The *Pigeonhole principle* (PHP) and *Least number principle* are perhaps the most popular fo principles in this area. The (negation of the) PHP asserts that there is an injection from a set of size n to a set of size $n-1$, and the (negation of the) LNP asserts that a strict partial order has no minimal element. Thus neither of these fo sentences has a finite model (though each has an infinite model).

A fairly recent sub-branch of Proof Complexity involves the studying of gap phenomena within propositional proof (refutation) systems. The first such result of this kind appeared in [6], where it was noted that the sequence of propositional contradictions derived from an fo sentence ϕ without finite models – as described above – either has 1.) a polynomially-sized refutation in tree-like Resolution, or 2.) requires fully exponential tree-like Resolution refutations. Moreover, the hard Case 2 prevails exactly when ϕ has some (infinite) model. Since the publication of

---

* The second author is supported by EPSRC grant EP/G020604/1.

F. Ferreira et al. (Eds.): CiE 2010, LNCS 6158, pp. 98–107, 2010.

[6], another gap has been discovered based on rank, and not size, for the integer linear programming-based refutation systems of Lovász-Schrijver and Sherali-Adams [1]. In these cases, the separating criterion is again whether or not an infinite model exists for ϕ, with the hard case – of polynomial instead of constant rank – prevailing if it does.

A gap for Resolution, if it exists at all, can not form along the the the same lines. It is known that the LNP – which clearly has infinite models – has polynomially-sized refutations in Resolution (while the PHP requires exponential refutations). Further, it has been argued that Resolution is not so suitable a system in which to search for a gap, because of its instability with respect to relativisation. When one considers certain relativisations of the LNP, the ensuing principles become hard for Resolution – requiring exponential-sized refutations [2]. Model-theoretically, the LNP and its relativisations are very similar, whence the argument that looking for a model-theoretic separation in the case of Resolution might be difficult. Perhaps a more robust system in which to search for a separation might be Resolution-with-bounded-conjunction, Res(k), introduced by Krajíček in [4] - for any relativisation of the LNP there exists a k such that it admits polynomially-sized Res(k) refutations [2].

In this paper we explore the boundary of tractability – polynomially-sized refutations – in the systems Res(k). We prove that any fo sentence with no finite models that admits a Σ_1 interpretation of the LNP, relativised to a set that is quantifier-free (qf) definable, generates a sequence of propositional contradictions that have polynomially-sized refutations in the system Res(k), for some k. When one considers the LNP with total order we demonstrate that a Π_1 interpretation of this is sufficient to allow polynomially-sized refutations in Res(k), for some k. On the other hand, we prove that a very simple fo sentence that admits a Π_1 interpretation of the LNP (with partial and not total order) – and without relativisation – requires exponentially-sized refutations in Resolution. This fo sentence is exactly a Π_1-variant of the LNP. We conjecture that this same fo sentence requires exponentially-sized refutation in Res(k), for all k. However, we defer the rather messy proof of this to the journal version of this paper. We briefly explore a sequence of Π_{d+1}-variants of the LNP, and conjecture that they (in fact their negations) may be used to separate depth d-Frege from depth $d + 1$-Frege.

The paper is organised as follows. After the preliminaries, we give in Section 3 our upper bounds for Res(k). In Section 4 we give our lower bound for Resolution. Finally, in Section 5 we give some final remarks as well as our conjectures. Owing to strict space restrictions, several proofs and diagrams are omitted. Please see the long version of this paper at http://www.dur.ac.uk/barnaby.martin

2 Preliminaries

Resolution and Res(k)

We denote by $\top$ and $\bot$ the Boolean values "true" and "false", respectively. A *literal* is either a propositional variable or a negated variable. We shall denote literals by small letters, usually ls. A k-*conjunction* (k-*disjunction*) is a conjunction (disjunction) of at most k literals. A *term* (k-*term*) is either a conjunction (k-conjunction) or a constant, $\top$ or $\bot$. We shall use capital letters to denote terms or k-terms, usually Cs for conjunctions and Ds for disjunctions. A k-*DNF* or k-*clause* (k-*CNF*) is a disjunction (conjunction) of an unbounded number of k-conjunctions (k-disjunctions). We shall use calligraphic capital letters to denote k-CNFs or k-DNFs, usually $\mathcal{C}s$ for CNFs and $\mathcal{D}s$ for DNFs. Sometimes, when clear from the context, we will say "clause" instead of "k-clause", even though, formally speaking, a clause is a 1-clause.

We can now describe the propositional refutation system Res(k), first introduced by Krajíček [3]. It is used *to refute* (i.e. to prove inconsistency) of a given set of k-clauses by deriving the empty clause from the initial clauses. There are four derivation rules:

1. The $\wedge$-*introduction rule* is

$$\frac{\mathcal{D}_1 \vee \bigwedge_{j \in J_1} l_j \quad \mathcal{D}_2 \vee \bigwedge_{j \in J_2} l_j}{\mathcal{D}_1 \vee \mathcal{D}_2 \vee \bigwedge_{j \in J_1 \cup J_2} l_j},$$

   provided that $|J_1 \cup J_2| \leq k$.

2. The *cut (or resolution) rule* is

$$\frac{\mathcal{D}_1 \vee \bigvee_{j \in J} l_j \quad \mathcal{D}_2 \vee \bigwedge_{j \in J} \neg l_j}{\mathcal{D}_1 \vee \mathcal{D}_2},$$

3. The two *weakening rules* are

$$\frac{\mathcal{D}}{\mathcal{D} \vee \bigwedge_{j \in J} l_j} \quad \text{and} \quad \frac{\mathcal{D} \vee \bigwedge_{j \in J_1 \cup J_2} l_j}{\mathcal{D} \vee \bigwedge_{j \in J_1} l_j},$$

   provided that $|J| \leq k$.

A Res(k)-proof can be considered as a directed acyclic graph (DAG), whose sources are the initial clauses, called also axioms, and whose only sink is the empty clause. We shall define *the size of a proof* to be the number of the internal nodes of the graph, i.e. the number of applications of a derivation rule, thus ignoring the size of the individual k-clauses in the refutation.

In principle the k from "Res(k)" could depend on n – an important special case is Res $(\log n)$. In the present paper, however, we shall be concerned only with Res (k) for some constant k.

Clearly, Res(1) is *(ordinary) Resolution*, working on 1-clauses, and using only the cut rule, which becomes the usual resolution rule, and the first weakening rule.

Equivalence between Res (k) and a Special Class of Branching Programs

If we turn a Res (k) refutation of a given set of k-clauses $\mathcal{D}$ upside-down, i.e. reverse the edges of the underlying graph and negate the k-clauses on the vertices, we get a special kind of restricted branching k-program. The restrictions are as follows.

Each vertex is labelled by a k-CNF which partially represents the information that can be obtained along any path from the source to the vertex (this is a *record* in the parlance of [5]). Obviously, the (only) source is labelled with the constant $\top$. There are two kinds of queries, which can be made by a vertex:

1. Querying a new k-disjunction, and branching on the answer, which can be depicted as follows.

$$
\begin{array}{ccc}
 & \mathcal{C} & \\
 & ?\bigvee_{j\in J} l_j & \\
\top\swarrow & & \searrow\bot \\
\mathcal{C}\wedge\bigvee_{j\in J} l_j & & \mathcal{C}\wedge\bigwedge_{j\in J}\neg l_j
\end{array}
\tag{1}
$$

2. Querying a known k-disjunction, and splitting it according to the answer:

$$
\begin{array}{ccc}
 & \mathcal{C}\wedge\bigvee_{j\in J_1\cup J_2} l_j & \\
 & ?\bigvee_{j\in J_1} l_j & \\
\top\swarrow & & \searrow\bot \\
\mathcal{C}\wedge\bigvee_{j\in J_1} l_j & & \mathcal{C}\wedge\bigvee_{j\in J_2} l_j
\end{array}
\tag{2}
$$

There are two ways of forgetting information,

$$
\begin{array}{ccc}
\mathcal{C}_1\wedge\mathcal{C}_2 & & \mathcal{C}\wedge\bigvee_{j\in J_1} l_j \\
\downarrow & \text{and} & \downarrow \\
\mathcal{C}_1 & & \mathcal{C}\wedge\bigvee_{j\in J_1\cup J_2} l_j
\end{array}
\tag{3}
$$

the point being that forgetting allows us to equate the information obtained along two different branches and thus to merge them into a single new vertex. A sink of the branching k-program must be labelled with the negation of a k-clause from $\mathcal{D}$. Thus the branching k-program is supposed by default to solve the *Search problem for $\mathcal{D}$*: given an assignment of the variables, find a clause which is falsified under this assignment.

The equivalence between a $\mathrm{Res}\,(k)$ refutation of $\mathcal{D}$ and a branching k-program of the kind above is obvious. Naturally, if we allow querying single variables only, we get branching 1-programs – decision DAGs – that correspond to Resolution. If we do not allow the forgetting of information, we will not be able to merge distinct branches, so what we get is a class of decision trees that correspond precisely to the tree-like version of these refutation systems.

Finally, we mention that the queries of the form (1) and (2) as well as forget-rules of the form (3) give rise to a Prover-Adversary game (see [5] where this game was introduced for Resolution). In short, Adversary claims that $\mathcal{D}$ is satisfiable, and Prover tries to expose him. Prover always wins if her strategy is kept as a branching program of the form we have just explained, whilst a good (randomised) Adversary's strategy would show a lower bound on the branching program, and thus on any $\mathrm{Res}\,(k)$ refutation of $\mathcal{D}$.

Translation of fo Sentences into Propositional CNF Formulae

We shall use the relational language of first-order logic with equality, but without function symbols.[1]

For the sake of explaining the translation, we assume that such an fo sentence ϕ is given in prenex normal form with quantifier-free part in r-CNF for some r. We start with the easy case of Π_1 sentences:

$$\forall x_1, x_2, \ldots x_l \; \mathcal{F}\left(x_1, x_2, \ldots x_l\right),$$

where $\mathcal{F}$ is quantifier-free, and thus can be considered as a propositional formula over propositional variables of two different kinds: $R\left(x_{i_1}, x_{i_2}, \ldots x_{i_p}\right)$, where R is a p-ary relation symbol, and $(x_i = x_j)$. We now take the union of the clauses of $\mathcal{F}$ as $x_1, x_2, \ldots x_l$ range over $[n]^l$ (we shall always use $[n] = \{1, 2, \ldots n\}$ as a finite universe). The variables of the form $(x_i = x_j)$ evaluate to either true or false, and we are left with variables of the form $R\left(x_{i_1}, x_{i_2}, \ldots x_{i_p}\right)$ only. The general case, a Π_l sentence ϕ,

$$\forall x_1 \exists y_1 \ldots \forall x_l \exists y_l \; \mathcal{F}\left(\overline{x}, \overline{y}\right),$$

can be reduced to the previous case by Skolemisation. We introduce Skolem relations $S_i(x_1, x_2, \ldots x_i, y_i)$ for $1 \leq i \leq l$. $S_i\left(x_1, x_2, \ldots x_i, y_i\right)$ witnesses y_i for any given $x_1, x_2, \ldots x_i$, so we need to add clauses stating that such a witness always exists, i.e.

$$\bigvee_{y_i=1}^{n} S_i\left(x_1, x_2, \ldots x_i, y_i\right) \tag{4}$$

for all $(x_1, x_2, \ldots x_i) \in [n]^i$. The original sentence can then be transformed into the following purely universal sentence

$$\forall \overline{x}, \overline{y} \left(\bigwedge_{i=1}^{l} S_i\left(x_1, \ldots x_i, y_i\right)\right) \rightarrow \mathcal{F}\left(\overline{x}, \overline{y}\right). \tag{5}$$

We shall call the clauses (4) "big" (or Skolem) clauses, and the clauses that result in the translation (5) "small" clauses in order to emphasise the fact that the former contain n literals while the latter contain only a constant number of literals, independent from n. Indeed, since $\mathcal{F}$ is assumed to be an r-CNF, we can see the small clauses have width $l + r$ – note the equivalence of $(\bigwedge_{i=1}^{l} S_i) \rightarrow \mathcal{F}$ and $\bigvee_{i=1}^{l} \neg S_i \vee \mathcal{F}$.

For the given fo sentence ϕ, we denote its CNF propositional translation obtained as explained above by $\mathcal{C}_{\phi,n}$ where n is the size of the (finite) model.

Given a (propositional) variable of the form $R_i\left(x_1, x_2, \ldots, x_p\right)$ or $S_j(x_1, x_2, \ldots, x_p, y)$, we call $x_1, x_2, \ldots, x_p$ *arguments* of R_i or S_j. We call y the *witness* of S_i. We also call $x_1, x_2, \ldots x_p$ and y the *elements* of R_i or S_j.

---

[1] This is for convenience only. Note that one may simulate constants with added outermost existentially quantified variables.

Finally, we point out that the Skolemisation also gives us a transformation of the original Π_k sentence ϕ into a Π_2 sentence ϕ':

$$\forall \overline{x}, \overline{y} \exists \overline{z} \; \bigwedge_{i=1}^{k} S_i \left(x_1, \ldots x_i, z_i \right) \wedge \left(\bigwedge_{i=1}^{k} S_i \left(x_1, \ldots x_i, y_i \right) \rightarrow \mathcal{F} \left(\overline{x}, \overline{y} \right) \right). \qquad (6)$$

Clearly ϕ' is equivalent to ϕ, i.e. ϕ' has the same set of countable models as ϕ except for the Skolem relations $S_i \left(x_1, x_2, \ldots x_i, y_i \right)$ that are explicit in ϕ' but not in ϕ.

Whenever we say that we refute an fo sentence ϕ in Res (k), we really mean that we first translate the sentence into a set of (1-)clauses, assuming a finite universe of size n, $\mathcal{C}_{\phi, n}$, and then refute $\mathcal{C}_{\phi, n}$ with Res(k). Naturally, the size of the refutation is then a function in n.

Least Number Principles

The *least number principle* is the assertion that every finite partial order has a minimal element. We will consider two versions of it:

LNP : $\forall x, y, z \exists w \; R(x, w) \wedge \neg R(x, x) \wedge (\neg R(x, y) \vee \neg R(y, z) \vee R(x, z))$
TLNP : $\forall x, y, z \exists w \; R(x, w) \wedge \neg R(x, x) \wedge (\neg R(x, y) \vee \neg R(y, z) \vee R(x, z))$
$\wedge (x = y \vee R(x, y) \vee R(y, x))$

the latter of which enforces that the order is total. The translation of these to propositional contradictions is a little verbose, involving as it does the introduction of an essentially unnecessary Skolem relation. We will therefore prefer the slightly more natural versions as follows (note that our results go through for the more verbose versions). Recall the variables are $R(i, j)$, $i, j \in [n]$; for LNP$_n$ we have the clauses:

$$\begin{array}{ll} \neg R(i, i) & \text{for } i \in [n] \\ \neg R(i, j) \vee \neg R(j, k) \vee R(i, k) & \text{for } i, j, k \in [n], \\ R(1, j) \vee \ldots \vee R(n, j), & \text{for } j \in [n] \end{array}$$

and for TLNP$_n$ we add the clauses $R(i, j) \vee R(j, i)$ for $i \neq j$.

3   Short Refutations in Res(k)

In this section we explore upper bounds in the systems Res(k). Owing to space restrictions, many of the diagrammatic explanations are omitted. It is well-known that the Least Number Principle has polynomially-sized Resolution refutations. We will now consider these as branching 1-programs (decision DAGs). Essentially, one maintains at each point a current minimal element m among the investigated $\{1, \ldots, j\}$, for $m \in [j]$. For the LNP$_n$, we consider major nodes, boxed below, of the program to be of the form $\boxed{\neg R(1,m) \wedge \ldots \wedge \neg R(j,m)}$, for $m \in [j]$. The refutation proceeds as follows.

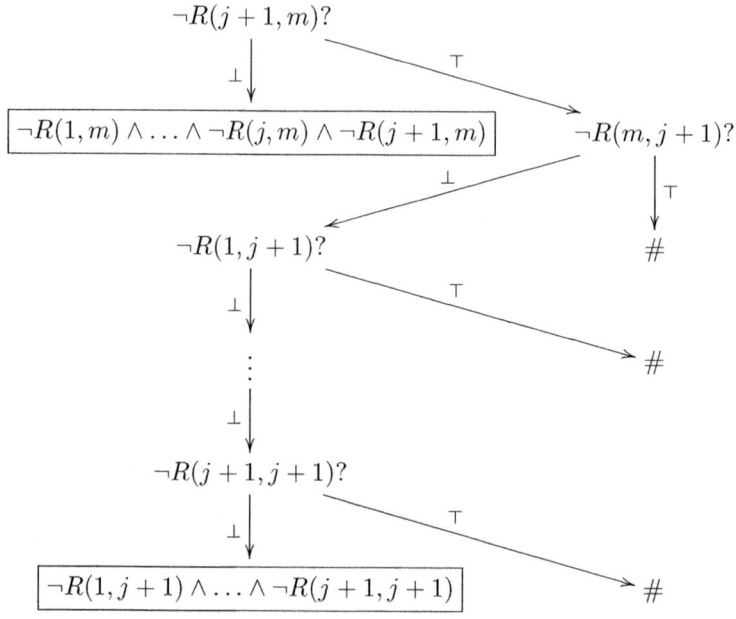

The total number of boxed nodes is bound by n^2 and the internal nodes in navigating between boxed nodes are fewer than, say, $2n$. It follows that the program is of size bound by $2n^3$. In the case of the TLNP_n we may go about our business in the dual fashion, with boxed nodes of the form

$$\boxed{R(m,1) \wedge \ldots \wedge R(m,m-1) \wedge R(m,m+1) \wedge \ldots \wedge R(m,j)},$$

for $m \in [j]$.

We say that an fo sentence ϕ admits a quantifier-free *interpretation of the relativised LNP* if there exist quantifier-free formulae $\mathcal{P}(\overline{x}, \overline{y})$ and $\mathcal{Q}(\overline{x})$ such that, in all models $\mathfrak{A}$ of ϕ, $\mathcal{P}(\overline{x}, \overline{y})$ defines a partial order without minimum on the set of tuples given by $\mathcal{Q}(\overline{x})$ (which is non-empty). We are now ready to state our first result, whose proof is omitted.

Proposition 1. *Let ϕ be an fo sentence with no finite models, but some infinite model, s.t. ϕ admits a quantifier-free interpretation of the relativised LNP. Then there exists k s.t. the sequence $\mathcal{C}_{\phi,n}$ has polynomially-sized Res(k) refutations.*

We say that an fo sentence ϕ admits a Σ_1 *interpretation of the relativised LNP* if there exist quantifier-free formulae $\mathcal{P}(\overline{z}, \overline{x}, \overline{y})$ and $\mathcal{Q}(\overline{x})$, where $\overline{x}, \overline{y}$ are l-tuples and $\overline{z}$ is an l'-tuple, such that, in all models $\mathfrak{A}$ of ϕ, $\exists \overline{z}\, \mathcal{P}(\overline{z}, \overline{x}, \overline{y})$ defines a partial order without minimum on the set of tuples given by $\mathcal{Q}(\overline{x})$ (which is non-empty). The proof of the following proposition is omitted.

Proposition 2. *Let ϕ be an fo sentence with no finite models, but some infinite model, s.t. ϕ admits a Σ_1 interpretation of the relativised LNP. Then there exists k s.t. the sequence $\mathcal{C}_{\phi,n}$ has polynomially-sized Res(k) refutations.*

We say that an fo sentence ϕ admits a Π_1 *interpretation of the relativised TLNP* if there exist quantifier-free formulae $\mathcal{P}(\overline{z}, \overline{x}, \overline{y})$ and $\mathcal{Q}(\overline{x})$, where $\overline{x}, \overline{y}$ are l-tuples and $\overline{z}$ is an l'-tuple, such that, in all models $\mathfrak{A}$ of ϕ, $\forall \overline{z} \; \mathcal{P}(\overline{z}, \overline{x}, \overline{y})$ defines a total order without minimum on the set of tuples given by $\mathcal{Q}(\overline{x})$ (which is non-empty). Proof of the following proposition is omitted.

Proposition 3. *Let ϕ be an fo sentence with no finite models, but some infinite model, s.t. ϕ admits a Π_1 interpretation of the relativised TLNP. Then there exists k s.t. the sequence $\mathcal{C}_{\phi,n}$ has polynomially-sized Res(k) refutations.*

4   Exponential Refutations in Resolution

In this section we will prove an exponential lower bound on a variant of the LNP in Resolution; a result that is somehow a counterpoint to those of the previous section. A similar result for a relativised version of the LNP has appeared in [2]. Our variant, which is not relativised, will be designated the Π_1-LNP. It will be specified by the conjunction of the following

$$\forall x, y, z \; \neg(\forall u R(u, x, y)) \vee \neg(\forall u R(u, y, z)) \vee (\forall u R(u, x, z))$$
$$\forall x \; \neg(\forall u R(u, x, x))$$
$$\forall x \exists y \; (\forall u R(u, y, x)),$$

and which more naturally appear as

$$\forall x, y, z \; (\exists u \neg R(u, x, y)) \vee (\exists u \neg R(u, y, z)) \vee (\forall u R(u, x, z))$$
$$\forall x \; (\exists u \neg R(u, x, x))$$
$$\forall x \exists y \; (\forall u R(u, y, x)).$$

It is transparent that this admits a Π_1 interpretation of the LNP (with partial and not total order) and no relativisation. We will translate this slightly differently from our stated procedure, with the addition of only a single Skolem relation $S(x, y)$ (this is in order to maintain some simplicity, our argument would work equally well in the normal circumstance).

$$\forall x, y, z \; (\exists u \neg R(u, x, y)) \vee (\exists u \neg R(u, y, z)) \vee (\forall u R(u, x, z))$$
$$\forall x \; (\exists u \neg R(u, x, x))$$
$$\forall x \forall y \forall u \; \neg S(x, y) \vee R(u, y, x).$$
$$\forall x \exists y \; S(x, y).$$

We can now give these naturally as the clauses

for each $w, x, y, z \in [n]$ $\bigvee_{u \in [n]} \neg R(u, x, y) \vee \bigvee_{u \in [n]} \neg R(u, y, z) \vee R(w, x, z)$
for each $x \in [n]$ $\qquad \bigvee_{u \in [n]} \neg R(u, x, x)$
for each $w, x, y, z \in [n]$ $\neg S(x, y) \vee R(w, y, x).$
for each $x \in [n]$ $\qquad \bigvee_{u \in [n]} S(x, y).$

The main result of this section is the following.

Theorem 1. *Any Resolution refutation of Π_1-LNP_n must be of size $\geq 2^{\frac{n}{64}}$.*

This will follow immediately from Lemma 6, below. We will derive our result through the probabilistic method, as appears, e.g., in [5]. Considering a decision DAG for the Π_1-LNP, we first prove that certain large records (conjunctions of facts) must appear. We then prove that a large - exponential - number of distinct large records ("bottlenecks") must appear, because otherwise there is a random restriction that generates a decision DAG for a (smaller) Π_1-LNP, that itself has no large records (a contradiction).

Let us imagine that n is divisible by four (we may handle the other cases in a similar manner). We will describe the following *random restrictions* to the Π_1-LNP. Consider a random partition of our universe into two equal parts A and B, with A again randomly divided into the two equal A_1 and A_2. Our random restriction will constrain the variables as follows.

- $S(x, y)$ and $R(w, y, x)$ for all w and $y \in A_1$ and $x \in A_2$.
- $\neg S(x, y)$ and $\neg R(w, y, x)$ for all w and $x \in A_1$ and $y \in A_2$.
- $\neg S(x, y)$ and $\neg R(w, y, x)$ for all w and $x \in A, y \in B$ or $x \in B, y \in A$.

Finally, we set all of the remaining variables of the form $R(w, y, x)$ [i.e. x, y both in A_1, A_2 or B], to $\top$ with probability $\frac{1}{4}$. Considering a decision DAG for the Π_1-LNP, we describe an element $x \in [n]$ as *busy* if either

$$S(x, y) \text{ holds} \qquad \text{(for some } y) \text{ or}$$
$$\bigwedge_{y \in Y} \neg S(x, y) \text{ holds (for a set of elements } Y \text{ of size } > \tfrac{n}{2}).$$

Further, we describe the pair (x, y) as *busy* if either

$$R(w, y, x) \text{ holds} \quad \text{(for some } w) \text{ or}$$
$$\neg R(w, y, x) \text{ holds (for some } w).$$

We will consider an $\frac{n}{2}$-*modified* variant of the Π_1-LNP_n in which there are, for each x, y, at most $\frac{n}{2}$ distinct w s.t. $R(w, y, x)$ is set to $\top$. Note that all technical proofs from this section are omitted.

Lemma 1. *Any decision DAG for an $\frac{n}{2}$-modified Π_1-LNP_n contains a record with $\geq \frac{n}{4}$ busy entities.*

Consider a set of clauses Γ obtained from Π_1-LNP_n by imposing the random restrictions. We describe Γ as *good* if, for all x, y, **both** of which are in either of A_1, A_2 or B, there are $\leq \frac{n}{2}$ distinct w s.t. $R(w, y, x)$ is set (to $\top$). If Γ is not good it is *bad*.

Lemma 2. *The probability that Γ is bad is $\leq \frac{3}{8} \cdot e^{\frac{-n}{12}}$.*

Lemma 3. *Consider a good Γ obtained from Π_1-LNP_n by imposing the random restrictions. Any decision DAG for Γ contains a record with $\geq \frac{n}{16}$ busy entities.*

Consider a decision DAG for Π_1-LNP_n. We describe a record involving $\geq \frac{n}{16}$ busy entities as a *bottleneck*.

Lemma 4. *With probability $> 1-(\frac{1}{2})^{\frac{n}{32}}$, any bottleneck is falsified by the random restrictions.*

Lemma 5. *If there are $< 2^{\frac{n}{64}}$ bottlenecks then there is a random restriction that falsifies all bottlenecks.*

Lemma 6. *Any decision DAG for Π_1-LNP_n must contain $\geq 2^{\frac{n}{64}}$ bottlenecks.*

5   Final Remarks

We believe that our proof of Theorem 1 can be canonically extended to prove that the lower bound holds not only in Resolution, but also in Res(k).

Conjecture 1. There exists $\epsilon_k > 0$ s.t. any Res(k) refutation of Π_1-LNP_n must be of size $\geq 2^{\epsilon_k \cdot n}$.

We have a proof of this statement which is both technical and messy and we would hope to have suitably distilled in the journal version of this paper.

We may define the following further variants of the least number principle which we will designate the Π_d-LNP. They may be specified by the conjunction of the following

$$\forall x, y, z \ \neg(\forall u_1 \exists u_2 \ldots, Qu_d \ R(u_1, .., u_d, x, y)) \vee \neg(\forall u_1 \exists u_2 \ldots, Qu_d \ R(u_1, .., u_d, y, z))$$
$$\vee (\forall u_1 \exists u_2 \ldots, Qu_d \ R(u_1, .., u_d, x, z))$$
$$\forall x \ \neg(\forall u_1 \exists u_2 \ldots, Qu_d \ R(u_1, .., u_d, x, x))$$
$$\forall x \exists y \ (\forall u_1 \exists u_2 \ldots, Qu_d \ R(u_1, .., u_d, y, x)),$$

where Q is $\forall$ if d is odd and is $\exists$ if d is even. We conjecture that the (negations of the) principles Π_d-LNP may be used to separate depth d-Frege from depth $d + 1$-Frege (for the definitions of these proof systems see [3]).

Conjecture 2. The negation of the Π_{d+1}-LNP gives rise to a sequence of propositional tautologies $\mathcal{C}_n$ that admits polynomially-sized proofs in depth $d + 1$-Frege but requires exponentially-sized proofs in depth d-Frege.

References

1. Dantchev, S.: Rank complexity gap for Lovász-Schrijver and Sherali-Adams proof systems. In: Proceedings of the 39th Annual ACM Symposium on Theory of Computing, San Diego, California, USA, June 11-13, pp. 311–317. ACM, New York (2007)
2. Dantchev, S., Riis, S.: On relativisation and complexity gap for resolution-based proof systems. In: Baaz, M., Makowsky, J.A. (eds.) CSL 2003. LNCS, vol. 2803, pp. 142–154. Springer, Heidelberg (2003)
3. Krajíček, J.: Bounded Arithmetic, Propositional Logic, and Complexity Theory. Cambridge University Press, Cambridge (1995)
4. Krajíček, J.: On the weak pigeonhole principle. Fundamenta Mathematica 170, 123–140 (2001)
5. Pudlák, P.: Proofs as games. American Mathematical Monthly (June-July 2000)
6. Riis, S.: A complexity gap for tree-resolution. Computational Complexity 10 (2001)

Haskell before Haskell: Curry's Contribution to Programming (1946–1950)*

Liesbeth De Mol[1], Maarten Bullynck[2], and Martin Carlé[3]

[1] Center for Logic and Philosophy of Science, University of Ghent,
Blandijnberg 2, 9000 Gent, Belgium
elizabeth.demol@ugent.be
[2] Université Paris VIII, Vincennes Saint-Denis
maarten.bullynck@kuttaka.org
[3] Κυψέλη, 15 Οδός Λευκάδος, 11362 Αθήνα, Greece
mc@aiguphonie.com

Abstract. This paper discusses Curry's work on how to implement the problem of inverse interpolation on the ENIAC (1946) and his subsequent work on developing a theory of program composition (1948-1950). It is shown that Curry anticipated automatic programming and that his logical work influenced his composition of programs.

Keywords: ENIAC, Haskell B. Curry, combinators, program composition, history of programming.

1 Introduction

In 1946, the ENIAC (Electronic Numerical Integrator and Computer) was revealed to the public. The machine was financed by the U.S. army and thus it was mainly used for military purposes like e.g. the computation of firing tables. It was the first U.S. electronic, digital and (basically) general-purpose computer. ENIAC did not have any programming interface. It had to be completely rewired for each new program. Hence, a principal bottleneck was the planning of a computation which could take weeks and it became obvious that this problem needed to be tackled at some point.

In the middle of 1945 a Computations Committee was set up at Aberdeen Proving Ground to prepare for programming the new computing machines [11]. The committee had four members: F.L. Alt, L.B. Cunningham, H.B. Curry and D.H. Lehmer. Curry and Lehmer prepared for testing the ENIAC, Cunningham was interested in the standard punch card section and Alt worked with Bell and IBM relay calculators. Lehmer's test program for the ENIAC was already

---

* This paper is a contribution to the ENIAC NOMOI project. We would like to thank G. Alberts for having given us the opportunity to present and discuss parts of this paper at CHOC (Colloquium History of Computing), University of Amsterdam. The first author is a postdoctoral fellow of the Fund for Scientific Research – Flanders (FWO).

F. Ferreira et al. (Eds.): CiE 2010, LNCS 6158, pp. 108–117, 2010.

studied in [9]. Here, we will deal with Haskell B. Curry's implementation of a very concrete problem on the ENIAC, i.e., inverse interpolation, and how this led him to develop what he called a theory of programming.

Of course, Curry is best known as a logician, especially for his work on combinators [1]. The fact that, on the one hand, a logician got involved with ENIAC, and, on the other hand, started to think about developing more efficient ways to program a machine, makes this a very early example of consilience between logic, computing and engineering. Curry's work in this context has materialized into three reports and one short paper. The first report [5], in collaboration with Willa Wyatt, describes the set-up of inverse interpolation for the ENIAC. The second and third report [2,3] develop the theory of program composition and apply it to the problem of inverse interpolation. A summary of these two reports is given in [4]. Despite the fact that the reports [2,3] were never classified, this work went almost completely unnoticed in the history of programming as well as in the actual history and historiography. Only Knuth and Pardo have discussed it to some extent [8, pp. 211–213].

2 A Study of Inverse Interpolation of the ENIAC

Together with Willa Wyatt, one of the female ENIAC programmers, Curry prepared a detailed technical report that presented a study of inverse interpolation of [sic] the ENIAC. The report was declassified in 1999. This problem is explained as follows: "Suppose we have a table giving values of a function $x(t)$ [...] for equally spaced values of the argument t. It is required to tabulate t for equally spaced values of x." [5, p.6] This problem is highly relevant for the computation of firing tables. Indeed, given the coordinates of the target, it allows to compute the angle of departure of a bomb as well as the fuse time.

The set-up is amply described in the report, with over 40 detailed figures of wirings for parts of the program and many technical notes on exploiting hardware specificities of the ENIAC. Notwithstanding this concreteness, Curry and Wyatt have more general ambitions, viz. "the problem of inverse interpolation is studied with reference to the programming on the ENIAC *as a problem in its own right* [m.i.]." [5, p.6] Thus, even though the practical goal was to compute firing tables, the aim was to consider the problem and its implementation in its generality: "[The] basic scheme was not designed specifically for a particular problem, but as a basis from which modifications could be made for various such problems." [5, p.54] For example, the scheme could be modified to provide for composite interpolation. In total, four modifications are discussed in detail and 21 more are suggested in the conclusion.

In their analysis Curry and Wyatt give a method for structuring a program. They distinguish between *stages* and *processes*. Processes are major subprograms that can be repeated over and over again. A process consists of pieces called stages. Each stage is a program sequence with an input and one or more outputs. About these stages Curry and Wyatt point out : "The stages can be programmed as independent units, with a uniform notation as to program lines, and then put together; and since each stage uses only a relatively small amount of the equipment the programming can be done on sheets of paper of ordinary size." [5, p.10] Thus, hand-writing replaces hand-wiring since modifications of the problem can be obtained by changing only these stages without the need to modify the complete (structure of the) program. In this sense, the concept of stages adds to the generality of their study.

Curry's experience with putting the inverse interpolation on the ENIAC was the starting point for a series of further investigations into programming. Moreover, he would use the interpolation problem as a prototypical example for developing and illustrating his ideas. As Curry wrote later: "This [interpolation] problem is almost ideal for the study of programming; because, although it is simple enough to be examined in detail by hand methods; yet it is complex enough to contain a variety of kinds of program compositions." [4, p. 102]

3   On the Composition of Programs

In 1949 and 1950 Curry wrote two lengthy reports for the Naval Ordnance that proposed a theory of programming different from the Goldstine-von Neumann (GvN hereafter) approach [6] (see Sec. 3.4) that "evolved with reference to inverse interpolation" [2, p.7]. The motivation and purpose of the reports is made absolutely clear [2, p.5]:

> In the present state of development of automatic digital computing machinery, a principal bottleneck is the planning of the computation [...] Ways of shortening this process, and of systemizing it, so that a part of the work can be done by the technically trained personnel or by machines, are much desired. The present report is an attack on this problem from the standpoint of composition of computing schedules [...] This problem is here attacked theoretically by using techniques similar to those used in some phases of mathematical logic.

Note that Curry explicitly considered the possibility of mechanizing the technique of program composition and thus made the firsts steps towards automatic programming, i.e., compiling. As G.W. Patterson stated in a 1957 (!) review on [4]: "automatic proramming is anticipated by the author" [10, p. 103].

3.1   Definitions and Assumptions

Unlike modern programming languages which are machine-independent, Curry chose to build up his theory on the basis of a concrete machine, viz. the IAS computer that had von Neumann architecture. He leaves "the question of ultimate generalization [i.e. machine-independence] until later", but does make an idealization of the IAS machine. Curry introduces several definitions related to his idealized IAS machine. Furthermore, he makes certain assumptions that are used to deal with practical problems of programming.

The machine has 3 main parts: a memory, an arithmetic unit (consisting mainly of accumulators A) and a control (keeping track of the active location in memory). The memory consists of locations and each location can store a word and is identified by its location number. There are two types of words: *quantities* and *orders*. An order consists of three main parts: an operator and two location numbers, a datum location and an exit location. There are four "species" of orders: arithmetical, transfer, control and stop orders. Roughly speaking, a transfer order moves a word from one position to another, a control order changes location numbers. A *program* is an assignment of $n + 1$ words to the first $n + 1$ locations. A program that exists exclusively of orders resp. quantities is called an order program resp. quantity program. A *normal program* X is a program where the orders and quantities are strictly separated into an order program A and a quantity program C with $X = AC$.

Note that it is impossible to tell from the appearance of a word, if it is a quantity or an order. Curry considers this as an important problem [2, p.98]:

> [...] from the standpoint of practical calculation, there is an absolute separation between [quantities and orders]. Accordingly, the first stage in a study of programming is to impose restrictions on programs in order that the words in all the configurations of the resulting calculation can be uniquely classified into orders and quantities.

Curry introduces a classification of the kinds of programs allowed. In this context, the *mixed arithmetic order* is crucial. This is an arithmetical operation that involves an order as datum. For example, an important use of mixed arithmetic orders is looking up consecutive data in a table . Here, it is employed to effectively calculate with location numbers. To enable this, Curry adds the *table condition*, i.e. one may add one to a location number to get a next value. Ultimately, this results in the notion of a *regular program*, which is either a primary program or a secondary program that satisfies the table condition whereby a primary program never involves mixed arithmetic orders, but a secondary program at least one. In any case, the calculation has to terminate.

3.2   Steps of Program Composition

Throughout the two main reports, Curry provides several steps for the process of program composition. As noted by Curry, these steps are suitable for mechanization. In this respect, the techniques provided in these steps can be considered as techniques for compiling.

Transformations and replacement. The first step in program composition as discussed by Curry concerns the definition of the different transformations needed in order to attain compositions on the machine level. Let $X = M_0 M_1 M_2 ... M_p; Y = N_0 N_1 N_2 ... N_q; Z = L_0 L_1 L_2 ... L_r$ be three regular programs with N_0, M_0, L_0 initiating orders, $T(k) = k'$ with $k \leq m$, $k' \leq n$ some numerical function. Given a program X such that $p = m$ then $\{T\}(X)$ computes the program Y with $q = n$ such that:

$$\{T\}(X) = \begin{cases} N_0 = M_0 \\ N_i = M_{k_i} & \text{if there are } \{k_1, ..., k_i, ..., k_t\} \text{ for which } T(k_j) = i, i > 0 \ (*) \\ & \text{and if } t > 1, \exists f \text{ such that } f(k_1, ..., k_t) = k_i \\ N_i = J & \text{if there is no } k \text{ such that } T(k) \text{ is defined} \end{cases}$$

where J is a blank. $\{T\}(X)$ is called a *transformation of the first kind*. It boils down to a reshuffling of the words in X, where it is not necessary that every word in X reappears in Y. Note that the function f is needed in order for $\{T\}$ to be uniquely defined.

A *transformation of the second kind* $(T)(X)$ gives the Y such that $q = p$ and every word $N_i \in Y$ is derived from a word $M_i \in X$ by replacing every location number k in every order M_i of X by $T(k)$. If M_i is a quantity then $N_i = M_i$. This results in changing the datum and exit numbers in the orders to correspond with the reshuffling from the transformation of the first kind.

Given program X and Y and θ a set of integers then the transformation $\frac{\theta}{Y}x$ called a *replacement* gives the program Z of length $r + 1$ where $r = p$ if $p \geq q$ else $r = q$ and for each $L_i \in Z$, $0 \leq i \leq r$:

$$L_i = \begin{cases} M_0 \text{ if } i = 0 M_i \text{ if } i \notin \theta, i \leq p \\ N_i \text{ if } i \leq q \text{ and } (i \in \theta \text{ or } i > p) \\ J \text{ if } i \in \theta, i > q \end{cases}$$

Thus, a replacement is a program made up from two programs by putting, in certain locations of one program, words from corresponding locations in the other program. Curry then gives definitions for transformations of the second kind with replacement, defined as $\{\frac{\theta,T}{Y}\} = \frac{\theta}{Y}(\{T\}(X))$ and transformations of the third kind. This last class concerns transformations that result from combining transformations of the first and second kind with replacements:

$$[T](x) = \{T\}(T)(x)$$
$$[\tfrac{T}{Y}] = \{\tfrac{T}{Y}\}(T)(x)$$
$$[\theta T](x) = \{\tfrac{\theta T}{0}\}(T)(x)$$
$$S(x) = [\tfrac{\theta T}{Y}](x) = \{\tfrac{\theta T}{Y}\}(T)(x)$$

Note that 0 is a void program.

Curry goes on to show that it is possible to convert every regular program into a normal program under certain restrictions.

Compositions. Using these three kinds of transformations, Curry embarks on the study of diverse program compositions. He considers the following compositions: simple substitution, multiple substitution, reduction to a single quantity program, loop programs and finally complex programs. This includes to find the right combination of the different transformations and to determine the numerical functions $T(k)$ used in the transformations. Here, we will just discuss how a simple substitution can be achieved.

Let $Y = BC$ and $X = AC$ be normal programs, with α, β, γ the respective lengths of A, B, C, m is the location number of a word M in A where the program Y is to be substituted and n the location number of the starting location of B. The following numerical functions are needed:

$$T_1(k) = \begin{cases} k & \text{for } 0 < k < m \\ m + n - 1 & \text{for } k = m \\ k + \beta - 1 & \text{for } m < k \leq \alpha + \gamma \end{cases}$$

$$T_2(k) = \begin{cases} m + k - n & \text{for } n \leq k \leq n + \beta \\ \alpha + k - n & \text{for } n + \beta < k \leq n + \beta + \gamma \end{cases}$$

Then with θ the set of k's with $n \leq k < m + \beta$, the simple substitution of Y in X at M is given by $Z = [\frac{\theta T_1}{[T_2](Y)}](x)$. Consider that, if M is an output of X, the simple substitution of Y in X results in the program composition Z, denoted as $Z = X \rightarrow Y$.

Basic programs. In [3], Curry starts with an analysis of programs into their simplest possible programs called basic programs. Curry gives several reasons for the significance of determining basic programs [4, p.100]:

(1) Experience in logic and in mathematics shows that an insight into principles is often best obtained by a consideration of cases too simple for practical use [...] (2) It is quite possible that the technique of program composition can completely replace the elaborate methods of Goldstine and von Neumann [...] (3) The technique of program composition can be mechanized; if it should prove desirable to set up programs [...] by machinery, presumably this may be done by analyzing them clear down to the basic programs

A basic program consists of a single order plus their necessary outputs and data [3, p.22]. Therefore, two important concepts are *locatum* and *term*. A locatum is a variable that designates a word in the machine, a term is a word constructed by the machine from its locata at any stage. If ξ is a term and λ a locatum, then $\{\xi : \lambda\}$ is a program that calculates the term ξ and stores it in the locatum λ, i.e. yielding what we would nowadays call an assignment. Given a predicate Φ constructed by logical connectives from equalities and inequalities of terms, then $\{\Phi\}$ designates a discrimination which tests whether Φ is true or not. The assignment program can be analyzed in more basic programs using the equation:

$$\{\phi(\xi) : \lambda\} = \{\xi : \mu\} \rightarrow \{\phi(\mu) : \lambda\}$$

where ϕ is some function.

Curry defines several functions and orders to list all the basic programs. First, he gives a list of arithmetic functions:

$$\pi_0(t) = +t \quad \pi_1(t) = -t$$
$$\pi_2(t) = +|t| \quad \pi_3(t) = -|t|$$

Besides these arithmetic functions, Curry defines two mixed arithmetic orders, i.e., $d(*)$ and $e(*)$ where $d(*)$ is an order that reads the location number of its own datum into the accumulator and $e(*)$ an order that reads the location number of its own exit number into the accumulator. Similarly, $d(x)$ designates the location number of the datum of order x, and $e(x)$ the location number of the exit number of the order x. Some more additional orders are the conditional and unconditional jump and the stop order.

On the basis of these most basic orders Curry defines his final set of basic orders, of which some are the result of simple compositions of the original set of basic orders. This is done for practical reasons. In this way, Curry replicates all orders of the GvN coding plus adds some more that he deems useful. E.g., 'clear', absent in the GvN coding, is $\{0 : A\}$; a conditional jump $\{A < 0\}$; $\{A + \pi_2(R) : A\}$, add the absolute value of the register R to the accumulator A; and the partial substitution order, S_p in the GvN coding, becomes $\{A : d(x)\}$.

Analysis and synthesis of expressions. How can one convert expressions such as $(x+1)(y+1)(z+1)$ to a composition of basic programs? This problem is treated in the last three chapters of [3]. Ch. III deals with arithmetic programs, Ch. IV with discrimination programs and Ch. V with secondary programs.

In Ch. III Curry gives all the necessary rules, which are based on inductive definitions, for transforming fairly straightforward arithmetic expressions (expressions that do not need control orders or discriminations) to a composition of basic programs. We will not go into the details of this here. However, to clarify what is done, the rules allow to convert the expression $(x + 1)(y + 1)(z + 1)$ to:

$$\{x : A\} \to \{A + 1 : A\} \to \{A : w\} \to \{y : A\} \to \{A + 1 : A\} \to \{A : R\} \to$$
$$\{wR : A\} \to \{A : w\} \to \{Z : A\} \to \{A + 1 : A\} \to \{A : R\} \to \{wR : A\}$$

Curry only gives a partial treatment of discrimination and secondary programs. For discrimination programs, this includes procedures to handle elementary propositions that are used as predicates. For secondary programs, some specific programs, such as 'tabulate', are synthesized.

3.3   Notation

As is clear from 3.2, Curry develops a notation for programs different from the flow charts discussed by Goldstine and von Neumann. We already introduced the notation for compositions and assignments. Another notation is that for conditionals, i.e., $X \to Y \ \& \ Z$ means that X is either followed by Y or Z depending on the output of X.

Curry's notation allows to code several expressions. Here is the notation for $n!$:

$$n! = \{1 : A\} \to \{A : x\} \to \{A : i\} \to (Y \to (It(m, i) \to I \ \& \ O_2))$$
$$Y = \{ix : x\}$$
$$It(m, i) = \{i : A\} \to \{A + 1 : A\} \to \{A : i\} \to \{m : A\} \to \{A - i : A\} \to \{A < 0\}$$

where I represents the input to Y, A is an accumulator of the arithmetical unit and O_2 is some output channel.

3.4   A Comparison with the Goldstine-von Neumann Approach

Originally, Curry feared that the third part of the GvN reports [6, Part III] *Combining routines* might overlap with his own work on composition of programs

[2, p. 6]. Yet by 1950, Curry had seen this report and promptly concluded that his approach differed significantly from GvN's [3, pp.3–4]. Perhaps, the chief difference in their respective approaches is the fact that Curry's is the more systematic one. There are several indications of this finding. For example, the classification of programs (Sec. 3.1) or their analysis into basic programs (Sec. 3.2) with the explicit goal of synthesizing and analyzing expressions. Curry also notes that the GvN approach is not suitable for automatic programming:

> [GvN give] a preparatory program for carrying out on the main machine a rather complicated kind of program composition. But one comment seems to be in order in regard to this arrangement. The scheme allows certain data to be inserted directly into the machine by means of a typewriter-like device. Such an arrangement is very desirable for trouble-shooting and computations of a tentative sort, but for final computations of major importance it would seem preferable to proceed entirely from a program or programs recorded in permanent form, not subject to erasure, such that the computation can be repeated automatically [...] on the basis of the record.' [3, p. 4]

To this Curry adds the following footnote in order to strengthen his point: "It is said that during the war an error in one of the firing tables was caused by using the wrong lead screw in the differential analyser. Such an error would have been impossible if the calculation had been completely programmed." Clearly, this remark indicates that Curry was not only highly aware of the significance of a digital approach but also of the possible merits of higher-level programming and the prospects of automated program optimization.

A second important point made by Curry, concerns the fact that he considered his notation as more suited for automatic programming than the flow chart notation of the GvN approach [2, p.7]:

> The present theory develops in fact a notation for program construction which is more compact than the "flow charts" of [6, Part I]. Flow charts will be used [...] primarily as an expository device. By means of this notation a composite program can be exhibited as a function of its components in such a way that the actual formation of the composite program can be carried out by a suitable machine.

In general, one may say that the cardinal difference of Curry's approach is due to the fact that it is based on a systematic logical theory, whereas this is not the case for the GvN-approach.

3.5 Logic and the Theory of Program Composition

In [3, p.5], Curry makes explicit that he regards his approach as a logician's:

> The objective was to create a programming technique based on a systematic logical theory. Such a theory has the same advantages here that it has in other fields of human endeavor. Toward that objective a beginning has been made.

Curry's guiding ideas while doing research in mathematical logic are the same as those that guided his theory of composing programs [1, p. 49]:

> [I]t is evident that one can formalize in various ways and that some of these ways constitute a more profound analysis than others. Although from some points of view one way of formalization is as good as any other, yet a certain interest attaches to the problem of simplification [...] In fact we are concerned with constructing systems of an extremely rudimentary character, which analyze processes ordinarily taken for granted.

Curry's analysis of basic orders and programs into the composition of more elementary orders (cfr. Sec. 3.2, pp. 113–114) is a clear example of this approach. In fact, Curry's analysis is at some points directly informed by his theory of combinators. In Sec. 3.2, p. 111 we explained the several kinds of transformations necessary to attain composition on the machine level. Remarkably, Curry makes use of functions rather than of combinators. However, he does rely on certain concepts of combinators in his explanation of how regular programs can be transformed into normal programs. This is related to the condition (∗) used in the definition of transformations of the first kind. There, a projection function is needed to select a value k_j. Then, the function $T(k_s) = i$ is called K-free, if it has at least one solution, W-free, if it has no multiple solutions, and C-free, if T is monotone increasing, whereby $K = \lambda.xy.x, W = \lambda xy.xyy$ and $C = \lambda xyz.xzy$.

On top of this analysis, Curry superposes a calculus of composition. While dealing with the synthesis of programs, Curry arrives at the problem that a program can be synthesized in more than one way. Hence, a calculus of equivalences becomes accessible. [4, p.100]:

> When these processes [of composition] are combined with one another, there will evidently be equivalences among the combinations. There will thus be a calculus of program composition. This calculus will resembles, in many respects the ordinary calculus of logic. It can be shown, for example, that the operation "→" is associative. But the exact nature of the calculus has not, so far as I know, been worked out.

Curry provides several equivalent programs and points out that finding shorter programs equivalent to longer ones is important with respect to memory which was very expensive at the time. His approach to study programs and their variations antedates by nearly a decade the Russian study of equivalent programs [7].

4   Discussion

The ENIAC was not just one of the first electronic computers, it was also the place where ideas originated that are still relevant today. While Lehmer's work on ENIAC announced the use of the computer as a tool for experimental mathematics, it was Curry's work to anticipate parts of a theory of programming.

Surely, Curry was aware that his suggestions would create a certain distance between man and machine, a distance which was almost completely absent with

ENIAC, and only marginally present in the GvN approach. Thus, he developed his theory of compositions in a way that allowed for mechanization and consequently automatic programming. As a result, a theory of programming emerged from the practical problems related to hand-wiring thereby also transgressing the limitations of a notation of programs still bound to hand-writing and manual composition. This, however, does not mean that Curry completely abstracted away from the concrete context he started from. On the contrary, his work is a good example of the mutual interactions between logic and the machinery of computing. To draw a conclusion, Curry's work reveals an interesting balance between, on the one hand, the highest awareness of the practicalities involved with the hardware and concrete problems such as inverse interpolation, and, on the other hand, the general character of a theory of compositions that becomes a calculus in its own right but still is evidently rooted in these practicalities.

References

1. Curry, H.B.: The combinatory foundations of mathematical logic. The Journal of Symbolic Logic 7(2), 49–64 (1942)
2. Curry, H.B.: On the composition of programs for automatic computing. Technical Report 9805, Naval Ordnance Laboratory (1949)
3. Curry, H.B.: A program composition technique as applied to inverse interpolation. Technical Report 10337, Naval Ordnance Laboratory (1950)
4. Curry, H.B.: The logic of program composition. In: Applications scientifiques de la logique mathématique, Actes du 2e Coll. Int. de Logique Mathématique, Paris, 25-30 août(1952), Institut Henri Poincaré, Paris, Gauthier-Villars, pp. 97–102 (1954)
5. Curry, H.B., Wyatt, W.A.: A study of inverse interpolation of the Eniac. Technical Report 615, Ballistic Research Laboratories, Aberdeen Proving Ground, Maryland (1946)
6. Goldstine, H.H., von Neumann, J.: Planning and coding of problems for an electronic computing instrument. vol. 2, part I,II and III, 1947-1948. Report prepared for U. S. Army Ord. Dept. under Contract W-36-034-ORD-7481
7. Ianov, I.I.: On the equivalence and transformation of program schemes. Communications of the ACM 1(10), 8–12 (1958)
8. Knuth, D.E., Pardo, L.T.: Early development of programming languages. In: Howlett, J., Metropolis, N., Rota, G.-C. (eds.) A History of Computing in the Twentieth Century, pp. 197–274. Academic Press, New York (1980)
9. De Mol, L., Bullynck, M.: A week-end off: The First Extensive Number-Theoretical computation on the ENIAC. In: Beckmann, A., Dimitracopoulos, C., Löwe, B. (eds.) CiE 2008. LNCS, vol. 5028, pp. 158–167. Springer, Heidelberg (2008)
10. Patterson, G.W.: Review of The logic of program composition by H.B. Curry. The Journal of Symbolic Logic 22(1), 102–103 (1957)
11. Tropp, H.S.: Franz Alt interview, September 12 (1972)

A Miniaturisation of Ramsey's Theorem

Michiel De Smet[*] and Andreas Weiermann[**]

Ghent University
Building S22 - Krijgslaan 281
9000 Gent - Belgium
{mmdesmet,weierman}@cage.ugent.be

Abstract. We approximate the strength of the infinite Ramsey Theorem by iterating a finitary version. This density principle, in the style of Paris, together with PA will give rise to a first-order theory which achieves a lot of the strength of ACA_0 and the original infinitary version. To prove our result, we use a generalisation of the results by Bigorajska and Kotlarski about partitioning α-large sets.

Keywords: Ramsey theory, α-largeness, density, unprovability, phase transitions.

1 Introduction

McAloon, in 1985, writes in [12]:

> It would be interesting to develop proofs of these results with the "direct" method of α-large sets of Ketonen-Solovay.

The results he speaks about concern Paris-Harrington incompleteness using finite Ramsey theorems. In that paper he gives a first order axiomatisation of the first order consequences of $ACA_0 + RT$, where RT stands for the infinite version of Ramsey's theorem.

Ketonen and Solovay used α-largeness in their paper on rapidly growing Ramsey functions ([11]), in which they extend the famous result of Paris and Harrington. They established sharp upper and lower bounds on the Ramsey function by purely combinatorial means. Bigorajska and Kotlarski generalised their ideas to ordinals below ε_0 and obtained several results on partitioning α-large sets (see [2,3,4]). We generalised those latter results in order to allow ordinals up to ε_ω (see [7]).

Since Paris introduced them in the late seventies densities turned out to be of interest for studying independence results (see [13,8,6]) . We define such a density notion to obtain a miniaturisation of the Ramsey Theorem. We consider the full infinite version with no restrictions on the dimensions or colours (RT). It turns

---

[*] Aspirant Fonds Wetenschappelijk Onderzoek (FWO) - Flanders.
[**] This authors research is supported in part by Fonds Wetenschappelijk Onderzoek (FWO) and the John Templeton Foundation.

F. Ferreira et al. (Eds.): CiE 2010, LNCS 6158, pp. 118–125, 2010.

out that our miniaturisation gives rise to transfinite induction up to ε_ω. Remark that McAloon proved that $\mathrm{ACA}_0 + \mathrm{RT}$ can be axiomatised by $\mathrm{ACA}_0 + \forall X \forall n$ $(\mathrm{TJ}(n, X)$, the nth Turing jump of X, exists). Analysing the latter theory one obtains that the proof-theoretic ordinal of $\mathrm{ACA}_0 + \mathrm{RT}$ is exactly ε_ω. For a more recent approach we refer to the PhD thesis of Afshari ([1]).

We would like to remark that an alternative way to miniaturising RT has been given by Ratajczyk in [14]. Ratajczyk's approach is based on relating iterated Ramseyan principles to iterated reflection principles.

2   Partitioning α-Large Sets

In this section we introduce all notions needed to state Lemmas 1, 2 and 3. The latter lemma will be crucial to obtain the results about the miniaturisation of Ramsey's theorem. As mentioned in the introduction, our results concerning partitioning α-large sets are generalisations of the ones of Bigorajska and Kotlarski ([2,4]). We only give definitions which are necessary here and state preliminary results without proofs, as this would lead us too far.

Define ε_α as the αth ordinal ξ, such that $\xi = \omega^\xi$. Henceforth, we sometimes slightly abuse notation and write ε_{-1} to denote ω, for the sake of generality of notation. The pseudonorm and fundamental sequence used by Bigorajska and Kotlarski can be extended in a natural way to ordinals below ε_ω. We introduce a standard notation for each ordinal and define the pseudonorm of an ordinal α as the maximum of the greatest natural number occurring in this notation and its height $ht(\alpha)$. The pseudonorm of an ordinal α will be denoted by $\mathrm{psn}(\alpha)$. In order to give a definition of the height $ht(\alpha)$, we need the function $\mathrm{tow}_n(\alpha)$ which is defined for ordinals α with $0 < \alpha < \varepsilon_\omega$, by $\mathrm{tow}_0(\alpha) = 1$ and $\mathrm{tow}_{k+1}(\alpha) = \alpha^{\mathrm{tow}_k(\alpha)}$.

Definition 1. *Let $\alpha < \varepsilon_\omega$.*

1. *$l(\alpha) := \min\{n \in \mathbb{N} : \alpha < \varepsilon_n\}$;*
2. *$ht(\alpha) := \min\{n \in \mathbb{N} : \alpha < \mathrm{tow}_n(\varepsilon_{l(\alpha)-1})\}$;*
3. *We say α is written in normal form to the base ε_m if*

$$\alpha = \varepsilon_m^{\alpha_0} \cdot \xi_0 + \varepsilon_m^{\alpha_1} \cdot \xi_1 + \ldots + \varepsilon_m^{\alpha_s} \cdot \xi_s,$$

for some $\alpha_0 > \alpha_1 > \ldots > \alpha_s$ and $0 < \xi_i < \varepsilon_m$, for $0 \leq i \leq s$. If $m = l(\alpha)-1$, then we say α is written in normal form and we write down $\alpha =_{NF} \varepsilon_m^{\alpha_0} \cdot \xi_0 + \varepsilon_m^{\alpha_1} \cdot \xi_1 + \ldots + \varepsilon_m^{\alpha_s} \cdot \xi_s$.
4. *If α is written in normal form, define*

$$\mathrm{psn}(\alpha) := \begin{cases} \max\left\{ht(\alpha), \mathrm{psn}(\alpha_0), \ldots, \mathrm{psn}(\alpha_s), \mathrm{psn}(\xi_0), \ldots, \mathrm{psn}(\xi_s)\right\} \\ \qquad\qquad\qquad\qquad\qquad\qquad\qquad\qquad\quad \textit{if } \alpha \geq \omega \\ \\ \alpha \\ \\ \qquad\qquad\qquad\qquad\qquad\qquad\qquad\qquad\quad \textit{if } \alpha < \omega \end{cases}.$$

We write $\beta \gg \alpha$ if either $\alpha = 0$ or $\beta = 0$ or all the exponents in the normal form of β are equal to or greater than all the exponents in the normal form of α. Observe that $\beta \gg \alpha$ does not imply $\beta \geq \alpha$, as one can verify in $\omega^3 \gg \omega^3 \cdot 2$.

For each ordinal α we define a sequence $\langle \alpha[n] : n \in \mathbb{N} \rangle$ of ordinals converging to α from below.

Definition 2. *Let α be an ordinal below ε_ω, written in normal form, and n be a natural number.*

1. *If $\alpha = \omega$, then $\alpha[n] = n$;*
2. *If $\alpha = \varepsilon_m$, with $m \geq 0$, then $\alpha[n] = \text{tow}_n(\varepsilon_{m-1})$;*
3. *If $\alpha = \varepsilon_m^{\alpha+1}$, with $m \geq -1$, then $\alpha[n] = \varepsilon_m^\alpha \cdot \varepsilon_m[n]$;*
4. *If $\alpha = \varepsilon_m^\psi$, with $m \geq -1$ and ψ a limit, then $\alpha[n] = \varepsilon_m^{\psi[n]}$;*
5. *If $\alpha = \varepsilon_m^\psi \cdot (\alpha+1)$, with $m \geq -1$, then $\alpha[n] = \varepsilon_m^\psi \cdot \alpha + \varepsilon_m^\psi[n]$;*
6. *If $\alpha = \varepsilon_m^\psi \cdot \xi$, with $m \geq 0$ and ξ limit, then $\alpha[n] = \varepsilon_m^\psi \cdot \xi[n]$;*
7. *If $\alpha = \varepsilon_m^{\alpha_0} \cdot \xi_0 + \ldots + \varepsilon_m^{\alpha_s} \cdot \xi_s$, with $m \geq 0$, then $\alpha[n] = \varepsilon_m^{\alpha_0} \cdot \xi_0 + \ldots + (\varepsilon_m^{\alpha_s} \cdot \xi_s)[n]$.*

We shall call the sequence $\alpha[n]$ the fundamental sequence for α.

Remark that one can find other definitions of a fundamental sequence in the literature, most of them differing only slightly. Apart from being interesting in its own right, the concept of a fundamental sequence will be used to introduce other notions, as one will see below.

Definition 3. *For $\alpha, \beta < \varepsilon_\omega$ we write $\beta \Rightarrow_n \alpha$ if there exists a finite sequence $\alpha_0, \ldots, \alpha_k$ of ordinals such that $\alpha_0 = \beta, \alpha_k = \alpha$ and $\alpha_{m+1} = \alpha_m[n]$ for every $m < k$.*

The next lemma will be needed later on.

Lemma 1. *For every $\beta < \varepsilon_\omega$ we have:*

$$(\forall \alpha < \beta)(\forall n > 1)((\text{psn}(\alpha) < n) \Rightarrow (\beta \Rightarrow_n \alpha)).$$

Proof. The proof goes by transfinite induction on β. A detailed proof can be found in [7] (Lemma 3).

Definition 4. *Let $h : \mathbb{N} \to \mathbb{N}$ be an increasing function such that $x < h(x)$ for all x. We define the Hardy hierarchy based on h as follows. For every $x \in \mathbb{N}$*

$$h_0(x) = x$$
$$h_{\alpha+1}(x) = h_\alpha(h(x))$$
$$h_\lambda(x) = h_{\lambda[x]}(x),$$

with λ a limit ordinal.

Often h will be the usual successor function $h(x) = x+1$. However, if we consider a specific set $A \subseteq \mathbb{N}$, we will work with the successor function in the sense of A. Thus h denotes the function defined on A if A is infinite and defined on $A \setminus \{\max A\}$ if A is finite, which associates with every a in its domain the next element of A. In this case we sometimes explicitly write h^A, instead of h. Similarly, if we want to stress the set A, we write h^A_α instead of h_α for the associated Hardy hierarchy.

For an ordinal $\alpha \leq \varepsilon_\omega$ we say that $A \subseteq \mathbb{N}$ is α-*large* if and only if $h^A_\alpha(\min A)$ is defined, i.e. $h^A_\alpha(\min A) \in A$. For example, in case A equals the interval $[a, b]$, then A will be α-large if $h^A_\alpha(a) \leq b$. A set which is not α-large is called α-*small*.

Lemma 2. *For every $\alpha < \varepsilon_\omega$:*

1. *h_α is increasing;*
2. *For every $\beta < \varepsilon_\omega$ and every b in the domain of h, if $\alpha \Rightarrow_b \beta$ and $h_\alpha(b)$ exists, then $h_\beta(b)$ exists and $h_\alpha(b) \geq h_\beta(b)$.*

Proof. By simultaneous induction on α. (For a detailed proof for ordinals below ε_0, see [10]).

Next we define the natural sum $\oplus$ for ordinals below ε_ω. This will be a clear generalisation of the natural sum for ordinals below ε_0 (which is exactly the case $m = 0$ in the definition below).

Definition 5. *Let α and β be ordinals below ε_ω. We define their natural sum $\alpha \oplus \beta$ by induction on $m = \max\{l(\alpha), l(\beta)\}$:*

If $m = 0$, then $\alpha, \beta < \varepsilon_0$ and we can write $\alpha = \omega^{\alpha_0} \cdot n_0 + \ldots + \omega^{\alpha_s} \cdot n_s$ and $\beta = \omega^{\beta_0} \cdot m_0 + \ldots + \omega^{\beta_u} \cdot m_u$, both in normal form to the base $\varepsilon_{-1} = \omega$. Permute items of both of these expansions so that we obtain a nonincreasing sequence of exponents. Then write this sequence as the normal form to the base ω of some ordinal which we denote $\alpha \oplus \beta$.

Now assume we have defined the natural sum for all ordinals γ, δ such that $\max\{l(\gamma), l(\delta)\} = m$ and let $\max\{l(\alpha), l(\beta)\} = m + 1$. Let $\alpha = \varepsilon_m^{\alpha_0} \cdot \xi_0 + \ldots + \varepsilon_m^{\alpha_s} \cdot \xi_s$ and $\beta = \varepsilon_m^{\beta_0} \cdot \eta_0 + \ldots + \varepsilon_m^{\beta_u} \cdot \eta_u$, both in normal form to the base ε_m. Permute items of both of these expansions so that we obtain a nonincreasing sequence of exponents. Then write this sequence as the normal form to the base ε_m of some ordinal which we denote $\alpha \oplus \beta$. While doing so, if $\alpha_i = \beta_j$ for some i, j, then use the induction hypothesis to obtain $\xi_i \oplus \eta_j$ as the coefficient of the term with exponent α_i.

The foregoing definition of natural sum enables us to introduce the function F.

Definition 6. *Define $F : (< \varepsilon_\omega) \to (< \varepsilon_\omega)$ by the following conditions:*

1. *$F(0) = 0$;*
2. *$F(\alpha + 1) = F(\alpha) + 1$;*
3. *$\beta \gg \alpha \Rightarrow F(\beta + \alpha) = F(\beta) \oplus F(\alpha)$;*
4. *$F(\omega^n) = \omega^n + \omega^{n-1} + \ldots + \omega^0$ for $n < \omega$;*
5. *$F(\omega^\alpha) = \omega^\alpha \cdot 2 + 1$ for $\alpha \geq \omega$.*

Clearly, these conditions determine exactly one function F from $(< \varepsilon_\omega)$ to $(< \varepsilon_\omega)$. Let us write an explicit formula for F. Let α be equal to

$$\omega^{\alpha_0} \cdot a_0 + \ldots + \omega^{\alpha_s} \cdot a_s + \omega^n \cdot m_n + \ldots + \omega^0 \cdot m_0,$$

written down in normal form to the base ω, with the difference that we allow some m_i's to be zero. Then $F(\alpha)$ is equal to

$$\omega^{\alpha_0} \cdot 2a_0 + \ldots + \omega^{\alpha_s} \cdot 2a_s$$

$$+\omega^n \cdot m_n + \omega^{n-1} \cdot (m_n + m_{n-1}) + \ldots + \omega^0 \cdot (m_n + \ldots + m_0)$$

$$+(a_0 + \ldots + a_s).$$

For our purpose here it suffices to remark that for $\omega^\omega \leq \alpha < \varepsilon_\omega$, $F(\alpha) \approx \alpha 2$. The meaning of the relation $A \to (\alpha)_m^r$ is as usual in Ramsey theory, with the slight adaptation that we demand the homogeneous subset to be α-large.

Now we are finally able to state the lemma which will lead to the unprovability result.

Lemma 3. Let $m, k \in \mathbb{N}$ and $\alpha = F(\varepsilon_m) + \frac{(k-2)(k-1)}{2} + 1$. Let A be such that $A \to (\alpha)_{3k-2}^k$, $3 \leq k \leq \min A$ and $3 < \min A$. Then A is $\mathrm{tow}_{k-2}(\varepsilon_m)$-large.

Proof. See Corollary 2 in [7].

3 Ramsey Density

Let us recall the infinite form of Ramsey's theorem (RT):

$$\mathrm{RT} \leftrightarrow (\forall n)(\forall k)(\forall G)(G : [\mathbb{N}]^n \to k \Rightarrow \exists H(H \text{ infinite and homogeneous for } G)).$$

In this section we give a miniaturisation of RT using the following density notion.

Definition 7. Let $A \subseteq \mathbb{N}$, with $\min A > 3$. Then A is called 0-dense if $|A| > \min A$. A is called $(n + 1)$-dense, if for every colouring $G : [A]^{\min A+2} \to 3^{\min A}$, there exists a subset $B \subseteq A$, such that B is homogeneous for G and B is n-dense.

We now investigate the logical strength of this density.

Theorem 1. For every $n \in \mathbb{N}$,

$$\mathrm{ACA}_0 + \mathrm{RT} \vdash (\forall a)(\exists b)([a, b] \text{ is } n\text{-dense}).$$

Proof. Let $n \in \mathbb{N}$. By applying König's lemma and the infinite version of Ramsey's theorem, we get the finite version of Ramsey's theorem. Then apply the latter n times to obtain a sufficiently large b.

Lemma 4. Let $A \subseteq \mathbb{N}$, such that $\min A > 2$. If A is $2n$-dense, then A is ε_{n-1}-large.

Proof. By induction on n. Put $a_0 = \min A$.

If n equals zero, then A is 0-dense and $|A| > a_0$. Thus $h_{a_0}^A(a_0) = h_{\varepsilon_{-1}}^A(a_0)$ exists, so A is ε_{-1}-large.

Assume the statement for $n = k$ and let A be $(2k+2)$-dense. Let $G : [A]^{a_0+2} \to 3^{a_0}$ be any function and B a subset of A such that B is homogeneous for G and B

is $(2k+1)$-dense. To invoke Lemma 3 we will show that B is $F(\varepsilon_{k-1})+\frac{a_0(a_0+1)}{2}+1$-large, with F as defined by Definition 6. Put $b_0 = \min B$ and

$$B_i = \{b \in B : h^B_{\varepsilon_{k-1}\cdot(i-1)}(b_0) \le b < h^B_{\varepsilon_{k-1}\cdot i}(b_0)\},$$

for $i \in \{1, \ldots, 3^{b_0} - 1\}$. Define $H : [B]^{b_0+2} \to 3^{b_0}$ by

$$H(x_1, \ldots, x_{b_0+2}) = \begin{cases} 1 & \text{if } x_1 \in B_1 \\ 2 & \text{if } x_1 \in B_2 \\ \vdots & \vdots \\ 3^{b_0} & \text{if } h^B_{\varepsilon_{k-1}\cdot(3^{b_0}-1)}(b_0) \le x_1 \end{cases},$$

for each $(x_1, \ldots, x_{b_0+2}) \in [B]^{b_0+2}$. Since B is $(2k + 1)$-dense, there must exist $C \subseteq B$, such that C is homogeneous for H and C is $2k$-dense. The induction hypothesis yields that C is ε_{k-1}-large. C cannot take any colour i with $i < 3^{b_0}$, since every B_i is ε_{k-1}-small. Thus $C \subseteq \{b \in B : h^B_{\varepsilon_{k-1}\cdot(3^{b_0}-1)}(b_0) \le b\}$. Since C is ε_{k-1}-large, $h^B_{\varepsilon_{k-1}\cdot 3^{b_0}}(b_0)$ exists and B is $\varepsilon_{k-1} \cdot 3^{b_0}$-large. Now, we obtain

$$\begin{aligned} h^B_{\varepsilon_{k-1}\cdot 3^{b_0}}(b_0) &= h^B_{\varepsilon_{k-1}\cdot 3^{b_0}-1+\varepsilon_{k-1}\cdot 3^{b_0}-1\cdot 2}(b_0) \\ &= h^B_{\varepsilon_{k-1}\cdot 3^{b_0}-1}(h^B_{\varepsilon_{k-1}\cdot 3^{b_0}-1\cdot 2}(b_0)) \\ &\ge h^B_{\varepsilon_{k-1}\cdot 3^{b_0}-1}(2^{3^{b_0-1}}b_0) \\ &\ge h^B_{F(\varepsilon_{k-1})+\frac{b_0(b_0+1)}{2}+1}(2^{3^{b_0-1}}b_0) \\ &\ge h^B_{F(\varepsilon_{k-1})+\frac{b_0(b_0+1)}{2}+1}(b_0) \\ &\ge h^B_{F(\varepsilon_{k-1})+\frac{a_0(a_0+1)}{2}+1}(b_0). \end{aligned}$$

The first and third inequality are due to Lemma 2 and the fact that

$$h^B_{\varepsilon_{k-1}\cdot 3^{b_0}-1\cdot 2}(b_0) \ge 2^{3^{b_0-1}}b_0 \ge b_0,$$

for all $k \in \mathbb{N}$. The second inequality holds because of Lemma 1 and Lemma 2 and the last one is due to $\min A \le \min B$. So B is $F(\varepsilon_{k-1}) + \frac{a_0(a_0+1)}{2} + 1$-large. Lemma 3 yields A is $\text{tow}_{a_0}(\varepsilon_{k-1})$-large, i.e. ε_k-large.

Corollary 1. *The following holds:*

$$\text{ACA}_0 \vdash (\forall n)(\forall a)(\exists b)([a, b] \text{ is } n\text{-dense}) \to (\forall a)(\exists b)([a, b] \text{ is } \varepsilon_\omega\text{-large}).$$

Theorem 2. *The following holds:*

$$\text{ACA}_0 + \text{RT} \nvdash (\forall n)(\forall a)(\exists b)([a, b] \text{ is } n\text{-dense})$$

Proof. Assume by contradiction that

$$\text{ACA}_0 + \text{RT} \vdash (\forall n)(\forall a)(\exists b)([a, b] \text{ is } n\text{-dense}).$$

Then

$$\text{ACA}_0 + \text{RT} \vdash (\forall a)(\exists b)([a, b] \text{ is } \varepsilon_\omega\text{-large}),$$

which states that h_{ε_ω} is a provably total function of $\text{ACA}_0 + \text{RT}$. This contradicts the claim that for each provably total function of $\text{ACA}_0 + \text{RT}$ there exists $k \in \mathbb{N}$, such that h_{ε_k} eventually dominates this function. The validity of this last claim can be roughly seen as follows. First combine an ordinal analysis of $\text{ACA}_0 + \text{RT}$ (see e.g. [1]) and results of Buchholz mentioned in the introduction of [5], to deduce that every arithmetical formula which is provable in $\text{ACA}_0 + \text{RT}$, is provable in $\text{PA} + \text{TI}(< \varepsilon_\omega)$, where $\text{TI}(< \varepsilon_\omega)$ denotes the scheme of transfinite induction over all ordinals strictly below ε_ω. Then apply results of Friedman and Sheard on provably total functions of systems of first-order arithmetic ([9]).

Theorem 3. $\text{PA} + \{(\forall a)(\exists b)([a, b] \text{ is } n\text{-dense})\}_{n \in \mathbb{N}}$ and $\text{ACA}_0 + \text{RT}$ have the same Π_2^0 consequences.

Proof. First remark that $\text{PA} + \{(\forall a)(\exists b)([a, b] \text{ is } n\text{-dense})\}_{n \in \mathbb{N}} \subseteq \text{ACA}_0 + \text{RT}$, because for every concrete $n \in \mathbb{N}$, Theorem 1 implies that $(\forall a)(\exists b)([a, b] \text{ is } n\text{-dense})$ can be proved in $\text{ACA}_0 + \text{RT}$.

Now suppose $\text{ACA}_0 + \text{RT}$ proves $\Phi = \forall x \exists y \varphi(x, y)$, where φ is a Δ_0 formula. This means that the function defined by φ is Δ_0-definable and provably total in $\text{ACA}_0 + \text{RT}$. Hence there exists $k \in \mathbb{N}$, such that h_{ε_k} eventually dominates this function (see the proof of Theorem 2). Due to Lemma 4, we can pick $n > 2(k+1)$, such that $\text{PA} + (\forall a)(\exists b)([a, b] \text{ is } n\text{-dense})$ proves the totality of h_{ε_k}, and so Φ. This completes the proof. ∎

We also conjecture that the unprovability result above will give rise to a phase transition as described below. Let us first introduce the parametrised version of n-density.

Definition 8. *Let* $A \subseteq \mathbb{N}$, *with* $\min A > 3$ *and* $f : \mathbb{N} \to \mathbb{N}$. *Then* A *is called* 0-dense(f) *if* $|A| > f(\min A)$. A *is called* $(n+1)$-dense(f), *if for every colouring* $F : [A]^{\min A + 2} \to 3^{\min A}$, *there exists a subset* $B \subseteq A$, *such that* B *is homogeneous for* F *and* B *is* n-dense(f).

We define a hierarchy F_k ($k \in \mathbb{N}$) of functions by putting

$$F_0(x) = 2^x$$
$$F_{k+1}(x) = \underbrace{F_k(\dots(F_k(x))\dots)}_{x \text{ times}} = F_k^x(x),$$

for all natural numbers x and k.

Whenever $G : \mathbb{N} \to \mathbb{N}$ is a function, the inverse function $G^{-1} : \mathbb{N} \to \mathbb{N}$ is defined by $G^{-1}(x) = \min\{y \mid G(y) \geq x\}$. $(F_2)^{-1}$ and $(F_1^k)^{-1}$ will denote the inverse function of F_2 and F_1^k, respectively. We conjecture the following.

Conjecture 1. *1. If* $f(x) = (F_2)^{-1}(x)$ *for every* $x \in \mathbb{N}$, *then*

$$\text{ACA}_0 + \text{RT} \vdash (\forall n)(\forall a)(\exists b)([a, b] \text{ is } n\text{-dense}(f)).$$

2. Let $k \in \mathbb{N}$. If $f(x) = (F_1^k)^{-1}(x)$ for every $x \in \mathbb{N}$, then

$$\text{ACA}_0 + \text{RT} \nvdash (\forall n)(\forall a)(\exists b)([a, b] \ \text{is } n\text{-dense}(f)).$$

The authors would like to thank the referees for their corrections and useful remarks.

This article is dedicated to the memory of Henryk Kotlarski who passed away on February 17, 2008. Kotlarski's work on α-largeness turned out to be very useful for proving the results of this paper.

References

1. Afshari, B.: Relative Computability and the Proof-theoretic Strength of some Theories. PhD thesis, School of Mathematics, University of Leeds (2009)
2. Bigorajska, T., Kotlarski, H.: A partition theorem for α-large sets. Fund. Math. 160(1), 27–37 (1999)
3. Bigorajska, T., Kotlarski, H.: Some combinatorics involving ξ-large sets. Fund. Math. 175(2), 119–125 (2002)
4. Bigorajska, T., Kotlarski, H.: Partitioning α-large sets: some lower bounds. Trans. Amer. Math. Soc. 358(11), 4981–5001 (2006) (electronic)
5. Buchholz, W.: An intuitionistic fixed point theory. Arch. Math. Logic 37(1), 21–27 (1997)
6. Bovykin, A., Weiermann, A.: The strength of infinitary ramseyan principles can be accessed by their densities. Annals of Pure and Applied Logic (to appear)
7. De Smet, M., Weiermann, A.: Partitioning α-large sets for $\alpha < \varepsilon_\omega$ (2010), http://arxiv.org/abs/1001.2437
8. Friedman, H.M., McAloon, K., Simpson, S.G.: A finite combinatorial principle which is equivalent to the 1-consistency of predicative analysis. In: Patras Logic Symposion (Patras, 1980). Stud. Logic Foundations Math., vol. 109, pp. 197–230. North-Holland, Amsterdam (1982)
9. Friedman, H., Sheard, M.: Elementary descent recursion and proof theory. Ann. Pure Appl. Logic 71(1), 1–45 (1995)
10. Kotlarski, H.: A model theoretic approach to proof theory of arithmetic, pp. 198–234. IOS, Amsterdam (2008)
11. Ketonen, J., Solovay, R.: Rapidly growing Ramsey functions. Ann. of Math. 113(2), 267–314 (1981)
12. McAloon, K.: Paris-Harrington incompleteness and progressions of theories. In: Recursion theory (Ithaca, NY, 1982). Proc. Sympos. Pure Math., vol. 42, pp. 447–460. Amer. Math. Soc., Providence (1985)
13. Paris, J.B.: Some independence results for Peano arithmetic. J. Symbolic Logic 43(4), 725–731 (1978)
14. Ratajczyk, Z.: Subsystems of true arithmetic and hierarchies of functions. Ann. Pure Appl. Logic 64(2), 95–152 (1993)

Graph Structures and Algorithms for Query-Log Analysis

Debora Donato

Yahoo! Labs
701 First Avenue, Sunnyvale CA
debora@yahoo-inc.com

Abstract. Query logs are repositories that record all the interactions of users with a search engine. This incredibly rich user behavior data can be modeled using appropriate graph structures. In the recent years there has been an increasing amount of literature on studying properties, models, and algorithms for query-log graphs. Understanding the structure of such graphs, modeling user querying patterns, and designing algorithms for leveraging the latent knowledge (also known as the *wisdom of the crowds*) in those graphs introduces new challenges in the field of graph mining. The main goal of this paper is to present the reader with an example of these graph-structures, i.e., the *Query-flow graph*. This representation has been shown extremely effective for modeling user querying patterns and has been extensively used for developing real time applications. Moreover we present graph-based algorithmic solutions applied in the context of problems appearing in web applications as query recommendation and user-session segmentation.

Keywords: graph mining, link mining, web mining, world wide web, query-log mining, query recommendation.

1 Introduction

A query log contains information about the interactions of users with a search engine — the queries that users make, the results returned by the search engines, and the documents that users click. The wealth of explicit and implicit information contained in the query logs can be valuable source of knowledge in a large number of applications.

Query-logs analysis have been introduced a number of challenges: the dimensions of these data sets are in the order of dozens of terabytes. The most of the algorithms applied to these data are not scalable. Even using advanced techniques as the map-reduce computational paradigm [13] and web-graphs compression algorithms [7], practical query-logs graphs are limited to few months of user-activity.

Nevertheless the range of applications that can be effectively addressed exploiting query-logs are huge and include the following: (*i*) analyzing the interests of users and their searching behavior, (*ii*) finding semantic relations between queries (which terms are similar to each other or which one is a specialization of

F. Ferreira et al. (Eds.): CiE 2010, LNCS 6158, pp. 126–131, 2010.

another) allowing to build taxonomies that are much richer than any human-built taxonomy, (*iii*) improving the results provided by search engines by analysis of the documents clicked by users and understanding the user information needs, (*iv*) fixing spelling errors and suggesting related queries, (*v*) improving advertising algorithms and helping advertisers select bidding keywords, and much more.

As a consequence of the wide range of applications and the importance of query logs, a large number of papers has recently been published on analyzing query logs and on addressing various data-mining problems related to them.

In the next sections we discuss a graph representation of query log data, i.e., the *Query-flow graph*, and consequently we present techniques for mining and analyzing the resulting graph structures.

2   Description of Query Logs

Query log. A typical query log $\mathcal{L}$ is a set of records $\langle q_i, u_i, t_i, V_i, C_i \rangle$, where q_i is the submitted query, u_i is an anonymized identifier for the user who submitted the query, t_i is a timestamp, V_i is the set of documents returned as results to the query, and C_i is the set of documents clicked by the user. We denote by Q, U, and D the set of queries, users, and documents, respectively. Thus $q_i \in Q$, $u_i \in U$, and $C_i \subseteq V_i \subseteq D$.

Sessions. A *user query session*, or just *session*, is defined as the sequence of queries of one particular user within a specific time limit. More formally, if t_θ is a timeout threshold, a user query session S is a *maximal* ordered sequence

$$S = \big\langle \langle q_{i_1}, u_{i_1}, t_{i_1} \rangle, \ldots, \langle q_{i_k}, u_{i_k}, t_{i_k} \rangle \big\rangle,$$

where $u_{i_1} = \cdots = u_{i_k} = u \in U$, $t_{i_1} \leq \cdots \leq t_{i_k}$, and $t_{i_{j+1}} - t_{i_j} \leq t_\theta$, for all $j = 1, 2, \ldots, k - 1$. Regarding the timeout threshold for splitting sessions, a typical value that is often used in query log analysis is $t_\theta = 30$ minutes [6,9,15,18].

Supersessions. The sequence of all the queries of a user in the query log, ordered by timestamp, is called a *supersession*. Thus, a supersession is a sequence of sessions in which consecutive sessions have time difference larger than t_θ.

Chains. A chain is a topically coherent sequence of queries of one user. Radlinski and Joachims [17] defined a chain as *"a sequence of queries with a similar information need"*. The concept of chain is also referred to in the literature with the terms *mission* [12] and *logical session* [1].

3   The Query-Flow Graph

Graphs might be used to provide a navigable, compact representation of the query-related information extracted by query-logs. Query graphs of different types have been extensively studied in literature [1,4,2,14,16]. In most of these studies query logs are represented by query-query graphs, where queries form the set of vertices and edges among queries capture various types of query information.

In order to capture the querying behavior of users, Boldi et al. [6] define the concept of the *query-flow graph*. The novelty of this representation with respect to the previous ones is that the timestamp information is directly taken into account and used for building the graph. A query-flow graph is a directed graph $G = (V, E, w)$ where:

- $V = Q \cup \{s, t\}$ is the set of distinct queries Q submitted to the search engine plus two special nodes s and t, representing a *starting state* and a *terminal state*, respectively, which can be seen as the start and the end of all the chains;
- $E \subseteq V \times V$ is the set of *directed edges*;
- $w : E \rightarrow (0..1]$ is a weighting function that assigns to every pair of queries $(q, q') \in E$ a weight $w(q, q')$.

In the query-flow graph, two queries q and q' are connected by an edge if there is at least one session of the query log in which a user issued the query q' immediately followed by the query q. The weight w is application-dependent; we can consider the weight to be the frequency of the transition in the original query log. That is, the transition probability of any edge (q, q') can be computed dividing its frequency by the sum of the frequencies of the edges leaving from q. In a more complex setting the weight w is the probability $\Pr[q, q']$ of transition from q to q', given that q and q' belong to the same session.

Session Segmentation. An interesting application of the query-flow graph is segmenting and assembling chains in user sessions. In this particular application, one complication is that for chains there is not necessarily some timeout constraint. Additionally, for the queries composing a chain it is not required to be consecutive. Thus, a session may contain queries from many chains, and inversely, a chain may contain queries from many sessions. The probabilities on the edges of the query-flow graph along with other data associated to each transition, are used to segment physical sessions (record of the activities of a single user before a time out of 30 minutes) in chains (i.e., records of activities that are topically related). This step is of key importance for applications aimed at improving the user search experience.

In [6] the problem of finding chains in query logs is modeled as an *Assymetric Traveling Salesman Problem* (ATSP) on the query-flow graph. The formal definition of the chain-finding problem is the following: Let $S = \langle q_1, q_2, \ldots, q_k \rangle$ be the supersession of one particular user, and assume that a query-flow graph has been built by processing a query log that includes S. Then we define a *chain cover* of S to be a partition of the set $\{1, \ldots, k\}$ into subsets $C_1, \ldots, C_h$. Each set $C_u = \{i_1^u < \cdots < i_{\ell_u}^u\}$ can be thought of as a chain $C_u = \langle s, q_{i_1^u}, \ldots, q_{i_{\ell_u}^u}, t \rangle$, which is associated with probability

$$\Pr[C_u] = \Pr[s, q_{i_1^u}] \Pr[q_{i_1^u}, q_{i_2^u}] \ldots \Pr[q_{i_{\ell_u-1}^u}, q_{i_{\ell_u}^u}] \Pr[q_{i_{\ell_u}^u}, t],$$

and we want to find a chain cover maximizing $\Pr[C_1] \ldots \Pr[C_h]$. is the probability for $q_{i_1^u}$ to be a starting query and similarly $\Pr[q_{i_{\ell_u}^u}, t]$ is the probability for $q_{i_{\ell_u}^u}$ to be the terminal query in the chain.

The chain-finding problem is then divided into two subproblems: *session re-ordering* and *session breaking*. The session reordering problem is to ensure that all the queries belonging to the same search mission are consecutive. Then, the session breaking problem is much easier as it only needs to deal with non-intertwined chains.

4   Query Recommendations

Query-log analysis is used to obtain insights on how users refine their queries, and what kind of search strategies they are using to locate the information they need. Often users find it hard to obtain relevant results because they can not express appropriately their information need or because the results of search algorithms are poor for certain queries [5]. Thus, *query-recommendation* systems are a fundamental feature of large-scale search systems, as they can help users reformulate their queries in a more effective way. Many of the algorithms for making query recommendations are based on defining similarity measures among queries then recommending the most popular queries in the query log among the similar ones to a give query [19,4,3,11].

A different set of methods for query recommendation [14,10] is based on graph structures.

Topical query decomposition. A common limitation of many query-recom-mendation algorithms is that they often provide recommendations that are very similar to each other. Bonchi et al. [8] formulate a new problem, which they call *topical query decomposition* and that partially overcomes the previous limitation. In this new framework the goal is to find a set of queries that cover different aspects of the original query. The intuition is that such a set of diverse queries can be more useful for a user whose query is typically too short (with 2.6 words on average) to receive recommendations only on a specific aspect of the query.

The problem statement of topical query decomposition is based on the click graph. A click graph $C = (V_q, V_d, E)$ is a bipartite graph, consisting of a node set $V_q \cup V_d$ and an edge set E. Nodes in V_q denote the set of distinct queries occurring in a query log $\mathcal{L}$, Nodes in V_d denote a set of distinct documents, and an edge $(q, d) \in E$ denotes that the query $q \in V_q$ has led some user to click on the document $d \in V_d$. In particular let q be a query and $D(q)$ be the result set of q, i.e., the neighbor nodes of q in the click graph. We denote with $\mathcal{Q}(q)$ the maximal set of queries p_i, where for each p_i, the set $D(p_i)$ has at least one document in common with the documents returned by q, this is,

$$\mathcal{Q}(q) = \{p_i | \langle p_i, D(p_i) \rangle \in \mathcal{L} \wedge D(p_i) \cap D(q) \neq \emptyset\}.$$

The goal is to compute a *cover*, i.e., selecting a sub-collection $\mathcal{C} \subseteq \mathcal{Q}(q_i)$ such that it covers almost all of $D(q_i)$. As stated before, the queries in $\mathcal{C}$ should represent coherent, conceptually well-separated set of documents: they should have small overlap, and they should not cover too many documents outside $D(q_i)$.

Query recommendations based on the query-flow graph. Boldi et al. [6] investigate the alternative approach of finding query recommendations using the query-flow graph instead of the click graph. In particular, given some original query q, the method computes the PageRank values of a random walk on the query-flow graph where the teleportation is always at the node of the graph that corresponds to query q. In this way, queries that are close to q in the graph are favored to be selected as recommendations. The advantage of using the query-flow graph instead of the click graph is that the method favors as recommendations for q queries q' that follow q in actual user sessions, thus it is likely that q' are natural continuations of q in an information seeking task performed by users.

Boldi et al. [6] explore various alternatives of using random walk on the query-flow graph for the query recommendation problem. One interesting idea is to use normalized PageRank. Here if $s_q(q')$ is the PageRank score for query q' on a random walk with teleportation to the original query q, instead of using the pure random-walk score $s_q(q')$, they consider the ratio $\hat{s}_q(q') = s_q(q')/r(q')$ where $r(q')$ is the absolute random-walk score of q' (i.e., the one computed using a uniform teleportation vector). The intuition behind this normalization is to avoid recommending very popular queries (like "ebay") that may easily get high PageRank scores even though they are not related with the original query. The experiments in [6] showed that in most cases $\hat{s}_q(q')$ produces rankings that are more reasonable, but sometimes tend to boost too much scores having a very low absolute score $r(q')$. To use a bigger denominator, they also tried dividing with $\sqrt{r(q')}$, which corresponds to the geometric mean between $s_q(q')$ and $\hat{s}_q(q')$.

Another interesting variant of the query-recommendation framework of Boldi et al. is providing recommendations that depend not only on the last query input by the user, but on some of the last queries in the user's history. This approach may help to alleviate the data sparsity problem — the current query may be rare, but among the previous queries there might be queries for which we have enough information in the query flow graph. Basing the recommendation on the user's query history may also help to solve ambiguous queries, as we have more informative suggestions based on what the user is doing during the current session. To take the recent queries of the user into account, one has to modify the random walk, in order to perform the teleportation into the set of last queries, instead of only the one last query. For more details on the method and various examples of recommendations see [6].

References

1. Baeza-Yates, R.: Graphs from search engine queries. In: van Leeuwen, J., Italiano, G.F., van der Hoek, W., Meinel, C., Sack, H., Plášil, F. (eds.) SOFSEM 2007. LNCS, vol. 4362, pp. 1–8. Springer, Heidelberg (2007)
2. Baeza-Yates, R., Tiberi, A.: Extracting semantic relations from query logs. In: Proceedings of the 13th ACM International Conference on Knowledge Discovery and Data Mining (KDD), pp. 76–85 (2007)
3. Baeza-Yates, R.A., Hurtado, C.A., Mendoza, M.: Query recommendation using query logs in search engines. In: Current Trends in Database Technology – EDBT Workshops, pp. 588–596 (2004)

4. Beeferman, D., Berger, A.: Agglomerative clustering of a search engine query log. In: Proceedings of the 6th ACM International Conference on Knowledge Discovery and Data Mining (KDD), pp. 407–416 (2000)
5. Belkin, N.J.: The human element: helping people find what they don't know. Communications of the ACM 43(8), 58–61 (2000)
6. Boldi, P., Bonchi, F., Castillo, C., Donato, D., Gionis, A., Vigna, S.: The query-flow graph: model and applications. In: Proceeding of the 17th ACM Conference on Information and Knowledge Management (CIKM), pp. 609–618 (2008)
7. Boldi, P., Vigna, S.: The WebGraph framework I: Compression techniques. In: Proc. of the 13th WWW Conf., Manhattan, USA, pp. 595–601. ACM Press, New York (2004)
8. Bonchi, F., Castillo, C., Donato, D., Gionis, A.: Topical query decomposition. In: Proceedings of the 14th ACM International Conference on Knowledge Discovery and Data Mining (KDD), pp. 52–60 (2008)
9. Catledge, L., Pitkow, J.: Characterizing browsing behaviors on the world wide web. Computer Networks and ISDN Systems 6, 1065–1073 (1995)
10. Craswell, N., Szummer, M.: Random walks on the click graph. In: Proceedings of the 30th Annual International ACM Conference on Research and Development in Information Retrieval (SIGIR), pp. 239–246 (2007)
11. Fonseca, B.M., Golgher, P.B., de Moura, E.S., Ziviani, N.: Using association rules to discover search engines related queries. In: LA-WEB, Washington, DC, USA, pp. 66–71 (2003)
12. Jones, R., Klinkner, K.L.: Beyond the session timeout: automatic hierarchical segmentation of search topics in query logs. In: Proceedings of the 16th ACM Conference on Conference on Information and Knowledge Management (CIKM), pp. 699–708 (2008)
13. Karloff, H., Suri, S., Vassilvitskii, S.: A model of computation for mapreduce. In: Proceedings of the Symposium on Discrete Algorithms (SODA), pp. 938–948 (2010)
14. Mei, Q., Zhou, D., Church, K.: Query suggestion using hitting time. In: Proceeding of the 17th ACM Conference on Information and Knowledge Management (CIKM), pp. 469–478 (2008)
15. Piwowarski, B., Zaragoza, H.: Predictive user click models based on click-through history. In: Proceedings of the 16th ACM Conference on Conference on Information and Knowledge Management (CIKM), pp. 175–182 (2007)
16. Poblete, B., Castillo, C., Gionis, A.: Dr. searcher and mr. browser: a unified hyperlink-click graph. In: Proceeding of the 17th ACM Conference on Information and Knowledge Management (CIKM), pp. 1123–1132 (2008)
17. Radlinski, F., Joachims, T.: Query chains: learning to rank from implicit feedback. In: Proceeding of the 11th ACM SIGKDD International Conference on Knowledge Discovery in Data Mining, pp. 239–248 (2005)
18. Teevan, J., Adar, E., Jones, R., Potts, M.A.S.: Information re-retrieval: repeat queries in yahoo's logs. In: Proceedings of the 30th Annual International ACM Conference on Research and Development in Information Retrieval (SIGIR), pp. 151–158 (2007)
19. Wen, J.-R., Nie, J.-Y., Zhang, H.-J.: Clustering user queries of a search engine. In: Proceedings of the 10th International Conference on World Wide Web (WWW), pp. 162–168 (2001)

On the Complexity of Local Search for
Weighted Standard Set Problems⋆

Dominic Dumrauf[1,2] and Tim Süß[2]

[1] Paderborn Institute for Scientific Computation, 33102 Paderborn, Germany
dumrauf@uni-paderborn.de
[2] Institute of Computer Science, University of Paderborn, 33102 Paderborn, Germany
tsuess@uni-paderborn.de

Abstract. In this paper, we study the complexity of computing *locally optimal* solutions for *weighted* versions of *standard set problems* such as SETCOVER, SETPACKING, and many more. For our investigation, we use the framework of $\mathcal{PLS}$, as defined in Johnson et al., [14]. We show that for most of these problems, computing a locally optimal solution is already $\mathcal{PLS}$-complete for a simple natural neighborhood of size one. For the local search versions of weighted SETPACKING and SETCOVER, we derive *tight bounds* for a simple neighborhood of size two. To the best of our knowledge, these are one of the very few $\mathcal{PLS}$ results about local search for weighted standard set problems.

1 Introduction

Set Problems and Their Approximation. In this paper, we study the complexity of computing *locally optimal* solutions for *weighted standard set problems* in the framework of $\mathcal{PLS}$, as defined in Johnson et al., [14]. In weighted set problems such as SETPACKING or SETCOVER, the input consists of a *set system* along with a *weight function* on the set system. The task is to compute a solution maximizing or minimizing some objective function on the set system while obeying certain constraints. Weighted set problems are *fundamental combinatorial optimization* problems with a wide range of applications spanning from *crew scheduling* in transportation networks and *machine scheduling* to *facility location* problems. Since they are of such fundamental importance on the one hand but of *computational intractability* on the other hand, [9], the *approximation* of weighted standard set problems has been extensively studied in the literature. Numerous heuristics have been developed for these problems, spanning from greedy algorithms and linear programming to local search.

Local Search and Set Problems. Local search is a standard approach to approximate solutions of hard combinatorial optimization problems. Starting from an arbitrary feasible solution, a sequence of feasible solutions is iteratively generated, such that each

---

⋆ This work is partially supported by German Research Foundation (DFG) Priority Programm 1307 Algorithm Engineering, DFG-Project DA155/31-1, ME872/11-1, FI14911-1 AVIPASIA and the DFG Research Training Group GK-693 of the Paderborn Institute for Scientific Computation. Some of the work was done while the first author visited Simon Fraser University.

F. Ferreira et al. (Eds.): CiE 2010, LNCS 6158, pp. 132–140, 2010.

solution is contained in the predefined *neighborhood* of its predecessor solution and strictly improves a given *cost function*. If no improvement within the neighborhood of a solution is possible, a *local optimum* (or *locally optimal solution*) is found. In practice, local search algorithms often require only a few steps to compute a solution. However, the running time is often pseudo-polynomial and even exponential in the worst case.

Polynomial Time Local Search. Johnson, Papadimitriou, and Yannakakis, [14], introduced the class $\mathcal{PLS}$ (polynomial-time local search) in 1988 to investigate the complexity of local search algorithms. Essentially, a problem in $\mathcal{PLS}$ is given by some *minimization* or *maximization problem* over instances with finite sets of feasible solutions together with a non-negative *cost function*. A *neighborhood structure* is superimposed over the set of feasible solutions, with the property that a local improvement in the neighborhood can be found in polynomial time. The objective is to find a *locally optimal* solution. The notion of a $\mathcal{PLS}$-reduction was defined in Johnson et al., [14], to establish relationships between $\mathcal{PLS}$-problems and to further classify them. In the recent past, *game theoretic approaches* re-raised the focus on the class $\mathcal{PLS}$ since in many games the computation of a Nash Equilibrium can be modeled as a local search problem, [8]. The knowledge about $\mathcal{PLS}$ is still very limited and not at all comparable with our rich knowledge about $\mathcal{NP}$.

In this paper, we show that for most weighted standard set problems, computing locally optimal solutions is $\mathcal{PLS}$-complete, even for very *small natural neighborhoods*. This implies that computing local optima for these problems via successive improvements may not yield a sufficient performance improvement over computing globally optimal solutions.

2   Notation and Contribution

In this section, we describe the notation, complexity classes, and problems considered throughout this paper. For all $k \in \mathbb{N}$, denote $[k] := \{1, \ldots, k\}$, and $[k]_0 := [k] \cup \{0\}$. Given a k-tuple T, let $P_i(T)$ denote the projection to the i-th coordinate for some $i \in [k]$. For some set S, denote by $2^{|S|}$ the power set of S.

$\mathcal{PLS}$, Reductions, and Completeness, [14]. A $\mathcal{PLS}$-problem

$$L = (D_L, F_L, c_L, N_L, \text{INIT}_L, \text{COST}_L, \text{IMPROVE}_L)$$

is characterized by seven parameters. The set of instances is given by $D_L \subseteq \{0,1\}^*$. Every instance $I \in D_L$ has a set of feasible solutions $F_L(I)$, where feasible solutions $\mathsf{s} \in F_L(I)$ have length bounded by a polynomial in the length of I. Every feasible solution $\mathsf{s} \in F_L(I)$ has a non-negative integer cost $c_L(\mathsf{s}, I)$ and a neighborhood $N_L(\mathsf{s}, I) \subseteq F_L(I)$. $\text{INIT}_L(I)$, $\text{COST}_L(\mathsf{s}, I)$, and $\text{IMPROVE}_L(\mathsf{s}, I)$ are polynomial time algorithms. Algorithm $\text{INIT}_L(I)$, given an instance $I \in D_L$, computes an initial feasible solution $\mathsf{s} \in F_L(I)$. Algorithm $\text{COST}_L(\mathsf{s}, I)$, given a solution $\mathsf{s} \in F_L(I)$ and an instance $I \in D_L$, computes the cost of the solution. Algorithm $\text{IMPROVE}_L(\mathsf{s}, I)$, given a solution $\mathsf{s} \in F_L(I)$ and an instance $I \in D_L$, finds a better solution in $N_L(\mathsf{s}, I)$ or returns that there is no better one.

A solution $s \in F_L(I)$ is *locally optimal*, if for every neighboring solution $s' \in N_L(s, I)$ it holds that $c_L(s', I) \leq c_L(s, I)$ in case L is a maximization $\mathcal{PLS}$-problem and $c_L(s', I) \geq c_L(s, I)$ in case L is a minimization $\mathcal{PLS}$-problem. A *search problem* R is given by a relation over $\{0, 1\}^* \times \{0, 1\}^*$. An algorithm "solves" R, when given $I \in \{0, 1\}^*$ it computes an $s \in \{0, 1\}^*$, such that $(I, s) \in R$ or it correctly outputs that such an s does not exist. Given a $\mathcal{PLS}$-problem L, let the according search problem be $R_L := \{(I, s) \mid I \in D_L, s \in F_L(I) \text{ is a local optimum}\}$. The *class* $\mathcal{PLS}$ is defined as $\mathcal{PLS} := \{R_L \mid L \text{ is a } \mathcal{PLS}\text{-problem}\}$. A $\mathcal{PLS}$-problem L_1 is $\mathcal{PLS}$-*reducible* to a $\mathcal{PLS}$-problem L_2 (written $L_1 \leq_{\text{pls}} L_2$), if there exist two polynomial-time computable functions $\Phi : D_{L_1} \mapsto D_{L_2}$ and Ψ defined for $\{(I, s) \mid I \in D_{L_1}, s \in F_{L_2}(\Phi(I))\}$ with $\Psi(I, s) \in F_{L_1}(I)$, such that for all $I \in D_{L_1}$ and for all $s \in F_{L_2}(\Phi(I))$ it holds that, if $(\Phi(I), s) \in R_{L_2}$, then $(I, \Psi(I, s)) \in R_{L_1}$. A $\mathcal{PLS}$-problem L is $\mathcal{PLS}$-*complete* if every $\mathcal{PLS}$-problem is $\mathcal{PLS}$-reducible to L.

We write limitations to a problem as a prefix and the size of the neighborhood as a suffix. For all $\mathcal{PLS}$-problems L studied in this paper, the algorithms INIT_L, COST_L, and IMPROVE_L are straightforward and polynomial-time computable.

2.1 Weighted Standard Set Problems

We next describe the $\mathcal{PLS}$-problems we study in this paper. All problems we present are local search versions of their respective decision problems. In the following, let $\mathcal{B}$ denote some finite set and let $\mathcal{C} := \{C_1, \ldots, C_n\}$ denote a collection of subsets over $\mathcal{B}$. Let $w_{\mathcal{C}} : \mathcal{C} \mapsto \mathbb{N}$ and $w_{\mathcal{B}} : \mathcal{B} \times \mathcal{B} \mapsto \mathbb{N}$. Denote by $m_{\mathcal{B}}$ and $m_{\mathcal{C}}$ positive integers with $m_{\mathcal{B}} \leq |\mathcal{B}|$ and $m_{\mathcal{C}} \leq |\mathcal{C}|$. Unless otherwise mentioned, we use the *k-differ-neighborhood* where two solutions are mutual neighbors if they differ in at most k-elements which describe a solution. Except for SETCOVER, all problems are maximization problems.

Definition 1 (W3DM-(p,q), [6]). $I \in D_{\text{W3DM}}$ of WEIGHTED-3-DIMENSIONALMAT-CHING *is a pair* (n, w) *with* $n \in \mathbb{N}$ *and* w *is a function* $w : [n]^3 \to \mathbb{R}_{\geq 0}$. *The components of triples are identified with boys, girls, and homes.* $F_{\text{W3DM}}(I)$ *are all matchings of boys, girls, and homes, i.e. all* $S \subseteq [n]^3$, *with* $|S| = n$, $P_k(T_i) \neq P_k(T_j)$, *for all* $T_i, T_j \in S$, $i \neq j$, $k \in [3]$. *For* $S \in F_{\text{W3DM}}(I)$ *the cost is* $c_{\text{W3DM}}(S, I) := \sum_{T_i \in S} w(T_i)$. $N_{\text{W3DM-(p,q)}}(S, I)$ *contains all feasible solutions where at most p triples are replaced and up to q boys or girls move to new homes.*

Definition 2 (X3C-(k)). $I \in D_{\text{X3C}}$ of EXACT-COVER-BY-3-SETS *is a collection* $\mathcal{C} := \{C_1, \ldots, C_n\}$ *of all 3-element sets over a finite set* $\mathcal{B}$, *with* $|\mathcal{B}| = 3q$ *for some* $q \in \mathbb{N}$, *and* $w : \mathcal{C} \mapsto \mathbb{N}$. $F_{\text{X3C}}(I)$ *are all* $S \subseteq \mathcal{C}$ *such that every* $b \in \mathcal{B}$ *is in exactly one* $C_i \in S$. *For* $S \in F_{\text{X3C}}(I)$ *the cost is*

$$c_{\text{X3C}}(S, I) := \sum_{C_i \in S} w(C_i).$$

Definition 3 (SP-(k)). $I \in D_{\text{SP}}$ of SETPACKING *is a triple* $(\mathcal{C}, w_{\mathcal{C}}, m_{\mathcal{C}})$. $F_{\text{SP}}(I)$ *are all sets* $S \subseteq \mathcal{C}$ *with* $|S| \leq m_{\mathcal{C}}$. *For* $S \in F_{\text{SP}}(I)$ *the cost is*

$$c_{\text{SP}}(S, I) := \sum_{C_i \in S \wedge \forall j \in [m], j \neq i : C_i \cap C_j = \emptyset} w(C_i).$$

Definition 4 (SSP-(k)). $I \in D_{\text{SSP}}$ of SETSPLITTING *is a tuple* $(\mathcal{C}, w_{\mathcal{C}})$. *Feasible solutions* $F_{\text{SSP}}(I)$ *are all partitionings* $S_1, S_2 \subseteq \mathcal{B}$ *of* $\mathcal{B}$. *For* $S \in F_{\text{SSP}}(I)$ *the cost is* $c_{\text{SSP}}(S, I) := \sum_{C_i \in \mathcal{C} \wedge \exists s_1 \in C_i : s_1 \in S_1 \wedge \exists s_2 \in C_i : s_2 \in S_2} w(C_i)$.

Definition 5 (SC-(k)). $I \in D_{\text{SC}}$ of SETCOVER *is a tuple* $(\mathcal{C}, w_{\mathcal{C}})$. $F_{\text{SC}}(I)$ *are all subsets* $S \subseteq \mathcal{C}$ *with* $\bigcup_{C_i \in S} C_i = \mathcal{B}$. *For* $S \in F_{\text{SC}}(I)$ *the cost is* $c_{\text{SC}}(S, I) := \sum_{C_i \in S} w(C_i)$.

Definition 6 (TS-(k)). $I \in D_{\text{TS}}$ of TESTSET *is a triple* $(\mathcal{C}, w_{\mathcal{B}}, m_{\mathcal{B}})$. *Feasible solutions* $F_{\text{TS}}(I)$ *are all sets* $S \subseteq \mathcal{C}$ *with* $|S| \in [m_{\mathcal{B}}]$. *For* $S \in F_{\text{TS}}(I)$ *the cost is* $c_{\text{TS}}(S, I) := \sum_{b_i, b_j \in \mathcal{B}; i < j \wedge \exists C_p, C_q \in S \text{ containing exactly one of } b_i \text{ and } b_j} w(b_i, b_j)$.

Definition 7 (SB-(k)). $I \in D_{\text{SB}}$ of SETBASIS *is a triple* $(\mathcal{C}, w_{\mathcal{C}}, m_{\mathcal{C}})$. $F_{\text{SB}}(I)$ *are all sets* $S = \{S_1, \ldots, S_{m_{\mathcal{C}}}\}$, *where* $S_i \in 2^{|\mathcal{B}|}$ *for all* $i \in [m_{\mathcal{C}}]$. *For* $S \in F_{\text{SB}}(I)$ *the cost is* $c_{\text{SB}}(S, I) := \sum_{C_i \in \mathcal{C} \wedge \exists S' \subseteq S : C_i = \bigcup_{S'_i \in S'} S'_i} w(C_i)$.

Definition 8 (HS-(k)). $I \in D_{\text{HS}}$ of HITTINGSET *is a triple* $(\mathcal{C}, w_{\mathcal{C}}, m_{\mathcal{B}})$. $F_{\text{HS}}(I)$ *are all sets* $S \subseteq \mathcal{B}$ *with* $|S| \leq m_{\mathcal{B}}$. *For* $S \in F_{\text{HS}}(I)$ *the cost is* $c_{\text{HS}}(S, I) := \sum_{C_i \in \mathcal{C} \wedge \exists s \in S : s \in C_i} w(C_i)$.

Definition 9 (IP-(k)). $I \in D_{\text{IP}}$ of INTERSECTIONPATTERN *are two symmetric* $n \times n$ *matrices* $A = (a_{ij})_{i,j \in [n]}$ *and* $B = (b_{ij})_{i,j \in [n]}$ *with positive integer entries and a collection* $\mathcal{D} := \{D_1, \ldots, D_l\}$ *with* $l \geq n$ *over a set* $\mathcal{B}$. $F_{\text{IP}}(I)$ *are all vectors* $\mathcal{C} := (C_1, \ldots, C_n)$ *with* $C_i \in \mathcal{D}$ *for all* $i \in [n]$. *For* $S \in F_{\text{IP}}(I)$ *the cost is* $c_{\text{IP}}(S, I) := \sum_{i \leq j \in [n], |C_i \cap C_j| = a_{ij}} b_{ij}$.

Definition 10 (CC-(k)). $I \in D_{\text{CC}}$ of COMPARATIVECONTAINMENT *are two collections* $\mathcal{C} := \{C_1, \ldots, C_n\}$, *and* $\mathcal{D} := \{D_1, \ldots, D_l\}$ *of sets over a set* $\mathcal{B}$, *and a function* $w : \mathcal{C} \cup \mathcal{D} \mapsto \mathbb{N}$. $F_{\text{CC}}(I)$ *are all sets* $S \subseteq \mathcal{B}$. *For* $S \in F_{\text{CC}}(I)$ *the cost is* $c_{\text{CC}}(S, I) := \sum_{C_i \in \mathcal{C}; S \subseteq C_i} w(C_i) - \sum_{D_i \in \mathcal{D}; S \subseteq D_i} w(D_i) + W$, *where* $W \geq \sum_{D_i \in \mathcal{D}} w(D_i)$.

2.2 Generalized Satisfiability Problems

The hardness results we present rely on known hardness results for the MCA-problem. We use the neighborhood where the value of one variable is changed. The task is to compute an assignment maximizing the sum of the weights.

Definition 11 ((p,q,r)-MCA, [5]). *An instance* $I \in D_{(p,q,r)\text{-MCA}}$ *of* (p, q, r)-MAX-CONSTRAINTASSIGNMENT *is a set of constraints* $\mathcal{C} := \{C_1, \ldots, C_m\}$ *over a set of variables* $\mathcal{X} := \{x_1, \ldots, x_n\}$. *Every constraint* $C_i(x_{i_1}, \ldots, x_{i_{p_i}}) \in \mathcal{C}$ *has length at most* p *and is a function* $w_{C_i} : [r]^{p_i} \mapsto \mathbb{R}_{\geq 0}$. *Every variable appears in at most* q *constraints and takes values from* $[r]$ *with* $r \in \mathbb{N}$. $F_{\text{MCA}}(I)$ *are all assignments* $a : \mathcal{X} \mapsto [r]$. *The cost of* $a \in F_{\text{MCA}}(I)$ *is*

$$c_{\text{MCA}}(a, I) := \sum_{C_i(x_{i_1}, \ldots, x_{i_{p_i}}) \in \mathcal{C}} w_{C_i}(a(x_{i_1}), \ldots, a(x_{i_{p_i}})).$$

2.3 Related Work

In this subsection, we mainly present related work about $\mathcal{PLS}$ and $\mathcal{PLS}$-completeness. The approximation of set problems has been intensively studied in the literature, [11,12,13,17]. Survey articles about local search algorithms can be found in several books, [1,2]. Local search for set problems has been applied in numerous papers, [4,10]. For a survey on the quality of solutions obtained via local search not only for set problems, confer [3]. $\mathcal{PLS}$ was defined in Johnson et al., [14], and the fundamental definitions and results are presented in [14,18]. The problem (p, q, r)-MCA is known to be $\mathcal{PLS}$-complete for triples (3,2,3), (2,3,8), and (6,2,2), [5,15]. An FPTAS for computing approximate local optima for every linear combinatorial optimization problem in $\mathcal{PLS}$ is presented in [16]. The book of Aarts et al., [1], contains a list of $\mathcal{PLS}$-complete problems known so far.

2.4 Our Contribution

In this paper, we show that for most of the weighted standard set problems given in Subsection 2.1, computing a locally optimal solution is $\mathcal{PLS}$-complete for the *1-differ-neighborhood*. This means, that the problems are already hard, when one element describing the solution is allowed to be added, deleted, or exchanged for another element which is not part of the solution. As our main result, we prove the following two theorems:

Theorem 1. *The problems* SSP-*(k),* TS-*(k),* HS-*(k),* SB-*(k),* IP-*(k), and* CC-*(k) are $\mathcal{PLS}$-complete for all $k \geq 1$. The problems* SP-*(k) and* SC-*(k) are $\mathcal{PLS}$-complete for all $k \geq 2$. The problems* W3DM-*(k,l) and* X3C-*(k) are $\mathcal{PLS}$-complete for all $k \geq 6$ and $l \geq 12$.*

Theorem 2. *The problems* SP-*(1) and* SC-*(1) are polynomial-time solvable.*

The proofs for SP-(k), SP-(1), SC-(k), and SC-(1) are given in Sections 3.1 and 3.2. Due to lack of space, we transfer the proofs for W3DM-(p,q), X3C-(k), SSP-(k), TS-(k), SB-(k), HS-(k), IP-(k), and CC-(k) to the full version of this paper, [7]. Let us remark that the reductions we present are *tight* in the sense of Schäffer and Yannakakis, [18]. To the best of our knowledge, these are one of the very few $\mathcal{PLS}$ results for local search on weighted standard set problems. Our reductions preserve the structure of the weights of the $\mathcal{PLS}$-complete generalized satisfiability problems we reduce from. Hence, in the reduced instances of the weighted standard set problems we investigate, weights are of exponential size. Our analysis also unveils that the hardness of the problems stems from the combination of a numerical problem on an underlying combinatorial problem.

3 The $\mathcal{PLS}$-Complexity of Weighted Standard Set Problems

In this section, we investigate the complexity of computing locally optimal solutions for the weighted standard set problems presented in Section 2.1.

Denote by (p,q,r)-MINCA the minimization version of (p,q,r)-MCA. Let us re-mark that results about (p,q,r)-MCA carry over to (p,q,r)-MINCA. For some given instance of (p,q,r)-MCA or (p,q,r)-MINCA, let integer

$$W \gg \sum_{C_i(x_{i_1},...x_{i_{p_i}}) \in C} \sum_{\mathsf{a}(x_{i_1}),...,\mathsf{a}(x_{i_{p_i}}) \in [r]} w_{C_i}(\mathsf{a}(x_{i_1}),...,\mathsf{a}(x_{i_{p_i}})).$$

3.1 The Exact Complexity of SETPACKING-(k)

In this subsection, we prove that SP-(k) is $\mathcal{PLS}$-complete for all $k \geq 2$ and polynomial-time solvable for $k = 1$. Given an instance $I \in D_{(3,2,r)\text{-MCA}}$, we construct a reduced instance $\Phi(I) = (\mathcal{M}, w) \in D_{\text{SP-}(2)}$, consisting of a collection $\mathcal{M}$ of sets over a finite set $\mathcal{B}$, a weight function $w : \mathcal{M} \mapsto \mathbb{N}$ that maps sets in collection $\mathcal{M}$ to positive integer weights, and a positive integer $m \leq |\mathcal{M}|$. W.l.o.g., we assume that in instance $I \in D_{(3,2,r)\text{-MCA}}$, every constraint $C_i \in \mathcal{C}$ has length 3 and the weight of every non-zero assignment is strictly larger than 1. Furthermore, we assume that every variable $x \in \mathcal{X}$ appears in 2 constraints and takes values from $[r]$. Additionally, we may assume that the variables are ordered by appearance.

The Reduction. We create a reduced instance of SP-(2) with $m := |\mathcal{C}|$. Sets in collec-tion $\mathcal{M}$ are defined on elements from the finite set $\mathcal{B} := \{e_i, c_i \mid i \in [m]\} \cup \{x_i \mid x \in \mathcal{X}, i \in [r]\}$. Collection $\mathcal{M}$ consists of the following sets: For all $i \in [m]$, we introduce sets $C_i^{SP} := \{e_i\}$ of weight $w(C_i^{SP}) := 1$ in $\mathcal{M}$. For every constraint $C_i(u,v,w) \in \mathcal{C}$ and every assignment $a,b,c \in [r]$, we introduce sets $C_i^{a,b,c}$ of weight $w(C_i^{a,b,c}) := w_{C_i}(a,b,c)$ in $\mathcal{M}$. Here, set $C_i^{a,b,c} := \{c_i, u'_a, v'_b, w'_c \mid a,b,c \in [r]\}$ where $u'_a := u_a$ if $u \in \mathcal{X}$ appears in $C_i(u,v,w)$ for the first time and $u'_a := u_1,\ldots,u_{a-1},u_{a+1},\ldots,u_r$ otherwise; analogously for v'_b and w'_c. We call an element x_j for some variable $x \in \mathcal{X}$ and assignment $j \in [r]$ enclosed in a set from $\mathcal{M}$ due to the first appearance of x *direct representative* of x. We say that a family of sets $C_i^{*,*,*} \in \mathcal{M}$ is *incident* to a family of sets $C_j^{*,*,*} \in \mathcal{M}$ if the clauses $C_i, C_j \in \mathcal{C}$ have a common variable.

Solution Mapping. We call a feasible solution $\mathsf{S} \in F_{\text{SP}}(\Phi(I))$ *set-consistent* if $|\mathsf{S}| = m$ and for every $i \in [m]$ there is exactly one set $C_i^{a,b,c}$ in S for some $a,b,c \in [r]$, which is pairwise disjoint from all other sets in S. For a feasible and set-consistent solution S, function $\Psi(I,\mathsf{S})$ returns for every set $C_i^{*,*,*} \in \mathsf{S}$ and every direct-representative of x_j the assignment j to variable $x \in \mathcal{X}$. If S is not set-consistent, then the assignment computed by $\text{INIT}_{(3,2,r)\text{-MCA}}(I)$ is returned.

Lemma 1. *Every locally optimal solution $\mathsf{S} \in F_{\text{SP}}(\Phi(I))$ is set-consistent.*

Proof. Assume there exists a locally optimal solution $\mathsf{S}' \in F_{\text{SP}}(\Phi(I))$ with $|\mathsf{S}'| < m$. By pigeonhole principle and construction of our reduction, this implies that there exists a set $C_j^{SP} \in \mathcal{M}$ with $j \in [m]$ which is not in S'. Adding C_j^{SP} to S' strictly improves the cost of S', since by construction C_j^{SP} is pairwise disjoint from all sets in $\mathcal{M}$. A contradiction to S' being locally optimal. Now, assume there exists a locally optimal solution $\mathsf{S}' \in F_{\text{SP}}(\Phi(I))$ with $|\mathsf{S}'| = m$ and there exists an $i \in [m]$ such that (1) at least two set $C_i^{a,b,c}, C_i^{d,e,f}$ are in S' for some $a, b, c, d, e, f \in [r]$ or (2) no set $C_i^{*,*,*}$ is in S'. First, consider case (1). Note that sets $C_i^{a,b,c}, C_i^{d,e,f} \in \mathsf{S}'$ are not pairwise disjoint, since by construction they both contain element $c_i \in \mathcal{B}$. By definition of SP-(k) this implies that they do not contribute to the cost of S'. By pigeonhole principle and construction of our reduction there exists a set $C_j^{SP} \in \mathcal{M}$ for some $j \in [m]$ which is not in S'. Replacing set $C_i^{d,e,f} \in \mathsf{S}'$ for set $C_j^{SP} \in \mathcal{M}$ strictly improves the cost of S', since $w(C_j^{SP}) = 1$ and C_j^{SP} is pairwise disjoint from all sets in $\mathcal{M}$. Set $C_i^{a,b,c}$ may only become pairwise disjoint and thus contribute to the cost of S'. A contradiction to S' being locally optimal. Now, consider case (2). Let $C_i^{*,*,*} \in \mathsf{S}'$ be incident to families of sets $C_o^{*,*,*}, C_p^{*,*,*}, C_q^{*,*,*} \in \mathcal{M}$. From case (1), we have that for every $j \in [m]$, at most one set $C_j^{*,*,*} \in \mathsf{S}'$. If present in S', assume that sets $C_o^{a,*,*}, C_p^{b,*,*}, C_q^{c,*,*}$ are in S' for some $a, b, c \in [r]$. Since there does not exist a set $C_i^{*,*,*} \in \mathsf{S}'$, this implies that there exists a set $C_j^{SP} \in \mathsf{S}'$ for some $j \in [m]$. Exchanging set $C_j^{SP} \in \mathsf{S}'$ for set $C_i^{a,b,c} \in \mathcal{M}$—if sets from incident families are not present in S', choose an arbitrary value for the respective variable—strictly increases the cost of S', since $C_i^{a,b,c}$ is pairwise disjoint from all sets in S' and $w(C_i^{a,b,c}) > w(C_j^{SP})$. A contradiction. □

Lemma 2. $(3, 2, r)$-MCA $\leq_{pls}$ SP-(k) *for all* $r \in \mathbb{N}$, $k \geq 2$.

Proof. Assume there exists a feasible solution $\mathsf{S} \in F_{SP}(\Phi(I))$ which is locally optimal for $\Phi(I)$, but is not locally optimal for I. By Lemma 1, S is set-consistent. This implies that $\Psi(I, \mathsf{S})$ is a legal assignment of values to variables $x \in \mathcal{X}$. Since $\Psi(I, \mathsf{S})$ is not locally optimal for I, there exists a variable $x \in \mathcal{X}$ from instance $I \in (3, 2, r)$-MCA, which can be set from value $i \in [r]$ to some value $j \in [r]$ such that the objective function strictly increases by some $z > 0$. Let variable x appear in constraints $C_p, C_q \in \mathcal{C}$. Exchanging the sets $C_p^{i,*,*}$ and $C_q^{i,*,*}$ by sets $C_p^{j,*,*}$ and $C_p^{j,*,*}$ in S yields a feasible and set-consistent solution and by construction this strictly increases the cost of S by z. A contradiction. □

Lemma 3. SP-(1) *is polynomial-time solvable.*

Proof. Given $I \in D_{SP\text{-}(1)}$, we use the following algorithm GREEDYPACKING: Starting from the feasible solution $\mathsf{S} := \emptyset$, process all sets in $\mathcal{M}$ by weight in descending order and add the heaviest yet unprocessed set to S, if it is disjoint from all sets $S_i \in \mathsf{S}$. It is obvious to see that all necessary operations can be performed in polynomial time. In order to prove that a solution $\mathsf{S} \in F_{SP\text{-}(1)}(I)$ computed by GREEDYPACKING is locally optimal, assume that GREEDYPACKING terminated and S is not locally optimal. This implies that there either exists a set $S_i \in \mathcal{M}$ that can be added, or a set $S_j \in \mathsf{S}$ that can be deleted, or exchanged for another set $S_\ell \in \mathcal{M}$ with $S_\ell \notin \mathsf{S}$. Assume there exists a set $S_i \in \mathcal{M}$ with $S_i \notin \mathsf{S}$ which can be added to S such that the cost strictly improves by some $z \in \mathbb{N}$. This implies that S_i is pairwise disjoint from all sets from S and thus, GREEDYPACKING would have included set S_i. A contradiction. Assume there exists a set $S_j \in \mathsf{S}$ which can be deleted from S such that the cost strictly improves by some $z \in \mathbb{N}$. This implies that S_j intersects with some set from S and GREEDYPACKING would have not included S_j. A contradiction. Assume there exists a set $S_j \in \mathsf{S}$ which can be exchanged for some set $S_\ell \in \mathcal{M}$ with $S_\ell \notin \mathsf{S}$ such that the cost strictly improves by some $z \in \mathbb{N}$. This implies that S_ℓ is pairwise disjoint from all sets in $\mathsf{S} \setminus S_j$ and has a larger weight than S_j. Thus, GREEDYPACKING would have included S_ℓ instead of S_j. A contradiction. □

3.2 The Exact Complexity of SETCOVER-(k)

In this subsection, we prove that SC-(k) is $\mathcal{PLS}$-complete for all $k \geq 2$ and polynomial-time solvable for $k = 1$.

The Reduction. Given an instance $I \in D_{(3,2,r)\text{-MINCA}}$, we construct a reduced instance $\Phi(I) = (\mathcal{M}, w) \in D_{\text{SC-(2)}}$, consisting of a collection $\mathcal{M}$ of sets over a finite set $\mathcal{B}$, and a weight function $w : \mathcal{M} \mapsto \mathbb{N}$ that maps sets in collection $\mathcal{M}$ to positive integer weights. As in proof of Lemma 2, we assume that in instance I, every constraint $C_i \in \mathcal{C}$ has length 3, every variable $x \in \mathcal{X}$ appears in 2 constraints and takes values from $[r]$. Denote $m := |\mathcal{C}|$. We create a reduced instance of SC-(2) over the finite set $\mathcal{B} := \{c_i \mid i \in [m]\} \cup \{x_i \mid x \in \mathcal{X}, i \in [r]\}$. For every constraint $C_i(u, v, w) \in \mathcal{C}$ and every assignment $a, b, c \in [r]$, we introduce sets $C_i^{a,b,c}$ of weight $w(C_i^{a,b,c}) := w_{C_i}(a, b, c) + W$ in $\mathcal{M}$. Here, sets $C_i^{a,b,c}$ are defined as in proof of Lemma 2. The definition of an incident family, a set-consistent solution, and the solution mapping $\Psi(I, \mathsf{S})$ is as in proof of Lemma 2, except that for a non-set-consistent solution, now the assignment computed by $\text{INIT}_{(3,2,r)\text{-MINCA}}(I)$ is returned.

Lemma 4. *Every locally optimal solution $\mathsf{S} \in F_{\text{SC}}(\Phi(I))$ is set-consistent.*

Proof. Assume there exists a locally optimal solution $\mathsf{S}' \in F_{\text{SC}}(\Phi(I))$ with $|\mathsf{S}'| < m$. By pigeonhole principle and construction of our reduction, this implies that there exists an element $c_i \in \mathcal{B}$ for some $i \in [m]$ which is not covered. A contradiction to S' being a feasible solution. Now, assume there exists a locally optimal solution $\mathsf{S}' \in F_{\text{SC}}(\Phi(I))$ with $|\mathsf{S}'| > m$. By pigeonhole principle, this implies that there are two sets $C_i^{a,b,c}$ and $C_i^{d,e,f}$ with total weight at least $2W$ in S for some $a, b, c, d, e, f \in [r]$ and $i \in [m]$. Since S' is a feasible solution, there exist sets $C_h^{o,*,*}, C_j^{p,*,*}, C_l^{q,*,*} \in \mathsf{S}'$ for some $o, p, q \in [r]$ and $h, j, l \in [m]$ from families incident to sets $C_i^{a,b,c}, C_i^{d,e,f} \in \mathsf{S}'$. Exchanging the two sets $C_i^{a,b,c}, C_i^{d,e,f}$ for set $C_i^{o,p,q}$ yields a feasible solution and strictly decreases the cost of S', since $w(C_i^{o,p,q}) < 2W$. A contradiction. □

Lemma 5. $(3, 2, r)$-MINCA $\leq_{pls}$ SC-(k) *for all* $r \in \mathbb{N}$, $k \geq 2$.

Proof. The proof of this lemma is similar to the proof of Lemma 2.

Lemma 6. SC-(1) *is polynomial-time solvable.*

Proof Sketch. Given $I \in D_{\text{SC-(1)}}$, we use the following algorithm GREEDYCOVER: Starting from the initial feasible solution $\mathsf{S} := \mathcal{M}$, process all sets in S by weight in descending order and remove the heaviest yet unprocessed set if S is still a legal cover of $\mathcal{B}$ after the removal. It is obvious to see that all necessary operations can be performed in polynomial time. The proof of correctness is similar to the proof of Lemma 3.

Acknowledgement. The first author thanks Petra Berenbrink for many fruitful discussions.

References

1. Aarts, E., Korst, J., Michiels, W.: Theoretical Aspects of Local Search. Springer, New York (2007)
2. Aarts, E., Lenstra, J. (eds.): Local Search in Combinatorial Optimization. John Wiley & Sons, Inc., New York (1997)
3. Angel, E.: A Survey of Approximation Results for Local Search Algorithms. In: Efficient Approximation and Online Algorithms, pp. 30–73 (2006)
4. Chandra, B., Halldórsson, M.: Greedy Local Improvement and Weighted Set Packing Approximation. In: SODA 1999, pp. 169–176 (1999)
5. Dumrauf, D., Monien, B.: On the PLS-complexity of Maximum Constraint Assignment (submitted), http://homepages.upb.de/dumrauf/MCA.pdf (preprint)
6. Dumrauf, D., Monien, B., Tiemann, K.: Multiprocessor Scheduling is PLS-complete. In: HICSS42, pp. 1–10 (2009)
7. Dumrauf, D., Süß, T.: On the Complexity of Local Search for Weighted Standard Set Problems. CoRR, abs/1004.0871 (2010)
8. Fabrikant, A., Papadimitriou, C., Talwar, K.: The Complexity of Pure Nash Equilibria. In: STOC 2004, pp. 604–612. ACM Press, New York (2004)
9. Garey, M., Johnson, D.: Computers and Intractability; A Guide to the Theory of NP-Completeness. Mathematical Sciences Series. W. H. Freeman & Co., New York (1990)
10. Halldórsson, M.: Approximating Discrete Collections via Local Improvements. In: SODA 1995, pp. 160–169 (1995)
11. Hochbaum, D.: Approximation Algorithms for the Set Covering and Vertex Cover Problems. SIAM Journal on Computing 11(3), 555–556 (1982)
12. Hoffman, K., Padberg, M.: Set Covering, Packing and Partitioning Problems. In: Encyclopedia of Optimization, 2nd edn., pp. 3482–3486. Springer, Heidelberg (2009)
13. Johnson, D.: Approximation Algorithms for Combinatorial Problems. In: STOC 1973, pp. 38–49 (1973)
14. Johnson, D., Papadimtriou, C., Yannakakis, M.: How Easy is Local Search? JCSS 37(1), 79–100 (1988)
15. Krentel, M.: Structure in Locally Optimal Solutions (Extended Abstract). In: FOCS 1989, pp. 216–221 (1989)
16. Orlin, J., Punnen, A., Schulz, A.: Approximate Local Search in Combinatorial Optimization. In: SODA 2004, pp. 587–596 (2004)
17. Paschos, V.: A Survey of Approximately Optimal Solutions to Some Covering and Packing Problems. ACM Comput. Surv. 29(2), 171–209 (1997)
18. Schäffer, A., Yannakakis, M.: Simple Local Search Problems That Are Hard to Solve. SIAM J. Comput. 20(1), 56–87 (1991)

Computational Interpretations of Analysis via Products of Selection Functions

Martín Escardó[1] and Paulo Oliva[2]

[1] University of Birmingham
[2] Queen Mary University of London

Abstract. We show that the computational interpretation of full comprehension via two well-known functional interpretations (dialectica and modified realizability) corresponds to two closely related infinite products of selection functions.

1 Introduction

Full classical analysis can be formalised using the language of finite types in Peano arithmetic PA^ω extended with the axiom schema of *full comprehension* (cf. [11])

$$\mathsf{CA} \quad : \quad \exists f^{\mathbb{N} \to \mathbb{B}} \forall n^{\mathbb{N}} (f(n) \leftrightarrow A(n)).$$

As $\forall n^{\mathbb{N}} (A(n) \vee \neg A(n))$ is equivalent to $\forall n^{\mathbb{N}} \exists b^{\mathbb{B}} (b \leftrightarrow A(n))$, full comprehension, in the presence of classical logic, follows from *countable choice* over the booleans

$$\mathsf{AC}_{\mathbb{B}}^{\mathbb{N}} \quad : \quad \forall n^{\mathbb{N}} \exists b^{\mathbb{B}} A(n, b) \to \exists f \forall n A(n, fn).$$

Finally, the negative translation of $\mathsf{AC}_{\mathbb{B}}^{\mathbb{N}}$ follows intuitionistically from $\mathsf{AC}_{\mathbb{B}}^{\mathbb{N}}$ itself together with the classical principle of *double negation shift*

$$\mathsf{DNS} \quad : \quad \forall n^{\mathbb{N}} \neg\neg A(n) \to \neg\neg \forall n A(n),$$

where $A(n)$ can be assumed to be of the form[1] $\exists y B^N(n, y)$. Therefore, full classical analysis can be embedded (via the negative translation) into $\mathsf{HA}^\omega + \mathsf{AC}_{\mathbb{B}}^{\mathbb{N}} + \mathsf{DNS}$, where HA^ω is Heyting arithmetic in the language of all finite types. It then follows that a computational interpretation of theorems in analysis can be obtained via a computational interpretation of the theory $\mathsf{HA}^\omega + \mathsf{AC}_{\mathbb{B}}^{\mathbb{N}} + \mathsf{DNS}$. The fragment $\mathsf{HA}^\omega + \mathsf{AC}_{\mathbb{B}}^{\mathbb{N}}$, excluding the double negation shift, has a very straightforward (modified) realizability interpretation [15], as well as a dialectica interpretation [1,10]. The remaining challenge is to give a computational interpretation of DNS.

A computational interpretation of DNS was first given by Spector [14], via the dialectica interpretation. Spector devised a form of recursion on well-founded trees, nowadays known as *bar recursion*, and showed that the dialectica interpretation of DNS can be witnesses by such recursion. A computational interpretation of DNS via realizability only came recently, first in [2], via a non-standard form of realizability, and then in [3,4], via Kreisel's modified realizability. The realizability interpretation of DNS makes use of a new form of bar recursion, termed *modified bar recursion*.

---

[1] B^N being the (Gödel-Gentzen) negative translation of B.

F. Ferreira et al. (Eds.): CiE 2010, LNCS 6158, pp. 141–150, 2010.

In this article we show that both forms of bar recursion used to interpret classical analysis, via modified realizability and the dialectica interpretation, correspond to two closely related infinite products of selection functions [9].

Notation. We use X, Y, Z for variables ranging over types. Although in HA^ω one does not have dependent types, we will develop the rest of the paper working with types such as $\Pi_{i \in \mathbb{N}} X_i$ rather than its special case X^ω, when all X_i are the same. The reason for this generalisation is that all results below go through for the more general setting of dependent types. Nevertheless, we hesitate to define a formal extension of HA^ω with dependent types, leaving this to future work. We often write $\Pi_i X_i$ for $\Pi_{i \in \mathbb{N}} X_i$. Also, we write $\Pi_{i \geq k} X_i$ for $\Pi_i X_{k+i}$, and $\mathbf{0}$ for the constant functional 0 of a particular finite type. If α has type $\Pi_{i \in \mathbb{N}} X_i$ we use the following abbreviations

$$[\alpha](n) \equiv \langle \alpha(0), \ldots, \alpha(n-1) \rangle, \quad \text{(initial segment of } \alpha \text{ of length } n)$$

$$\alpha[k, n] \equiv \langle \alpha(k), \ldots, \alpha(n) \rangle, \quad \text{(finite segment from position } k \text{ to } n)$$

$$\overline{\alpha, n} \equiv \langle \alpha(0), \ldots, \alpha(n-1), \mathbf{0}, \mathbf{0}, \ldots \rangle, \quad \text{(infinite extension of } [\alpha](n) \text{ with } \mathbf{0}\text{'s)}$$

$$\hat{s} \equiv \langle s_0, \ldots, s_{|s|-1}, \mathbf{0}, \mathbf{0}, \ldots \rangle. \quad \text{(infinite extension of finite seq. } s \text{ with } \mathbf{0}\text{'s)}$$

If x has type X_n and s has type $\Pi_{i=0}^{n-1} X_i$ then $s * x$ is the concatenation of s with x, which has type $\Pi_{i=0}^{n} X_i$. Similarly, if x has type X_0 and α has type $\Pi_{i=1}^{\infty} X_i$ then $x * \alpha$ has type $\Pi_{i \in \mathbb{N}} X_i$. Finally, by q_s or ε_s we mean the partial evaluation of q or ε on the finite string $s \colon \Pi_{i=0}^{n-1} X_i$, e.g. if q has type $\Pi_{i=0}^{\infty} X_i \to R$ then $q_s \colon \Pi_{i=n}^{\infty} X_i \to R$ is the functional $q_s(\alpha) = q(s * \alpha)$.

1.1 Background: Selection Functions and Their Binary Product

In our recent paper [9] we showed how one can view any element of type $(X \to R) \to R$ as a *generalised quantifier*. The particular case when $R = \mathbb{B}$ corresponds to the types of the usual logical quantifiers $\forall, \exists$. We also showed that some generalised quantifiers $\phi \colon (X \to R) \to R$ are *attainable*, in the sense that for some *selection function* $\varepsilon \colon (X \to R) \to X$, we have

$$\phi p = p(\varepsilon p)$$

for all (generalised) predicates p. In the case when ϕ is the usual existential quantifier, for instance, ε corresponds to Hilbert's epsilon term. Since the types $(X \to R) \to R$ and $(X \to R) \to X$ shall be used quite often, we will abbreviate them as $K_R X$ and $J_R X$, respectively. Moreover, since R will be a fixed type, we often simply write KX and JX, omitting the subscript R. In [9] we also defined the following products of quantifiers and selection functions.

Definition 1. *Given a quantifier $\phi \colon KX$ and a family of quantifiers $\psi \colon X \to KY$, define a new quantifier $\phi \otimes \psi \colon K(X \times Y)$ as*

$$(\phi \otimes \psi)(p^{X \times Y \to R}) \overset{R}{:=} \phi(\lambda x^X . \psi(x, \lambda y^Y . p(x, y))).$$

Also, given a selection function $\varepsilon\colon JX$ and a family of selection functions $\delta\colon X \to JY$, define a new selection function $\varepsilon \otimes \delta\colon J(X \times Y)$ as

$$(\varepsilon \otimes \delta)(p^{X \times Y \to R}) \overset{X \times Y}{:=} (a, b(a))$$

where $b(x) := \delta(x, \lambda y^Y.p(x,y))$ and $a := \varepsilon(\lambda x^X.p(x, b(x)))$.

One of the results we obtained is that the product of attainable quantifiers is also attainable. This follows from the fact that the product of quantifiers corresponds to the product of selection functions, as made precise in the following lemma:

Lemma 1 ([9], lemma 3.1.2). *Given a selection function $\varepsilon \colon JX$, define a quantifier $\overline{\varepsilon}\colon KX$ as*

$$\overline{\varepsilon}p := p(\varepsilon p).$$

Then for $\varepsilon\colon JX$ and $\delta\colon X \to JY$ we have $\overline{\varepsilon \otimes \delta} = \overline{\varepsilon} \otimes \lambda x.\overline{\delta_x}$.

It is well known that the construction K can be given the structure of a strong monad, called the *continuation monad*. We have shown in [9] that J also is a strong monad, with the map $(\bar{\cdot})\colon J \to K$ defined above playing the role of a monad morphism. Any strong monad T has a canonical morphism $TX \times TY \to T(X \times Y)$ (and a symmetric version). We have also shown in *loc. cit.* that for the monads $T = K$ and $T = J$ the canonical morphism turns out to be the product of quantifiers and of selection functions respectively. For further details on the connection between strong monads, products, and the particular monads J and K, see [9]. In the following we explore the concrete structure of J and K and their associated products considered as binary versions of bar recursion, which are then infinitely iterated to obtain countable versions.

2   Two Infinite Products of Selection Functions

Given a finite sequence of selection functions, the binary product defined above can be iterated so as to give rise to a finite product. We have shown that such construction appears in a variety of areas such as game theory (backward induction), algorithms (backtracking), and proof theory (interpretation of the infinite pigeon-hole principle). In the following we describe two possible ways of iterating the binary product of selection functions an infinite (or unbounded) number of times.

2.1   Explicitly Controlled Iteration

The first possibility for iterating the binary product of selection functions we consider here is via an "explicitly controlled" iteration, which we will show to correspond to Spector's bar recursion. In the following subsection we also define an "implicitly controlled" iteration, which we will show to correspond to modified bar recursion.

Definition 2. *Let $\varepsilon\colon \Pi_{k\in\mathbb{N}}((\Pi_{j<k}X_j) \to JX_k)$ be a family of selection functions. Define the* explicitly controlled infinite product of the selection functions ε *as*

$$\mathsf{EPS}_s(\omega)(\varepsilon) \stackrel{J(\Pi_{i=|s|}^\infty X_i)}{=} \begin{cases} \mathbf{0} & \text{if } \omega_s(\mathbf{0}) < |s| \\ \varepsilon_s \otimes \lambda x^{X_{|s|}}.\mathsf{EPS}_{s*x}(\omega)(\varepsilon) & \text{otherwise,} \end{cases}$$

where $s\colon \Sigma_{k\in\mathbb{N}}(\Pi_{j<k}X_j)$. (Note that $\omega_s(\mathbf{0}) = \omega(\hat{s})$)

We refer to this infinite iteration of the product $\otimes$ as "explicitly controlled" because we have an explicit test $\omega_s(\mathbf{0}) < |s|$ for when the iteration stops. As we will see in Section 2.2 (next), we could also iterate the product without using the functional ω.

As with Spector's bar recursion, we consider extensions of Gödel's T with the EPS-schema above. It is then natural to ask what are the models for the calculus of functionals T + EPS. It will follow from our result that EPS is primitive recursively equivalent to Spector's bar recursion, that EPS is validated both in the model of continuous functionals [13] and in the model of strongly majorizable functionals [6]. The same will be true for the functional IPS defined in Section 2.2. For further discussion on the models validating EPS and IPS see [9].

Lemma 2. *Let $q\colon \Pi_{i=|s|}^\infty X_i \to R$ and $\omega\colon \Pi_i X_i \to \mathbb{N}$. EPS can be equivalently defined as*

$$\mathsf{EPS}_s(\omega)(\varepsilon)(q) \stackrel{\Pi_{i=|s|}^\infty X_i}{=} \begin{cases} \mathbf{0} & \text{if } \omega_s(\mathbf{0}) < |s| \\ c * \mathsf{EPS}_{s*c}(\omega)(\varepsilon)(q_c) & \text{otherwise,} \end{cases}$$

where $c = \varepsilon_s(\lambda x.q_x(\mathsf{EPS}_{s*x}(\omega)(\varepsilon)(q_x)))$.

Although we will only need to work with EPS, it will be useful (for the sake of clarity) to define also the explicitly controlled infinite product of *quantifiers*:

Definition 3. *Let $\phi\colon \Pi_{k\in\mathbb{N}}((\Pi_{j<k}X_j) \to KX_k)$ be a family of quantifiers. The* explicitly controlled infinite product of the quantifiers ϕ *is defined as*

$$\mathsf{EPQ}_s(\omega)(\phi) \stackrel{K(\Pi_{i=|s|}^\infty X_i)}{=} \begin{cases} \lambda q.q(\mathbf{0}) & \text{if } \omega_s(\mathbf{0}) < |s| \\ \phi_s \otimes \lambda x^{X_{|s|}}.\mathsf{EPQ}_{s*x}(\omega)(\phi) & \text{otherwise.} \end{cases}$$

The following lemma explains why EPQ can be defined from EPS if we are working with attainable quantifiers.

Lemma 3. *Assuming $\forall\alpha\exists n(\omega_{[\alpha](n)}(\mathbf{0}) \leq n)$ we have $\mathsf{EPQ}_s(\omega)(\bar{\varepsilon}) = \overline{\mathsf{EPS}_s(\omega)(\varepsilon)}$.*

2.2   Implicitly Controlled Iteration

The binary product of selection functions can also be infinitely iterated without the need for the "control functional" ω as follows:

Definition 4. *Let $\varepsilon\colon \Pi_{k\in\mathbb{N}}((\Pi_{j<k}X_j) \to (JX_k))$ and $s\colon \Sigma_{k\in\mathbb{N}}(\Pi_{j<k}X_j)$. Define the* implicitly controlled infinite product of selection functions IPS *as*

$$\mathsf{IPS}_s(\varepsilon) \stackrel{J(\Pi_{i=|s|}^\infty X_i)}{=} \varepsilon_s \otimes \lambda x^{X_{|s|}}.\mathsf{IPS}_{s*x}(\varepsilon),$$

where $s\colon \Sigma_{k\in\mathbb{N}}(\Pi_{j<k}X_j)$.

Again, by unwinding the definition of the binary product of selection functions (and using course-of-values induction) one can show that IPS is equivalent to the following:

Lemma 4. *Let* $q\colon \Pi_{i=|s|}^{\infty} X_i \to R$. *IPS can be equivalently defined as*

$$\mathsf{IPS}_s(\varepsilon)(q)(n) \overset{X_{|s|+n}}{=} \varepsilon_{s*t_{s,n}}(\lambda x^{X_{|s|+n}}.q_{t_{s,n}*x}(\mathsf{IPS}_{s*t_{s,n}*x}(\varepsilon)(q_{t_{s,n}*x})))$$

where $t_{s,n} := [\mathsf{IPS}_s(\varepsilon)(q)](n)$.

The functional IPS generalises Escardó's [7] construction that selection functions for a sequence of spaces can be combined into a selection function for the product space.

Proposition 1. *IPS (with $R = \mathbb{B}$ and ε_s dependent only on $|s|$) is primitive recursively equivalent to Escardo's Π functional of [7]:*

$$\Pi(\varepsilon)(q)(n) \overset{X_n}{=} \varepsilon_n(\lambda x^{X_n}.q_{n,x}(\Pi(\lambda i.\varepsilon_{n+i+1})(q_{n,x})))$$

where

$$q_{n,x}(\alpha^{\Pi_{i=n+1}^{\infty} X_i}) \overset{\mathbb{B}}{:=} q\left(\lambda i. \begin{cases} \Pi(\varepsilon)(q)(i) & i < n \\ x & i = n \\ \alpha(i-n-1) & i > n \end{cases}\right).$$

Proof. For one direction we take

$$\Pi(\varepsilon)(q) := \mathsf{IPS}_{\langle\rangle}(\varepsilon)(q),$$

for the other

$$\mathsf{IPS}_s(\{\varepsilon_n\}_{n\in\mathbb{N}})(q) := \Pi(\{\varepsilon_{|s|+n}\}_{n\in\mathbb{N}})(q).$$

We omit the details of the verification.   □

3   Dialectica Interpretation of Classical Analysis

We now show how EPS can be used to solve Spector's equations (which arise from the dialectica interpretation of full classical analysis).

Theorem 1 (cf. lemma 11.5 of [12]). *Let* $q\colon \Pi_{i=0}^{\infty} X_i \to R$ *and* $\omega\colon \Pi_{i=0}^{\infty} X_i \to \mathbb{N}$ *and* $\varepsilon\colon \Pi_{i=0}^{\infty} JX_i$ *be given. Define*

$$\alpha \quad := \mathsf{EPS}_{\langle\rangle}(\omega)(\varepsilon)(q)$$

$$p_n(x) := \mathsf{EPQ}_{[\alpha](n)*x}(\omega)(\overline{\varepsilon})(q_{[\alpha](n)*x}),$$

identifying ε_s with $\varepsilon_{|s|}$. The functionals α and p_n are a solution to Spector's system of equations, i.e. for $n \leq \omega(\alpha)$ we have

$$\alpha(n) \overset{X_n}{=} \varepsilon_n(p_n)$$

$$p_n(\alpha(n)) \overset{Y}{=} q\alpha.$$

Proof. First, let us show by induction that for all n the following holds:

(i) $\quad \alpha = [\alpha](n) * \mathsf{EPS}_{[\alpha](n)}(\omega)(\varepsilon)(q_{[\alpha](n)})$.

If $n = 0$ this follows by definition. Assume this holds for n we wish to show it for $n + 1$. Consider two cases.

(a) If $\omega(\overline{\alpha, n}) < n$ then $\mathsf{EPS}_{[\alpha](n)}(\omega)(\varepsilon)(q_{[\alpha](n)}) = \mathbf{0}$ and hence $\alpha \overset{(\mathrm{IH})}{=} \overline{\alpha, n} = \overline{\alpha, n + 1}$. Therefore, $\omega(\overline{\alpha, n + 1}) = \omega(\overline{\alpha, n}) < n < n + 1$. So,

$$[\alpha](n + 1) * \mathsf{EPS}_{[\alpha](n+1)}(\omega)(\varepsilon)(q_{[\alpha](n+1)}) = \overline{\alpha, n + 1} = \overline{\alpha, n} = \alpha.$$

(b) If, on the other hand, $\omega(\overline{\alpha, n}) \geq n$, then

$$\alpha \overset{(\mathrm{IH})}{=} [\alpha](n) * \mathsf{EPS}_{[\alpha](n)}(\omega)(\varepsilon)(q_{[\alpha](n)}) = [\alpha](n) * c * \mathsf{EPS}_{[\alpha](n)*c}(\omega)(\varepsilon)(q_{[\alpha](n)*c}),$$

where $c = \alpha(n)$. Hence $\alpha = [\alpha](n + 1) * \mathsf{EPS}_{[\alpha](n+1)}(\omega)(\varepsilon)(q_{[\alpha](n+1)})$.

Now, let $n := \omega(\alpha)$. We argue that $(ii)\ \omega(\overline{\alpha, n}) \geq n$. Otherwise, assuming $\omega(\overline{\alpha, n}) = \omega_{[\alpha](n)}(\mathbf{0}) < n$ we would have, by (i), that $\alpha = \overline{\alpha, n}$. Hence[2], $n > \omega_{[\alpha](n)}(\mathbf{0}) = \omega(\alpha) = n$, a contradiction.

Then, it follows easily that, if $n \leq \omega(\alpha)$,

$$
\begin{aligned}
\alpha(n) &\overset{(i)}{=} \mathsf{EPS}_{[\alpha](n)}(\omega)(\varepsilon)(q_{[\alpha](n)})(0) \\
&\overset{(ii)}{=} (\varepsilon_n \otimes \lambda x.\mathsf{EPS}_{[\alpha](n)*x}(\omega)(\varepsilon))(q_{[\alpha](n)})(0) \\
&= \varepsilon_n(\lambda x.q_{[\alpha](n)*x}(\mathsf{EPS}_{[\alpha](n)*x}(\omega)(\varepsilon)(q_{[\alpha](n)*x}))) \\
&= \varepsilon_n(\lambda x.\overline{\mathsf{EPS}_{[\alpha](n)*x}(\omega)(\varepsilon)}(q_{[\alpha](n)*x})) \\
&= \varepsilon_n(\lambda x.\mathsf{EPQ}_{[\alpha](n)*x}(\omega)(\overline{\varepsilon})(q_{[\alpha](n)*x})) = \varepsilon_n(p_n).
\end{aligned}
$$

For the second equality, we have

$$
\begin{aligned}
p_n(\alpha(n)) &= \mathsf{EPQ}_{[\alpha](n+1)}(\omega)(\overline{\varepsilon})(q_{[\alpha](n+1)}) \\
&= \overline{\mathsf{EPS}_{[\alpha](n+1)}(\omega)(\varepsilon)}(q_{[\alpha](n+1)}) \\
&= q_{[\alpha](n+1)}(\mathsf{EPS}_{[\alpha](n+1)}(\omega)(\varepsilon)(q_{[\alpha](n+1)})) \\
&= q([\alpha](n + 1) * \mathsf{EPS}_{[\alpha](n+1)}(\omega)(\varepsilon)(q_{[\alpha](n+1)})) \overset{(i)}{=} q(\alpha).
\end{aligned}
$$

That concludes the proof.   □

---

[2] Note that extensionality is used here. It turns out that only a weak form of extensionality rule is necessary (cf. [12, Lemma 11.5]). We recall that the dialectica interpretation does not validate the axiom of extensionality, but it validates the aforementioned rule. We are obviously allowed, however, to appeal to extensionality when *verifying* that the dialectica interpretation of a certain principle (e.g. DNS) is correct. So, the effort to avoid extensionality is purely philosophical.

Remark 1. The theorem above has a very natural game theoretic reading. Following the nomenclature of [9], each ε_n can be viewed as the selection function defining an outcome quantifier for round n. The functional q is the outcome functional, mapping infinite plays (in $\Pi_i X_i$) to the outcome of the game (in R). The construction used in the theorem for α and p_n calculates an infinite play α of the game which is optimal up to the point $n = \omega(\alpha)$. If ω is thought of as deciding when the game is terminated, then we have in fact an optimal play in the game.

Remark 2. Note that we are only using EPQ for the sake of clarity. As shown in Lemma 3, any use of EPQ above can be replaced by an instance of EPS. Therefore, the recursion schema EPS alone can be used to solve Spector's equations.

3.1 Relation to Spector's Bar Recursion

As we have shown above, EPS solves the computational interpretation of classical analysis via the dialectica interpretation. Spector, however, describing the recursion schema used in his solution, formulated first the general "construction by bar recursion" as

$$\mathsf{BR}_s(\omega)(\phi)(g) \overset{R}{=} \begin{cases} g(s) & \text{if } \omega_s(\mathbf{0}) < |s| \\ \phi_s(\lambda x^{X_{|s|}}.\mathsf{BR}_{s*x}(\omega)(\phi)(g)) & \text{otherwise.} \end{cases}$$

Then, Spector explicitly says that only a "restricted form" of this is used. It is this restricted form that we shall from now on call "Spector's bar recursion":

Definition 5. *Let* $R = \Pi_{i=0}^{\infty} X_i$ *and* $\varepsilon_s \colon JX_{|s|}$ *and* $\omega \colon \Pi_{i=0}^{\infty} X_i \to \mathbb{N}$. *Spector's bar recursion [14] is the following recursion schema*

$$\mathsf{SBR}_s(\omega)(\varepsilon) \overset{R}{=} \begin{cases} \hat{s} & \text{if } \omega_s(\mathbf{0}) < |s| \\ \mathsf{SBR}_{s*c}(\omega)(\varepsilon) & \text{otherwise,} \end{cases}$$

where $c \overset{X_{|s|}}{=} \varepsilon_s(\lambda x.\mathsf{SBR}_{s*x}(\omega)(\varepsilon))$.

We showed above how EPS can be used to solve Spector's equations. In fact, we have:

Proposition 2. EPS *and* SBR *are primitive recursively equivalent.*

4 Realizability Interpretation of Classical Analysis

We have seen (Section 3 above) that EPS solves the dialectica interpretation of classical analysis. In this section we show that when interpreting DNS via modified realizability, an *unrestricted* iterated product of selection functions naturally arises. Assuming continuity[3], for instance, one may say that the infinite iterated product is *implicitly* controlled, by the continuity of q.

---

[3] By continuity of $q \colon \Pi_i X_i \to R$ we mean that for all $\alpha \colon \Pi_i X_i$ there exists a point n (called 'point of continuity') such that the value $q(\alpha)$ is determined by $[\alpha](n)$, i.e. for any β extending $[\alpha](n)$ we have $q\alpha = q\beta$.

As discussed in the introduction, only a restricted form of DNS is used for the interpretation of full comprehension, namely, DNS for formulas $A \equiv \exists y B^N(n, y)$. For such formulas we have that $\bot \to \forall n A(n)$, and hence this restricted form of DNS is equivalent to

$$\forall n((A(n) \to \bot) \to A(n)) \to (\forall n A(n) \to \bot) \to \forall n A(n).$$

Moreover, since the negative translation brings us into minimal logic, falsity $\bot$ can be replaced by an arbitrary formula[4] R. In practice, however, because we will require a continuity assumption we restrict R to be a Σ_1^0 formula. As such, recalling that $J_R A \equiv (A \to R) \to A$, the resulting principle we obtain is what we shall call J-shift

$$J\text{-shift} \quad : \quad \forall n J_R A(n) \to J_R \forall n A(n),$$

where $A(n)$ is an arbitrary formula and R is a Σ_1^0 formula.

Theorem 2 (cf. [3], theorem 3). $\mathsf{IPS}_{\langle \rangle}$ *modified realizes* J-shift.

Proof. Let

$$\varepsilon_n \; \mathrm{mr} \; (A(n) \to R) \to A(n)$$

$$q \; \mathrm{mr} \; \forall n A(n) \to R.$$

As in [3], we shall assume continuity of q. We show $\forall s \in S \, \forall n P(s, n)$ by relativised bar induction (see [3] for precise formulation), where

$$P(s, n) \equiv (s * \mathsf{IPS}_s(\varepsilon)(q_s))(n) \, \mathrm{mr} \, A(n)$$

and the predicate used in the relativisation is

$$s \in S \equiv \forall n < |s| \, (s_n \, \mathrm{mr} \, A(n)).$$

We write $\alpha \in S$ as an abbreviation for $\forall n([\alpha](n) \in S)$. We now prove the two assumptions of the bar induction:

(i) $\forall \alpha \in S \, \exists k \forall t \succeq [\alpha](k) \, \forall n P(t, n)$, where $t \succeq s$ means that t is an extension of the finite sequence s. Given α we pick k to be a point of continuity of q on α. The result follows simply unfolding the definition of IPS.

(ii) $\forall s \in S(\forall t, x(s * t * x \in S \to \forall n P(s * t * x, n)) \to \forall n P(s, n))$. Fix $s \in S$ and assume

$$\text{(a)} \; \forall t, x(s * t * x \in S \to \forall n P(s * t * x, n)).$$

We prove $\forall n P(s, n)$ by course-of-values induction. Assume $\forall k < n \, P(s, k)$, i.e.

$$\text{(b)} \; \forall k < n \, ((s * \mathsf{IPS}_s(\varepsilon)(q_s))(k) \, \mathrm{mr} \, A(k)).$$

We want to show $(s * \mathsf{IPS}_s(\varepsilon)(q_s))(n) \, \mathrm{mr} \, A(n)$. If $n < |s|$ we are done, since in this case $(s * \mathsf{IPS}_s(\varepsilon)(q_s))(n) = s_n$ (and $s \in S$). Assume $n \geq |s|$. Then, our goal becomes

---

[4] This is known as the (refined) A-translation [5], and is useful to analyse proofs of Π_2^0 theorems in analysis.

$$\varepsilon_n(\lambda x^{X_n}.q_{s*t_{s,n}*x}(\mathsf{IPS}_{s*t_{s,n}*x}(\varepsilon)(q_{s*t_{s,n}*x}))) \text{ mr } A(n),$$

where $t_{s,n} = [\mathsf{IPS}_s(\varepsilon)(q_s)](n - |s|)$. That follows from

$$\lambda x^{X_n}.q_{s*t_{s,n}*x}(\mathsf{IPS}_{s*t_{s,n}*x}(\varepsilon)(q_{s*t_{s,n}*x})) \text{ mr } A(n) \to R$$

which, by definition, is

$$\forall x^{X_n}(x \text{ mr } A(n) \; \to \; q_{s*t_{s,n}*x}(\mathsf{IPS}_{s*t_{s,n}*x}(\varepsilon)(q_{s*t_{s,n}*x})) \text{ mr } R).$$

Fix x such that x mr $A(n)$. By our assumption (b) we have that $s * t_{s,n} * x \in S$. And by assumption (a) we get $(s * t_{s,n} * x * \mathsf{IPS}_{s*t_{s,n}*x}(\varepsilon)(q_{t_{s,n}*x}))$ mr $\forall n A(n)$. The proof is then concluded by the assumption that q mr $\forall n A(n) \to R$.   □

Remark 3. We analyse the J-shift in more detail in the companion paper [8], where a proof translation based on the construction JX is also defined.

4.1   Relation to Modified Bar Recursion

We now show that IPS and modified bar recursion are in fact primitive recursively interdefinable. Modified bar recursion [3], when generalised to the language of dependent types, can be formulated as

$$\mathsf{MBR}_s(\varepsilon)(q)(n) \overset{X_n}{=} \begin{cases} s_n & \text{if } n < |s| \\ \varepsilon_s(\lambda x^{X_{|s|}}.q(\mathsf{MBR}_{s*x}(\varepsilon)(q)))(n - |s|) & \text{otherwise,} \end{cases}$$

where $\varepsilon_s \colon (X_{|s|} \to R) \to \Pi_{j \geq |s|} X_j$. The following lemma says that MBR is equivalent to a variant which can make use of any value bar recursively computed, and not just the immediate children $s * x$ of the node s. We are assuming that types are restricted so that finite sequences of X_k's can be coded as single elements.

Lemma 5   ([3], lemma 2). MBR *is primitive recursively equivalent to*

$$\mathsf{MBR}_s^0(\varepsilon)(q)(n) \overset{X_n}{=} \begin{cases} s_n & \text{if } n < |s| \\ \varepsilon_s(\lambda r^{\Pi_{k=|s|}^{j-1} X_k} \lambda x^{X_j}.q(\mathsf{MBR}_{s*r*x}^0(\varepsilon)(q)))(n - |s|) & \text{otherwise.} \end{cases}$$

The next theorem essentially says that MBR is also equivalent to a variant which makes use of course-of-values recursion to access values previously computed, i.e. in order to define the point n of the infinite sequence $\mathsf{MBR}_s^1(\varepsilon)(q)$ we are allowed to use $\mathsf{MBR}_s^1(\varepsilon)(q)(k)$ for $k < n$.

Lemma 6. MBR^0 *is primitive recursively equivalent to*

$$\mathsf{MBR}_s^1(\varepsilon)(q)(n) \overset{X_n}{=} \begin{cases} s_n & \text{if } n < |s| \\ \varepsilon_s(r_{s,n}, \lambda r^\eta, x^{X_j}.q(\mathsf{MBR}_{s*r*x}^1(\varepsilon)(q)))(n - |s|) & \text{otherwise,} \end{cases} \tag{1}$$

where $r_{s,n} := \mathsf{MBR}_s^1(\varepsilon)(q)[|s|, n - 1]$ *and* $\eta = \Pi_{k=|s|}^{j-1} X_k$.

Corollary 1. MBR *primitive recursively defines* IPS.

Theorem 3. IPS *primitive recursively defines* MBR.

Corollary 2. *The equation defining* IPS *has a solution in the type structure* $\mathcal{M}$ *of the strongly majorizable functionals.*

Proof. This follows from the result in [4] that MBR lives in the model $\mathcal{M}$.   □

We have also recently shown the following:

Theorem 4. *The iterated product of selection functions* $\otimes$ *defined in [9] (which is clearly a particular case of* IPS*) is primitive recursively equivalent to* IPS.

Acknowledgements. The second author gratefully acknowledges support of the Royal Society (grant 516002.K501/RH/kk).

References

1. Avigad, J., Feferman, S.: Gödel's functional ("Dialectica") interpretation. In: Buss, S.R. (ed.) Handbook of proof theory. Studies in Logic and the Foundations of Mathematics, vol. 137, pp. 337–405. North Holland, Amsterdam (1998)
2. Berardi, S., Bezem, M., Coquand, T.: On the computational content of the axiom of choice. The Journal of Symbolic Logic 63(2), 600–622 (1998)
3. Berger, U., Oliva, P.: Modified bar recursion and classical dependent choice. Lecture Notes in Logic 20, 89–107 (2005)
4. Berger, U., Oliva, P.: Modified bar recursion. Mathematical Structures in Computer Science 16, 163–183 (2006)
5. Berger, U., Schwichtenberg, H.: Program extraction from classical proofs. In: Leivant, D. (ed.) LCC 1994. LNCS, vol. 960, pp. 77–97. Springer, Heidelberg (1995)
6. Bezem, M.: Strongly majorizable functionals of finite type: a model for bar recursion containing discontinuous functionals. The Journal of Symbolic Logic 50, 652–660 (1985)
7. Escardó, M.H.: Infinite sets that admit fast exhaustive search. In: Proceedings of LICS, pp. 443–452 (2007)
8. Escardó, M.H., Oliva, P.: The Peirce translation and the double negation shift. In: Ferreira, F., Lowe, B., Mayordomo, E., Gomes, L.M. (eds.) CiE 2010. LNCS, vol. 6158, pp. 151–161. Springer, Heidelberg (2010)
9. Escardó, M.H., Oliva, P.: Selection functions, bar recursion, and backward induction. Mathematical Structures in Computer Science 20(2), 127–168 (2010)
10. Gödel, K.: Über eine bisher noch nicht benützte Erweiterung des finiten Standpunktes. Dialectica 12, 280–287 (1958)
11. Kohlenbach, U.: Higher order reverse mathematics. In: Simpson, S.G. (ed.) Reverse Mathematics 2001, ASL, A.K. Peters. Lecture Notes in Logic, vol. 21, pp. 281–295 (2005)
12. Kohlenbach, U.: Applied Proof Theory: Proof Interpretations and their Use in Mathematics. In: Monographs in Mathematics. Springer, Heidelberg (2008)
13. Normann, D.: The continuous functionals. In: Griffor, E.R. (ed.) Handbook of Computability Theory, ch. 8, pp. 251–275. North Holland, Amsterdam (1999)
14. Spector, C.: Provably recursive functionals of analysis: a consistency proof of analysis by an extension of principles in current intuitionistic mathematics. In: Dekker, F.D.E. (ed.) Recursive Function Theory: Proc. Symposia in Pure Mathematics, vol. 5, pp. 1–27. American Mathematical Society, Providence (1962)
15. Troelstra, A.S.: Metamathematical Investigation of Intuitionistic Arithmetic and Analysis. LNM, vol. 344. Springer, Berlin (1973)

The Peirce Translation and the Double Negation Shift

Martín Escardó[1] and Paulo Oliva[2]

[1] School of Computer Science, University of Birmingham,
Birmingham B15 2TT, United Kingdom
[2] School of Electronic Engineering and Computer Science, Queen Mary,
University of London, Mile End Road, London E1 4NS, United Kingdom

Abstract. We develop applications of selection functions to proof theory and computational extraction of witnesses from proofs in classical analysis. The main novelty is a translation of classical minimal logic into minimal logic, which we refer to as the *Peirce translation*, and which we apply to interpret both a strengthening of the double-negation shift and the axioms of countable and dependent choice, via infinite products of selection functions.

1   Introduction

In previous work [5,7], we investigated *selection functions*

$$\varepsilon \in JA \equiv ((A \to R) \to A)$$

for *generalised quantifiers*

$$\phi \in KA \equiv ((A \to R) \to R),$$

where R is a fixed object of generalised truth values. Moreover, we developed various applications to higher-type computability, algorithms, game theory, and proof theory, among others. In this paper, we develop further applications to proof theory. Before discussing the new applications, we introduce background from the above work.

Selection functions in higher-type computability. In [5], the first author considered the particular case where the object A is a domain, and the object R is the domain of boolean values. The particular quantifier ϕ studied was the bounded existential quantifier $\exists_S$ for a subset S of A, with the requirement that $\varepsilon(p)$ be an element of S such that if $p(s)$ holds for some $s \in S$, then $p(\varepsilon(p))$ holds, that is:

$$\phi(p) = p(\varepsilon(p)), \tag{1}$$

for all $p \in A \to R$. The set $S \subseteq A$ is called *exhaustible* if the quantifier $\phi = \exists_S$ is computable, and *searchable* if additionally there is a computable functional $\varepsilon \in JA$ satisfying (1).

It turns out that any searchable set (of total elements) is topologically compact, and, mimicking the Tychonoff theorem from topology, it was shown that

F. Ferreira et al. (Eds.): CiE 2010, LNCS 6158, pp. 151–161, 2010.

searchable sets are closed under countable products. This relies on countable-product functionals of type

$$(JA)^n \to JA^n \qquad (n \leq \omega).$$

In [7], we considered much more general choices for A and R (objects of a cartesian closed category), and for ϕ (e.g. supremum functional when R are the reals in the category of sets, or in suitable categories of spaces). Again, we required that the selection function ε be related to the quantifier ϕ as in Equation (1). Moreover, we considered the above product in more generality, allowing the object A to vary:

$$\otimes : \prod_{i<n} JA_i \to J\left(\prod_{i<n} A_i\right) \qquad (n \leq \omega).$$

The case $n = \omega$ is restricted to a category of continuous maps of certain spaces, which include Kleene–Kreisel spaces of continuous functionals, and requires that R be discrete (e.g. the natural numbers or more generally the types defined in [5, Definition 4.12]) to be well defined.

Selection functions in game theory. Let A_i be the set of possible moves of a sequential game at round i, and let R be the set of possible outcomes of the game. Moreover, let $p \colon \prod_i A_i \to R$ be a function that gives the outcome of a play (or payoff of a profile), and consider quantifiers $\phi_i \in KA_i$ for each round i defining the "goal" for that round (see [7] for details). Finally, assume the quantifiers ϕ_i have associated selection functions $\varepsilon_i \in JA_i$ that choose moves to locally optimise the play, which may be regarded as policy functions. It turns out that product of selection functions calculate optimal plays, profiles in Nash equilibrium, and optimal strategies. As a simple example, for Abelard and Eloise playing in alternating rounds, we take R to be the booleans, and we use universal quantifiers for Abelard, and existential quantifiers for Eloise. If a draw is possible, we instead consider $R = \{-1, 0, 1\}$, and we replace these quantifiers by infimum and supremum functionals respectively. For Nash equilibria, consider $R = \mathbb{R}^n$, with supremum functionals in all rounds (cf. [7]).

Selection functions in bar recursion. Moreover, we showed in [7] that:

1. The infinite case $n = \omega$ of the product of selections functions is the iteration of the finite case $n = 2$,

    $$\otimes \colon JA \times JB \to J(A \times B),$$

    in the sense that

    $$\bigotimes_i \varepsilon_i = \varepsilon_0 \otimes \bigotimes_i \varepsilon_{i+1}.$$

2. This iteration is an instance of the bar recursion scheme.

In the companion paper [6], we establish relations to traditional instances of bar recursion, such as Spector's bar recursion [11] and modified bar recursion [2,3].

Selection functions in category theory. We also showed in [7] that the construction J over any cartesian closed category gives rise to a strong monad, with a monad morphism into the well-known continuation monad K [9]. The morphism $J \to K$ assigns the quantifier $\phi \in KA$ defined by Equation (1) to any given selection function $\varepsilon \in JA$. Moreover, the case $n = 2$ of the product of selection functions turns out to be simply the canonical map that makes any strong monad into a monoidal monad.

Selection functions in proof theory. We now move to the results developed in this paper. We interpret the objects A and R as logical formulae, and the morphisms as proofs in intuitionistic or minimal logic, or as computable realisers of entailments. For $T = J$ or $T = K$, or more generally any strong monad T, one has the intuitionistic laws

$$A \to TA \qquad \text{(unit)} \qquad\qquad T(A \to B) \to TA \to TB \quad \text{(functor)}$$

$$TTA \to TA \quad \text{(multiplication)} \qquad A \wedge TB \to T(A \wedge B) \qquad \text{(strength)}.$$

In the terminology of [1], the construction T is a lax modal operator.

Application to the double negation shift. It turns out that the infinite product of selection functions realises, in the sense of formalised modified realisability, the following shift principle for $T = J$, assuming that R has a discrete type of realisers:

$$T\text{-shift} \quad : \quad \forall n TA(n) \to T \forall n A(n).$$

The well-known double negation shift is the case $T = K$ with $R = \bot$, but it is realised only for special types of formulae A, including those in the image of the negative translation, whereas the J-shift is realised for *all* formulae A. We also show that the double negation shift for formulas A in the image of a negative translation follows from the J-shift. With this, we will get an alternative way of interpreting classical analysis and extracting computational witnesses via infinite products of selection functions.

Application to the elimination of Peirce's law. It is well known that several forms of the negative translation can be understood in terms of the continuation monad K. It is also well known that any monad T gives rise to a translation (see e.g. [1]). Here we consider the *T-translation* inductively defined as

$$P^T \quad = TP \qquad (A \wedge B)^T = A^T \wedge B^T \qquad (A \vee B)^T \ = T(A^T \vee B^T)$$

$$(\exists x A)^T = T(\exists x A^T) \quad (\forall x A)^T \ = \forall x A^T \qquad (A \to B)^T = A^T \to B^T.$$

That is, we prefix T in front of atomic formulae, disjunctions and existential quantifications. For $T = K$ and $R = \bot$, this amounts to the standard Gödel-Gentzen negative translation [13], and for $R = A$, with A a Σ_1^0-formula, this corresponds to Friedman's A-translation [8] of the negative translation. From well-known properties of monads on cartesian closed categories, one sees by induction that any C in the image of the T-translation is a T-algebra and in particular $TC \to C$ is provable. Putting this together:

1. $TC \to C$ is provable in minimal logic ML for formulae C in the image of the T-translation.

2. For $T = K$ and $R = \bot$ this principle amounts to double negation elimination.

3. For $T = J$ this is the instance $((C \to R) \to C) \to C$ of Peirce's law, and hence we also refer to the J-translation as the *Peirce translation*.

Because there is a monad morphism $J \to K$, any K-algebra is a J-algebra, which gives the standard fact that the usual negative translations also eliminate Peirce's law. Notice that the implication $JA \to KA$ can be reversed if and only if $R \to A$. In fact, a main difference between the K-translation and the J-translation is that the former also eliminates ex-falso-quodlibet EFQ ($\bot \to A$), whereas the latter is sound with respect to EFQ but does not eliminate it.

Notation. We use X, Y, Z for variables ranging over types. Although in HA^ω one does not have dependent types, we will develop the rest of the paper working with types such as $\Pi_{i \in \mathbb{N}} X_i$ rather than the special case X^ω, when all X_i are the same. The reason for this generalisation is that all results below go through for the more general setting of dependent types. Nevertheless, we hesitate to define a formal extension of HA^ω with dependent types, leaving this to future work. We often write $\Pi_i X_i$ for $\Pi_{i \in \mathbb{N}} X_i$. If x has type X_n and s has type $\Pi_{i=0}^{n-1} X_i$ then $s * x$ is the concatenation of s with x, which has type $\Pi_{i=0}^n X_i$.

2   Products of Selection Functions

In this background section we briefly recall some functionals defined and studied in more detail in [6,7]. Given selection functions $\varepsilon \in JX$ and $\delta \in JY$, define their product $\varepsilon \otimes \delta \in J(X \times Y)$ by

$$(\varepsilon \otimes \delta)(p) = (a, b(a)) \text{ where } b(x) = \delta(\lambda y.p(x, y)) \text{ and } a = \varepsilon(\lambda x.p(x, b(x))).$$

Then, the infinite product functional is defined in [7] by the equation

$$\bigotimes_i \varepsilon_i = \varepsilon_0 \otimes \bigotimes_i \varepsilon_{i+1}.$$

Also, given a selection function $\varepsilon \in JX$ and a family of selection functions $\delta \in X \to JY$, define their *dependent product* $\varepsilon \otimes_d \delta \in J(X \times Y)$ as

$$(\varepsilon \otimes_d \delta)(p) = (a, b(a)) \text{ where } b(x) = \delta(x)(\lambda y.p(x, y)) \text{ and } a = \varepsilon(\lambda x.p(x, b(x))).$$

For $\varepsilon \colon \Pi_{k \in \mathbb{N}}((\Pi_{j<k} X_j) \to (JX_k))$ and $s \colon \Sigma_{k \in \mathbb{N}}(\Pi_{j<k} X_j)$, define the *iterated dependent product of selection functions* IPS as

$$\mathsf{IPS}_s(\varepsilon) \overset{J(\Pi_{i \overset{\infty}{=} k} X_i)}{=} \varepsilon_s \otimes_d \lambda x^{X_k}.\mathsf{IPS}_{s*x}(\varepsilon).$$

The recursive definitions for $\bigotimes$ and IPS uniquely define functionals in the models of partial and total continous functionals (cf. [7]). Finally, we remark that $\bigotimes$ and IPS are actually inter-definable over system T, as stated in [6].

3   Shift Principles, Countable Choice, and Dependent Choice

In this section we investigate Heyting arithmetic (HA) and classical extensions of HA induced by the monads $T = J$ and $T = K$, as discussed in the introduction.

Given a formal system S we write S^ω for the finite type generalisation of S with a neutral treatment of equality (cf. [12]). Before specialising to the cases of interest, we consider the general T-translation for an arbitrary strong monad T.

We refer as T-*logic* to the extension of intuitionistic logic with the T-*elimination axiom* $TA \to A$. As such, classical logic amounts to K-logic if we choose $R = \perp$ in the definition of K. Using this language, the discussion of the introduction shows that the T-translation eliminates the T-elimination axiom, thereby mapping T-logic into intuitionistic logic (see Section 4 below for a sharper analysis of this fact). As is well known, in the case $T = K$ with $R = \perp$, this mapping actually lands in minimal logic ML, that is, intuitionistic logic without EFQ.

We refer as T-*arithmetic* (TA) to the extension of HA with T-logic. Then Peano arithmetic (PA) is K-arithmetic for $R = \perp$. If a formula does not have occurrences of disjunction or existential quantification, its T-translation only prefixes T to atomic formulae, and hence the T-translations of the Peano axioms follow from the Peano axioms. This shows that the T-translation maps TA into HA.

However, the T-translation does not map $TA^\omega + AC_\mathbb{N}$ into $HA^\omega + AC_\mathbb{N}$, where $AC_\mathbb{N}$ is the axiom of countable choice

$AC_\mathbb{N}$: $\forall n^\mathbb{N} \exists x^X A(n, x) \to \exists f \forall n\, A(n, fn),$

and this failure applies to the particular cases $T = J$ and $T = K$ too. In fact, the T-translation of $AC_\mathbb{N}$ is

$AC_\mathbb{N}^T$: $\forall n T \exists x A^T(n, x) \to T \exists f \forall n A^T(n, fn),$

which is not an instance of $AC_\mathbb{N}$. In order to overcome this, the following was first observed by Spector [11] for the special case $T = K$ and $R = \perp$, where

T-shift(A) : $\forall n^\mathbb{N} TA(n) \to T \forall n A(n).$

Proposition 1. $AC_\mathbb{N}$ *and* T-shift *together imply* $AC_\mathbb{N}^T$. *Hence, the T-translation maps* $TA^\omega + AC_\mathbb{N}$ *into* $HA^\omega + AC_\mathbb{N} + T$-shift.

Proof. Applying T-shift to the premise $\forall n T \exists x A^T(n, x)$ of $AC_\mathbb{N}^T$, we deduce that $T \forall n \exists x A^T(n, x)$. Functoriality of T applied to $AC_\mathbb{N}$ with A instantiated to A^T gives that $T \forall n \exists x A^T(n, x)$ implies $T \exists f \forall n\, A^T(n, fn)$, and hence we get $T \exists f \forall n A^T(n, fn)$ by modus ponens, which is the conclusion of $AC_\mathbb{N}^T$.   □

Spector, in the context of the dialectica interpretation, showed that a form of bar recursion, now known as *Spector bar recursion*, realises the *double negation shift* (DNS), which amounts to the T-shift for $T = K$ and $R = \perp$. Moreover, via different forms of bar recursion with R a Σ_1^0 formula, it is shown in [2,3] how computational information can also be extracted via (modified) realisability from proofs in classical analysis in the presence of countable choice. But the K-shift is established only for formulae $\exists x A^K$ where A^K is in the image of the K-translation. Now notice that any formula A^K we have $\perp \to \exists x A^K$.

Proposition 2. *Over minimal logic, if* $R \to A$ *then* J-shift$(A) \to K$-shift(A).

Proof. We know that $JA \rightarrow KA$ for any A, and the assumption $R \rightarrow A$ is easily seen to give the converse, and hence $JA \leftrightarrow KA$. (Moreover, notice that if $KA \rightarrow JA$ holds then $R \rightarrow A$, and hence the assumption $R \rightarrow A$ is optimal.)

Hence the following gives an alternative way of realising the K-shift for the purposes of extracting witnesses from classical proofs with countable choice. The notions in the assumptions of the following theorem are defined in [3,12].

Theorem 1. *Assuming continuity and relativised bar induction, the infinite product of selection functions $\bigotimes$ modified-realises J-shift(A) for any A, provided the parameter R in the definition of J is a formula with a discrete type of realisers.*

We omit the proof for lack of space, but we fully prove a stronger result in Section 5. The restriction on R is needed for the infinite product to be well-defined [7], and notice that it is fulfilled if R is Σ_1^0 or a Harrop formula.

We emphasise that the above theorem states that the infinite product functional *itself* realises the shift principle. This is in contrast with the work discussed above, where the bar recursive functionals in question are *used in order to define* functionals that realise shift principles, but do not realise the shift principles themselves as they do not have the required types. We regard as rather striking the fact that a functional that was originally introduced to mimic a theorem from topology in a computational setting, as discussed in the introduction, turns out to have a natural logical reading related to traditional work in proof theory, and we think that this deserves further investigation. In summary, the J-shift turns out to be a logical analogue of the Tychonoff theorem from topology.

We now compare TA^ω and HA^ω with respect to the axiom of *dependent choice*

$$\mathsf{DC}_X \quad : \quad \forall n^{\mathbb{N}}, x^X \exists y^X A_n(x,y) \rightarrow \forall x_0 \exists \alpha(\alpha_0 = x_0 \wedge \forall n A_n(\alpha_n, \alpha_{n+1})).$$

Proposition 3. $\mathsf{DC}_{\mathbb{N}}$ *and T-shift together imply $\mathsf{DC}_{\mathbb{N}}^T$. Hence, the T-translation maps $\mathsf{TA} + \mathsf{DC}_{\mathbb{N}}$ into $\mathsf{HA} + \mathsf{DC}_{\mathbb{N}} + T$-shift.*

Proof. The argument is essentially the same as that of Proposition 1, but one applies the T-shift twice, to move T outside two numerical universal quantifiers.

In general, however, when X is a higher-type, the situation is subtler, because the T-shift will not be available for $T = J$ (let alone $T = K$). The case $T = K$ has been addressed in [2,3], and in Section 5 below we address the case $T = J$ (which has the case $T = K$ as a corollary).

Proposition 4. *The T-shift principle is equivalent to*
$$\forall n (\forall k < n\, A(k) \rightarrow TA(n)) \rightarrow T \forall n A(n),$$
which we will refer to as the course-of-values T-shift.

Proof. It is straightforward that this condition implies the T-shift. Conversely, assume $\forall n (\forall k < n\, A(k) \rightarrow TA(n))$. By the extension law $(B \rightarrow TC) \rightarrow (TB \rightarrow TC)$ of strong monads in a cartesian closed category and induction on n, we deduce that $\forall n (\forall k < n\, TA(k) \rightarrow TA(n))$. Hence $\forall n TA(n)$ by course-of-values induction, and the T-shift gives the desired result. $\qquad\square$

Theorem 2. $\mathsf{IPS}_{\langle\rangle}$ *modified-realises the course-of-values J-shift(A) for any A, provided the parameter R in the definition of J has a discrete type of realisers.*

In Section 5 we show that IPS also realises a more general logical principle that implies DC_X^J for any type X, not just $X = \mathbb{N}$ as above.

4   Extraction of Witnesses via the J-Translation

Let ML stand for *minimal logic*[1] and consider the T-elimination scheme

T-elim   : $TA \to A$.

As discussed above, for $T = J$, this is the instance $((A \to R) \to A) \to A$ of Peirce's law. If $R = \bot$, the proof system $\mathsf{ML} + J\text{-elim} + \mathsf{EFQ}$ amounts to full first-order classical logic CL. For R arbitrary, the J-translation is such that the translated instance of Peirce's law $JA \to A$ becomes provable in minimal logic. More generally:

Lemma 1. *For any strong monad T:*

1. $\mathsf{ML} \vdash TA^T \to A^T$.
2. $\mathsf{ML} + T\text{-elim} \vdash A^T \to A$.
3. $\mathsf{ML} + T\text{-elim} \vdash A$ *if and only if* $\mathsf{ML} \vdash A^T$.

These facts are well known (see e.g. [1]) and are easily proved by induction on formulae, although they are usually stated for intuitionistic logic rather than minimal logic.

Theorem 3. *Assume that $P(x,y) \to R$ and that the variable y is not free in R. If*

$\mathsf{ML} + J\text{-elim} \vdash \forall x \exists y P(x,y)$

then also $\mathsf{ML} \vdash \forall x \exists y P(x,y)$.

Proof. First notice that under the assumption $P(x,y) \to R$ we have

(i) $\mathsf{ML} \vdash JP(x,y) \to P(x,y)$,

(ii) $\mathsf{ML} \vdash J\exists y P(x,y) \to \exists y P(x,y)$.

If $\mathsf{ML} + J\text{-elim} \vdash \forall x \exists y P(x,y)$ then $\mathsf{ML} + J\text{-elim} \vdash \exists y P(x,y)$, and hence Lemma 1 gives $\mathsf{ML} \vdash J\exists y JP(x,y)$, which by (i) and (ii), implies that $\mathsf{ML} \vdash \exists y P(x,y)$.

The first part of the next proposition shows that if multiple instances of J-elimination are used in a proof, for different parameters R, one can reduce to a single instance with the conjunction of all the parameters. For example, this can be applied to the above theorem if one needs to use several instances of Peirce's law. The second part shows that the J- and K-translations coincide over intuitionistic logic.

---

[1] Intuitionistic logic without the ex-falso-quodlibet axiom scheme $\mathsf{EFQ}\colon \bot \to A$ (see e.g. [13]).

Proposition 5

1. $\mathsf{ML} + J_{R_0 \wedge R_1}$-elim $\vdash J_{R_0}$-elim $\wedge J_{R_1}$-elim.
2. *For* $R \equiv \bot$ *we have that* $\mathsf{ML} + \mathsf{EFQ} \vdash A^K \leftrightarrow A^J$.

Proof. The first part is routine verification. The second part follows from Proposition 2.

The following theorem (cf. Proposition 1 of [3]) shows how one can extract witnesses from proofs of Π^0_2-statements in classical analysis via the J-translation and the J-shift (as opposed to via the negative translation and the double negation shift).

Theorem 4. *If* $\mathsf{PA}^\omega + \mathsf{AC}_\mathbb{N} + \mathsf{DC}_\mathbb{N} \vdash \forall x^X \exists n^\mathbb{N} P(x, n)$ *then one can extract a term* t *in system* T *extended with the product functional* $\bigotimes$ *such that* $P(x, tx)$.

Proof. Write $C = \mathsf{AC}_\mathbb{N} \wedge \mathsf{DC}_\mathbb{N}$. By prefixing each atomic formula with a double negation, EFQ is eliminated. Hence the assumption of the theorem implies that $\mathsf{MA}^\omega + J_\bot$-elim $+ C \vdash \forall x \exists n \neg\neg P(x, n)$. Because the proof is in ML, we can replace $\bot$ by any formula, which we take to be R:

$$\mathsf{MA}^\omega + J_R\text{-elim} + C \vdash \forall x \exists n (P(x, n) \to R) \to R.$$

If we now take $R \equiv \exists n P(x, n)$, we conclude that $\mathsf{MA}^\omega + J_R$-elim $+ C \vdash \forall x \exists n P(x, n)$. By the J-translation we have $\mathsf{MA}^\omega + C^J \vdash \forall x J \exists n J P(x, n)$, and, by the choice of R, $\mathsf{MA}^\omega + C^J \vdash \forall x \exists n P(x, n)$. We are now done because C^J follows, in $\mathsf{MA}^\omega + C$, from J-shift, which, by Theorem 1, is realised by $\bigotimes$, and because $\mathsf{AC}_\mathbb{N}$ and $\mathsf{DC}_\mathbb{N}$, and hence C, are also realised.

5   Full Dependent Choice

Recall that it has been standard to interpret the axiom of *countable* choice computationally by reducing it to the computational interpretation of the double negation shift (cf. [2,3,11] and Theorem 4 above). When it comes to the computational interpretation of the *dependent* choice, one normally does it directly, as it is seems not possible to reduce the negative translation of DC using the simple double negation shift. In this section, continuing the discussion started in Section 3, we show that what is needed in order to approach this from a logical point of view is a *dependent* variant of the shift principle. The binary version of the course-of-values J-shift (cf. Proposition 4) is

$$JA \wedge (A \to JB) \to J(A \wedge B).$$

Now, let us consider a dependent version of this, where JB depends on the witness for A:

$$J\exists x A(x) \wedge \forall x \in A \, J \exists y B(x, y) \to J \exists x, y (A(x) \wedge B(x, y)).$$

Using finite sequences, this can be generalised to an arbitrary finite number of predicates:

$$\bigwedge_{i=0}^{n} \forall s \in (\bigwedge_{j=0}^{i-1} A_j) \, J \exists x_i A_i(s * x_i) \rightarrow J \exists t \bigwedge_{i=0}^{n} A_i(t_0, \ldots, t_i),$$

where $\bigwedge_{i=0}^{n}$ and $\bigwedge_{j=0}^{i-1}$ stand for bounded universal quantifications. Generalising this further to the case of infinitely many predicates, we have

$$J^d\text{-shift} \quad : \quad \forall s \in (\bigwedge_{j=0}^{|s|-1} A_j) \, J \exists x A_{|s|}(s * x) \rightarrow J \exists \alpha \forall n A_n([\alpha](n+1)),$$

which we call the *dependent J-shift*. Note that $[\alpha](n)$ stands for the initial segment of the infinite sequence α of length n, i.e. $[\alpha](n) = \langle \alpha(0), \alpha(1), \ldots, \alpha(n-1)\rangle$.

We show now how the computational content of (the classical) dependent choice can be reduced to that of the dependent J-shift. More precisely, the dependent J-shift together with dependent choice proves the J-translation of dependent choice (as with countable choice). We use the following variant of *dependent choice* based on finite sequences [10, Section 2.3]:

$$\mathsf{SDC} \; : \; \forall s (\forall j < |s| \, A_j([s](j)) \rightarrow \exists x A_{|s|}(s * x)) \rightarrow \exists \alpha \forall n A_n([\alpha](n+1)).$$

Lemma 2. *The following are provable in* MA^ω:

1. $\mathsf{SDC} \vdash \mathsf{DC}$.

2. $\mathsf{SDC} + J^d\text{-shift} \vdash \mathsf{SDC}^J$.

Theorem 5. *Let R be a Σ_1^0-formula. Assuming continuity and relativised bar induction, $\mathsf{IPS}_{\langle\rangle}$ modified-realises J^d-shift.*

Proof. Assume the realiser for $\exists y^{Y_n} A_n(s * y)$ has type $X_n(s) \equiv Y_n \times Z_n(s)$. Also, assume

$$\varepsilon_s \; \mathrm{mr} \; \forall j < |s| \, A_j([s](j)) \rightarrow J \exists y A_{|s|}(s * y)$$
$$q \; \mathrm{mr} \; \exists \alpha \forall n A_n([\alpha](n+1)) \rightarrow R.$$

Then ε_s and q have types

$$\Pi_{j < |s|} \Pi_{z_j \in Z_j(s_0, \ldots, s_j)} J X_{|s|}(s) \quad \text{and} \quad \Sigma_{\alpha \in \Pi_k Y_k} \Pi_n Z_n([\alpha](n+1)) \rightarrow R,$$

respectively. We prove $\mathsf{IPS}_{\langle\rangle}(\varepsilon)(q) \, \mathrm{mr} \, \exists \alpha \forall n A_n([\alpha](n+1))$. We proceed by relativised bar induction (cf. [3]) to prove $\forall s P(s)$, where

$$P(s) \equiv \mathsf{IPS}_s(\varepsilon)(q_s) \, \mathrm{mr} \, \exists \alpha \forall n \, A_{|s|+n}(s^0 * [\alpha](n+1)).$$

Note that s is a finite sequence of pairs. We write s^0 (respectively, s^1) for the finite sequence consisting of the first (respectively, second) component of each pair. The bar induction will be relativised to the predicate

$$R(s) \equiv \forall j < |s| \, (s_j^1 \, \mathrm{mr} \, A_j(\langle s_0^0, \ldots, s_{j-1}^0\rangle)).$$

We now prove the two hypothesis (i) and (ii) of the bar induction.

(i) $\forall \alpha^R \exists k P([\alpha](k))$. Given α, pick k to be a point of continuity of q on α. We must show $P([\alpha](k))$, i.e.

$$\mathsf{IPS}_{[\alpha](k)}(\varepsilon)(q_{[\alpha](k)}) \text{ mr } \exists \beta \forall n\, A_{k+n}(([\alpha](k))^0 * [\beta](n+1)).$$

Abbreviate $\gamma, \delta \equiv \mathsf{IPS}_{[\alpha](k)}(\varepsilon)(q_{[\alpha](k)})$. The above follows from, for all n,

$$\delta(n) \text{ mr } A_{k+n}(([\alpha](k))^0 * [\gamma](n+1)).$$

Unfolding the definition of IPS, this is equivalent to, for all n,

$$(\varepsilon_{[\alpha](k)*r}(\lambda x.q_{[\alpha](k)*r*x}(\mathsf{IPS}_{[\alpha](k)*r*x}(\varepsilon)(q_{[\alpha](k)*r*x}))))_1 \text{ mr } A_{k+n}(([\alpha](k))^0 * [\gamma](n+1)),$$

where $r = \mathsf{IPS}_{[\alpha](k)}(\varepsilon)(q_{[\alpha](k)})[k, k+n-1]$ (computed by course-of-values). By the fact that k is a point of continuity of q on α, this is equivalent to

$$(\varepsilon_{[\alpha](k)*r}(\lambda x.q_{[\alpha](k)*r*x}(\mathbf{0})))_1 \text{ mr } A_{k+n}(([\alpha](k))^0 * [\gamma](n+1)).$$

By course-of-values induction $[\alpha](k) * r \in R$, Hence, by the assumption on ε it remains to show that

$$\lambda x.q_{[\alpha](k)*r*x}(\mathbf{0}) \text{ mr } \exists y A_{k+n}(([\alpha](k))^0 * [\gamma](n) * y) \to R$$

which follows from the assumptions on q, and that $\alpha \in R$.

(ii) $\forall s^R(\forall t, x(R(s*t*x) \to P(s*t*x)) \to P(s))$. Let $s \in R$ be given, and assume
(1) $\forall t, x(R(s*t*x) \to P(s*t*x))$. We must show $P(s)$, i.e.

$$\mathsf{IPS}_s(\varepsilon)(q_s) \text{ mr } \exists \alpha \forall n A_{|s|+n}(s^0 * [\alpha](n+1)).$$

Again let $\gamma, \delta \equiv \mathsf{IPS}_s(\varepsilon)(q_s)$. Then, $P(s)$ follows from

$$(\mathsf{IPS}_s(\varepsilon)(q_s)(n))_1 \text{ mr } A_{|s|+n}(s^0 * [\gamma](n+1)),$$

which, by the definition of IPS is

$$(\varepsilon_{s*r}(\lambda x.q_{s*r*x}(\mathsf{IPS}_{s*r*x}(\varepsilon)(q_{s*r*x}))))_1 \text{ mr } A_{|s|+n}(s^0 * [\gamma](n+1)),$$

where $r = \mathsf{IPS}_s(\varepsilon)(q_s)[|s|, |s|+n-1]$. This follows from

$$(2)\ \lambda x.q_{s*r*x}(\mathsf{IPS}_{s*r*x}(\varepsilon)(q_{s*r*x})) \text{ mr } \exists x_n A_{|s|+n}(s^0 * [\gamma](n) * x_n) \to R.$$

Now, assume x is such that $R(s*r*x)$. Then, by (1) we have, $P(s*r*x)$, i.e.

$$(3)\ \mathsf{IPS}_{s*r*x}(\varepsilon)(q_{s*r*x}) \text{ mr } \exists \alpha \forall n A_{|s*r*x|+n}((s*r*x)^0 * [\alpha](n+1)).$$

By the assumption on q we have that (3) implies (2), which concludes the proof.

Corollary 1. *If* $\mathsf{PA}^\omega + \mathsf{AC_N} + \mathsf{SDC} \vdash \forall x^X \exists n^{\mathbb{N}} P(x,n)$ *then one can extract a term* t *in system* T *extended with* $\bigotimes$ *such that* $P(x, tx)$.

Proof. $\mathsf{AC_N}$ and SDC are modified-realizable in system T. The result follows because $\bigotimes$ is inter-definable with IPS (cf. [6]), and hence the J^d-shift is modified-realizable in $T + \bigotimes$.

6   Concluding Remarks

We have developed a proof translation based on the selection monad J, and shown how to realise a corresponding J-shift principle, which is more general than the double negation shift. We plan to investigate the use of the product of selection functions $\bigotimes$ for extraction of computational content from proofs involving countable/dependent choice, as done by Seisenber [10] with modified bar recursion. Based on the experimental results and theoretical conjectures of [4] and [5, Section 8.10], we wish to investigate whether $\bigotimes$ would give rise to more efficient computational extraction of witnesses.

Acknowledgements. The second author gratefully acknowledges support of the Royal Society (grant 516002.K501/RH/kk).

References

1. Aczel, P.: The Russell-Prawitz modality. Math. Structures Comput. Sci. 11(4), 541–554 (2001); Modalities in type theory, Trento (1999)
2. Berardi, S., Bezem, M., Coquand, T.: On the computational content of the axiom of choice. The Journal of Symbolic Logic 63(2), 600–622 (1998)
3. Berger, U., Oliva, P.: Modified bar recursion and classical dependent choice. Lecture Notes in Logic 20, 89–107 (2005)
4. Escardó, M.H.: Infinite sets that admit fast exhaustive search. In: Proceedings of LICS, pp. 443–452 (2007)
5. Escardó, M.H.: Exhaustible sets in higher-type computation. Logical Methods in Computer Science 4(3) (2008)
6. Escardó, M.H., Oliva, P.: Computational interpretations of analysis via products of selection functions. In: Ferreira, F., Lowe, B., Mayordomo, E., Gomes, L.M. (eds.) CiE 2010. LNCS, vol. 6158, pp. 141–150. Springer, Heidelberg (2010)
7. Escardó, M.H., Oliva, P.: Selection functions, bar recursion, and backward induction. Mathematical Structures in Computer Science 20(2), 127–168 (2010)
8. Friedman, H.: Classically and intuitionistically provably recursive functions. In: Scott, D., Müller, G. (eds.) Higher Set Theory. LNM, vol. 669, pp. 21–28. Springer, Berlin (1978)
9. Griffin, T.G.: A formulae-as-types notion of control. In: 17th Annual ACM Symp. on Principles of Programming Languages, POPL 1990, San Francisco, CA, USA, pp. 17–19 (1990)
10. Seisenberger, M.: Programs from proofs using classical dependent choice. Annals of Pure and Applied Logic 153(1-3), 97–110 (2008)
11. Spector, C.: Provably recursive functionals of analysis: a consistency proof of analysis by an extension of principles in current intuitionistic mathematics. In: Dekker, F.D.E. (ed.) Recursive Function Theory: Proc. Symposia in Pure Mathematics, vol. 5, pp. 1–27. American Mathematical Society, Providence (1962)
12. Troelstra, A.S.: Metamathematical Investigation of Intuitionistic Arithmetic and Analysis. LNM, vol. 344. Springer, Berlin (1973)
13. Troelstra, A.S., Schwichtenberg, H.: Basic Proof Theory, 2nd edn. Cambridge University Press, Cambridge (2000)

Counting the Changes of Random Δ_2^0 Sets

Santiago Figueira[1,*], Denis Hirschfeldt[2,**], Joseph S. Miller[3,***],
Keng Meng Ng[3], and André Nies[4,†]

[1] Dept. of Computer Science, FCEyN, University of Buenos Aires and CONICET
[2] Dept. of Mathematics, The University of Chicago
[3] Dept. of Mathematics, University of Wisconsin—Madison
[4] Dept. of Computer Science, The University of Auckland

Abstract. Consider a Martin-Löf random Δ_2^0 set Z. We give lower bounds for the number of changes of $Z_s \restriction_n$ for computable approximations of Z. We show that each nonempty Π_1^0 class has a low member Z with a computable approximation that changes only $o(2^n)$ times. We prove that each superlow ML-random set already satisfies a stronger randomness notion called balanced randomness, which implies that for each computable approximation and each constant c, there are infinitely many n such that $Z_s \restriction_n$ changes more than $c2^n$ times.

1   Introduction

A *computable approximation* of a set $Z \subseteq \mathbb{N}$ is a computable sequence $(Z_s)_{s \in \mathbb{N}}$ of finite sets such that $Z(x) = \lim_s Z_s(x)$ for each x. The Shoenfield Limit Lemma states that a set $Z \subseteq \mathbb{N}$ is Δ_2^0 iff Z has a computable approximation.

In Sections 3 to 5 we are interested in the number of changes of $Z_s \restriction_n$ for computable approximations of a Martin-Löf random Δ_2^0 set Z. We give some lower bounds. Next, we obtain a hierarchy theorem saying that allowing more changes yields new ω-c.e. ML-random sets. Thereafter, we prove the "$o(2^n)$ changes" low basis theorem which says that each nonempty Π_1^0 class has a low member Z with a computable approximation such that $Z \restriction_n$, the initial segment of length n, changes only $o(2^n)$ times. We conclude that there is a computable approximation of a low ML-random set which changes only $o(2^n)$ times.

In [1, Sect. 8.6] evidence was presented for the following thesis: Among ML-random sets, being computationally less complex is equivalent to being more

---

* Figueira was partially supported by UBA (grant UBACyT X615) and CONICET (grant PIP 370).

** Hirschfeldt was partially supported by the National Science Foundation of the United States, grants DMS-0801033 and DMS-0652521.

*** Miller was partially supported by the National Science Foundation of the United States under grants DMS-0945187 and DMS-0946325, the latter being part of a Focused Research Group in Algorithmic Randomness.

† Nies was partially supported by the Marsden Fund of New Zealand, grant no. 08-UOA-187.

F. Ferreira et al. (Eds.): CiE 2010, LNCS 6158, pp. 162–171, 2010.

random. For instance, a ML-random set forms a minimal pair with $\emptyset'$ iff it is weakly 2-random. In the final section we use the foregoing results to give some more evidence for this when the ML-random set is Δ_2^0.

To specify what we mean by being more random, we consider variants of Demuth randomness, a notion that strengthens ML-randomness but is still compatible with being Δ_2^0. Demuth tests (see [1, Def. 3.6.24]) generalize Martin-Löf tests $(G_m)_{m \in \mathbb{N}}$ in that one can exchange the m-th component a computably bounded number of times. A set $Z \subseteq \mathbb{N}$ passes a Demuth test if Z is in only finitely many final versions of the G_m.

The passing condition that Z is not in at least one of the G_m yields weak Demuth randomness. In this case, we can require as well that $G_m \supseteq G_{m+1}$ for each m, since we can replace G_m by $\bigcap_{i \le m} G_i$ if necessary. A test with this property will be called *monotonic*. Note that the number of version changes is still computably bounded. Thus Z is weakly Demuth random iff it passes all monotonic Demuth tests (where passing the test can be taken in either sense).

We introduce balanced randomness, an even more restricted form of weak Demuth randomness where the bound on the number of changes of the m-th version is $O(2^m)$. Each balanced random set is ML-random and Turing incomplete.

For evidence of the direction from left to right in the thesis above, we show that a ML-random set that is superlow is already balanced random. Being ω-c.e. tracing is a highness property due to Greenberg and Nies [2] that is incompatible with superlowness (see Sect. 6). In fact we show that a ML-random set that is not ω-c.e. tracing is already balanced random.

Evidence for the direction from right to left in the thesis above is given by the fact that a Demuth random set bounds only generalized low$_1$ sets, and the result of [3] that a c.e. set Turing below a Demuth random set must be strongly jump-traceable. In [3] further evidence for this direction is given by showing that a weakly Demuth random set Z is not superhigh, namely, $Z' \not\ge_{tt} \emptyset''$. (However, it can be high.) We conjecture that a balanced random set is not LR-complete, and prove a result in that direction.

2   Counting Changes of a Δ_2^0 Set

For a computable approximation $(Z_s)_{s \in \mathbb{N}}$, unless otherwise stated, we will assume that $Z_s(x) = 0$ for each $x \ge s$. Given such an approximation, for a number n and a stage number $s > 0$, to say that $Z \upharpoonright_n$ *changes* at stage s means that $Z_s \upharpoonright_n \ne Z_{s-1} \upharpoonright_n$.

When we say that we bound the number of changes for a Δ_2^0 set Z from above, we mean that the changes of *some* approximation can be bounded from above.

Definition 1. Let $f \colon \mathbb{N} \to \mathbb{N}$. We say a set $Z \subseteq \mathbb{N}$ is f-c.e. if there is a computable approximation $(Z_s)_{s \in \mathbb{N}}$ of Z such that for each n, $Z_s \upharpoonright_n$ changes at most $f(n)$ times via this approximation. Terminology such as $O(f)$-c.e. set, $o(f)$-c.e. set and so on has the obvious meaning. For instance, Z is $o(f)$-c.e. if there is a function $g \in o(f)$ such that Z is g-c.e.

Each left-c.e. set is $o(2^n)$-c.e.:

Fact 2. *Let Z be a left-c.e. set as shown by the computable approximation $(Z_s)_{s \in \mathbb{N}}$. Then Z is $o(2^n)$-c.e. via this computable approximation.*

Proof. Given k, let t be the least stage such that $Z_t \upharpoonright_{k+1}$ has the final value. Let $n \geq t + k + 1$. By our convention that $Z_s(x) = 0$ for each $x \geq s$, $Z \upharpoonright_n$ changes at no more than $2^t \leq 2^{n-k-1}$ stages that are $\leq t$. Furthermore, since the approximation cannot return to previous states, $Z \upharpoonright_n$ changes at no more than 2^{n-k-1} stages that are greater than t. Thus $Z \upharpoonright_n$ changes at no more than 2^{n-k} stages.   □

Actually, the fact still holds if we require only that the approximation to $Z \upharpoonright_n$ can never return to a previous value.

3   Some Lower Bounds on the Number of Changes of a ML-Random Set

In this section we assume that Z is a ML-random Δ_2^0 set with a fixed computable approximation $(Z_s)_{s \in \mathbb{N}}$. We give some lower bounds on the number of times $Z \upharpoonright_n$ can change. We confirm the intuition that the number of changes cannot be far below 2^n.

First we look at computable functions bounding the number of changes of $Z \upharpoonright_n$ for only infinitely many n.

Proposition 3. *Let $q : \mathbb{N} \to \mathbb{Q}^+$ be computable. If $Z \upharpoonright_n$ changes fewer than $\lfloor 2^n q(n) \rfloor$ times for infinitely many n, then $\sum_n q(n) = \infty$.*

Proof. Assume for contradiction that $\sum_n q(n) < \infty$. We define an effective sequence $(\mathcal{S}_i)_{i \in \mathbb{N}}$ of Σ_1^0 classes in the following way. For each n, we put into $\mathcal{S}_n$ the first $\lfloor 2^n q(n) \rfloor$ versions of $[Z \upharpoonright_n]$. Clearly $(\mathcal{S}_i)_{i \in \mathbb{N}}$ is a sequence of uniformly c.e. open sets and $\lambda \mathcal{S}_n \leq q(n)$ for all n, where λ is Lebesgue measure. Thus $(\mathcal{S}_i)_{i \in \mathbb{N}}$ is a Solovay test. By hypothesis $Z \in \mathcal{S}_n$ for infinitely many n. This means that Z fails the test $(\mathcal{S}_i)_{i \in \mathbb{N}}$ and therefore is not ML-random.   □

Example 4. $Z \upharpoonright_n$ changes at least $2^n n^{-2}$ times for almost every n.

The proof of the foregoing proposition can easily be extended to the case that the function q is effectively approximable from below, that is, $q(n) = \sup_s q_s(n)$ for an effective sequence of rationals that is nondecreasing in s. For instance, we can let $q(n) = 2^{-K(n)}$, where K is prefix-free Kolmogorov complexity. Thus, in the example above, in fact we have a lower bound of $2^{n-K(n)}$.

If for almost every n the number of changes of $Z \upharpoonright_n$ is bounded above by $2^n q(n)$, then the function q is in fact bounded away from 0.

Proposition 5. *Let $q : \mathbb{N} \to \mathbb{Q}^+$ be computable. If $Z \upharpoonright_n$ changes fewer than $\lfloor 2^n q(n) \rfloor$ times for almost every n, then $\inf_n q(n) > 0$.*

Proof. Let n^* be a number such that the bound holds from n^* on. Assume for a contradiction that $\inf_n q(n) = 0$. We show that $\exists^\infty n \, K(Z \upharpoonright_n) \leq^+ n$, contrary to the assumption that Z is ML-random. To do so we build a bounded request (aka Kraft-Chaitin) set L. Let $(n_i)_{i>0}$ be a computable sequence of numbers greater than n^* such that $q(n_i) < 2^{-i}$ for each i. For each s, we put the request

$$\langle n_i, Z_s \upharpoonright_{n_i} \rangle$$

into L. For each $i > 0$, the weight put into L is at most $2^{-n_i} 2^{n_i} q(n_i) \leq 2^{-i}$. Thus L is a bounded request set. Hence by the usual machine existence theorem (aka Kraft-Chaitin Theorem), we have $\exists^\infty n \, K(Z \upharpoonright_n) \leq^+ n$ as required.   □

The proof of the foregoing proposition can easily be extended to the case that the function q is effectively approximable from *above*. For each i, we can search for an s and an n_i such that $q_s(n_i) < 2^{-i}$.

We remark that neither Proposition 3 nor Proposition 5 can be extended to reals of positive effective Hausdorff dimension. It is easy to construct a real Z with effective Hausdorff dimension 1, and a computable function q such that $\sum_n q(n) < \infty$, and $Z \upharpoonright_n$ changes fewer than $\lfloor 2^n q(n) \rfloor$ times for almost every n.

It is natural to ask what else we can say about the number of times $Z \upharpoonright_n$ can change for a Δ_2^0 ML-random Z. In particular, we consider strengthening Propositions 3 and 5 simultaneously: whenever $Z \upharpoonright_n$ changes fewer than $\lfloor 2^n q(n) \rfloor$ times for infinitely many n, then $q(n)$ is bounded away from zero on these n. By the following proposition this is true if q is a computable nonincreasing function, but by Corollary 11 this fails in general.

Proposition 6. *Let* $q : \mathbb{N} \to \mathbb{Q}^+$ *be computable and nonincreasing. If* $Z \upharpoonright_n$ *changes fewer than* $\lfloor 2^n q(n) \rfloor$ *times for infinitely many* n, *then* $\inf_n q(n) > 0$.

Proof. Suppose the contrary, that $\inf_n q(n) = 0$. Let $(n_i)_{i \in \mathbb{N}}$ be a computable sequence of natural numbers such that for every i, n_i is the least number larger than n_{i-1} such that $q(n_i) < 2^{-i-1}$. We build a Solovay test $(\mathcal{S}_i)_{i \in \mathbb{N}}$ by the following. For each i enumerate into $\mathcal{S}_i$ the first $2^{n_i - i}$ different versions of $[Z \upharpoonright_{n_i}]$. Then $\lambda \mathcal{S}_i \leq 2^{-i}$ for every i. Since Z is ML-random and $Z \upharpoonright_n$ changes fewer than $\lfloor 2^n q(n) \rfloor$ times for infinitely many n, we fix $m > n_0$ and $i > 0$ such that $Z \upharpoonright_m$ changes fewer than $\lfloor 2^m q(m) \rfloor$ times, $Z \notin \mathcal{S}_i$ and i is the least such that $n_i \geq m$. Since $Z \notin \mathcal{S}_i$, there must be at least $2^{n_i - i} + 1$ many distinct elements in the set $\{Z_s \upharpoonright_{n_i} : s \in \mathbb{N}\}$. Now since $n_{i-1} < m$ we have $q(m) \leq q(n_{i-1}) < 2^{-i}$. Hence $Z \upharpoonright_m$ changes fewer than 2^{m-i} times. This is a contradiction.   □

4   A Hierarchy Theorem for ML-Random ω-c.e. Sets

Using a method of Kučera one can code a given set into a path on a Π_1^0 class of positive measure. The method rests on the following lemma (see [1, Lem. 3.3.1]), where $\lambda(\mathcal{C}|x)$ denotes $2^{|x|} \lambda(\mathcal{C} \cap [x])$.

Lemma 7. *Let $C \subseteq 2^\omega$ be measurable and $\lambda(C|x) \geq 2^{-(r+1)}$. Then for every $n \geq |x| + r + 2$ there are distinct strings $y_0, y_1 \succ x$ with $|y_i| = n$ such that $\lambda(C|y_i) > 2^{-(r+2)}$ for $i = 0, 1$.*

An *order function* is a nondecreasing unbounded computable function.

Theorem 8. *Let b be an order function such that $\forall n\, b(n) \geq \epsilon 2^n$ for some positive real ϵ. Then for each order function s there is a ML-random Z which is $s \cdot b$-c.e. but not b-c.e.*

We can restate Proposition 5 as follows: if the ML-random set Z is b-c.e. for some computable function b, then there is $\epsilon > 0$ such that $\forall n\, b(n) \geq \epsilon 2^n$. This shows that the additional hypothesis $\forall n\, b(n) \geq \epsilon 2^n$ in this hierarchy theorem does not restrict its generality.

Proof (Theorem 8). The idea is the following. To make Z ML-random, we ensure that it belongs to an appropriate Π^0_1-class. To make Z non b-c.e., let $(f_e)_{e \in \mathbb{N}^+}$ be an enumeration of all total computable functions f mapping pairs of natural numbers to strings such that for all n, $\#\{t : f(n,t) \neq f(n, t+1)\} \leq b(n)$, $|f(n,t)| = n$, and $f(n,t) \prec f(n+1, t)$. Each such f is the approximation of some b-c.e. set. Conversely, if a set is b-c.e. then there is some f giving the set in the limit. Thus it suffices to ensure that for every e there is an n such that $\lim_t f_e(n,t) \neq Z{\restriction}_n$.

Here are the details. Recall that s is the given order function. Choose a computable sequence $(n_e)_{e \in \mathbb{N}^+}$ such that $n_1 = 0$,

$$s(n_e) > e + 1/\epsilon, \quad \text{and} \quad n_{e+1} \geq n_e + e + 2.$$

Let $\mathcal{P}$ be a Π^0_1-class such that $\mathcal{P} \subseteq \mathsf{MLR}$, where MLR is the class of ML-random sets, and $\lambda\mathcal{P} > 1/2$. Let $\hat{\mathcal{P}}$ be the Π^0_1 class of paths through the Π^0_1 tree

$$T = \{y : (\forall i)[n_i \leq |y| \to \lambda(\mathcal{P}|(y{\restriction}_{n_i})) \geq 2^{-(i+1)}]\}.$$

Note that $\hat{\mathcal{P}} \subseteq \mathcal{P}$. Since $\lambda\mathcal{P} \geq 1/2$, by Lemma 7, $\hat{\mathcal{P}}$ is nonempty.

We define $z_0 \prec z_1 \prec z_2 \prec \cdots$ in such a way that $|z_e| = n_e$ and $z_e \neq \lim_t f_e(n_e, t)$. We also define $Z = \bigcap_e [z_e]^\prec$. In this way, we ensure that for all $e \geq 1$, $Z{\restriction}_{n_e} \neq \lim_t f_e(n_e, t)$ and therefore Z is not b-c.e. At the same time, we ensure that $Z \in \hat{\mathcal{P}}$, and hence Z is ML-random.

The definition of z_e proceeds by steps. Let $z_{0,s} = \emptyset$ and for $e > 0$ let

$$z_{e+1,s} = \min\{[z] \subseteq \hat{\mathcal{P}}_s : |z| = n_{e+1} \wedge z \succ z_{e,s} \wedge f_e(n_e, s) \neq z\}. \quad (1)$$

Recall that each Π^0_1 class $\mathcal{P}$ has an effective approximation by descending clopen sets $\mathcal{P}_s$; see [1, Sect. 1.8].

Suppose $z_{e,s}$ has already been defined. By Lemma 7 and the definition of $\hat{\mathcal{P}}$, there are two distinct strings $y_0, y_1 \succ z_{e,s}$ such that $|y_i| = n_e$ and $[y_i] \cap \hat{\mathcal{P}} \neq \emptyset$. Hence $z_{e+1,s}$ is well defined in equation (1).

To show that Z is $s \cdot b$-c.e., define a computable approximation $(Z_i)_{i \in \mathbb{N}^+}$ with $Z_s = z_{s,s}$. Suppose $n_e \leq n < n_{e+1}$.

If $Z_{s+1}\restriction_n \neq Z_s \restriction_n$ then

$$[Z_s \restriction_n] \not\subseteq \hat{\mathcal{P}}_{s+1} \text{ or } \exists i \leq e \, f_i(n, s+1) \neq f_i(n, s).$$

The former may occur at most 2^n many times, and the latter at most $e \cdot b(n)$ times. For all $e \geq 1$, the number of changes of $Z \restriction_n$ is at most

$$2^n + e \cdot b(n) \leq b(n)/\epsilon + e \cdot b(n)$$
$$\leq b(n)(e + 1/\epsilon)$$
$$\leq b(n) \cdot s(n_e) \leq b(n) \cdot s(n).$$

5 Counting Changes for Sets Given by the (Super)Low Basis Theorem

The low basis theorem of Jockusch and Soare [4] says that every nonempty Π_1^0 class has a member Z that is low, that is, $Z' \leq_T \emptyset'$. The proof actually makes Z *superlow*, that is, $Z' \leq_{tt} \emptyset'$. Here we study possible bounds on the number of changes for a low member of the class. Surprisingly, we find that to make the member superlow will in general take more changes, not fewer.

Theorem 9. *Let $\mathcal{P}$ be a nonempty Π_1^0 class. For each order function h, the class $\mathcal{P}$ has a superlow $2^{n+h(n)}$-c.e. member.*

Proof. The idea is to run the proof of the superlow basis theorem with a c.e. operator W^X that codes X' only at a sparse set of positions, and simply copies X for the other bit positions. Let R be the infinite computable set $\{n\colon h(n+1) > h(n)\}$. Define the c.e. operator W by

$$W^X(n) = \begin{cases} X(i) & \text{if } n \text{ is the } i\text{-th smallest element in } \mathbb{N} - R \\ X'(j) & \text{if } n \text{ is the } j\text{-th smallest element in } R \end{cases} \tag{2}$$

By the proof of the superlow basis theorem as in [1, Thm. 1.8.38], there is a $Z \in \mathcal{P}$ such that $B = W^Z$ is left-c.e. via some approximation (B_s). Let Z_s be the computable approximation of Z given by $Z_s(i) = B_s(n)$ where n is the i-th smallest element in $\mathbb{N} - R$. If $Z_s \restriction_n$ changes then $B_s \restriction_{n+h(n)}$ changes. Thus $Z_s \restriction_n$ changes at most $2^{n+h(n)}$ times. Furthermore, $Z' \leq_m B$. Since B is ω-c.e. we have $B \leq_{tt} \emptyset'$, so Z is superlow. $\square$

Theorem 18 below shows that if $\mathcal{P} \subseteq \mathsf{MLR}$, no superlow member can be $O(2^n)$-c.e. On the other hand, if we merely want a low member, we can actually get away with $o(2^n)$ changes. For the case $\mathcal{P} \subseteq \mathsf{MLR}$, this shows that $o(2^n)$-c.e. ML-random sets can be very different from the Turing complete ML-random set Ω, even though Ω is also $o(2^n)$-c.e. by Fact 2.

Theorem 10. *Each nonempty Π_1^0 class $\mathcal{P}$ contains a low $o(2^n)$-c.e. member.*

Proof. We combine the construction in the proof of Theorem 9 with a dynamic coding of the jump. At each stage we have movable markers γ_k at the positions where $X'(k)$ is currently coded. Thus, the positions where X' is coded become sparser and sparser as the construction proceeds.

Construction. At stage 0 let $\gamma_{0,0} = 1$ and B_0 be the empty set.

Stage $t > 0$.

(i). Let $W^X[t]$ be the c.e. operator such that

$$
W^X[t](v) = \begin{cases} X(i) & \text{if } v \text{ is the } i\text{-th smallest element} \\ & \text{not of the form } \gamma_{k,t-1} \\ X'(k) & \text{if } v = \gamma_{k,t-1}. \end{cases} \tag{3}
$$

We define a sequence of Π^0_1 classes $\mathcal{Q}_n[t]$ ($n \in \mathbb{N}$) according to the proof of the low basis theorem as in [1, Thm. 1.8.38], but at stage t we use the operator $W[t]$ instead of the jump operator.

Let $\mathcal{Q}_0[t] = \mathcal{P}$. If $\mathcal{Q}_n[t]$ has been defined, let

$$
\mathcal{Q}_{n+1}[t] = \begin{cases} \mathcal{Q}_n[t] & \text{if for all } X \in \mathcal{Q}_{n,t}[t], \\ & \text{we have } n \in W^X[t] \\ \{X \in \mathcal{Q}_n[t] : n \notin W^X[t]\} & \text{otherwise.} \end{cases}
$$

In the first case, define $B_t(n) = 1$; in the second case, define $B_t(n) = 0$.

(ii). Let k be least such that $k = t$ or $B_t \upharpoonright 2k \neq B_{t-1} \upharpoonright 2k$. Define $\gamma_{r,t} = \gamma_{r,t-1}$ for $r < k$, and $\gamma_{r,t} = t + 2r$ for $t \geq r \geq k$.

Verification

Claim 1. *B is left-c.e. via the computable approximation $(B_t)_{t \in \mathbb{N}}$.*

Suppose i is least such that $B_t(i) \neq B_{t-1}(i)$. Since $\gamma_{r,t-1} > 2r$ for each r, this implies that $\gamma_{r,t} = \gamma_{r,t-1}$ for all r such that $\gamma_{r,t-1} \leq i$. Thus the construction up to $\mathcal{Q}_i[t]$ behaves like the usual construction to prove the low basis theorem, whence we have $B_{t-1}(i) = 0$ and $B_t(i) = 1$.

We conclude that $\gamma_k = \lim_t \gamma_{k,t}$ exists for each k, and therefore $\mathcal{Q}_n = \lim_t \mathcal{Q}_n[t]$ exists as well.

By the compactness of 2^ω there is $Z \in \bigcap_n \mathcal{Q}_n$. Clearly Z is low because $Z'(k) = B(\gamma_k)$ and the expression on the right can be evaluated by $\emptyset'$. It remains to show the following.

Claim 2. *Z is $o(2^n)$-c.e.*

We have a computable approximation to Z given by

$Z_t(i) = B_t(v)$ where v is the i-th smallest number not of the form $\gamma_{k,t}$.

Given n let k be largest such that $\gamma_k \leq n$. We show that $Z \upharpoonright n$ changes at most 2^{n-k+1} times.

For $n \geq r \geq k$ let t_r be least stage t such that $\gamma_{r+1,t} > n$. Then $B_t \upharpoonright 2r$ is stable for $t_r \leq t < t_{r+1}$. Since $(B_t)_{t \in \mathbb{N}}$ is a computable approximation via which B is left-c.e., $B \upharpoonright_{n+r}$ changes at most 2^{n-r} times for $t \in [t_r, t_{r+1})$. Hence $Z \upharpoonright_n$ changes at most 2^{n-r} times for such t. The total number of changes is therefore bounded by $\sum_{k \leq r \leq n} 2^{n-r} < 2^{n-k+1}$.   □

Corollary 11. *There is a ML-random Z and a computable $q : \mathbb{N} \rightarrow \mathbb{Q}^+$ such that $Z \upharpoonright_n$ changes fewer than $\lfloor 2^n q(n) \rfloor$ times for infinitely many n, and $\lim_n q(n) = 0$.*

Proof. Follow the proof of Theorem 10, and let $\mathcal{P}$ be a Π_1^0 class containing only ML-randoms. We define $q(m) = 2^{-r+1}$, where r is the least such that $\gamma_{r,m} \geq m$. Then q is computable and $\lim_n q(n) = 0$ because each marker reaches a limit. Also, $q(\gamma_r) = 2^{-r+1}$ for every r. By the proof of Theorem 10, for every r, $Z \upharpoonright_{\gamma_r}$ changes fewer than $2^{\gamma_r - r+1} = \lfloor 2^{\gamma_r} q(\gamma_r) \rfloor$ times.   □

6   Balanced Randomness

Basics on balanced randomness. We study a more restricted form of weak Demuth randomness (which was defined in the introduction). The bound on the number of changes of the m-th version is now $O(2^m)$.

Definition 12. A *balanced test* is a sequence of c.e. open sets $(G_m)_{m \in \mathbb{N}}$ such that $\forall m \, \lambda G_m \leq 2^{-m}$; furthermore, there is a function f such that G_m equals the Σ_1^0 class $[W_{f(m)}]^{\prec}$ and $f(m) = \lim_s g(m, s)$ for a computable function g such that the function mapping m to the size of the set $\{s : g(m, s) \neq g(m, s-1)\}$ is in $O(2^m)$.

A set Z passes *the test if $Z \notin G_m$ for some m. We call Z balanced random *if it passes each balanced test.*

We denote $[W_{g(m,s)}]^{\prec}$ by $G_m[s]$ and call it the *version of G_m at stage s.*

Example 13. No $O(2^n)$-c.e. set is balanced random.

To see this, simply let $G_m[s] = [Z_s \upharpoonright_m]$; then Z fails the balanced test $(G_m)_{m \in \mathbb{N}}$.

Again, we may monotonize a test and thus assume $G_m \supseteq G_{m+1}$ for each m, because the number of changes of $\bigcap_{i \leq m} G_i[s]$ is also $O(2^m)$.

Let $(\alpha_i)_{i \in \mathbb{N}}$ be a nonincreasing computable sequence of rationals that converges effectively to 0, for instance $\alpha_i = 1/i$. If we build monotonicity into the definition of balanced randomness, we can replace the bound 2^{-m} on the measure of the m-th component by α_m, and bound the number of changes by $O(1/\alpha_m)$. Thus, the important condition is being balanced in the sense that the measure bound times the bound on the number of changes is $O(1)$. We emulate a test $(G_m)_{m \in \mathbb{N}}$ by a test $(H_i)_{i \in \mathbb{N}}$ as in Definition 12 by letting $H_i[s] = G_m[s]$, where m is least such that $2^{-i} \geq \alpha_m > 2^{-i-1}$.

Difference randomness and Turing incompleteness. Franklin and Ng have recently introduced *difference randomness*, where the m-th component of a test is a class of the form $A_m - B_m$ with measure at most 2^{-m}, for uniformly given Σ_1^0 classes

A_m, B_m. To pass such a test means not to be in $A_m - B_m$ for some m. (We could replace the individual B_m in each component by $B = \bigcup B_m$. We may also assume that the test is monotonic after replacing $A_m - B_m$ by $\bigcap_{i \le m} A_i - B$ if necessary.)

Proposition 14. *Each balanced random set is difference random.*

Proof. Given a test $(A_m - B_m)_{m \in \mathbb{N}}$, we may assume that $\lambda(A_{m,t} - B_{m,t}) \le 2^{-m}$ for each t (these are the clopen sets effectively approximating A_m, B_m). At stage t let i be greatest such that $\lambda B_{m,t} \ge i2^{-m}$, and let $t^* \le t$ be least such that $\lambda B_{m,t^*} \ge i2^{-m}$. Let $G_m[t] = A_m - B_{m,t^*}$. Then G_m changes at most 2^m times. Clearly $A_m - B_m$ is contained in the last version of G_m. For each t we have $\lambda G_m[t] \le 2^{-m+1}$, so after omitting the first component we have a balanced test.   □

Franklin and Ng [5] show that for ML-random sets, being difference random is equivalent to being Turing incomplete. Nonetheless, it is instructive to give a direct proof of this fact for balanced randomness.

Proposition 15. *Each balanced random set is Turing incomplete.*

Proof. Suppose Z is ML-random and Turing complete. Then $\Omega = \Gamma(Z)$ for some Turing functional Γ. By a result of Miller and Yu (see [1, Prop. 5.1.14]), there is a constant c such that $2^{-m} \ge \lambda\{Z: \Omega \restriction_{m+c} \prec \Gamma(Z)\}$ for each m. Now let the version $G_m[t]$ copy $\{Z: \Omega_t \restriction_{m+c} \prec \Gamma_t(Z)\}$ as long as the measure does not exceed 2^{-m}. Then Z fails the balanced test $(G_m)_{m \in \mathbb{N}}$.   □

Balanced randomness and being ω-c.e. tracing. The following (somewhat weak) highness property was introduced by Greenberg and Nies [2]; it coincides with the class $\mathcal{G}$ in [1, Proof of 8.5.17].

Definition 16. Z *is called ω-c.e. tracing if each function $f \le_{wtt} \emptyset'$ has a Z-c.e. trace T_x^Z such that $|T_x^Z| \le 2^x$ for each x.*

Since we trace only total functions, by a method of Terwijn and Zambella (see [1, Thm. 8.2.3]), the bound 2^x can be replaced by any order function without changing the class. Greenberg and Nies [2] show that there is a single benign cost function such that each c.e. set obeying it is Turing below each ω-c.e. tracing ML-random set. In particular, each strongly jump-traceable, c.e. set is below each ω-c.e. tracing set.

Fact 17. *No superlow set Z is ω-c.e. tracing.*

To prove this, one notes that T_x^Z is truth-table below $\emptyset'$ uniformly in x.

Theorem 18. *Let Z be a ML-random set. If Z is not balanced random then Z is ω-c.e. tracing. In particular, Z is not superlow.*

Proof. Fix a balanced test $(G_m)_{m \in \mathbb{N}}$ such that $Z \in \bigcap_m G_m$. Suppose we are given a function $f \le_{wtt} \emptyset'$ with computable use bound h. Thus there is a computable approximation $f(x) = \lim_s f_s(x)$ with at most $h(x)$ changes. Let $(m_i)_{i \in \mathbb{N}}$ be a computable sequence of numbers such that

$$\sum_i h(i) 2^{-m_i} < \infty,$$

for instance $m_i = \lfloor \log h(i) + 2 \log(i+1) \rfloor$.

To obtain the required trace for f, we define an auxiliary Solovay test $\mathcal{S}$ of the form $\bigcup \mathcal{S}_i$. We put $[\sigma]$ into $\mathcal{S}_i$ if there are 2^i versions $G_{m_i}[t]$ such that $[\sigma] \subseteq G_{m_i}[t]$. Clearly $\mathcal{S}_i$ is uniformly Σ_1^0. We show that $\lambda \mathcal{S}_i = O(2^{-i})$ for each i. Let (σ_k) be a prefix free set of strings such that $\bigcup_k [\sigma_k] = \mathcal{S}_i$. Then

$$O(1) \geq \sum_s \lambda G_{m_i}[s] \geq \sum_s \sum_k \lambda(G_{m_i}[s] \cap [\sigma_k]) \geq 2^i \sum_k \lambda[\sigma_k] = 2^i \lambda \mathcal{S}_i.$$

To define the c.e. operators T_i^Z, when Z enters $G_{m_i}[s]$, put $f_s(i)$ into T_i^Z. Since Z passes the Solovay test $\mathcal{S}$, for almost every i we put at most 2^i numbers into T_i^Z.

To show that f is traced, we define a further Solovay test $\mathcal{R}$. When $f_s(i) \neq f_{s-1}(i)$, put the current version $G_{m_i}[s]$ into $\mathcal{R}$. Note that $\mathcal{R}$ is a Solovay test because $\sum_i h(i) 2^{-m_i} < \infty$. Since Z passes the test $\mathcal{R}$ but fails $(G_m)_{m \in \mathbb{N}}$, we have $f(i) \in T_i^Z$ for almost every i. For, if $f_s(i) \neq f_{s-1}(i)$ then Z must enter a further version $G_m[t]$ for some $t \geq s$, so we can put the new value $f_s(i)$ into T_i^Z. $\qquad \square$

By Theorem 10 we have the following result.

Corollary 19. *There is an ω-c.e. tracing low ML-random set.*

Proof. Applying Theorem 10 to a Π_1^0 class $\mathcal{P} \subseteq \mathsf{MLR}$, we obtain a low ML-random set that is $o(2^n)$-c.e. This set is not balanced random. Then, by Theorem 18 the set is ω-c.e. tracing. $\qquad \square$

Recall that any incomplete ML-random set is difference random. So the proof also shows that some difference random set is not balanced random.

We do not know at present whether the converse of Theorem 18 holds: if Z is balanced random, does it fail to be ω-c.e. tracing? Any LR-complete set is ω-c.e. tracing by [1, Thm. 8.4.15]. So this would imply that a balanced random set is not LR-complete; in particular, no K-trivial set can be cupped above $\emptyset'$ via a balanced random set. By the method of [1, Lem. 8.5.18], we have the following somewhat weaker converse of Theorem 18.

Theorem 20. *Suppose Z is an $O(h(n)2^n)$-weak Demuth random set for some order function h. Then Z is not ω-c.e. tracing.*

References

1. Nies, A.: Computability and Randomness. Oxford Logic Guides, vol. 51. Oxford University Press, Oxford (2009)
2. Greenberg, N., Nies, A.: Benign cost functions and lowness properties (to appear)
3. Kučera, A., Nies, A.: Demuth randomness and computational complexity (to appear 20xx)
4. Jockusch Jr., C., Soare, R.: Degrees of members of Π_1^0 classes. Pacific J. Math. 40, 605–616 (1972)
5. Franklin, J.N., Ng, K.M.: Difference randomness. In: Proceedings of the American Mathematical Society (to appear)

Boole: From Calculating Numbers to Calculating Thoughts*

Michèle Friend

Department of Philosophy, George Washington University,
801 22nd St NW, Washington,
DC 20052, United States of America

We are often taught in a first course in the history of logic or in the philosophy of mathematics that Frege singlehandedly invented second-order logic, and that there was no one close to his achievements before him. In a slightly more sophisticated version of the same course, a qualifying nod is made in the direction of Boole, who did "bring quantifiers to logic".[1] But the student is given the impression that Boole, like Frege, rose *ex nihilo* from the weedy wasteland of Aristotelian syllogistic reasoning.[2] While this is not a wholly mistaken impression, it is misleading, and should be corrected. Boole was working in the context of a revival of logic—where "logic", especially as it was being taught in England, was gradually being prised away from the *doxa* of Aristotelian syllogistic reasoning. It is in this context that Boole made innovative contributions by bringing together a number of ideas. The combination brought us closer to our modern conception of logic.

We shall only discuss some of Boole's contributions in this paper. They can be summarized in the following way. Boole noticed, treated systematically and extended:

---

* I should like to thank the reviewers of this article for many helpful comments on the draft version.

[1] Of course, this is already misleading since Aristotle has "all" and "some" in his logic. It is the translation of quantifiers "all" and "some" into mathematical quantifiers, deployed as algebraic operators, which was important. This move is absent in [7].

[2] This impression is not completely unfounded. Indeed one of Boole's recent predecessors, Lord Dudley, in his letters of 1814 to Lord Copleston writes concerning "the study [of the then Aristotelian syllogistic] logic, that it was sedulously cultivated during the dark periods in which in which the intellectual powers of mankind seemed nearly paralysed,—when no discoveries were made, and when various errors were wide-spread and deep-rooted: and that when the mental activity of the world revived, and philosophical inquiry flourished and bore its fruits, logical studies fell into decay and contempt... " From [7, p. ix], Whatley was intent on reviving logic. However, "The *revival* of a study which had for a long time been regarded as an obsolete absurdity, would probably have appeared to many persons, thirty years ago [1800] as an undertaking far more difficult than the undertaking of some *new* study;- as resembling rather the attempt to restore to life one of the antediluvian fossil-plants, than the rearing of a young seedling into a tree." [7, p. ix]

F. Ferreira et al. (Eds.): CiE 2010, LNCS 6158, pp. 172–179, 2010.

1. the similarity between algebraic calculating and logical reasoning and
2. the separation of logic from some aspects of epistemology and ontology, bringing us to an extensional / predicative notion of semantics.

In this paper, we shall see that many small pieces in his work contribute in separate ways to our modern conception of logic. We shall look at nine pieces. We should be quite clear about the nature of the claims being made here. It is not that Boole invented or noticed for the first time each of the pieces we shall discuss below, rather, it was Boole who brought them together in a system of logic: where it is not numbers which are being calculated, but thoughts quite generally. Furthermore, the nine pieces are not exhaustive of Boole's contribution, for example I make no mention of probability theory. Nor, is the numbering of the nine pieces significant, it is merely heuristically convenient for this paper.

We see the first idea at least in Leibniz, but Boole made this more explicit by giving a method of translating from propositions to equations which are purely symbolic, where there are no natural language words. This is our first piece. Roughly speaking, (true) propositions represent a fact in the world. In English, a declarative sentence refers to a proposition. Once propositions are taken as a unit, and represented with a symbol, we can add to them the usual algebraic operations (which we now often think of as Boolean connectives): union, intersection and complement. In declarative sentences these more or less correspond to the words "or", "and" and "not", respectively. Once translated, our propositions and connectives are now susceptible to being treated as parts of algebraic equations. We can simply manipulate the symbols. So we learn, for example, from Boole's contemporary, De Morgan, that $\sim(P \vee Q) = \sim P \wedge \sim Q$.[3]

A major influence in this prising of logic away from the confining grips of syllogistic reasoning came from Leibniz, but more immediately, from [3, 6, 7], at least, and there were undoubtedly other lesser known contributors to the movement to modernise logic. Whatley rehearses through syllogistic reasoning, but started the modernising process by adding more general considerations about disjunctions [7, p. 75] for example, which are given a separate treatment from the syllogism. Peacock reversed the direction, and extended the idea of algebra to work not only with numbers but with arbitrary term symbols as well: so bringing us closer to our modern conception of logic. However, having introduced the more general idea of a term-symbol algebra, he did not do much work with it, preferring to work on arithmetical algebra. Gregory was also extending the notion of the symbolic algebra to solve problems with differential equations. De Morgan was, perhaps, the closest to Boole, since he was making direct contributions to the calculation of symbols.[4]

---

[3] I do not use Boole's own notation. $\sim$ is the symbol for negation. P and Q are proposition variables. $\wedge$ is conjunction or intersection. $\vee$ is disjunction or union.

[4] I should like to thank the reviewer for pointing out to me these aspects of the history before Boole.

Under the influence of these logicians and mathematicians, Boole started to really develop the technical machinery needed to make an algebra of natural language terms, where propositions are one sort of term, amongst others. When we think of an algebra of terms, as opposed to a syllogistic logic of terms, we can have multiple premises, so we cleanly break away from the syllogism. This is the second piece. Previous to Boole, in logic, we were allowed to unpack, or string together several syllogistic arguments, but each syllogism was considered to be a self-contained unit. Boole did not just develop a propositional, or term, logic. He also allowed logical treatment within propositions and terms. This is the third piece. Thus, even the terms have a logical, formal structure. Instead of Aristotle's term logic, where the terms refer to subjects and predicates, and are impenetrable units, Boole allows for predicates, relations and functions found in English language propositions to form terms. These can then be treated algebraically, as relations and function in algebra. For example, see [2, p. 140]. When we allow symbolic representations of predicates relations and functions, we can calculate out thoughts, much as we calculate out numbers.

One of the interesting points about Boole's importing of algebraic ideas into logic is that the terms, like numbers, are abstract, in the sense of removed from an ontology of objects being calculated. Similarly, if we calculate thoughts, we calculate while disregarding the content of the thoughts. As Frege comments: "In all this [Boole's calculus] there is no concern about content whatsoever." [4, p. 12]. We set up premises which we think are interesting, (or not—they can be quite arbitrary, or un-interpreted).[5] We then move to the calculation, or manipulation, of the symbols according to some Boolean algebraic rules to come up with a conclusion. We are then free to re-inject the conclusion with content, i.e., we interpret the conclusion. See Figure 1.

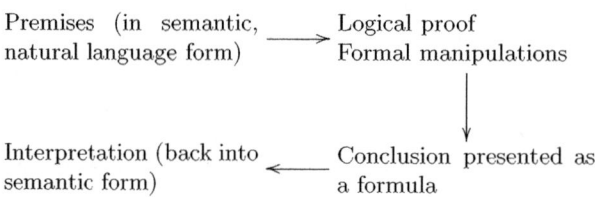

Fig. 1.

As Bar-Am emphasises, the separation of logical calculation from content, and so, from epistemology and ontology, was a huge conceptual leap because the formal calculation is "mindless" in some sense. It is automatic, and takes place without interpretations, or without requiring understanding of the content or meaning of the symbols. The leap was only possible in the light of Boole's commitment to extensionalism: the idea that logic is not intensional but

---

[5] That is, whether or not a premise is interesting or informative is of no concern to the Boolean logician. This is where he parts company with his predecessors.

extensional.[6] That is, the mode of presentation (intension) of a term is irrelevant in logical calculation, what matters is what is picked out by a term or predicate in a formal language. Boole's extensionalism is the fourth piece. This makes the logic completely general over interpretations. The generality is possible, since we add to the traditional interpretations the notion that there are different universes of discourse, and the extreme limit cases of these are the null class: 0, and the universe: 1. The introduction of the null class and the universe of discourse to logic is the fifth piece. With this contribution, we are then free to interpret these as falsity and truth respectively, or as other types of valuation. We then notice that non-referring terms can also be calculated, they just refer to the null class. Similarly, we can calculate terms referring to the complement of the null-class, i.e., the universe. As Bar-Am argues, once we have these ingredients, we are released from the traditional constraint on only using terms which have referents, such as natural kinds. Instead, we are free to choose, and calculate with, any grammatically determined term in the language. The calculations become rich when our grammar is extended to include relations and functions as separate grammatical items.

> The freedom to study [reason over] all terms, those that seem to us natural and even those that do not make any sense to us, and the freedom to study necessarily or possibly false propositions is essential to modern logic. It is the freedom from premises about what is allegedly known *a priori* and about the conditions of knowledge. The discovery of this freedom was a gradual and laborious process. [1, p. 6]

The calculations can proceed unchecked by reference to content (ontology) or to epistemology to justify inferences. This is important. For, in conceptions of logic before Boole (and some logic after Boole (see Russell and Frege))[7] we insist on, or cannot see past, using only proper definitions: definitions of terms which pick

---

[6] [1, pp. 97–123]. Bar-Am claims that Boole's extensionalism has been overlooked by other commentators, and that Boole's extensionalism resulted in a liberating of logic from the previous Aristotelian and Scholastic attachments to epistemology and ontology. I shall be drawing on his paper extensively. This is not to say that Boole invented extensionalism, it was already present in [7]. Rather, it is the inheritance of extensionalism to his calculation of thoughts which makes the syntactical manipulations "mindless".

[7] Gottlob Frege, "Boole's Logical Formula Language and my Concept Script" in [5, p. 47]. Here, Frege explicitly says that Boole's symbols were chosen in order to ignore content altogether, this is not at all the case with his own symbols and script. Frege might have been merely defending his script—since it is quite awkward. But it is more likely that he is giving priority to logic over arithmetic—logic was meant to justify arithmetic. So the symbols used in logic should be quite different from those found in arithmetic, in order not to beg any questions. Frege repeatedly insists on this difference between him and Boole: that Boole sought to bring arithmetic to bear on logic, whereas Frege does the reverse, he wants to justify arithmetic by appeal to the more basic logic.

out natural kinds or concepts which are useful, "interesting", or already known,[8] or going back further in history to Aristotle and the Mediaevals: essences.[9] With Boole's contributions, the notion of interpretation is completely divorced from the calculation. When we think of terms grammatically, when we determine that a string of symbols is a term only by reference to the grammar of the language, then the interpretation of the premises and conclusion of an argument is completely divorced from the calculation. We can now logically calculate using terms which fail to refer in the real world, and terms which refer to the entire universe of discourse. It is then a small step to admit all of the grammatically possible terms and allow them into the calculations, which gives us our modern conception of logic, and is related to the conception used in computer science. In computer science, while definitions have to be constructed effectively, the meaning of the defined term is a quite separate matter. Effectiveness is a methodological constraint, not an ontological or epistemological one. The computer takes the term and makes calculations. In particular, the computer is not in any way tied to intended interpretations, or any interpretation at all. It occupies the right hand side of Figure 1.

For further contributions, let us examine some of Boole's calculations more closely. Boole's calculations of thoughts are algebraic calculations. Begin with negation. Picking up on De Morgan, Boole noticed that negation acts like the notion of complement. This is the sixth piece. If there are no donkeys on the farm, then this tells us that the animals (if any) which are on the farm are to be found in the complement set: animals which are not donkeys.

Disjunction is like set theoretic union: the two classes are merged together, and any overlap is only mentioned once. This is the seventh piece. Let us illustrate the meaning of disjunction by appeal to an example. Say that the extension of the concept: "member of the poetry club" is: {Benedikt, Olga, Michele, Thomas and Norma}. The extension of the concept: "is presenting in the section on Reasoning as Computation from Leibniz to Boole at the CiE 2010 conference" is: {Olga, Michele, Volker and Sara}. The disjunction, or union, of the two sets is: {Benedikt, Olga, Michele, Thomas, Norma, Volker and Sara}.

Conjunction in natural language is also not like arithmetic addition, it is more like the set theoretic notion of intersection. This is the eighth piece. If you are not convinced by the examples, then consider the connectives as applied to De Morgan's laws: $\sim(P \,\&\, Q) \equiv \sim P \vee \sim Q$.

---

[8] For Frege, proper, and new, definitions have to be shorthand for a concept already proved in the logic. This issue is tricky because Frege also included Basic Law V and Hume's Principle in his logic, and these are sometimes (but not by Frege) called "impredicative definitions". They are impredicative because we have to already understand the terms on each side of the definitional biconditional in order to understand the concept. Russell was leery of impredicativity, and outlawed it from his formal system: the outlawing is called the "vicious circle principle": that circular definitions are not allowed.

[9] Bar-Am points out that the syllogistic logic was really developed as an epistemological tool for going from truths about certain issues in science to new scientific facts. [1, p. 8]

The importance of giving formal definitions for connectives in English, is that we can now make axioms governing the connectives. Where "A" is a term which has an extension, $\mathbf{U}$ is the universe and $\emptyset$ is the null class, the Boolean axioms include the following:

1. $A \,\&\, \overline{A} = \emptyset$.

2. $A \vee \overline{A} = \mathbf{U}$.

3. $A \,\&\, A = A$.

What is important is that Boole axiomatized logic. This is the ninth piece. Once we define the connectives by giving axioms and rules of inference, then we can calculate thoughts or concepts. This, again, allows us to entirely separate the calculation from the interpretations, since we bring the universe of discourse to the calculation only at the stage of interpretation. The ontology is no longer part of the logic, or the reasoning. What about epistemology and methodology?

In conclusion, Boole is very modern in this respect. Boole separated episte- mology from the methodology of reasoning. Consider the well accepted claim that logic inference is truth preserving. Today, this is quite liberally interpreted, since we are allowed to discharge assumptions (as in conditional proofs), or even falsehoods (as in *reductio ad absurdum* proofs). In a conditional proof, we assume the antecedent of a conditional statement we want to prove, prove the conse- quent of the conditional from the assumed antecedent, and then we are entitled, logically, to assert the conditional statement. When we do this we discharge the initial assumption, since we are only asserting a conditional. So the antecedent could well be false. Nevertheless, the conditional statement is logically derived. In *reductio ad absurdum* proofs, we assume the opposite of what we want to prove. We make the assumption temporarily, and for the sake of argument. We then prove a contradiction from the (faulty) assumption, thus demonstrating that the assumption was faulty. We then reject it, by concluding the negation of the assumption. Again we have been reasoning with falsehoods. The proofs preserve truth, but that is not the same as their only containing truths, which is what the predecessors of Boole insisted on.

To be quite clear, Boole's reasoning does not contain conditional proofs or *reductio* proofs. Rather, what I should like to claim is that it is his approach to reasoning which made conditional proofs and *reductio* proofs possible later additions to logic. This is part of the significance of his axiomatization of logic. The addition was only subsequently possible because he allowed us to reason completely rigorously and logically with falsehoods. Boole's system does not allow us to discharge assumptions in the elegant way that we do today, but he did allow us to manipulate terms which failed to refer, or were devoid of content—this much Frege recognised in Boole; but Frege did not adopt the idea. To appreciate the significance of Boole's more algebraic axioms, consider Frege's later, but less epistemologically liberal system. Frege's formal system is methodologically and epistemologically more restrictive than Boole's since the only rule of inference is modus ponens. For this reason, the axioms have to be necessary truths. The

Basic Laws of Frege's system were introduced as indubitable truths. Some were just tautologies, others were just meant to be quite obvious analytic truths.[10] Contrast this to Boole's axioms above. These are better thought of as convenient and intuitive truths, once one is thinking in the right sort of way. For example, the last axiom, above, is not at all obvious as a logical truth. We have to convince ourselves of its truth, and use it to define the symbol "&" together with the notion of idempotence.

To appreciate why the divorce of epistemology from methodology was such a grand conceptual advance, we should revisit briefly the history. Traditionally, before Boole, epistemology was part of logic. We learned logic in order to bring us from truths to other truths. Logic is supposed to be truth-preserving. But, traditionally we interpret this to mean that we may only reason with truths. Before, and even subsequent to Boole, with Frege, we could only start logical arguments with truths, and were not allowed to bring in hypothetical assumptions for the sake of argument. This sounds quite reasonable since we want to reason about the real world after all, and we thought of logic as an epistemological tool. Going further back—to Aristotle—we find even more surprising constraints. For Aristotle, the premises of a syllogism did not only have to be truths. They had to be truths about the essences of things. Aristotle's syllogistic reasoning was developed in order to advance science. There is a passage in Aristotle which is often overlooked because it is embarrassing to the modern reader (and even the Mediaeval reader). Aristotle insists that, for the purposes of scientific demonstration, the conclusions of syllogistic arguments should never be used as premises for further arguments. [11] Thus, we are not allowed to string syllogisms together! The reason for this lies in Aristotle's conceived purpose for developing logic: to help in the development of a system of biology—or of biological taxonomy, according to the essence of biological entities—hence all of the horses, man and swan examples. It is clear that this intended application or reason for developing logic was quickly generalised by subsequent philosophers. They (and probably even Aristotle) noticed that syllogistic logic could be useful in other areas of research as well, especially if we drop the clause which says that conclusions can never figure as premises. What works for biology also works for other sciences and more mundane reasoning. Nevertheless, its primary purpose was to educate us, to bring us new, and reliable, information. We can learn from the other panellists that after the Mediaevals had had their run with the syllogism, deductive logic (as opposed to inductive logic) fell into disrepute in some circles. Some critics of syllogistic reasoning thought of it as epistemologically misleading. Others thought of it as trivial (so epistemologically neutral, since there were no possible advances in knowledge made through syllogistic reasoning). For this reason alone it was important to breath new life into logic, and this was possible in light of Boole's separation of epistemology from logical calculation.

The point we want to retain from this paper is that Boole was a significant contributor to our modern computer science through his contribution to logic,

---

[10] Of course there was a fatal problem with Basic Law V, but we'll ignore this here.

[11] Aristotle *Posterior Analytics* 71b 26-36. Referred to in [1, n.11].

although we can only appreciate this in hindsight. Boole divorced the old scientific view of epistemology from the methodology of reasoning. He gave the methodology in the form of a formal system. This is something which Leibniz had alluded to but had never fully accomplished. To repeat some of the quotation from Bar Am: "The discovery of this freedom [to use un-interpreted terms] was a gradual and laborious process," [1, p. 6 of the web manuscript]. We identified nine pieces in that process. How the pieces are counted is not what is important. Nor are the missing pieces—I said nothing about Boole's bringing probability theory to logic, for example. Nor are all of the pieces inventions by Boole. Most of the pieces were there in other fields of research. They were ripe for the picking and placing in a system of logic. Further, to really follow Boole's system is not a completely easy matter, his system had to be smoothed out, added to and modified to resemble our modern logical systems. Rather, the point is to appreciate that the development of modern logic was not a sudden process from Aristotle, jumping to Boole, jumping to Frege. Nor was it a smooth gradual process, since Boole's lesson of the distinction between epistemology and methodology was not fully appreciated during his life time (possibly not even by himself), and was even overlooked shortly after Boole's major publications. Even Frege (50 years after Boole) insisted that we begin our arguments with truths: no hypotheses and no falsehoods—even temporarily for the sake of argument!

References

1. Bar-Am, N.: Extensionalism and Induction in Boole. Physics 41, 97–123 (2004); Also on the author's webpage (page references are taken from the web version)
2. Boole, G.: An Investigation of the Laws of Thought. Barnes & Noble, New York (2005); Originally (1854)
3. De Morgan, A.: Formal Logic. In: Taylor, A.E. (ed.), Open Court, London (1927); Originally (1847)
4. Frege, Gottlob "Boole's Logical Calculus and the Concept Script". In: Frege, Gottlob Posthumous Writings, Trans. Peter Long, Roger White. Eds. Hans Hermes, Friedrich Kambartel, Friedrich Kaulbach, Basil Blackwell, Oxford (1976); Originally submitted for publication in 1881. pp. 9–46 (1881)
5. Frege, "Boole's Logical Formula Language and my Concept Script". In: Frege, Gottlob Posthumous Writings, Trans. Peter Long, Roger White. Eds. Hans Hermes, Friedrich Kambartel, Friedrich Kaulbach, Basil Blackwell, Oxford (1976); Originally written 1882, pp. 47–52 (1882)
6. Peacock, G.: A Treatise of Algebra. J.&J.J. Deighton, Cambridge (1830)
7. Whatley, R.: Elements of Logic. Reprinted from the 9th (Octavio) edn., Longmans Green & Co., London (1913)

Approximability and Hardness in Multi-objective Optimization

Christian Glaßer[1], Christian Reitwießner[1],
Heinz Schmitz[2], and Maximilian Witek[1]

[1] Lehrstuhl für Theoretische Informatik, Julius-Maximilians-Universität Würzburg,
Am Hubland, 97074 Würzburg, Germany
{glasser,reitwiessner,witek}@informatik.uni-wuerzburg.de
[2] Fachbereich Informatik, Fachhochschule Trier, Schneidershof, 54293 Trier, Germany
schmitz@informatik.fh-trier.de

Abstract. We study the approximability and the hardness of combinatorial multi-objective NP optimization problems (multi-objective problems, for short). Our contributions are:

- We define and compare several solution notions that capture reasonable algorithmic tasks for computing optimal solutions.
- These solution notions induce corresponding NP-hardness notions for which we prove implication and separation results.
- We define approximative solution notions and investigate in which cases polynomial-time solvability translates from one to another notion. Moreover, for problems where all objectives have to be minimized, approximability results translate from single-objective to multi-objective optimization such that the relative error degrades only by a constant factor. Such translations are not possible for problems where all objectives have to be maximized (unless P = NP).

As a consequence we see that in contrast to single-objective problems (where the solution notions coincide), the situation is more subtle for multiple objectives. So it is important to exactly specify the NP-hardness notion when discussing the complexity of multi-objective problems.

1 Introduction

Many technical, economical, natural- and socio-scientific processes contain multiple optimization objectives in a natural way. For instance, in logistics one is interested in routings that simultaneously minimize transportation costs and transportation time. For typical instances there does not exist a single solution that is optimal for both objectives, since they are conflicting. Instead one will encounter trade-offs between both objectives, i.e., some routings will be cheap, others will be fast. The *Pareto set* captures the notion of optimality in this setting. It consists of all solutions that are optimal in the sense that there is no solution that is strictly better. For decision makers the Pareto set is very useful as it reveals all trade-offs between all optimal solutions for the current instance.

In practice, multi-objective problems are often solved by turning the problem into a single-objective problem first and then solving that problem. This

F. Ferreira et al. (Eds.): CiE 2010, LNCS 6158, pp. 180–189, 2010.
© Springer-Verlag Berlin Heidelberg 2010

approach has the benefit that one can build on known techniques for single-objective problems. However, it comes along with the disadvantage that the translation of the problem significantly changes its nature, i.e., sometimes the problem becomes harder to solve and sometimes certain optimal solutions are not found anymore. In general, single-objective problems cannot adequately represent multi-objective problems, so optimization of multi-objective problems is studied on its own. The research area of multi-objective combinatorial optimization has its origins in the late 1980s and has become increasingly active since that time [5]. For a general introduction we refer to the survey by Ehrgott and Gandibleux [4] and the textbook by Ehrgott [3].

Typically, the exact solution of a multi-objective problem is not easier than the exact solution of an underlying single-objective problem. Therefore, polynomial-time approximation with performance guarantee is a reasonable approach also to multi-objective problems. In this regard, Papadimitriou and Yannakakis [10] show the following important result: Every Pareto set has a $(1+\varepsilon)$-approximation of size polynomial in the size of the instance and $1/\varepsilon$. Hence, even though a Pareto set might be an exponentially large object, there always exists a polynomial-sized approximative set. This clears the way for a general investigation of the approximability of Pareto sets. However, complexity issues raised by multi-objective problems have not been addressed systematically yet [10]. We consider our paper as a first step to a systematic investigation of hardness and approximability of multi-objective problems. Our contribution is as follows:

Solution Notions (Section 3): We define several notions that capture reasonable algorithmic tasks for computing *optimal* solutions of multi-objective problems. On the technical side, we see that these notions can be uniformly and precisely described by refinements of total, multivalued functions. This yields the suitable concept of polynomial-time Turing reducibility for solution notions. It turns out that the relationships shown in Figure 1 hold for arbitrary multi-objective problems.

NP-Hardness Notions (Section 4): Solution notions for multi-objective problems induce corresponding NP-hardness notions for these problems. We provide examples of multi-objective problems that are hard with respect to some notion, but polynomial-time solvable with respect to some other notion. So in contrast to the single-objective case where all notions coincide, we see a more subtle picture in case of multiple objectives: our separation results show that NP-hardness notions in fact differ, unless P = NP. As a consequence, we suggest to exactly specify the solution notion when talking about complexity of multi-objective problems and when comparing problems in these terms.

Approximation Notions (Section 5): We also define and investigate various approximative solution notions for multi-objective problems. As a summary, Figure 2 shows for arbitrary multi-objective problems in which cases polynomial-time solvability of one such notion implies polynomial-time solvability of another notion, and what quality of approximation can be preserved at least. Moreover, we reveal a significant dichotomy between approximation of minimization and maximization problems in this context. For problems where all objectives have to be minimized, approximability results translate from single-objective to multi-objective optimization such that the relative error degrades only by a constant factor (the number of objectives). With this general result we provide a procedure how to translate (already known) single-objective approximation results to the multi-objective case and show its application to some example problems. In contrast to this result we prove that such translations are not possible for problems where all objectives have to be maximized, unless P = NP.

The proofs omitted in this paper can be found in the technical report [7].

2   Preliminaries

Let $k \geq 1$. A *combinatorial k-objective* NP *optimization problem* (*k-objective problem*, for short) is a tuple $(S, f, \leftarrow)$ where

- $S \colon \mathbb{N} \to 2^{\mathbb{N}}$ maps an instance $x \in \mathbb{N}$ to the set of feasible solutions for this instance, denoted as $S^x \subseteq \mathbb{N}$. There must be some polynomial p such that for every $x \in \mathbb{N}$ and every $s \in S^x$ it holds that $|s| \leq p(|x|)$ and the set $\{(x, s) \mid x \in \mathbb{N}, s \in S^x\}$ must be polynomial-time decidable.
- $f \colon \{(x, s) \mid x \in \mathbb{N}, s \in S^x\} \to \mathbb{N}^k$ maps an instance $x \in \mathbb{N}$ and $s \in S^x$ to its value, denoted by $f^x(s) \in \mathbb{N}^k$. f must be polynomial-time computable.
- $\leftarrow \subseteq \mathbb{N}^k \times \mathbb{N}^k$ is the partial order relation specifying the direction of optimization. It must hold that $(a_1, \ldots, a_k) \leftarrow (b_1, \ldots, b_k) \iff a_1 \leftarrow_1 b_1 \wedge \cdots \wedge a_k \leftarrow_k b_k$, where $\leftarrow_i \in \{\leq, \geq\}$ depending on whether the i-th objective is minimized or maximized.

For instances and solutions we relax the restriction to integers and allow other objects (e.g., graphs) where a suitable encoding is assumed, possibly setting $S^x = \emptyset$ if x is not a valid code. We write $\leq$ and $\geq$ also for their multidimensional variants. Our notation allows concise definitions of problems which we exemplify by defining two well-known problems on graphs with edge-labels from $\mathbb{N}^k$.

Example 1 (k-Objective Minimum Matching)

    k-MM $= (S, f, \leq)$ where instances are $\mathbb{N}^k$-labeled graphs $G = (V, E, l)$, $S^G = \{M \mid M \subseteq E \text{ is a perfect matching on } G\}$, and $f^G(M) = \sum_{e \in M} l(e)$.

Example 2 (2-Objective Minimum Traveling Salesman)

    2-TSP $= (S, f, \leq)$ where instances are $\mathbb{N}^2$-labeled graphs $G = (V, E, l)$, $S^G = \{H \mid H \subseteq E \text{ is a Hamiltonian circuit}\}$, and $f^G(H) = \sum_{e \in H} l(e)$.

The superscript x of f and S can be omitted if it is clear from context. The projection of f^x to the i-th component is f_i^x. If $a \leftarrow b$ we say that a *weakly dominates* b (i.e., a is at least as good as b). If $a \leftarrow b$ and $a \neq b$ we say that a *dominates* b. Note that $\leftarrow$ always points in direction of the better value. If f and x are clear from the context, then we extend $\leftarrow$ to combinations of values and solutions, and we write $s \leftarrow t$ if $f^x(s) \leftarrow f^x(t)$, $s \leftarrow c$ if $f^x(s) \leftarrow c$, and so on, where $s, t \in S^x$ and $c \in \mathbb{N}^k$. Furthermore, we define $\mathrm{opt}_{\leftarrow}(M) = \{y \in M \mid \forall z \in M[z \leftarrow y \Rightarrow z = y]\}$ as a function that maps sets of values to sets of optimal values. The operator $\mathrm{opt}_{\leftarrow}$ is also applied to subsets $S' \subseteq S^x$ as $\mathrm{opt}_{\leftarrow}(S') = \{s \in S' \mid f^x(s) \in \mathrm{opt}_{\leftarrow}(f^x(S'))\}$. If even $\leftarrow$ is clear from the context, we write $S^x_{\mathrm{opt}} = \mathrm{opt}_{\leftarrow}(S^x)$ and $\mathrm{opt}_i(S') = \{s \in S' \mid f^x_i(s) \in \mathrm{opt}_{\leftarrow_i}(f^x_i(S'))\}$.

For approximations we need to relax the notion of dominance by a factor of α. For $a \geq 1$ define $u \overset{a}{\leq} v \iff u \leq a \cdot v$ and $u \overset{a}{\geq} v \iff a \cdot u \geq v$. Let $x = (x_1, \ldots, x_k)$, $y = (y_1, \ldots, y_k) \in \mathbb{N}^k$, and let $\alpha = (a_1, \ldots, a_k) \in \mathbb{R}^k$ where $a_1, \ldots, a_k \geq 1$. We say that x *weakly α-dominates* y, $x \overset{\alpha}{\leftarrow} y$ for short, if $x_i \overset{a_i}{\leftarrow_i} y_i$ for $1 \leq i \leq k$. For all $x, y, z \in \mathbb{N}^k$ it holds that $x \overset{\alpha}{\leftarrow} x$, and $x \overset{\alpha}{\leftarrow} y \overset{\beta}{\leftarrow} z \Rightarrow x \overset{(\alpha \cdot \beta)}{\leftarrow} z$. Again we extend $\overset{\alpha}{\leftarrow}$ to combinations of values and solutions.

Let A and B be sets. $\mathcal{F}$ is a *multivalued function* from A to B, if $\mathcal{F} \subseteq A \times B$. The *set of values* of x is set-$\mathcal{F}(x) = \{y \mid (x, y) \in \mathcal{F}\}$. $\mathcal{F}$ is called *total*, if for all x, set-$\mathcal{F}(x) \neq \emptyset$. In order to compare solution notions of optimization problems we need an appropriate reducibility notion. All solution notions $\mathcal{F}$ considered in this paper have in common that each instance x specifies a non-empty set of suitable outputs set-$\mathcal{F}(x) = \{y \mid y$ solves x in terms of solution notion $\mathcal{F}\}$. In this sense, a solution notion $\mathcal{F}$ is a total multivalued function that maps an instance x to all $y \in$ set-$\mathcal{F}(x)$. Therefore, solution notions can be compared by means of a reducibility for total multivalued functions. We use Selman's [12] definition of polynomial-time Turing reducibility for multivalued functions, restricted to *total* multivalued functions. It can also be applied to decision problems, since characteristic functions are total (multivalued) functions. A solution notion $\mathcal{F}$ is called *polynomial-time solvable*, if there is a total, polynomial-time computable function f such that f is a refinement [12] of $\mathcal{F}$. A solution notion $\mathcal{F}$ is called NP-*hard*, if all problems in NP are polynomial-time Turing-reducible to $\mathcal{F}$.

3   Multi-objective Solution Notions

For a k-objective problem $\mathcal{O} = (S, f, \leftarrow)$ we discuss several reasonable concepts of "solving $\mathcal{O}$". We investigate their relationships and conclude this section with a taxonomy of these concepts.

Apparently a dominated solution s is not optimal for $\mathcal{O}$, since solutions exist that are at least as good as s in all objectives and better than s in at least one objective. So we are only interested in non-dominated solutions, which are called *(Pareto-)optimal solutions*. Note that the set S^x_{opt} of non-dominated solutions may contain several solutions with identical values. Since these solutions cannot be distinguished, it suffices to find one solution for each optimal value, as it is usual in single-objective optimization. This motivates the following definition.

E-$\mathcal{O}$ **Every-optimum notion**
> Compute a set of optimal solutions that generate all optimal values.
> Input: instance x
> Output: some $S' \subseteq S^x_{\mathrm{opt}}$ such that $f^x(S') = f^x(S^x_{\mathrm{opt}})$

Although E-$\mathcal{O}$ formalizes the canonical notion of solving multi-objective problems, this is far too ambitious in many cases, since every set S' can be of exponential size. We call $\mathcal{O}$ *polynomially bounded* if there is some polynomial p such that $\#f^x(S_{\mathrm{opt}}^x) \leq p(|x|)$ for all x. If $\mathcal{O}$ is not polynomially bounded, then E-$\mathcal{O}$ is not polynomial-time solvable. The earlier defined problems 2-MM and 2-TSP are examples that show this effect. So E-$\mathcal{O}$ is infeasible in general, and hence more restricted concepts of solving multi-objective problems are needed.

A-$\mathcal{O}$ Arbitrary-optimum notion

Compute an arbitrary optimal solution.

Input: instance x

Output: some $s \in S_{\mathrm{opt}}^x$ or report that $S^x = \emptyset$

S-$\mathcal{O}$ Specific-optimum notion

Compute an optimal solution that weakly dominates a given cost vector.

Input: instance x, cost vector $c \in \mathbb{N}^k$

Output: some $s \in S_{\mathrm{opt}}^x$ with $f^x(s) \leftarrow c$ or report that there is no such s

D-$\mathcal{O}$ Dominating-solution notion

Compute a solution that weakly dominates a given cost vector.

Input: instance x, cost vector $c \in \mathbb{N}^k$

Output: some $s \in S^x$ with $f^x(s) \leftarrow c$ or report that there is no such s

If no additional information is available (including no further criteria, no prior knowledge, and no experience by decision makers), then it is not plausible to distinguish non-dominated solutions. In these cases it suffices to consider A-$\mathcal{O}$, since all elements in S_{opt}^x are "equally optimal". The notion S-$\mathcal{O}$ additionally allows us to specify the minimal quality c that an optimal solution s must have. With D-$\mathcal{O}$ we relax the constraint that s must be optimal.

There exist several well-established approaches that turn $\mathcal{O}$ into a single-objective problem first and then treat it with methods known from single-objective optimization. The following definitions are motivated by such methods. Later we will show that these approaches differ with respect to their computational complexity (cf. Figure 1). We consider W-$\mathcal{O}$ only for multi-objective problems where all objectives have to be minimized (resp., maximized).

W-$\mathcal{O}$ Weighted-sum notion

Single-objective problem that weights the objectives in a given way.

Input: instance x, weight vector $w \in \mathbb{N}^k$

Output: some $s \in S^x$ that optimizes $\sum_{i=1}^{k} w_i f_i^x(s)$ or report that $S^x = \emptyset$

C$_i$-$\mathcal{O}$ Constraint notion for the i-th objective

Single-objective problem that optimizes the i-th objective while respecting constraints on the remaining objectives.

Input: instance x, constraints $b_1, \ldots, b_{i-1}, b_{i+1}, \ldots, b_k \in \mathbb{N}$

Output: for $S_{\mathrm{con}}^x = \{s \in S^x \mid f_j^x(s) \leftarrow_j b_j \text{ for all } j \neq i\}$ return some $s \in \mathrm{opt}_i(S_{\mathrm{con}}^x)$ or report that $S_{\mathrm{con}}^x = \emptyset$

L-$\mathcal{O}$ Lexicographical notion for a fixed order of objectives
Single-objective problem with a fixed order of objectives (here $1, 2, \ldots, k$).
Input: instance x
Output: some $s \in \mathrm{opt}_k(\ldots(\mathrm{opt}_2(\mathrm{opt}_1(S^x)))\ldots)$ or report that $S^x = \emptyset$

Strictly speaking, W-$\mathcal{O}$, $\mathrm{C_i}$-$\mathcal{O}$, and L-$\mathcal{O}$ are not only solution notions for $\mathcal{O}$, but in fact they are single-objective problems. In the literature, L-$\mathcal{O}$ is also known as *hierarchical optimization*. Moreover, W-$\mathcal{O}$ is a particular *normalization approach*, since a norm is used to aggregate several cost functions into one.

We observe that if E-$\mathcal{O}$ is polynomial-time solvable, then so are all other notions. Moreover, if $\mathcal{O}$ is a single-objective problem, then all notions defined so far are polynomial-time Turing equivalent.

E-$\mathcal{O}$ plays a special role among the solution notions for $\mathcal{O}$, since solutions of E-$\mathcal{O}$ are typically of exponential size. So in general, E-$\mathcal{O}$ is not polynomial-time Turing reducible to any of the other notions. On the other hand, polynomial-time Turing reductions to E-$\mathcal{O}$ are problematic as well, since answers to oracle queries can be exponentially large which makes the reduction very sensitive to encoding issues (as the reduction machine can only read the left-most polynomial-size part of the answer). Therefore, we will not compare E-$\mathcal{O}$ with the other notions by means of polynomial-time Turing reductions, and we will not consider the NP-hardness of E-$\mathcal{O}$. However, in some sense E-$\mathcal{O}$ is covered by D-$\mathcal{O}$ (resp., S-$\mathcal{O}$), since the latter can be used in a binary search manner to find solutions for arbitrary optimal values. In Section 5, E-$\mathcal{O}$ will become important again in the context of approximate solutions.

Next we show that the remaining notions are closely related (see Figure 1 for a summary). SAT denotes the NP-complete set of all satisfiable Boolean formulas.

Theorem 1. *Let $\mathcal{O} = (S, f, \leftarrow)$ be some k-objective problem.*

1. A-$\mathcal{O} \leq_{\mathrm{T}}^{\mathrm{P}}$ L-$\mathcal{O} \leq_{\mathrm{T}}^{\mathrm{P}}$ S-$\mathcal{O}$
2. S-$\mathcal{O} \equiv_{\mathrm{T}}^{\mathrm{P}}$ D-$\mathcal{O} \equiv_{\mathrm{T}}^{\mathrm{P}}$ $\mathrm{C_1}$-$\mathcal{O} \equiv_{\mathrm{T}}^{\mathrm{P}}$ $\mathrm{C_2}$-$\mathcal{O} \equiv_{\mathrm{T}}^{\mathrm{P}} \ldots \equiv_{\mathrm{T}}^{\mathrm{P}}$ $\mathrm{C_k}$-$\mathcal{O}$
3. L-$\mathcal{O} \leq_{\mathrm{T}}^{\mathrm{P}}$ W-$\mathcal{O}$ *if all objectives have to be minimized (resp., maximized)*
4. W-$\mathcal{O} \leq_{\mathrm{T}}^{\mathrm{P}}$ SAT *and* D-$\mathcal{O} \leq_{\mathrm{T}}^{\mathrm{P}}$ SAT

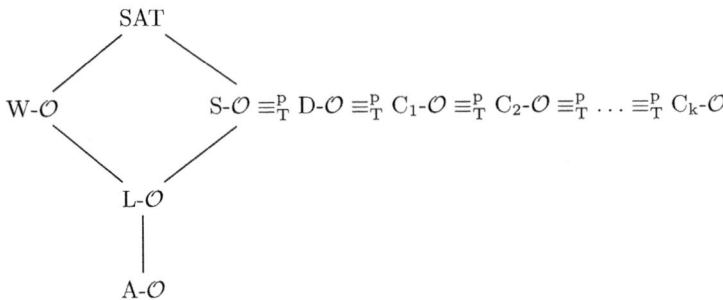

Fig. 1. Polynomial-time Turing reducibility among different solution notions for any multi-objective problem $\mathcal{O}$ (corresponding separations are shown in Section 4)

We will see that Theorem 1 is complete in the sense that no further reductions among the solution notions are possible in general and the complexity of different notions can be separated, unless P = NP. Therefore, even if multi-objective problems are not polynomially bounded it is worthwhile to classify them according to the taxonomy given in Figure 1.

4   Multi-objective NP-Hardness

Each solution notion for a multi-objective problem induces a corresponding NP-hardness notion. In this section we study the relationships between these NP-hardness notions and present separation results which imply the strictness of the reductions shown in Theorem 1 and Figure 1.

From Theorem 1 we obtain the following relationships between NP-hardness notions.

Theorem 2. *Let $\mathcal{O} = (S, f, \leftarrow)$ be some k-objective problem.*

1. *A-$\mathcal{O}$ NP-hard $\implies$ L-$\mathcal{O}$ NP-hard $\implies$ D-$\mathcal{O}$ NP-hard and W-$\mathcal{O}$ NP-hard*
2. *D-$\mathcal{O}$ NP-hard $\iff$ S-$\mathcal{O}$ NP-hard $\iff$ C$_i$-$\mathcal{O}$ NP-hard*

We can show that no further implications hold between the NP-hardness notions, unless P = NP. Such differences concerning the NP-hardness are not unusual. Below we give several examples of natural problems that are NP-hard with respect to one notion and that are polynomial-time solvable with respect to another one. This shows the importance of an exact specification of the NP-hardness notion that is used when discussing the complexity of multi-objective problems.

Let 2-LWF denote the scheduling problem of minimizing both the maximum lateness and the weighted flowtime of a schedule for some number of jobs with given processing times, due dates, and weights. This problem is one of the most studied 2-objective scheduling problems in the literature [8]. Let 2-LF denote the variant where all weights are set to 1. We further consider the 2-objective minimum quadratic diophantine equation problem.

2-QDE $= (S, f, \leq)$ where instances are triples $(a, b, c) \in \mathbb{N}^3$, $S^{(a,b,c)} = \{(x, y) \in \mathbb{N}^2 \mid ax^2 + by^2 - c \geq 0\}$, and $f^{(a,b,c)}(x, y) = (x^2, y^2)$.

The following table shows complexity and hardness results we obtained for all solution notions of the given problems, where 'P' indicates that this solution notion is polynomial-time solvable and L$_i$-$\mathcal{O}$ denotes the lexicographical problem where the i-th objective is the primary one. As a consequence, under the assumption P $\neq$ NP we can separate the NP-hardness notions and hence prove the strictness of the Turing-reduction order shown in Theorem 1 and Figure 1.

Problem $\mathcal{O}$	A-$\mathcal{O}$	L$_1$-$\mathcal{O}$	L$_2$-$\mathcal{O}$	W-$\mathcal{O}$	S-$\mathcal{O}$, D-$\mathcal{O}$, C$_i$-$\mathcal{O}$
2-LF	P	P	P	P	P
2-MM	P	P	P	P	NP-hard
2-QDE	P	P	P	NP-hard	P
2-LWF	P	NP-hard	P	NP-hard	NP-hard
2-TSP	NP-hard	NP-hard	NP-hard	NP-hard	NP-hard

5   Multi-objective Approximation

We define several approximation notions for multi-objective problems and study their relationships. Moreover, we show that if all objectives have to be minimized, then approximability results translate from single-objective to multi-objective optimization. Such translations are not possible for problems where all objectives have to be maximized, unless P = NP.

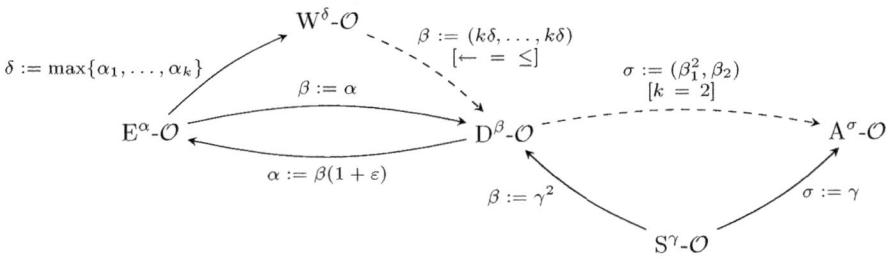

Fig. 2. Implications between polynomial-time solvability of approximate solution notions for any k-objective problem $\mathcal{O}$ ($\varepsilon > 0$ can be chosen arbitrary close to zero). Dashed lines indicate that the implication is shown only for the case that the additional condition [...] holds.

Let $\mathcal{O} = (S, f, \leftarrow)$ be a k-objective problem. We start with the α-approximate version of E-$\mathcal{O}$, where $\alpha = (a_1, \ldots, a_k)$ for $a_1, \ldots, a_k \geq 1$.

E$^\alpha$-$\mathcal{O}$ α-Approximate every-solution notion
Compute a set of solutions that α-dominates every solution.
Input: instance x
Output: some $S' \subseteq S^x$ such that $\forall s \in S^x \exists s' \in S'[s' \overset{\alpha}{\leftarrow} s]$

In Section 3 we argued that E-$\mathcal{O}$ is not an appropriate solution notion due to its large solutions. In contrast, approximations of E-$\mathcal{O}$ are very useful and feasible. Papadimitriou and Yannakakis [10] show that E$^\alpha$-$\mathcal{O}$ has polynomial-size outputs for every $\alpha = (1 + \varepsilon, \ldots, 1 + \varepsilon)$ with $\varepsilon > 0$.

A$^\alpha$-$\mathcal{O}$ α-Approximate arbitrary-optimum notion
Compute a solution that weakly α-dominates an arbitrary optimal solution.
Input: instance x
Output: an $s \in S^x$ such that $s \overset{\alpha}{\leftarrow} t$ for some $t \in S^x_{\mathrm{opt}}$ or report that $S^x = \emptyset$

S$^\alpha$-$\mathcal{O}$ α-Approximate specific-optimum notion
Compute a solution that weakly α-dominates an optimal solution specified by a given cost vector.
Input: instance x, cost vector $c \in \mathbb{N}^k$
Output: an $s \in S^x$ such that $s \overset{\alpha}{\leftarrow} t \overset{\alpha}{\leftarrow} c$ for some $t \in S^x_{\mathrm{opt}}$
            or report that there is no $s \in S^x$ such that $s \leftarrow c$

D^α-$\mathcal{O}$ α-**Approximate dominating-solution notion**

Compute a solution that weakly α-dominates a given cost vector.

Input: instance x, cost vector $c \in \mathbb{N}^k$

Output: some $s \in S^x$ such that $s \overset{\alpha}{\leftarrow} c$ or report that there is no $s \in S^x$ such that $s \leftarrow c$

We consider the following δ-approximations for $\delta \geq 1$. Again we define W^δ-$\mathcal{O}$ only for multi-objective problems where all objectives have to be minimized (resp., maximized). In this case, $\leftarrow_1$ below can be replaced by any of the $\leftarrow_i$.

W^δ-$\mathcal{O}$ δ-**Approximate weighted-sum notion**

Single-objective problem that weights the objectives in a given way.

Input: instance x, weight vector $w \in \mathbb{N}^k$

Output: some $s \in S^x$ such that $\sum_{i=1}^k w_i f_i^x(s) \overset{\delta}{\leftarrow}_1 \sum_{i=1}^k w_i f_i^x(s')$ for all $s' \in S^x$ or report that $S^x = \emptyset$

C_i^δ-$\mathcal{O}$ δ-**Approximate constraint notion for the i-th objective**

Single-objective problem that approximates the i-th objective while respecting constraints on the remaining objectives.

Input: instance x, constraints $b_1, \ldots, b_{i-1}, b_{i+1}, \ldots, b_k \in \mathbb{N}$

Output: for $S_{con}^x = \{s \in S^x \mid f_j^x(s) \leftarrow_j b_j \text{ for all } j \neq i\}$ return an $s \in S_{con}^x$ with $s \overset{\delta}{\leftarrow}_i \text{opt}_i(S_{con}^x)$ or report that $S_{con}^x = \emptyset$

We disregard approximations for the lexicographical problem L-$\mathcal{O}$, since here it is not clear how to measure the performance of the approximation. Observe that each of these notions coincides with its exact version if $\alpha = (1, \ldots, 1)$ or $\delta = 1$, respectively. Note that if $\mathcal{O}$ is a single-objective problem, then A^δ-$\mathcal{O}$, S^δ-$\mathcal{O}$, D^δ-$\mathcal{O}$, W^δ-$\mathcal{O}$, and C_1^δ-$\mathcal{O}$ are polynomial-time Turing equivalent to computing a δ-approximation for the single-objective problem.

Theorem 3 ([10]). *Let $\mathcal{O} = (S, f, \leftarrow)$ be a k-objective problem and $\alpha \in \mathbb{R}^k$. D^α-$\mathcal{O}$ polynomial-time solvable $\implies$ $E^{\alpha(1+\varepsilon)}$-$\mathcal{O}$ polynomial-time solvable*

Theorem 4. *Let $\mathcal{O} = (S, f, \leftarrow)$ be a k-objective problem and $\alpha \in \mathbb{R}^k$.*

1. *D^{α^2}-$\mathcal{O} \leq_T^P S^\alpha$-$\mathcal{O}$ and A^α-$\mathcal{O} \leq_T^P S^\alpha$-$\mathcal{O}$*
2. *$A^{(\alpha_1^2, \alpha_2)}$-$\mathcal{O} \leq_T^P D^\alpha$-$\mathcal{O}$ if $k = 2$*
3. *E^α-$\mathcal{O}$ polynomial-time solvable $\implies$ $W^{\max_i(\alpha_i)}$-$\mathcal{O}$ polynomial-time solvable*

We now show that for k-objective problems $\mathcal{O}$ where all objectives have to be minimized, approximability results translate from single-objective to multi-objective optimization where the relative error degrades only by the constant factor k.

Theorem 5. *For any k-objective problem $\mathcal{O} = (S, f, \leq)$ and any $\delta \geq 1$ it holds that $D^{(k\delta, \ldots, k\delta)}$-$\mathcal{O} \leq_T^P W^\delta$-$\mathcal{O}$.*

The following table shows examples for multi-objective approximability results obtained from known single-objective results by applying Theorem 5.

Problem $\mathcal{O}$	solvable in P	translation from
2-objective minimum vertex cover	$D^{(4,4)}$-$\mathcal{O}$	[1,9]
2-objective minimum TSP with repetitions	$D^{(3,3)}$-$\mathcal{O}$	[2]
2-objective minimum k-spanning tree	$D^{(6,6)}$-$\mathcal{O}$	[6]
2-objective minimum k-cut	$D^{(4,4)}$-$\mathcal{O}$	[11]

We argue that Theorem 5 does not hold for maximization. Let 2-CLIQUE$_{\text{restr}}$ denote the 2-objective maximum clique problem restricted to instances consisting of a graph G with labels $(1,1)$ and two additional nodes x, y with labels $(2|G| + 1, 0)$ and $(0, 2|G| + 1)$ that have no connections to other nodes.

Theorem 6. *For $\mathcal{O} = $ 2-CLIQUE$_{\text{restr}}$, W-$\mathcal{O}$ is polynomial-time solvable, but there is no $\alpha \in \mathbb{R}^2$ such that D^{α}-$\mathcal{O}$ is polynomial-time solvable, unless P $=$ NP.*

References

1. Bar-Yehuda, R., Even, S.: A local ratio theorem for approximating the weighted vertex cover problem. Analysis and Design of Algorithms for Combinatorial Problems. Annals of Discrete Mathematics 25, 27–46 (1985)
2. Christofides, N.: Worst-case analysis of a new heuristic for the travelling salesman problem. Tech. Rep. 388, Graduate School of Industrial Administration, Carnegie-Mellon University, Pittsburgh, PA (1976)
3. Ehrgott, M.: Multicriteria Optimization. Springer, Heidelberg (2005)
4. Ehrgott, M., Gandibleux, X.: A survey and annotated bibliography of multiobjective combinatorial optimization. OR Spectrum 22(4), 425–460 (2000)
5. Ehrgott, M., Gandibleux, X. (eds.): Multiple Criteria Optimization: State of the Art Annotated Bibliographic Survey. Kluwer's International Series in Operations Research and Management Science, vol. 52. Kluwer Academic Publishers, Dordrecht (2002)
6. Garg, N.: A 3-approximation for the minimum tree spanning k vertices. In: 37th Annual Symposium on Foundations of Computer Science, pp. 302–309. IEEE Computer Society Press, Los Alamitos (1996)
7. Glaßer, C., Reitwießner, C., Schmitz, H., Witek, M.: Hardness and approximability in multi-objective optimization. Tech. Rep. TR10-031, Electronic Colloquium on Computational Complexity (2010)
8. Lee, C.Y., Vairaktarakis, G.L.: Complexity of single machine hierarchical scheduling: A survey. In: Pardalos, P.M. (ed.) Complexity in Numerical Optimization, pp. 269–298. World Scientific, Singapore (1993)
9. Monien, B., Speckenmeyer, E.: Ramsey numbers and an approximation algorithm for the vertex cover problem. Acta Informatica 22(1), 115–123 (1985)
10. Papadimitriou, C.H., Yannakakis, M.: On the approximability of trade-offs and optimal access of web sources. In: FOCS 2000: Proceedings of the 41st Annual Symposium on Foundations of Computer Science, pp. 86–95. IEEE Computer Society, Washington (2000)
11. Saran, H., Vazirani, V.V.: Finding k-cuts within twice the optimal. In: 32nd Annual Symposium on Foundations of Computer Science, pp. 743–751. IEEE Computer Society Press, Los Alamitos (1991)
12. Selman, A.L.: A taxonomy on complexity classes of functions. Journal of Computer and System Sciences 48, 357–381 (1994)

$\mathcal{P}_w$ Is Not a Heyting Algebra

Kojiro Higuchi

Mathematical institute, Tohoku University, 6-3, Aramaki Aoba,
Aoba-ku, Sendai, Miyagi, Japan
sa7m24@math.tohoku.ac.jp

Abstract. Let $\mathcal{P}_w$ denote the set of weak degrees of nonempty Π_1^0 classes in the Cantor space 2^ω. We show that $\mathcal{P}_w$ is not a Heyting algebra. This is a solution to a question presented by Simpson [3].

Keywords: Π_1^0 classes, weak reducibility, Heyting algebras.

1 Introduction

A distributive lattice $\mathcal{L}$ with the top $\mathbf{1}$ and the bottom $\mathbf{0}$ is said to be a *Heyting algebra* if for any $a, b \in \mathcal{L}$ there exists the maximal element c satisfying that $\inf(a, c) \leq b$. And a distributive lattice is said to be a *Brouwer algebra* if the dual lattice is a Heyting algebra. Terwijn[5] has shown that the lattice $\mathcal{P}_s$ of strong degrees of nonempty Π_1^0 classes in the Cantor space 2^ω is not a Heyting algebra. Simpson[3] has shown that the lattice $\mathcal{P}_w$ of weak degrees of nonempty Π_1^0 classes in 2^ω is not a Brouwer algebra. We show that $\mathcal{P}_w$ is not a Heyting algebra. This is a solution to a problem presented in Simpson[3].

Let us recall the definitions of $\mathcal{P}_s$ and $\mathcal{P}_w$. Let ω denote the set of natural numbers and let 2^ω denote the Cantor space, $2^\omega = \{f | f : \omega \to \{0, 1\}\}$. For $P, Q \subset 2^\omega$ we say that P is *strongly reducible* to Q, written $P \leq_s Q$, if there exists a computable functional Φ such that $\Phi(g) \in P$ for any $g \in Q$. We say that P is *weakly reducible* to Q, written $P \leq_w Q$, if for any $g \in Q$ there exists $f \in P$ such that $f \leq_T g$, where $\leq_T$ denotes the Turing reducibility. For a reducibility $\leq_r$ we write $P \equiv_r Q$ if $P \leq_r Q$ and $Q \leq_r P$. $P \subset 2^\omega$ is said to be a Π_1^0 *class* if there exists a computable relation R such that $P = \{f \in 2^\omega | (\forall n) R(n, f)\}$. The relations $\equiv_s$ and $\equiv_w$ are equivalence relations among nonempty Π_1^0 classes. The equivalence class of a nonempty Π_1^0 class P under the relation $\equiv_s$ is called the *strong degree* of P and is denoted by $\deg_s(P)$. The equivalence class of a nonempty Π_1^0 class P under the relation $\equiv_w$ is called the *weak degree* of P and is denoted by $\deg_w(P)$. Let $\mathcal{P}_s$ and $\mathcal{P}_w$ denote the order structures of strong degrees and weak degrees, respectively. In fact, the degree of 2^ω is the bottom element and the degree of **CPA**, the set of completions of Peano Arithmetic, is the top element in either structure. $\mathcal{P}_s$ and $\mathcal{P}_w$ are, indeed, distributive lattices under the following operations. For $f_0, f_1 \in 2^\omega$ we define the function $f_0 \oplus f_1$ by $(f_0 \oplus f_1)(2n + i) = f_i(n)$ for any $n \in \omega$ and any $i < 2$. $2^{<\omega}$ denotes the set of binary strings. For $\sigma \in 2^{<\omega}$, let $\sigma^\frown P = \{\sigma^\frown f | f \in P\}$, where $^\frown$ denotes concatenation. We define $P \vee Q$ and $P \wedge Q$ by

F. Ferreira et al. (Eds.): CiE 2010, LNCS 6158, pp. 190–194, 2010.
© Springer-Verlag Berlin Heidelberg 2010

$$P \vee Q = \{f \oplus g | f \in P \ \& \ g \in Q\}$$

and

$$P \wedge Q = \langle 0 \rangle ^\frown P \cup \langle 1 \rangle ^\frown Q.$$

The equivalence relation $\equiv_s$ and $\equiv_w$ are congruent to these operations $\vee$ and $\wedge$. So we can consider these operations $\vee$ and $\wedge$ as the operations on $\mathcal{P}_s$ and $\mathcal{P}_w$ and one can easily show that the operation $\vee$ is the supremum operation and the operation $\wedge$ is the infimum operation in either structure.

2 $\mathcal{P}_w$ Is Not a Heyting Algebra

Let $\mathcal{L}$ be a distributive lattice with the top and the bottom and let $\vee$ and $\wedge$ be the supremum operation and the infimum operation of $\mathcal{L}$ respectively. In order to show that $\mathcal{L}$ is not a Heyting algebra, we should find $a, b \in \mathcal{L}$ such that for any $c \in \mathcal{L}$ satisfying $a \wedge c \leq b$, $\mathcal{L}$ has an element $\widehat{c} > c$ satisfying $a \wedge \widehat{c} \leq b$. But, in fact, the following lemma holds.

Lemma 1. *The following are equivalent:*
(i) $\mathcal{L}$ is not a Heyting algebra,
(ii) There exist $a, b \in \mathcal{L}$ such that for any $c \geq b$ satisfying $a \wedge c \leq b$, $\mathcal{L}$ has an element $\widehat{c} \not\leq c$ satisfying $a \wedge \widehat{c} \leq b$.

Proof. Trivially, (i) implies (ii). Assume (ii) holds and let $a, b \in \mathcal{L}$ satisfying the condition (ii). It suffices to show that

$$\forall c \in \mathcal{L}[a \wedge c \leq b \Longrightarrow \exists d \in \mathcal{L}[d > c \ \& \ a \wedge d \leq b]].$$

Let $c \in \mathcal{L}$ such that $a \wedge c \leq b$. If $c \not\geq b$, define $d = b \vee c$. Then $d = b \vee c > c$ and $a \wedge d = a \wedge (b \vee c) = (a \wedge b) \vee (a \wedge c) \leq b$. If $c \geq b$, $\mathcal{L}$ has an element $\widehat{c} \not\leq c$ such that $a \wedge \widehat{c} \leq b$ because of (ii). Define $d = \widehat{c} \vee c$. $d = \widehat{c} \vee c > c$ and $a \wedge d = a \wedge (\widehat{c} \vee c) = (a \wedge \widehat{c}) \vee (a \wedge c) \leq b$. Therefore, $\mathcal{L}$ is not a Heyting algebra. $\square$

We want to construct nonempty Π_1^0 classes P and Q which weak degrees satisfy the condition (ii) of the previous lemma. We need two lemmas which have been proved by Simpson[2].

Lemma 2. *Let P and Q be nonempty Π_1^0 classes such that $P \leq_w Q$. Then there exists a nonempty Π_1^0 class $\widehat{Q} \subset Q$ such that $P \leq_s \widehat{Q}$.*

Proof. See [2, Lemma 6.9.]. $\square$

Lemma 3 (The Embedding Lemma). *Let P be a nonempty Π_1^0 class and let Q be a Σ_3^0 class. Then there exists a nonempty Π_1^0 class R such that $R \equiv_w P \cup Q$.*

Proof. See [4, Lemma 3.3.]. $\square$

In the proof of our main theorem 2, we will use the embedding lemma for **CPA** and $(Q \setminus \{g\})$, where Q is a nonempty Π_1^0 class and g is a Δ_2^0 set in Q. Note that $(Q \setminus \{g\})$ is a Σ_2^0 class. The following theorem is proved in the next section.

Theorem 1. *There exist nonempty Π_1^0 classes P and Q such that*
(a) $\mathbf{CPA} \cup (Q \setminus \{g\}) \not\leq_w \{g\}$ for any $g \in Q$,
(b) $P \not\leq_s \widehat{Q}$ for any nonempty Π_1^0 class $\widehat{Q} \subset Q$,
(c) for any nonempty Π_1^0 class $\widehat{Q} \subset Q$, there exists a Δ_2^0 $g \in \widehat{Q}$ such that
$P \leq_w \{g\}$.

Now we can prove our main theorem.

Theorem 2. $\mathcal{P}_w$ *is not a Heyting algebra.*

Proof. Let P and Q be nonempty Π_1^0 classes of theorem 1. The condition (ii) of lemma 1 holds via $\deg_w(P)$ and $\deg_w(Q)$ if and only if for any nonempty Π_1^0 class S which satisfies $S \geq_w Q$ and $P \wedge S \leq_w Q$, there exists a nonempty Π_1^0 class $\widehat{S}$ such that $\widehat{S} \not\leq_w S$ and $P \wedge \widehat{S} \leq_w Q$. We show the latter statement of this equivalence.

Let S be a nonempty Π_1^0 class such that $S \geq_w Q$ and $P \wedge S \leq_w Q$. From lemma 2, there exists a nonempty Π_1^0 class $\widehat{Q} \subset Q$ such that $P \wedge S \leq_s \widehat{Q}$. Since $P \not\leq_s \widehat{Q}$, there exists a nonempty Π_1^0 class $R \subset \widehat{Q}$ such that $S \leq_s R$. From (c), $R \subset \widehat{Q} \subset Q$ implies that there exists a Δ_2^0 $g \in R$ such that $P \leq_w \{g\}$. Let g be such an element of R. From lemma 3, there exists a nonempty Π_1^0 class $\widehat{S}$ such that $\widehat{S} \equiv_w \mathbf{CPA} \cup (Q \setminus \{g\})$. We show that $\widehat{S} \not\leq_w S$ and $P \wedge \widehat{S} \leq_w Q$. First, there exists a $h \in S$ such that $h \leq_T g$ since $S \leq_s R$. $S \geq_w Q$ implies there exists a $g_0 \in Q$ such that $g_0 \leq_T h$. From (a), we conclude that $g_0 = g \equiv_T h$ and therefore $\widehat{S} \equiv_w \mathbf{CPA} \cup (Q \setminus \{g\}) \not\leq_w \{g\} \equiv_w \{h\}$. Hence $\widehat{S} \not\leq_w S$ since $h \in S$. Second, let $f \in Q$. If $f = g$, then $\{f\} = \{g\} \geq_w P \geq_w P \wedge \widehat{S}$. Otherwise, $\{f\} \geq_w Q \setminus \{g\} \geq_w \mathbf{CPA} \cup (Q \setminus \{g\}) \equiv_w \widehat{S} \geq_w P \wedge \widehat{S}$. Thus $P \wedge \widehat{S} \leq_w \{f\}$ for any $f \in Q$. This means that $P \wedge \widehat{S} \leq_w Q$. $\square$

3   The Proof of Theorem 1

To obtain nonempty Π_1^0 classes P, Q of theorem 1, we use two theorems which have been proved by Jockusch and Soare[1]. The first one is a special case of Theorem 2.5. of Jockusch and Soare[1].

Theorem 3. *Let Q be a nonempty Π_1^0 class and let $S \subset 2^\omega$ be a countable class with no computable element. Then $S \not\leq_w Q$.*

Proof. See [1, Theorem 2.5.]. $\square$

For $f, g \in 2^\omega$, we denote $f|_T g$ if $f \not\leq_T g$ and $f \not\geq_T g$.

Theorem 4. *There exists a nonempty Π_1^0 class Q with no computable element such that $f|_T g$ for any distinct elements $f, g \in Q$.*

Proof. See [1, Theorem 4.7.]. $\square$

Lemma 4. *Let Q be a nonempty Π_1^0 class with no computable element such that $f|_T g$ for any distinct elements $f, g \in Q$. Then $\mathbf{CPA} \cup (Q \setminus \{g\}) \not\leq_w \{g\}$ for any $g \in Q$.*

Proof. Let $g \in Q$. From the assumption, $f \not\leq_T g$ for any $f \in (Q \setminus \{g\})$. So $(Q \setminus \{g\}) \not\leq_w \{g\}$. Since g is not computable, $(Q \setminus \{g\}) \neq \emptyset$. Let $f \in (Q \setminus \{g\})$ and $\sigma = f \cap g$. Then the nonempty Π_1^0 class $S = \{h \in Q \mid \sigma^\frown \langle f(|\sigma|) \rangle \subset h\}$ satisfies that $S \subset (Q \setminus \{g\})$ and $S \leq_w \mathbf{CPA}$. Since $(Q \setminus \{g\}) \leq_w S \leq_w \mathbf{CPA}$ and $(Q \setminus \{g\}) \not\leq_w \{g\}$, $\mathbf{CPA} \not\leq_w \{g\}$. Hence, $\mathbf{CPA} \cup (Q \setminus \{g\}) \not\leq_w \{g\}$ for any $g \in Q$. $\qquad\square$

Let $\{W_e\}_{e \in \omega}$ be a standard computable listing of all c.e. sets. We may view W_e as a set of strings since there is a computable bijection between ω and $2^{<\omega}$. For $\Sigma \subset 2^{<\omega}$, $P \subset 2^\omega$ and $n \in \omega$, we define $[\Sigma] = \{\sigma^\frown f \in 2^\omega \mid \sigma \in \Sigma\}$ and $P \upharpoonright n = \{f \upharpoonright n \mid f \in P\}$. Let $R_e = 2^\omega \setminus [W_e]$ and $R_{e,s} = 2^\omega \setminus [W_{e,s}]$ for any $e, s \in \omega$. Then $\{R_e\}_{e \in \omega}$ is a computable listing of all Π_1^0 classes and $R_e = \bigcap_{s \in \omega} R_{e,s}$ for any $e \in \omega$. We may safely assume that $R_{e,s} = [R_{e,s} \upharpoonright n]$. Let Q be a nonempty Π_1^0 class and let $g \in Q$. We say g is the *left most path* of Q if for any $f \in (Q \setminus \{g\})$ there exists a string $\sigma \in 2^{<\omega}$ such that $\sigma^\frown \langle 0 \rangle \subset g$ and $\sigma^\frown \langle 1 \rangle \subset f$. Note that the left most path of $R_{e,s}$ is computable uniformly in e and s.

Lemma 5. *The left most path of a nonempty Π_1^0 class is Δ_2^0.*

Proof. Assume $R_e \neq \emptyset$. Let g_s be the left most path of $R_{e,s}$ and g be the left most path of R_e. Then $g(x) = \lim_{s \to \infty} g_s(x)$ for any $x \in \omega$. Hence g is Δ_2^0. $\qquad\square$

Let $\sigma \in 2^{<\omega}$ and let $f \in 2^\omega$. We say that f is *on* σ if $\sigma \subset f$.

Lemma 6. *Let Q be a nonempty Π_1^0 class. Then*

$$\bigcup_{e \in \omega} \{g \mid g \text{ is the left most path of } R_e \cap Q\}$$

is Σ_3^0.

Proof. Let h be a computable function such that $R_{h(e)} = R_e \cap Q$ for any $e \in \omega$. Then

$$f \in \bigcup_{e \in \omega} \{g \mid g \text{ is the left most path of } R_e \cap Q\}$$

$$\Longleftrightarrow (\exists e)[f \in R_{h(e)} \ \& \ (\forall \sigma \subset f)(\exists s)[\text{the left most path of } R_{h(e),s} \text{ is on } \sigma]].$$

$\qquad\square$

We restate our main theorem 5 of this section and prove it.

Theorem 5. *There exist nonempty Π_1^0 classes P and Q such that*
(a) $\mathbf{CPA} \cup (Q \setminus \{g\}) \not\leq_w \{g\}$ for any $g \in Q$,
(b) $P \not\leq_s \widehat{Q}$ for any nonempty Π_1^0 class $\widehat{Q} \subset Q$,
(c) for any nonempty Π_1^0 class $\widehat{Q} \subset Q$, there exists a Δ_2^0 $g \in \widehat{Q}$ such that $P \leq_w \{g\}$.

Proof. Let Q be a nonempty Π_1^0 class with no computable element such that $f|_T g$ for any distinct elements $f, g \in Q$. From lemma 4, Q satisfies the condition (a). Since $\bigcup_{e \in \omega} \{g \mid g$ is the left most path of $R_e \cap Q\}$ is Σ_3^0, we can use lemma 3 to obtain a nonempty Π_1^0 class P such that

$$P \equiv_w \mathbf{CPA} \cup \bigcup_{e \in \omega} \{g \mid g \text{ is the left most path of } R_e \cap Q\}.$$

Clearly, P and Q satisfy the condition (c). We claim that P and Q satisfy the condition (b).

Suppose that there exists a nonempty Π_1^0 class $\widehat{Q} \subset Q$ such that $P \leq_s \widehat{Q}$. $\bigcup_{e \in \omega} \{g \mid g$ is the left most path of $R_e \cap Q\} \leq_w \widehat{Q}$, since $\mathbf{CPA} \not\leq_w \{g\}$ for any $g \in \widehat{Q}$. This contradicts to theorem 3. Hence the condition (b) holds.   □

References

1. Jockusch Jr., C.G., Soare, R.I.: Π_1^0 classes and degrees of theories. Transactions of the American Mathematical Society 173, 35–56 (1972)
2. Simpson, S.G.: Mass problems and randomness. Bulletin of Symbolic Logic 11, 1–27 (2005)
3. Simpson, S.G.: Mass problems and intuitionism. Notre Dame Journal of Formal Logic 49, 127–136 (2008)
4. Simpson, S.G.: An extension of the recursively enumerable Turing degrees. Journal of the London Mathematical Society 75, 287–297 (2007)
5. Terwijn, S.A.: The Medvedev lattice of computably closed sets. Archive for Mathematical Logic 45, 179–190 (2006)

Lower Bounds for Reducibility to the Kolmogorov Random Strings[*]

John M. Hitchcock

Department of Computer Science
University of Wyoming

Abstract. We show the following results for polynomial-time reducibility to R_C, the set of Kolmogorov random strings.
1. If P $\neq$ NP, then SAT does not dtt-reduce to R_C.
2. If PH does not collapse, then SAT does not n^α-tt-reduce to R_C for any $\alpha < 1$.
3. If PH does not collapse, then SAT does not n^α-T-reduce to R_C for any $\alpha < \frac{1}{2}$.
4. There is a problem in E that does not dtt-reduce to R_C.
5. There is a problem in E that does not n^α-tt-reduce to R_C, for any $\alpha < 1$.
6. There is a problem in E that does not n^α-T-reduce to R_C, for any $\alpha < \frac{1}{2}$.

These results hold for both the plain and prefix-free variants of Kolmogorov complexity and are also independent of the choice of the universal machine.

1 Introduction

Because the Kolmogorov complexity function $C(x)$ is noncomputable, the set

$$R_C = \{x \mid C(x) > |x|\}$$

of Kolmogorov random strings is undecidable. In fact, R_C has no infinite computably enumerable subset. From this and the fact that the complement R_C^c is computably enumerable, Arslanov's completeness criterion implies that R_C is hard for the c.e. sets under Turing reductions. Kummer [7] showed a stronger result: $H^c \leq_{\text{dtt}} R_C$, where H^c is the complement of the halting problem and $\leq_{\text{dtt}}$ denotes a disjunctive truth-table reduction. Neither of these reductions from the halting problem to R_C is efficient. This raises the question: what can be efficiently reduced to R_C?

Recall that the Kolmogorov complexity [11] of a binary string x is the length of a shortest program that prints x on a universal Turing machine U:

$$C_U(x) = \min\{|p| \mid U(p) \text{ prints } x\}.$$

[*] This research was supported in part by NSF grant 0652601 and by an NWO travel grant. Part of this research was done while the author was on sabbatical at CWI.

F. Ferreira et al. (Eds.): CiE 2010, LNCS 6158, pp. 195–200, 2010.

For the most part, the theory of Kolmogorov complexity does not depend on the choice of the universal machine U: for any two universal machines U and V, C_U and C_V are within an additive constant of each other. As usual, we fix a universal machine U and omit it from the notation, writing $C(x)$ instead of $C_U(x)$. There will, however, be situations when the choice of universal machine matters and then we will be explicit with the subscript. We use the notation $P_\tau(A)$ to denote the class of problems that reduce to A by $\leq_\tau^P$-reductions.

Kummer's result [7] implies there is a computable time bound $t(n)$ such that for every decidable A, $A \leq_{\mathrm{dtt}}^{t(n)} R_C$. Kummer's proof is nonconstructive and does not yield any information about the function $t(n)$. In fact, Allender et al. [1] show that some uncertainty about the time bound $t(n)$ is inevitable. They show that the $t(n)$ in Kummer's theorem may be arbitrarily large, depending on the choice of the universal machine U. Formally, for every computable time bound $t(n)$, there exists a universal machine U and a decidable set A such that A does not $\leq_{\mathrm{dtt}}^{t(n)}$-reduce to R_{C_U}. On the other hand, independent of U, there exist sets with arbitrarily high time complexity that reduce to R_{C_U} via a polynomial-time dtt-reduction: for every computable time bound $t(n)$ and every universal machine U, there is a set $A \in \mathrm{DEC} - \mathrm{DTIME}(t(n))$ such that $A \leq_{\mathrm{dtt}}^{P} R_{C_U}$. While this result shows $P_{\mathrm{dtt}}(R_C)$ contains sets of high time complexity, the set A in this theorem is constructed via padding, which makes A very sparse. Thus while A has high time complexity, A is very simple in other terms. We show that this simplicity is inherent: any such A is highly predictable in the sense of polynomial-time dimension. From this it follows that R_C is not hard for E under $\leq_{\mathrm{dtt}}^{P}$-reductions. This holds for every universal machine, i.e. $\mathrm{E} \not\subseteq P_{\mathrm{dtt}}(R_{C_U})$ for every U. We also show that R_C is not hard for NP unless $P = NP$. Both of these results follow from showing that if a decidable set $\leq_{\mathrm{dtt}}^{P}$-reduces to R_C, then the set $\leq_{\mathrm{dtt}}^{P}$-reduces to a tally set. These results complement the result of Allender et al. [1] that

$$P = \mathrm{DEC} \cap \bigcap_U P_{\mathrm{dtt}}(R_{C_U}),$$

where the intersection is over all universal machines. While the class $\mathrm{DEC} \cap P_{\mathrm{dtt}}(R_{C_U})$ contains arbitrarily complex sets, it is intuitively "close" to P for every U, in that it has small dimension and cannot contain NP unless $P = NP$.

Allender et al. [2] showed that R_C is hard for PSPACE under polynomial-time Turing reductions: $\mathrm{PSPACE} \subseteq P_T(R_C)$. Recently, Buhrman et al. [3] showed that R_C is hard for BPP under polynomial-time truth-table reductions: $\mathrm{BPP} \subseteq P_{\mathrm{tt}}(R_C)$. We consider bounded query Turing and truth-table reductions. Based on the Winnow algorithm [8] and polynomial-time dimension [6], we show that R_C is not $\leq_{n^\alpha\text{-tt}}^{P}$-hard for E, for any $\alpha < 1$. This is an improvement of a result in [1] which obtained the same consequence for EE. Also, we use the techniques of [5,4] to show that R_C is not $\leq_{n^\alpha\text{-tt}}^{P}$-hard for NP unless $\mathrm{NP} \subseteq \mathrm{coNP/poly}$ and the polynomial-time hierarchy collapses by Yap's theorem [13]. Finally, we obtain the same consequences for $\leq_{n^\alpha\text{-T}}^{P}$-reductions, for all $\alpha < \frac{1}{2}$.

2   Preliminaries

We use standard notions of polynomial-time reducibilities [9]. We also need the following two notions of reducibility.

Definition 1. *Let $\mathcal{B} = (B_n \mid n \geq 0)$ be a family of subsets of $\{0,1\}^*$. We say that A NP-reduces to $\mathcal{B}$ if there is an NPMV function N such that for all n, for all $x \in \{0,1\}^n$, $x \in A$ iff at least one output of $N(x)$ is in B_n.*

Definition 2. *Let $\mathcal{B} = (B_n \mid n \geq 0)$ be a family of subsets of $\{0,1\}^*$. We say that A disjunctively reduces to $\mathcal{B}$ in $t(n)$ time if there is an algorithm M such that for all n, for all $x \in \{0,1\}^n$, $M(x)$ outputs a list of strings in $t(n)$ time and $x \in A$ iff at least one output of $M(x)$ is in B_n.*

The following lemma is from [4], based on a technique of [5]. An AND-function (of order 1) for a set A is a polynomial-time computable function g such that for all strings $x_1, x_2, \ldots, x_n$, $|g(x_1, \ldots, x_n)| = O\left(\sum_{i=1}^{n} |x_i|\right)$ and $g(x_1, x_2, \ldots, x_n) \in A$ iff $x_i \in A$ for all i.

Lemma 1. *Let A have an AND-function and let $\alpha < 1$. Let $\mathcal{B} = (B_n \mid n \geq 0)$ be a family of sets with $|B_n| \leq 2^{n^\alpha}$ for sufficiently large n. If A NP-reduces to $\mathcal{B}$, then $A \in \mathrm{NP/poly}$.*

The p-dimension [10] of a complexity class is a real number in $[0, 1]$. The p-dimension of P is 0 and the p-dimension of E is 1. For this paper, we do not need the full details of p-dimension; all we require is the fact that a p-dimension 0 class cannot contain E and the following lemma. The proof of this lemma relies on the Winnow online learning algorithm [8] and is straightforward to prove using the approach of [6].

Lemma 2. *Let $\alpha < 1$ and let $c \geq 1$. Let X be the class of all A for which there exists a family $\mathcal{B} = (B_n \mid n \geq 0)$ with $|B_n| \leq 2^{n^\alpha}$ such that A disjunctively reduces to $\mathcal{B}$ in 2^{cn} time. Then X has p-dimension 0. In particular, X does not contain E.*

3   Disjunctive Reductions

Theorem 1. *If A is decidable and $A \leq_{\mathrm{dtt}}^{\mathrm{P}} R_C$, then $A \leq_{\mathrm{dtt}}^{\mathrm{P}} B$ for some $B \in$ TALLY.*

Proof. We use the proof technique from [1] that A is decidable and $A \leq_{\mathrm{mtt}}^{\mathrm{P}} R_C$ (monotone truth-table) implies $A \in \mathrm{P/poly}$, observing that we can encode in a tally set to obtain the stronger result.

Suppose A is decidable and $A \leq_{\mathrm{dtt}}^{\mathrm{P}} R_C$ via a reduction computable in time n^d. Let the queries on input x be denoted by $Q(x)$. For some constant c, we claim only the queries of length at most $l(n) = c \log n$ "matter."

For any x, we have $x \in A$ iff $Q(x) \cap R_C \neq \emptyset$. Define $Q'(x) = Q(x) \cap \Sigma^{\leq l(n)}$, where $n = |x|$. We claim that for each $x \in A$, there is some $q \in Q'(x)$ such that for all y with $|y| = |x|$, $q \in Q'(y)$ implies $y \in A$.

Suppose the claim is false. Then given n, we can find the first string x of length n such that $x \in A$ and each query $q \in Q'(x)$ belongs to $Q'(y)$ for some $y \notin A$. This implies that $Q'(x) \cap R_C = \emptyset$. Since $x \in A$, it follows that $Q(x) - Q'(x)$ contains a string $r \in R_C$. This string r has $C(r) > l(n)$ because $r \notin Q'(x)$. We can describe r by describing n and the index of r in $Q(x)$. Since $|Q(x)| \leq n^d$, this takes at most $(d+3)\log n$ bits, a contradiction if we choose $c = d + 4$.

Let $\{w_1, \ldots, w_N\}$ be an enumeration of $\Sigma^{\leq l(n)}$. Let I_n be the collection of all i where for all y of length n, $w_i \in Q(y)$ implies $y \in A$. Our desired tally set is $\{0^{\langle n, i \rangle} \mid n \geq 0 \text{ and } i \in I_n\}$, where $\langle \cdot, \cdot \rangle$ is a pairing function on the natural numbers.   □

Corollary 1. *If* $\mathrm{NP} \subseteq \mathrm{P_{dtt}}(R_C)$*, then* $\mathrm{P} = \mathrm{NP}$*.*

Proof. By Theorem 1, SAT $\leq^{\mathrm{P}}_{\mathrm{dtt}} B$ for a tally set B. Then $\mathrm{SAT}^c \leq^{\mathrm{P}}_{\mathrm{ctt}} B^c \cap 0^*$. Ukkonen [12] showed that $\mathrm{P} = \mathrm{NP}$ if coNP has a sparse $\leq^{\mathrm{P}}_{\mathrm{ctt}}$-hard set.   □

Because $\mathrm{P_{dtt}}(\mathrm{SPARSE})$ has p-dimension 0 [6], we obtain the following.

Corollary 2. *The class* $\mathrm{P_{dtt}}(R_C) \cap \mathrm{DEC}$ *has p-dimension 0.*

Corollary 3. $\mathrm{E} \not\subseteq \mathrm{P_{dtt}}(R_C)$*.*

4   Truth-Table Reductions

Theorem 2. *Let* $\alpha < 1$*.*

1. *If* A *is decidable,* A *has an AND-function, and* $A \leq^{\mathrm{P}}_{n^\alpha\text{-tt}} R_C$*, then* $A \in$ NP/poly.
2. *The class* $\mathrm{P}_{n^\alpha-\mathrm{tt}}(R_C) \cap \mathrm{DEC}$ *has p-dimension 0.*

Proof. The main idea of the proof is from [1]. We reproduce the argument here and show how to apply Lemmas 1 and 2.

Let A be decidable such that $A \leq^{\mathrm{P}}_{n^\alpha-\mathrm{tt}} R_C$. Write $Q(x)$ for the truth-table reduction's queries on input x and $Z_x \subseteq \Sigma^{n^\alpha}$ for the query answer sequences that cause the reduction to accept x. That is, if $Q(x) = \{q_1, \ldots, q_{n^\alpha}\}$ in lexicographic order, then $x \in A$ if and only if $R_C[q_1] \cdots R_C[q_{n^\alpha}] \in Z_x$.

Let $l(n) = n^\epsilon$, where $0 < \epsilon < 1 - \alpha$. We claim that the truth-table reduction is still correct if we only use the queries of length at most $l(n)$. Formally, let $Q'(x) = Q(x) \cap \Sigma^{\leq l(n)}$ and let Z'_x be the restriction of Z_x with bits corresponding to strings in $Q(x) - Q'(x)$ removed.

Call two strings x and y equivalent if $Q'(x) = Q'(y)$. We claim that for each $x \in A$, there is some $z_x \in Z'(x)$ such that for all y equivalent to x, $z_x \in Z'(y)$ iff $y \in A$.

Suppose the claim is false. We can find the least $x \in A$ such that for all $z \in Z'(x)$, there is some y_z equivalent to x such that $z \in Z'(y)$ iff $y_z \notin A$. Let v be the correct answer sequence for $Q'(x) \cap R_C$ and let r be the number of 1's in v; that is, $r = |Q'(x) \cap R_C|$. Given x and r, we can enumerate R_C^c to compute $Q'(x) \cap R_C$ and obtain v. Then we can compute y_v such that query answers v are incorrect for y_v. This means that $Q(y_v) - Q'(y_v)$ must contain a string in R_C with length $> l(n)$. However, we can describe this string by describing n, r, and its index in $Q(y_v)$, which takes $O(\log n)$ bits, a contradiction.

We define a family of sets $\mathcal{B} = (B_n \mid n \geq 0)$ as follows. For each equivalence class $[x]$ with queries $Q'(x) = \{w_1, \ldots, w_{n^\alpha}\}$ and $z_x \in Z'(x)$ the answer sequence that is correct for all strings in the equivalence class, we put the tuple $\langle w_1, \ldots, w_{n^\alpha}, z_x \rangle$ in B_n. Note that $|B_n| < 2^{n^\gamma}$ where $\alpha + \epsilon < \gamma < 1$. By the claim, A NP-reduces to $\mathcal{B}$. It follows from Lemma 1 that $A \in \text{NP}/\text{poly}$ if A has an AND-function.

We also have that A is disjunctively reducible in 2^n time to $\mathcal{B}$. Therefore Lemma 2 applies to show $P_{n^\alpha - \text{tt}}(R_C) \cap \text{DEC}$ has p-dimension 0.   □

Corollary 4. *If* $\text{NP} \subseteq P_{n^\alpha - \text{tt}}(R_C)$ *for some* $\alpha < 1$, *then* $\text{NP} \subseteq \text{coNP}/\text{poly}$.

Proof. This follows from Theorem 2 because the hypothesis implies $\text{SAT}^c \leq^p_{n^\alpha - \text{tt}} R_C$ and SAT^c has an AND-function.   □

Corollary 5. *For* $\alpha < 1$, $\text{E} \nsubseteq P_{n^\alpha - \text{tt}}(R_C)$.

5   Turing Reductions

Theorem 3. *Let* $\alpha < \frac{1}{2}$.

1. *If* A *is decidable,* A *has an AND-function, and* $A \leq^p_{n^\alpha - \text{T}} R_C$, *then* $A \in \text{NP}/\text{poly}$.
2. *The class* $P_{n^\alpha - \text{T}}(R_C) \cap \text{DEC}$ *has* p-*dimension 0.*

Proof. Let $\alpha < \beta < \frac{1}{2}$. Suppose that $A \in \text{DEC}$ and $A \leq^p_{n^\alpha - \text{T}} R_C$ via M. Let M' be the Turing machine that simulates M and whenever M makes a query of length at least n^β, M' makes no query and proceeds as if the answer to the query were no. We use the following concepts:

- An *advice* is a tuple $(z, w_1, \ldots, w_{n^\alpha})$ such that $z \in \Sigma^{n^\alpha}$ and each $w_i \in \Sigma^{< n^\beta}$.
- A string y is *accepted with advice* $(z, w_1, \ldots, w_{n^\alpha})$ if $M'(y)$ queries $w_1, \ldots, w_{n^\alpha}$ and accepts y when M' is given $z[1], \ldots, z[n^\alpha]$ as the query answers.
- An advice $(z, w_1, \ldots, w_{n^\alpha})$ is *safe* if for all $y \in \Sigma^n$, y is accepted with advice $(z, w_1, \ldots, w_{n^\alpha})$ implies $y \in A$.

We claim that for all $x \in A_{=n}$, there is a safe advice $(z, \boldsymbol{w})$ such that x is accepted with advice $(z, \boldsymbol{w})$.

Suppose the claim is false. Then we can find the least $x \in A_{=n}$ that does not have a safe advice. We can specify the correct answer sequence $z \in \Sigma^{n^\alpha}$ for

$M(x)$ when querying oracle R_C. With this correct answer sequence z, M must query some string in R_C that is not in $\Sigma^{<n^\beta}$. Therefore we can describe a string r with $C(r) \geq n^\beta$ by describing n, z, and the index of r in $M(x)$'s query set on query answer sequence z. Thus $C(r) \leq n^\alpha + O(\log n)$, which is a contradiction since $\alpha < \beta$.

We define a family of sets $\mathcal{B}$ by putting into B_n all advices $(z, w_1, \ldots, w_{n^\alpha})$ that are safe. Let $1 > \gamma > \alpha + \beta$. The total number of possible advices is at most $2^{n^\alpha} \cdot (2^{n^\beta})^{n^\alpha} < 2^{n^\gamma}$, so $|B_n| < 2^{n^\gamma}$. We have that A NP reduces to $\mathcal{B}$ and A disjunctively reduces in 2^n time to $\mathcal{B}$, so the theorem follows from Lemmas 1 and 2. ☐

Corollary 6. *If* $\mathrm{NP} \subseteq \mathrm{P}_{n^\alpha - \mathrm{T}}(R_C)$ *for some* $\alpha < \frac{1}{2}$, *then* $\mathrm{NP} \subseteq \mathrm{coNP/poly}$.

Corollary 7. *For* $\alpha < \frac{1}{2}$, $\mathrm{E} \not\subseteq \mathrm{P}_{n^\alpha - \mathrm{T}}(R_C)$.

References

1. Allender, E., Buhrman, H., Koucký, M.: What can be efficiently reduced to the Kolmogorov-random strings? Annals of Pure and Applied Logic 138, 2–19 (2006)
2. Allender, E., Buhrman, H., Koucký, M., van Melkebeek, D., Ronneburger, D.: Power from random strings. SIAM Journal on Computing 35, 1467–1493 (2006)
3. Buhrman, H., Fortnow, L., Koucký, M., Loff, B.: Derandomizing from random strings. Technical Report 0912.3162, arXiv.org e-Print archive (2009)
4. Buhrman, H., Hitchcock, J.M.: NP-hard sets are exponentially dense unless NP ⊆ coNP/poly. In: Proceedings of the 23rd Annual IEEE Conference on Computational Complexity, pp. 1–7. IEEE Computer Society, Los Alamitos (2008)
5. Fortnow, L., Santhanam, R.: Infeasibility of instance compression and succinct PCPs for NP. In: Proceedings of the 40th Annual ACM Symposium on Theory of Computing, pp. 133–142 (2008)
6. Hitchcock, J.M.: Online learning and resource-bounded dimension: Winnow yields new lower bounds for hard sets. SIAM Journal on Computing 36(6), 1696–1708 (2007)
7. Kummer, M.: On the complexity of random strings. In: Proceedings of the 13th Annual Symposium on Theoretical Aspects of Computer Science, pp. 25–36. Springer, Heidelberg (1996)
8. Littlestone, N.: Learning quickly when irrelevant attributes abound: A new linear-threshold algorithm. Machine Learning 2(4), 285–318 (1988)
9. Ladner, R.E., Lynch, N.A., Selman, A.L.: A comparison of polynomial-time reducibilities. Theoretical Computer Science 1(2), 103–123 (1975)
10. Lutz, J.H.: Dimension in complexity classes. SIAM Journal on Computing 32(5), 1236–1259 (2003)
11. Li, M., Vitányi, P.M.B.: An Introduction to Kolmogorov Complexity and its Applications, 3rd edn. Springer, Heidelberg (2008)
12. Ukkonen, E.: Two results on polynomial time truth-table reductions to sparse sets. SIAM Journal on Computing 12(3) (1983)
13. Yap, C.K.: Some consequences of non-uniform conditions on uniform classes. Theoretical Computer Science 26, 287–300 (1983)

Spatial Models for Virtual Networks

Jeannette Janssen

Dalhousie University, Halifax, NS, B3H 3J5, Canada

Abstract. This paper discusses the use of spatial graph models for the analysis of networks that do not have a direct spatial reality, such as web graphs, on-line social networks, or citation graphs. In a spatial graph model, nodes are embedded in a metric space, and link formation depends on the relative position of nodes in the space. It is argued that spatial models form a good basis for link mining: assuming a spatial model, the link information can be used to infer the spatial position of the nodes, and this information can then be used for clustering and recognition of node similarity. This paper gives a survey of spatial graph models, and discusses their suitability for link mining.

Keywords: spatial graph models, virtual networks.

1 Introduction

Through the advent of the Internet and especially the World Wide Web, huge repositories of data have become available in a naturally linked form. Examples are: on-line social networks, thematically coherent or domain-restricted segments of the World Wide Web, and electronic libraries of scientific papers. The links connecting entities in such data collections form a virtual network. The link structure of this network encodes information about the data represented by the nodes. *Link mining* is the process of extracting that information. Link mining can give information about the data collection when node-specific data is unavailable or private, as in on-line social networks, or can be combined with text mining in web graphs or citation networks to gain a better understanding of the data.

In order to interpret the structure of the network it helps to model the process that led to the formation of the network. The virtual networks that are the subject of this paper are *self-organizing*: they are not governed by central control or design, but formed by individual actions of autonomous agents: Facebook users, Web page designers, authors of scientific papers. Moreover, since the networks are virtual and link creation is free, there are no physical constraints that limit the link structure.

The first studies of the link structure of the Web revealed that virtual self-organizing networks exhibit a characteristic structure. Initially, models for such networks mainly aimed to generate graphs with a similar substructure. For a survey of such models, see [4,3]. The principal properties observed were:

F. Ferreira et al. (Eds.): CiE 2010, LNCS 6158, pp. 201–210, 2010.

1. *A heavy tail degree distribution.* A characteristic of almost all virtual networks is that high degree nodes are relatively common. That is, the degree distribution $P(k)$, where $P(k)$ is the proportion of nodes of degree k, does not fall off exponentially as k grows large. Often, the tail of the distribution follows a *power law*: $P(k) \sim k^{-\gamma}$ for some exponent γ which is usually between 2 and 3.

2. *The graphs are globally sparse, but locally dense.* The average degree is $O(\log n)$, where n is the size of the network, or can even be constant. On the other hand, locally the graph is denser than one would expect if the links were distributed randomly. This can be measured by the value of the *clustering coefficient*, which is the average density of the subgraphs induced by the neighbourhood of a node.

3. *Small distances between nodes.* The local density might lead one to expect that it can take a large number of hops to go from one node to the other, since many links remain local. However, this is not the case: the average distance between nodes is $O(\log n)$ or smaller.

While many of the graph models were succesful in reproducing some or all of the observed graph properties, they are based on an assumption that is incompatable with the concept of link mining. Generally, nodes in such models are indistinguishable. That is, the stochastic process that determines the link neighbourhood of a node is *only* based on the existing network, not on any individual property of the node. Link mining is based on the premise that nodes are *not* equal. The premise that it is possible to extract information about the nodes from the link structure implies that this information is present but hidden, and further, that this information influences the formation of the link structure. Thus, the link structure is a visible manifestation of an underlying, hidden reality.

Two principal tasks of link mining are those of detecting *similarity* between nodes, and of identifying *communities* of related nodes. Thus, the hidden reality should give information about node similarity. A natural way to model similarity is to assume that the nodes are embedded in a metric space, where the metric distance between nodes is a measure of the similarity between nodes. If two nodes are similar the metric distance between them will be small, and communities will correspond to spatial clusters.

Graph models where the nodes are embedded in a metric space and link formation is influenced by the metric distance between nodes are called *spatial models*. The main principle of spatial models is that nodes that are metrically close are more likely to link to each other. This is a formal expression of the intuitive notion we hold about virtual networks: Web links are likely to point to similar pages, people that share similar interests are more likely to become friends on Facebook, and a scientific paper mostly refers to papers on a similar topic. This paper gives an overview of spatial models and their suitability for link mining.

2   Spatial Models with Network-Based Link Formation

In this section we review models where the link formation is directly influenced by the distance between the nodes. Precisely, let d be the metric of the space in which the nodes are located. The probability that node v_i links to node v_j is a decreasing function of $d(v_i, v_j)$. A review of these models can also be found in Section 2.5 of [3]. The models have in common that link formation is *network based*, which means that the stochastic process according to which the links attaching to a node are generated depends on the entire network, and the node can potentially link to any node in the network, albeit with increasingly smaller probability as the distance increases.

Most of the models presented in this Section were proposed as models for *spatial networks*, characterized in [3] as "networks whose nodes occupy a precise position in two- or three-dimensional Euclidean space, and whose edges are physical connections". Examples of such networks are: the Internet (where the nodes are the routers) and other physical communication networks, railway and road networks, electric power grids, and neural networks in the brain. In spatial networks, there are limitations on the network posed by the physical reality; in virtual networks such limitations are largely absent. Still, the principle that nodes that are close in space are more likely to link to each other holds for both types. We will argue that some of these spatial models fit the reality of virtual models.

Unless otherwise mentioned, we consider the space S in which the nodes are embedded to be the hypercube $[0, 1]^D$, where D is the dimension. In order to eliminate boundary effects, we use the *torus metric* derived from any of the L_p norms. Formally, this means that for any two points x and y in S,

$$d(x, y) = \min\{||x - y + u||_p \, : \, u \in \{-1, 0, 1\}^D\}.$$

The torus metric thus "wraps around" the boundaries of the unit cube, so every point in S is equivalent.

In the models of spatial networks, the dimension D must equal 2 or 3, and the metric is the Euclidean metric (derived from the L_2 norm). For a realistic model of virtual networks we expect D to be higher, while the metrics derived from the L_∞ norm (determined by the largest coordinate) and the L_1 norm (determined by the sum of the coordinates) can be reasonable alternatives.

The nodes are embedded in S according to a given probability distribution. For ease of analysis, this distribution is often assumed to be uniform. However, a distribution that aims to model real data should contain clusters of closely spaced points.

The most straightforward spatial models are those where links between nodes are formed independently, and a link between two nodes that have distance r to each other is formed with probability $p(r)$. An early model for the Internet by Waxman [27] takes $p(r) = \beta \exp(-r/\alpha c)$, where $\alpha, \beta \in (0, 1)$ are parameters, and c is the maximum distance between any pair of points. A similar model in studied in [17]. The exponential decay in $p(r)$ implies that links substantially longer than αc are highly unlikely to occur. The graph distance between two nodes that are far apart in S depends on the maximum metric distance that can be spanned by a link. A good model for virtual networks will need to include *long links*, i.e. links between nodes that are far apart in the space, in order to achieve the property of small average distance between nodes.

In order to generate networks that have the desired properties, especially a suitable degree distribution, the link neighbourhood should also depend on the link structure of the network. This can be done if the graph is generated node by node. Formally, starting from a small initial graph, at each time step t a new node v_t is generated and embedded in S according to the predetermined probability distribution. The new node is given a fixed number m of initial links. For each of the links, the probability $p(i, t)$ that an existing node v_i is chosen as the other endpoint is a function of the metric distance between v_i and v_t, and graph properties of v_i in the existing network; usually the node degree. The probability $p(i, t)$ is called the *link probability*.

In [20,23,30], the link probability is given as:

$$p(i, t) = \frac{deg(v_i)}{c(t)\, d(v_i, v_t)^\alpha},$$

where $\alpha > 0$ is a parameter, and $c(t)$ is a normalizing constant. In [30], this model is studied in one dimension ($D = 1$). Since we use the torus metric, we can imagine the nodes as being located in a circle. Thus, the model is a natural extension of the "small-world" network of Watts and Strogatz [26]. For this model, it as been determined from simulations that, for $\alpha < 1$ the degree distribution follows a power law, $P(k) \sim k^{-3}$, and for $\alpha > 1$ it is a stretched exponential, $P(k) \sim \exp(-bk^\gamma)$, where γ depends on α. For all values of α, the average distance between nodes is of order $\log n$.

In [20], the model is studied numerically in all dimensions, and a critical value α_c is determined so that for all $\alpha < \alpha_c$, the degree distribution follows a power law, while for $\alpha > \alpha_c$, it is a stretched exponential. It is also shown that the *link length distribution* $\ell(k)$, where $\ell(k)dk$ is the proportion of links that have length between k and $k + dk$, follows a power law.

The papers discussed above derive their results from a uniform distribution of points in space; in [23], a non-uniform distribution is studied. Here $D = 2$. Any new node is embedded randomly in S at distance r from the barycenter of the existing set of nodes, where r is chosen probabilistically according to a probability distribution $P(r) \sim r^{-(2+\beta)}$, where $\beta > 0$ is a parameter. Using methods from statistical mechanics, the authors numerically determine the values of the parameters for which the graph has a power law degree distribution.

In [2], the case where $p(i, t) \sim deg(v_i) \exp(-d(v_i, v_j)/r_c)$ is studied for $D = 2$. In this case, links significantly longer than r_c will be rare. In [28], a model for the Internet is proposed where $D = 2$, and

$$p(i, t) = \frac{deg(v_i)^\sigma}{c(t)\, d(v_i, v_t)^\alpha},$$

where both σ and α are parameters. Moreover, a non-uniform distribution of nodes in space is used which mimicks actual human population densities.

In [18], nodes are placed on a grid pattern in a 2-dimensional space. Each node is connected to its four nearest neighbours in the grid, and to one other node. The destination of the extra link is chosen from all other nodes with probability $r^{-\alpha}$, where r is the distance between the origin and destination node. Thus, this model can be considered as a grid graph superimposed on a spatial graph model with $p(i,t) = d(v_i, v_j)^{-\alpha}$, $D = 2$ and $m = 1$.

A spatial model developed with the explicit aim to make inferences about the underlying *latent* metric space was proposed for the modelling of social networks by Hoff, Raftery and Handcock [14]. Here, nodes represent individuals in a social network, and these nodes are assumed to embedded in a latent space representing simularities between nodes. The dimension is assumed to be low: in the examples given in the paper the dimension does not exceed 3. In the model, the relation between link probability p and metric distance d between nodes is given as:

$$\frac{p}{1-p} = \log a + bd,$$

where a and b are parameters of the model. Note that this link probability function can be seen as a smoothing of a threshold function. If a threshold function is used instead, the graph model corresponds to the random geometric graph model discussed in the next section. Extensions of this model can be found in [11,15,24].

Finally, we mention three spatial models that are based on different principles. In [29], nodes are linked independently with probability proportional to the dot product of the vectors reprenting the nodes. In [21], and later also [7], nodes are assigned random weights w_i, and two nodes v_i and v_j are linked precisely when a function of the weights and the distance, for example $(w_i + w_j)/d(v_i, v_j)$, exceeds a given threshold θ. In [8], each new node v_t links to the node v_i that minimizes a function which is the convex combination of the graph distance of v_i to the center of the graph, and the metric distance between v_i and v_t.

3   Spatial Models with Node-Based Link Formation

Virtual networks can be very large, so it is reasonable to assume that any user is only familiar with a small fraction of the network, and that this fraction consists of the nodes that are similar to the node associated with the user. In the context of spatial models, this implies that a new node can only see the part of the network that corresponds to nodes that are metrically close. The second class of models we survey are based on this principle.

The simplest model is the *random geometric graph* [22]. In this model, nodes are embedded in a metric space according to a given probability distribution, and two nodes are linked precisely when their distance is smaller than a given threshold value θ. The random geometric graph was proposed as a model for spatial networks, wireless multi-hop networks, and for biological networks, see for example [12,13,25]. In [19], a variant of the model is presented, where the metric space is the hyperbolic space. In random geometric graphs, a link between two nodes gives exact binary information about whether the nodes are within distance θ of each other or not. Thus the graph distance should be highly correlated with the metric distance, up to multiples of θ. In [5], this relationship is confirmed for infinite random geometric graphs. In the same paper, a variation of the random geometric graph model is proposed, where nodes that are within distance θ of each other are linked independently with probability p.

A variation of the random geometric graph model where the link probability is partially determined by the network was proposed by Flaxman *et al.* in [9]. Here, nodes join the network one by one, and each new node receives m links, and chooses its neighbours from among the nodes that are within distance θ of it. The probability that a node v_i, which is within distance θ of the new node, receives a link, is proportional to its degree. As in the previous model, the threshold θ limits the length of a link, so that long links become impossible. In [10], the same authors extend the model, so that the hard threshold θ is replaced by a link probability function which makes it less likely that a node receives a link if it is far removed from the new node.

The random geometric graph can also be represented as follows: each node is the centre of an associated sphere of radius θ. A new node can link to an existing node only when it falls within its associated sphere. The SPA model, proposed in [1], is based on the same principle, except that the radii of the spheres are not uniform, but depend on the in-degree of the node (the SPA model generates directed graphs). Precisely, each node v_i is the center of a sphere whose radius is chosen so that its volume equals

$$A(v_i, t) = \min\{\frac{A_1 \deg^-(v_i, t) + A_2}{t + A_3}, 1\},$$

where A_1, A_2 and A_3 are parameters of the model.

At each time step t, a new node v_t is created and embedded in S uniformly at random. For each node v_i so that v_t falls inside the sphere of v_i, independently, a link from v_t to v_i is created with probability p. Thus, a node with high degree is "visible" over a larger area of S than a node with small degree. It was shown in [1] that the SPA model produces graphs with a power law degree distribution, with exponent $1 + 1/pA_1$. Moreover, because of the unequal size of the spheres, long links can occur, but only to nodes of high degree. It is shown in the next section that the SPA model can be used to retrieve the metric distances between nodes from the network with high precision.

A related model, the geo-protean model, was proposed in [6]. Here, the size of the network is constant, but at each step a node is deleted and a new node is added. Nodes are ranked from 1 to n, and the sphere around a node is such that its volume is proportional to its rank raised to the power $-\alpha$. The link neighbourhood of the new nodes is determined in a similar way as in the SPA model. It is shown that the degree distribution, link density, average distance between nodes, and clustering behaviour are consistent with those observed in social networks.

4   Estimating Distance from the Number of Common Neighbours

In this section, we show how the metric distances between nodes can be estimated from their number of common neighbours. The model we use is the SPA model, described in the previous section. The work presented here can be found in more detail in [16]. In this section, the dimension of S is assumed to be 2, and the parameters of the SPA model are $A_1 = A_2 = A_3 = 1$.

The SPA model produces directed graphs. The term "common neighbour" here refers to common in-neighbours. Precisely, a node w is a common neighbour of nodes u and v if there exist directed links from w to u and from w to v. In the SPA model, this can only occur if w is younger than u and v, and, at its birth, w lies in the intersection of the spheres of influence of u and v. We use $cn(u, v, t)$ to denote the number of common in-neighbours of u and v at time t.

First of all, we show that a blind approach to using the co-citation measure does not work. From the description of the SPA model it is clear that there exists a correlation between the spatial distance and number of common in-neighbours of a given pair of nodes. However, as shown in Figure 1, when we plot spatial distance versus number of common neighbours without further processing, no relation between the two is apparent.

The data presented in Figure 1 was obtained from a graph with 100K nodes. The graph was generated from points randomly distributed in the unit square in $\mathbb{R}^2$ according to the SPA model, with $n = 100,000$ and $p = 0.95$.

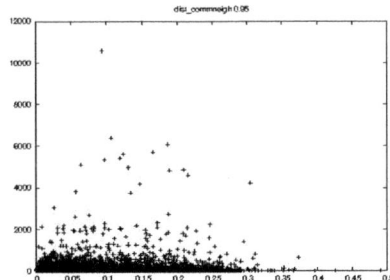

Fig. 1. Actual distance vs. number of common neighbours

By analyzing the model, we can extract the relationship between the number of common neighbours and the metric distance. First, we make a simplifying assumption. In [1], it was shown that the expected degree, at time t, of node v_i born at time i is proportional to $(t/i)^p$. In the following, assuming that the degree of each node is close to its expected degree, we will set the the volume of the sphere of a node equal to its expected value:

$$A(v_i, t) = \frac{\left(\frac{t}{i}\right)^p}{t}. \tag{1}$$

Before we continue, it is important to note that the simulation data was generated using the *original* SPA model as described in the previous section, and not using the simplifying assumption given above. The simplified model, where the volume of the sphere of influence is deterministic instead of a random variable, is only used for the analysis. The simulation results given later will show that this simplification is justified.

The radius $r(v_i, t)$ of the sphere of influence of node v_i at time t can now be derived from its volume. Note that, under the simplifying assumption, the spheres of influence shrink in every time step.

Consider the relationship between the number of common neighbours and the metric distance of two nodes v_i and v_j at distance $d(v_i, v_j) = d$, where $i < j$, so v_i is older than v_j. Three distinct cases can be distinguished:

1. If $d > r(v_i, j+1) + r(v_j, j+1)$, then the spheres of influence of v_i and v_j are disjoint at the time of v_j's birth. Since spheres of influence shrink over time, this means that these spheres remain disjoint, so v_i and v_j can have no common neighbours.
2. If $d \leq r(v_i, n) - r(v_j, n)$, then the sphere of influence of v_j is contained in the sphere of v_i at the end of the process (time n). Thus, at any time during its existence, the sphere of v_j was completely contained in the sphere of v_i, and thus any neighbour of v_j has a probability p of also becoming a neighbour of v_i. Thus, the expected number of common neighbours is asymptotically equal to p times the degree of v_j, which is $p(n/j)^p$.
3. If $r(v_i, n) - r(v_j, n) < d \leq r(v_i, j+1) + r(v_j, j+1)$, then the sphere of v_j is initially contained inside the sphere of v_i, but as the spheres shrink, eventually the spheres will become disjoint. The time at which this happens depends on the distance d between v_i and v_j, and this time, in turn, will determine the number of common neighbours. Ignoring a lower order error term, the expected number of common neighbours is given by:

$$\mathbb{E}\, cn(v_i, v_j, n) = p\pi^{-\frac{p}{1-p}} \left(i^{-\frac{p^2}{1-p}} \right) \left(j^{-p} \right) \left(d^{-\frac{2p}{1-p}} \right) \tag{2}$$

These formulas lead to an estimate $\hat{d}$ of the metric distance between two nodes, based on the number of common neighbours of the pair. Note that from case 1 and 2, we can only obtain a lower and upper bound on the distance, respectively. Only in case 3 we obtain a formula which explicitly links the number of common neighbours with the distance d.

In our simulation, in order to eliminate case 1, we consider only pairs that have at least 20 common neighbours (19.2K pairs). To eliminate case 2, we require that the number of common neighbours should be less than $p/2$ times the lowest degree of the pair. This reduces the data set to 2.4K pairs.

When we are likely in case 3, we can derive a precise estimate of the distance. We base our estimate on Equation (2). The estimated distance between nodes v_i and v_j, given that their number of common neighbours equals k, is then given by

$$\hat{d} = \left(\pi^{-1/2} p^{\frac{1-p}{2p}} \right) \left(i^{-p/2} \right) \left(j^{-\frac{1-p}{2}} \right) \left(k^{-\frac{1-p}{2p}} \right).$$

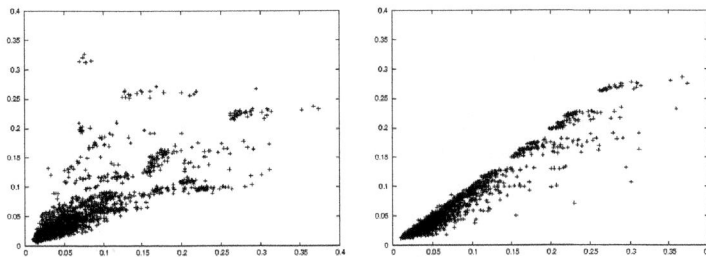

Fig. 2. Actual distance vs. estimated distance for eligible pairs from simulated data, calculated using the age (left) and estimated age from degree (right) of both nodes

Note that i and j appear in the formula above, so the estimated distance depends not only on the number of common neighbours of the two nodes, but also on their age. In our simulation data, the age of the nodes is known, and used in the estimate of $\hat{d}$. Figure 2 shows estimated distance vs. real distance between all pairs of nodes that are likely to be in case 3.

While there is clearly some agreement between estimated and real distance, the results can be improved if we use the *estimated age*, instead of the real age. The estimated time of birth $\hat{a}(v)$ of a node v which has in-degree k at time n will be:

$$\hat{a}(v) = nk^{-1/p}.$$

Thus, we can compute $\hat{d}$ again, but this time based on the estimated birth times. This method has the added advantage that it can be more conveniently applied to real-life data, where the degree of a node is much easier to obtain than its age. Figure 2 again shows estimated vs. real distance for the exact same data set, but now estimated age is used in its calculation. This time, we see almost perfect agreement between estimate and reality.

References

1. Aiello, W., Bonato, A., Cooper, C., Janssen, J., Prałat, P.: A spatial web graph model with local influence regions. Internet Mathematics 5(1-2), 175–196 (2009)
2. Barthelemy, M.: Crossover from scale-free to spatial networks. Europhys. Lett. 63(6), 915–921 (2003)
3. Boccaletti, S., et al.: Complex networks: Structure and dynamics. Physics Reports 424, 175–308 (2006)
4. Bonato, A.: A Course on the Web Graph. American Mathematical Society, Providence (2008)
5. Bonato, A., Janssen, J.: Infinite random geometric graphs (2010) (preprint)
6. Bonato, A., Janssen, J., Prałat, P.: A geometric model for on-line social networks (2010) (preprint)
7. Bradonjic, M., Hagberg, A., Percus, A.G.: The structure of geographical threshold graphs. Internet Mathematics 4(1-2), 113–139 (2009)

8. Fabrikant, A., Koutsoupias, E., Papadimitriou, C.H.: Heuristically Optimized Trade-offs: a new paradigm for power laws in the Internet. In: Widmayer, P., Triguero, F., Morales, R., Hennessy, M., Eidenbenz, S., Conejo, R. (eds.) ICALP 2002. LNCS, vol. 2380, pp. 110–122. Springer, Heidelberg (2002)
9. Flaxman, A., Frieze, A.M., Vera, J.: A geometric preferential attachment model of networks. Internet Mathematics 3(2), 187–206 (2006)
10. Flaxman, A., Frieze, A.M., Vera, J.: A geometric preferential attachment model of networks II. Internet Mathematics 4(1), 87–111 (2008)
11. Handcock, M.S., Raftery, A.E., Tantrum, J.M.: Model-based clustering for social networks. J. R. Statist. Soc. A 170(2), 301–354 (2007)
12. Herrmann, C., Barthélemy, M., Provero, P.: Connectivity distribution of spatial networks. Phys. Rev. E 68, 026128 (2003)
13. Higham, D.J., Rasajski, M., Przulj, N.: Fitting a geometric graph to a protein-protein interaction network. Bioinformatics 24(8), 1093–1099 (2008)
14. Hoff, P.D., Raftery, A.E., Handcock, M.S.: Latent space approaches to social network analysis. J. Am. Stat. Assoc. 97, 1090–1098 (2002)
15. Hoff, P.D.: Bilinear mixed-effects models for dyadic data. J. Am. Stat. Assoc. 100, 286–295 (2005)
16. Janssen, J., Prałat, P., Wilson, R.: Estimating node similarity from co-citation in a spatial graph model. In: phProceedings of 2010 ACM Symposium on Applied Computing–Special Track on Self-organizing Complex Systems, 5 p. (2010)
17. Kaiser, M., Hilgetag, C.C.: Spatial growth of real-world networks. Phys. Rev. E 69, 036103 (2004)
18. Kleinberg, J.: Navigation in a small world. Nature 406, 845 (2000)
19. Krioukov, D., Papadopoulos, F., Vahdat, A., Boguña, M.: Curvature and Temperature of Complex Networks. arXiv0903.2584v2 (2009)
20. Manna, S.S., Sen, P.: Modulated scale-free network in Euclidean space. Phys. Rev. E 66, 066114 (2002)
21. Masuda, N., Miwa, M., Konno, N.: Geographical threshold graphs with small-world and scale-free properties. Phys. Rev. E 71(3), 036108 (2005)
22. Penrose, M.: Random Geometric Graphs. Oxford University Press, Oxford (2003)
23. Soares, D.J.B., Tsallis, C., Mariz, A.M., da Silva, L.R.: Preferential attachment growth model and nonextensive statistical mechanics. Europhys. Lett. 70(1), 70–76 (2005)
24. Schweinberger, M., Snijders, T.A.B.: Settings in social networks: a measurement model. Sociol. Methodol. 33, 307–341 (2003)
25. Ta, X., Mao, G., Anderson, B.D.O.: On the Properties of Giant Component in Wireless Multi-hop Networks. In: Proceedings of 28th IEEE INFOCOM (2009)
26. Watts, D.J., Strogatz, S.H.: Collective dynamics of "small-world" networks. Nature 393(6684), 409–410 (1998)
27. Waxman, B.M.: Routing of multipoint connections. IEEE J. Sel. Areas in Comm. 6(9), 1617–1622 (1988)
28. Yook, S.-H., Jeong, H., Barabási, A.-L.: Modeling the Internet's large-scale topology. Proc. Natl. Acad. Sci. USA 99, 13382 (2002)
29. Young, S.J., Scheinerman, E.R.: Random dot product graph models for social networks. In: Bonato, A., Chung, F.R.K. (eds.) WAW 2007. LNCS, vol. 4863, pp. 138–149. Springer, Heidelberg (2007)
30. Xulvi-Brunet, R., Sokolov, I.M.: Evolving networks with disadvantaged long-range connections. Phys. Rev. E 66, 026118 (2002)

DNA Rearrangements through Spatial Graphs*

Nataša Jonoska and Masahico Saito

Department of Mathematics and Stataistics, University of South Florida, USA
jonoska@math.usf.edu, saito@math.usf.edu

Abstract. The paper is a short overview of a recent model of homologous DNA recombination events guided by RNA templates that have been observed in certain species of ciliates. This model uses spatial graphs to describe DNA rearrangements and show how gene recombination can be modeled as topological braiding of the DNA. We show that a graph structure, which we refer to as an assembly graph, containing only 1- and 4-valent rigid vertices can provide a physical representation of the DNA at the time of recombination. With this representation, 4-valent vertices correspond to the alignment of the recombination sites, and we model the actual recombination event as smoothing of these vertices.

Keywords: DNA rearrangements, spatial graphs, vertex smoothing, ciliates.

1 Introduction

Theoretical models for DNA recombination have been proposed for both DNA sequence reorganization (e.g.,[5,6,9]), and topological processes of knotting (e.g., [15]). Several species of ciliates are often taken as model organisms to study gene rearrangements. In *Oxytricha*, for example, DNA deletion reduces a 1Gb germline genome of an archival micronucleus to a somatic macronuclear genome of only 50Mb, nearly devoid of non-coding DNA. These massively occurring recombination processes involve DNA processing events that effectively eliminates *all* so-called "junk" DNA, including intergenic DNA, as well as hundreds of thousands of intervening DNA segments (internal eliminated sequences, IESs) that interrupt genes. Each macronuclear gene may appear as several nonconsecutive segments (macronuclear destined sequences, MDSs) in the micronucleus. During macronuclear development, the IESs that interrupt MDSs in the micronucleus are all deleted. Moreover, the MDS segment order for thousands of genes in the micronucleus can be permuted or sequences reversed, with no coding strand asymmetry. Formation of the macronuclear genes in these ciliates thus requires any combination of the following three events: unscrambling of segment order, DNA inversion, and IES removal. Figure 1 shows an example of a typical scrambled gene requiring all three events.

---

* Research supported in part by the NSF grants CCF-0726396 and DMS-0900671.

F. Ferreira et al. (Eds.): CiE 2010, LNCS 6158, pp. 211–218, 2010.

Pointer-like sequences that are repeated at the end of each nth MDS and the beginning of each $(n + 1)$st MDS in the micronucleus have been observed such that each pointer sequence is retained as exactly one copy in the macronuclear sequence [10,14]. Such repetition of sequences suggests "pointer guided" homologous recombination. Several models for these processes have been pro-

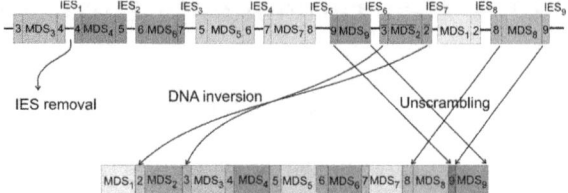

Fig. 1. Schematic representation of the scrambled Actin I micronuclear germline gene in *Oxytricha nova* (top) and the correctly assembled macronuclear gene (bottom)[14]. Each block represents an MDS, and each line between blocks is an IES. The numbers at the beginning and at the end of each segment represent the pointer sequences. The bars above MDS_2 and its pointers indicate that this block is inverted in the germline relative to the others.

posed, including the models in [5,7] which all assume that a correct pair of pointers align and splice. Although the general mechanism that guides this process of assembly is still not known, an epigenetic model has been proposed in which a DNA template (Prescott et al. [13]), or an RNA template (Angeleska et al. [1]) derived from the maternal macronucleus guides assembly of the new macronuclear chromosomes. Recently, the RNA model was supported by several experimental observations showing that maternal RNA templates guide DNA rearrangement in the early development of the macronucleus [12]. A schematic representation of the steps of this model for a double stranded RNA guided DNA recombination is depicted in Figure 2. The computational power of the template guided model has also been studied by many authors (see for ex. [4]) and it has been observed that its power is not greater than the power of a finite state machine. In this paper we describe a theoretical model based on spatial graphs that depicts the DNA structure at the time of recombination. This model was initiated in [1] with more detailed analysis found in [3]. An overview of the experimental results from [12,11] in conjunction with the theoretical model can be found in [2].

2   Mathematical Models by Spatial Graphs

The model in [1] utilizes graphs as a physical representation of the DNA at the time of recombination. Schematically, the moment of homologous recombination, alignment of the pointers, can be represented as a vertex in a graph as depicted in Figure 3(left). The pointer alignment and recombination can be seen as a 4-valent rigid vertex v made of two DNA segments; for each edge e incident to the vertex v, there is a predefined "predecessor" edge, and a predetermined "successor" edge with respect to v constituting the *neighbors* of e. The homologous recombination corresponds to removal of the crossing (vertex) called "smoothing" (see Figure 3(right)). The RNA or DNA template enables alignment of pointers and, as shown in [1], this alignment (being parallel or

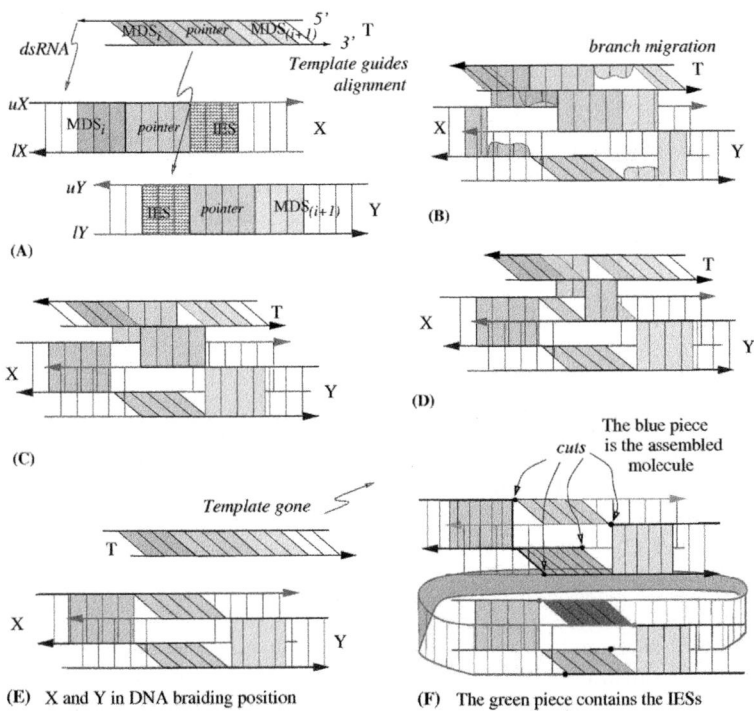

Fig. 2. Step by step model of a DNA recombination guided by a dsRNA template. (A) Three molecules, two micronuclear segments X, Y and a dsRNA template T interact, permitting the template strands to find their corresponding complements in X and Y. (B) Through branch migration, the template displaces base-pairs between X and Y where substrings of X and Y might be transiently unpaired during this thermodynamically driven process. A portion of the "lower" strand, lX, containing the pointer sequence becomes single stranded and pairs with a portion of lY containing the corresponding complementary sequence. (C) Branch migration begins and pairing between the template strands reinstates releasing the "upper" strands uX and uY. (D) As uX and uY dissociate from the template T, pairing between uX and uY develops, via strand complementarity. (E) *DNA Braiding:* uX completely binds to uY through the pointer sequence. Template base-pairs are restored, leaving the template unchanged. (E) resulting molecules obtained after cuts are introduced. The blue portion of the molecule indicates the newly recombined molecule. The green IES portion is excised.

antiparallel) determines the type of recombination and the type of smoothing performed at a vertex.

The graph depicted in Figure 4(A) is a schematic planar representation of the MDS-IES structure of the scrambled gene in Figure 1 at the time of rearrangement. Joining the neighboring MDSs forms a path in the graph which in some sense determines the smoothings of the vertices as depicted in Figure 4(B). The resulting graph, after smoothing of every vertex, is depicted in Figure 4(C). As shown, this result is composed of three connected components, although in reality these

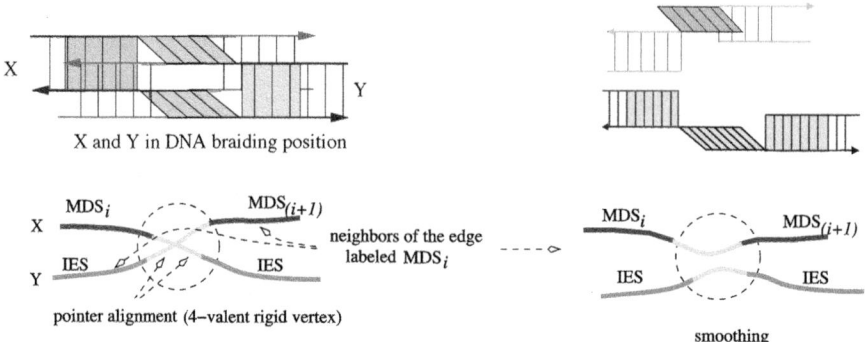

Fig. 3. (Left) Schematic representation of the pointer alignment shown as a 4-valent vertex. Two micronuclear DNA segments exchange pointer nucleotides through branch migration as in Figure 2(E) with MDS segments indicated in blue (top). This alignment is represented as a 4-valent vertex in a graph (bottom). **(Right)** Two molecules after the recombination (top), with the MDSs joined on the same molecule. Schematic representation of the finished recombination as smoothing of the vertex (bottom).

molecules are likely non-circular. Two of them are labeled only with IESs, indicating IES excisions, while one of the components contains $MDS_1 - MDS_2 - \cdots - MDS_9$, i.e., the assembled macronuclear gene in correct MDS order. In [1] and subsequently in [3] it was shown that every MDS-IES micronuclear gene structure can be modeled by a spatial graph, and the assembly of a macronuclear gene can be viewed as a smoothing of every vertex in the graph.

Therefore, a micronuclear sequence is modeled with an *assembly graph* which is a finite connected graph with 4-valent rigid vertices. Instead of circular molecules, we may allow assembly graphs to have two end-points (1-valent vertices) representing the initial and terminal points of a DNA molecule. An *open path*, i.e., the image of an open interval in an assembly graph models a subsequence of DNA. A macronuclear gene consisting of ordered MDS segments is modeled with a *polygonal path*, an open path such that consecutive edges are neighbors with respect to the joint incident vertex. In other words, a polygonal path makes a "90 degree" turn at every rigid 4-valent vertex. A *(transverse) component* of an assembly graph is a path of maximal length containing non-repeating edges, such that no two consecutive edges are neighbors. The graph in Figure 4 (A) consists of a single component made of edges labeled with the micronuclear sequence of MDSs and IESs.

In an assembly graph, two open paths are *disjoint* if they do not have a vertex in common. A set of pairwise disjoint open paths with non-repeating vertices $\{\gamma_1, \ldots, \gamma_k\}$ is called *Hamiltonian* for a graph Γ if their union contains all 4-valent vertices of Γ. An open path γ is called *Hamiltonian* if the set $\{\gamma\}$ is Hamiltonian. Hamiltonian polygonal sets are of special interest, because they trace the correct order of the MDSs in the macronucleus. In Figure 4(B) the assembly graph that models the micronuclear Actin I gene from Figure 1 con-

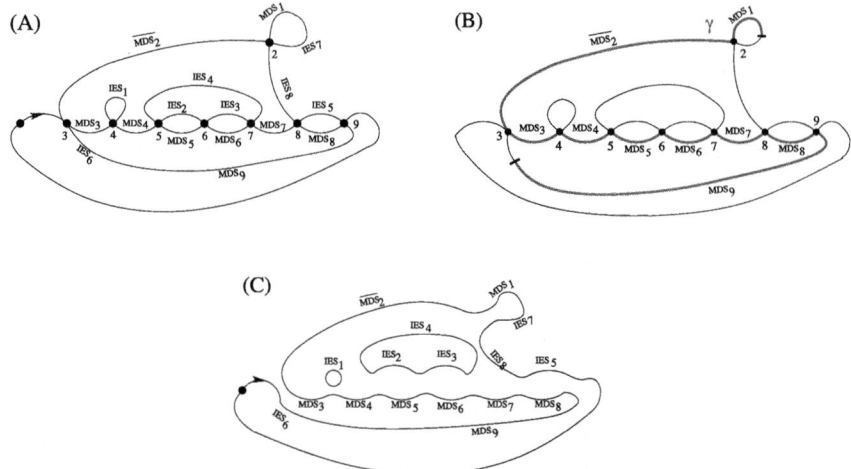

Fig. 4. (A) Graph structure of simultaneous recombination for Actin I gene from Fig 1, at the moment of recombination; (B) polygonal path in the graph containing the MDSs for the macronuclear gene; (C) smoothing of the vertices relative the polygonal path in (B) and the resulting molecules after recombination

tains the Hamiltonian polygonal path γ that represents the macronuclear (after assembly) Actin I gene.

Two types of smoothings are depicted in Figure 5. To distinguish them, we fix an orientation of an assembly graph to obtain an *oriented* assembly graph. An orientation of an assembly graph is a fixed direction of each component of the graph. These orientations are defined abstractly, independent from the $5' - 3'$ orientation of DNA. For an oriented assembly graph, each smoothing of a vertex is either orientation preserving (*parallel smoothing*, or *p*-smoothing, as in Figure 5(left), modeling the case when pointers are aligned in parallel) or non-preserving (*non-parallel smoothing*, or *n*-smoothing as in Figure 5(right), modeling the case when pointers are aligned anti-parallel).

The sets of Hamiltonian paths can be related to sets of smoothings as follows. Let Γ be an assembly graph and $\gamma = \{\gamma_1, \ldots, \gamma_k\}$ be a Hamiltonian set of polygonal paths for Γ. A *smoothing of a 4-valent vertex v in Γ with respect to a*

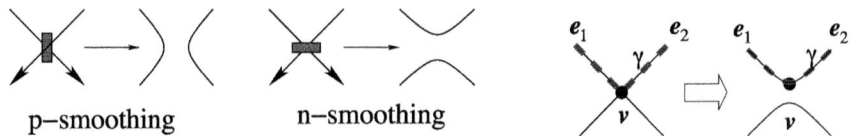

Fig. 5. Two types of smoothings, parallel (*p*-) smoothing (left) and non-parallel (*n*-) smoothing (right)

Fig. 6. Smoothing at a vertex v with respect to a polygonal path γ

polygonal path γ_i is a smoothing such that the arcs traversing γ_i at vertex v are connected as depicted in Figure 6. Each polygonal path corresponds to a given MDS order in a macronuclear gene.

3   Using Knot and Spatial Graph Theories Techniques

Recall from Section 2, an assembly graph is a connected graph with rigid 4-valent vertices (possibly with pairs of uni-valent vertices); a polygonal path in an assembly graph is a path with open ends that makes "90 degree" turn at every vertex; and a set of open paths is Hamiltonian if they visit every vertex once. A connected Hamiltonian polygonal path models a single MDS sequence to be assembled. Two assembly graphs with end-points, Γ_1 and Γ_2, can be composed into $\Gamma_1 \circ \Gamma_2$ by joining one end-point of Γ_1 with one end-point of Γ_2. Such composed graphs are called *reducible*. If an assembly graph cannot be obtained as a composition of two other assembly graphs, we say that the graph is *irreducible*.

Define the *assembly number* of Γ, denoted by $\mathrm{An}(\Gamma)$, for a given assembly graph Γ, to be $\mathrm{An}(\Gamma) = \min\{\ k\ |\ \text{there exists a Hamiltonian set of polygonal}$ paths $\{\gamma_1, \ldots, \gamma_k\}$ in $\Gamma\}$. Thus an assembly graph Γ with $\mathrm{An}(\Gamma) = 1$ models a single scrambled macronuclear gene (micronuclear sequence) whose structure during the recombination process has a possible spatial graph structure isomorphic to Γ. In general, the assembly number could be seen as a measure of failure for the given assembly graph to represent a structure modeling a single scrambled macronuclear gene. On the other side, a graph Γ with $\mathrm{An}(\Gamma) = n$ models a DNA sequence which contains at least k scrambled macronuclear genes. In [3], the following properties are proved. For any positive integer n, there exists

  (i) a reducible assembly graph Γ such that $\mathrm{An}(\Gamma) = n$,
 (ii) an irreducible assembly graph Γ such that $\mathrm{An}(\Gamma) = n$, and
(iii) an assembly graph Γ with no endpoints such that $\mathrm{An}(\Gamma) = n$.

Several other properties related to realization of the assembly numbers are studied in [3] as well.

Although the Hamiltonian polygonal path definition is motivated by DNA recombination, it is also of independent mathematical interest. In particular, the smoothing of vertices along such a polygonal path seems similar to the colored smoothing defined by Kauffman [8]. Smoothings of knot diagrams corresponding to so called, proper colorings have been introduced by Kauffman in [8]. A *proper coloring* is an assignment of two colors, 0 and 1, to the edges of an assembly graph such that non-neigboring edges are colored distinctly. The smoothing with respect to a proper coloring is performed such that the neighboring edges colored with the same color remain connected (see left of Figure 7 (A)). For a given proper coloring of an assembly graph, there is a unique smoothing at every crossing such that the resulting circles have consistent colors, 0 or 1, as depicted in the right of Figure 7 (A). In Figure 7 (B), such a smoothing is depicted for a projection of the trefoil knot. An invariant for virtual knots is defined through "counting" of the resulting circles relative the set of all proper colorings [8].

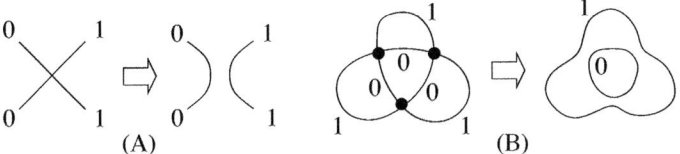

Fig. 7. (A) Proper coloring of a vertex and smoothing corresponding to the proper coloring (Kauffman [8]). (B) An example of a proper coloring for an assembly graph "representing" the trefoil and the resulting components after the smoothing.

Let $\mathcal{C}$ be the set of proper colorings for an assembly graph Γ with one or more transverse components and no end points. Given a coloring $\xi \in \mathcal{C}$ let ξ_0 and ξ_1 denote the number of components colored 0, respectively 1, after smoothing corresponding to ξ. The minimal number of components of smoothed circles with one color, is defined to be $K_{\min} = \min_{\xi \in \mathcal{C}} \{\xi_0, \xi_1\}$. In [3], the following relations between the assembly numbers and the number $K_{\min}$ related to Kauffman's colored smoothings were given: For any assembly graph Γ with no end points, $\mathrm{An}(\Gamma) \le K_{min}(\Gamma)$, and for any positive integer m and n with $m \le n$, there is an assembly graph Γ with no end points such that $\mathrm{An}(\Gamma) = m$ and $K_{min}(\Gamma) = n$.

4   Concluding Remarks

The RNA-guided DNA recombination model initiated description of DNA rearrangements with spatial graphs which we call assembly graphs. Some recombination events may appear within a narrow time window to be considered simultanous [11]. We can assume that such simultaneous recombination events do not disturb the natural MDS order in the macronuclear gene, therefore we can consider subsets of pointers that allow such recombination together with smoothing of the vertices that corresponds to those pointers. This is achieved by requiring that after smoothing of vertices the polygonal paths remain undisturbed. The component of the resulting graphs that contains an undisturbed polygonal path is a potential intermediate sequence that may be expected, and experimentally extracted, at certain part of the development. Successful sets (vertices for simultaneous smoothing) of a single molecule with a single gene are characterized in [3]. However, even in this simple case, experimental results [11] suggest that the unscrambling process is not necessarily just intramolecular, but there might be instances in which this process is also intermolecular. Therefore, performing similar investigations for multicomponent graphs remains important in understanding unscrambling of a genome full of thousands of scrambled genes. Furthermore, the notion of smoothing along a polygonal path is new in mathematics and its study provides new research pathways for knot and graph theory.

References

1. Angeleska, A., Jonoska, N., Saito, M., Landweber, L.F.: RNA-guided DNA assembly. J. of Theoretical Biology 248(4), 706–720 (2007)
2. Angeleska, A., Jonoska, N., Saito, M., Landweber, L.F.: Strategies for RNA-guided DNA recombination. In: Kok, J., et al. (eds.) Algorithmic Bioprocesses, pp. 83–98. Springer, Heidelberg (2009)
3. Angeleska, A., Jonoska, N., Saito, M.: DNA recombinations through assembly graphs. Discrete and Applied Math. 157, 3020–3037 (2009)
4. Daley, M., McQuillan, I.: Template-guided DNA recombination. Theor. Comput. Sci. 330, 237–250 (2005)
5. Ehrenfeucht, A., Harju, T., Petre, I., Prescott, D.M., Rozenberg, G.: Computing in Living Cells. Springer, Heidelberg (2005)
6. Head, T.: Formal language theory and DNA: an analysis of the generative capacity of specific recombinant behaviors. Bull. Math. Biology 49, 737–759 (1987)
7. Kari, L., Landweber, L.F.: Computational power of gene rearrangement. In: Winfree, E., Gifford, D.K. (eds.) DNA Based Computers V, pp. 207–216. AMS (1999)
8. Kauffman, L.H.: A self-linking invariant of virtual knots. Fund. Math. 1(84), 135–158 (2004)
9. Jonoska, N., Saito, M.: Algebraic and topological models for DNA recombinant processes. In: Calude, C.S., Calude, E., Dinneen, M.J. (eds.) DLT 2004. LNCS, vol. 3340, pp. 49–62. Springer, Heidelberg (2004)
10. Landweber, L.F., Kuo, T.-C., Curtis, E.A.: Evolution and assembly of an extremely scrambled gene. Proc. Nat. Acad. Sci. USA 97(7), 3298–3303 (2000)
11. Mollenbeck, M., Zhou, Y., Cavalcanti, A.R.O., Jonsson, F., Chang, W.-J., Juranek, S., Doak, T.G., Rozenberg, G., Lipps, H.J., Landweber, L.F.: The pathway for detangling a scrambled gene. PLoS ONE 3(6), e2330 (2008),
    http://www.plosone.org/article/info:doi/10.1371/journal.pone.0002330
12. Nowacki, M., Vijayan, V., Zhou, Y., Doak, T., Swart, E., Landweber, L.F.: RNA-template guided DNA recombination: epigenetic reprogramming of a genome rearrangement pathway. Nature 451, 153–158 (2008)
13. Prescott, D.M., Ehrenfeucht, A., Rozenberg, G.: Template-guided recombination for IES elimination and unscrambling of genes in stichotrichous ciliates. J. of Theoretical Biology 222, 323–330 (2003)
14. Prescott, D.M., Greslin, A.F.: Scrambled actin I gene in the micronucleus of Oxytricha nova. Dev. Genet. 13(1), 66–74 (1992)
15. Sumners, D.W.: Lifting the curtain: using topology to probe the hidden action of enzymes. Notices of AMS 42(5), 528–537 (1995)

On Index Sets of Some Properties of Computable Algebras

Bakhadyr Khoussainov[1] and Andrey Morozov[2,*]

[1] University of Auckland, Auckland, New Zealand
bmk@cs.auckland.ac.nz
[2] Sobolev Institute of Mathematics, Novosibirsk, Russia
morozov@math.nsc.ru

Abstract. We study the index sets of the following properties of computable algebras: to have no nontrivial congruences, to have a finite number of congruences, to have infinite decreasing or increasing chains of congruence relations. We prove completeness of these index sets in arithmetic and hyperarithmetic hierarchies.

1 Introduction and Basic Definitions

Structures derived from algebras, such as congruence lattices, lattice of subalgebras, automorphism groups, etc., often tell something important about the underlying algebras. In 1963, Grätzer and Schmidt proved that every algebraic lattice is isomorphic to the congruence lattice of an algebra [1]. In 1972, Lampe [2] proved that for any two algebraic lattices L_1 and L_2 with more than one element each and any group G there exists an algebra whose congruence lattice is isomorphic to L_1, the subalgebra lattice is isomorphic to L_2, and the automorphisms group is isomorphic to G. For the current problems on representation of lattices as congruence lattices of algebras we refer the reader to the survey paper by W. Lampe [3].

When algebras are given effectively (in some formalization of the notion of effectiveness) one would like to investigate the interactions between the algebra and its derivative structures. For example, one would like to investigate the effective content of Grätzer and Schmidt theorem. This has recently been investigated by Brodhead and Kjos-Hanssen in [4]. There has also been an intensive study of a relationship between effectively given algebras and their automorphism groups (see [5]). There is a large body of work on the study of subalgebra lattices of particular classes of algebras, such as effectively given Boolean algebras and vector spaces. In this paper we continue this line of research work and study algorithmic complexities of known algebraic properties of the congruence lattices of computable algebras.

Let $\nu : \omega \to S$ be an onto mapping called a *numbering* of S. For a subset $S' \subseteq S$, the set $\nu^{-1}(S') = \{n \mid \nu(n) \in S'\}$ is called the *index set of* S' under

---

* Partially supported by RFBR grants 08-01-00336-a and 09-01-12140-OFI-M.

F. Ferreira et al. (Eds.): CiE 2010, LNCS 6158, pp. 219–228, 2010.

the numbering ν. The study of complexity of various index sets in computability theory (e.g. index sets of infinite and co-infinite sets) as well as in computable model theory (e.g. index sets of computable rigid structures) is a rather standard topic of research. One can find some facts and properties of index sets in computability theory in [6] or [7]. For some results on index sets in computable model theory we refer the reader to [8,9].

In this paper our goal is to study index sets of universal algebras whose congruence lattices satisfy certain natural algebraic properties. In particular, we investigate complexities of the index sets of universal algebras that (1) have no nontrivial congruences, (2) possess finite number of congruences, and (3) have no infinite decreasing or increasing chains of congruences. We prove completeness of these index sets in arithmetic and hyperarithmetic hierarchies. We assume that the reader has basic knowledge of computability, universal algebra, and computable model theory. For instance, see [6,?,10,11]. For self-containment of this paper, we will present some of the basic definitions and notations.

We fix some computable coding of tuples of natural numbers. If a pair of natural numbers $\langle m, n \rangle$ is coded by k then the functions $(\cdot)_0$ and $(\cdot)_1$ compute the left and the right components of the pair, that is $m = (k)_0$ and $n = (k)_1$. Cardinality of the set A will be denoted by $|A|$.

An *algebra* $\mathcal{A}$ is a tuple $(A; f_0^{n_0}, \ldots, f_k^{n_k}, \ldots)$, where A is called the domain of the algebra and each $f_i^{n_i}$ is a total operation on A of arity n_i. Sometimes we refer to these operations as *basic operations* of the algebra. The sequence $f_0^{n_0}$, $\ldots, f_k^{n_k}, \ldots$ is called the signature of this algebra. In this paper we assume that the signature is given effectively, that is, the function $i \to n_i$ is computable.

Let $\mathcal{A}$ be an algebra. The *diagram* of this algebra consists of all sentences of the form $f(a_1, \ldots, a_k) = a_{k+1}$, $f(a_1, \ldots, a_k) \neq a_{k+1}$, $a = b$, $a \neq b$ true in $\mathcal{A}$. Denote the diagram of the algebra $\mathcal{A}$ by $D(\mathcal{A})$. The *positive part of* $D(\mathcal{A})$, i.e., its sentences of kind $A = B$ is denoted by $D^+(\mathcal{A})$. We say that the algebra $\mathcal{A}$ is *computable* if its domain is an initial segment of ω and its diagram is a computable set. Clearly, computability of $\mathcal{A}$ is equivalent to saying that the domain of $\mathcal{A}$ is an initial segment of ω and that there is an algorithm that given i and $m_1, \ldots, m_{n_i}$ all in ω produces the value of $f_i^{n_i}(m_1, \ldots, m_{n_i})$.

A *congruence relation* of an algebra $\mathcal{A}$ is an equivalence relation θ such that all the operations of $\mathcal{A}$ are well-behaved with respect to θ. In other words, for any n-ary basic operation f and elements $a_1, \ldots, a_n, b_1, \ldots, b_n$ with $(a_i, b_i) \in \theta$ for $i = 1, \ldots, n$, it should be the case that $(f(a_1, \ldots, a_n), f(b_1, \ldots, b_n)) \in \theta$. Given a congruence relation θ one can define the quotient algebra $\mathcal{A}/\theta$ in a standard way. The set of all congruences of the algebra forms an algebraic lattice [11].

Let f be an n-ary basic operation of the algebra $\mathcal{A}$. Any function $\phi : A \to A$ defined as $\phi(x) = f(a_1, \ldots, a_{i-1}, x, a_{i+1}, \ldots, a_n)$, where the elements $a_1, \ldots, a_{i-1}, a_{i+1}, \ldots, a_n$ are fixed parameters, is called a *transition* of the algebra $\mathcal{A}$.

Let a, b be two elements of the algebra $\mathcal{A}$. Consider the smallest congruence relation that contains the pair (a, b). Denote this congruence relation by $\theta(a, b)$. This is called the *principal congruence relation* generated by the pair (a, b). A known characterization of the principal congruence relations $\theta(a, b)$ states that

$(x, y) \in \theta(a, b)$ if and only if there exists a sequence $z_0, z_1, \ldots, z_n$ of elements and transitions $\phi_0, \ldots, \phi_{n-1}$ such that $x = z_0$, $y = z_n$, and $\{\phi_i(a), \phi_i(b)\} = \{z_i, z_{i+1}\}$ for all $i \leq n - 1$ (e.g. see [11]). Basically, this tells us that $\theta(a, b)$ consists of all the pairs (x, y) such that $D^+(\mathcal{A}), a = b \vdash x = y$, that is the diagram $D^+(\mathcal{A})$ together with $a = b$ can formally prove the equality $x = y$.

By $W_0, W_1, \ldots$ we denote the standard enumeration of all computably enumerable (c.e.) sets. The above characterization of the principal congruence $\theta(a, b)$ tells us that the congruence $\theta(a, b)$ is uniformly c.e. from a and b in case the algebra $\mathcal{A}$ is computable. In other words, there is an algorithm that given (a, b) produces an index of a computably enumerable set that equals $\theta(a, b)$.

Let $A, B \subseteq \omega$. Recall that A *m-reduces to* B if there exists a computable function f such that for all $x \in \omega$ holds $x \in A \Leftrightarrow f(x) \in B$. Let $\mathfrak{I}$ be a class of sets. We say that *a set $A \in \mathfrak{I}$ is m-complete in $\mathfrak{I}$* if each set in $\mathfrak{I}$ m-reduces to A.

We give the main definition of our paper. Let $\mathfrak{I}$ be a class of subsets of ω closed under m–reducibility. Let $\mathfrak{P}$ be an abstract property of algebras from an abstract class $\mathcal{K}$, i.e., if an algebra $\mathcal{A} \in \mathcal{K}$ satisfies $\mathfrak{P}$ then all its isomorphic copies also satisfy $\mathfrak{P}$.

Definition 1 (Main Definition). *We say that the property $\mathfrak{P}$ of algebras in a class $\mathcal{K}$ is $\mathfrak{I}$–complete if the following properties are true:*

1. *For all computable sequences of algebras $\{\mathcal{A}_i\}_{i<\omega}$ in $\mathcal{K}$, the set $\{i \mid \mathcal{A}_i$ satisfies $\mathfrak{P}\}$ is in $\mathfrak{I}$;*
2. *There exists a computable sequence of algebras $\{\mathcal{A}_i\}_{i<\omega}$ in $\mathcal{K}$ for which the set $\{i \mid \mathcal{A}_i$ satisfies $\mathfrak{P}\}$ is m-complete in $\mathfrak{I}$.*

If the class $\mathcal{K}$ coincides with the class of all algebras then we will omit mentioning $\mathcal{K}$ when speaking about the completeness of a property.

2 Basic Results and Their Proofs

We begin with the following theorem showing that the set of indexes of computable algebras with two element congruence lattices is Π_2^0–complete.

Theorem 1. *The property "to have no nontrivial congruences" is Π_2^0–complete.*

Proof. We prove that p. 1 of the main definition is satisfied. Assume $\{\mathcal{A}_i\}_{i<\omega}$ is a computable family of algebras. Then $\mathcal{A}_i$ has no nontrivial congruences if and only if the following is true:

$$\forall a \forall b \ (a \neq b \rightarrow \theta(a, b) = \mathcal{A}_i \times \mathcal{A}_i). \tag{1}$$

The property $\theta(a, b) = \mathcal{A}_i \times \mathcal{A}_i$ is equivalent to $\forall x \forall y \ (\langle x, y \rangle \in \theta(a, b))$ which is Π_2^0. Therefore the property "to have no nontrivial congruence relations" is a Π_2^0-set.

Now we show that p. 2 of the main definition is true. For this, we define two computable sequences of natural numbers $(a_n)_{n<\omega}$ and $(b_n)_{n<\omega}$ as follows:

$$a_0 = 2; \quad b_0 = 4; \quad a_{n+1} = b_n; \quad b_{n+1} = 2a_{n+1}.$$

Let $I_n = [1, a_n]$, $J_n = [a_n + 1, b_n]$. One can easily check by induction that these intervals satisfy the following properties:

$$I_n \cap J_n = \varnothing;$$
$$I_n \cup J_n = I_{n+1};$$
$$\bigcup_{n<\omega} I_n = \omega.$$

Recall now that A_S (where S is a nonempty set) denotes the set of all even permutations on S. This group is known as the alternating group of degree $|S|$. We now define the finite group H_n as the group that consists of all permutations on ω, that are presentable as product of an even permutation on I_n extended by identity mapping outside I_n and of an even permutation on J_n extended by identity mapping outside J_n. Clearly, this group is isomorphic to $A_{I_n} \times A_{J_n}$. These groups form a uniformly computable increasing chain whose union is A_ω:

$$H_0 \subseteq H_1 \subseteq \ldots \subseteq H_k \subseteq \ldots \bigcup_{i<\omega} H_i = A_\omega.$$

Note that the function $n \to |H_n|$ is a computable function, and hence A_ω is a computable group. All these groups $H_n \cong A_{I_n} \times A_{J_n}$ have nontrivial normal subgroups isomorphic to A_{I_n} and hence they possess nontrivial congruences.

Take now a Π_2^0–complete set $B \subseteq \omega$. It is known that each Π_2^0–set A can be defined as follows:

$$x \in A \Leftrightarrow \exists^\omega t P(t, x),$$

for some computable predicate $P(t, n)$ (for instance, see [7]). Fix some such predicate $P(t, n)$ for the set B.

For each n, we define a non-decreasing computable sequence of natural numbers m_i^n as follows: $m_i^n = |\{t < i \mid P(t, n)\}|$. Now consider the uniformly computable sequence of groups $(\mathfrak{G}_n)_{n<\omega}$, where each $\mathfrak{G}_n$ is defined as:

$$\mathfrak{G}_n = \bigcup_{i<\omega} H_{m_i^n}.$$

One can easily check that

$$\exists^\omega t P(t, n) \Leftrightarrow \lim_{i \to \infty} m_i^n = \omega \Leftrightarrow \mathfrak{G}_n \cong A_\omega$$

and that

$$\neg(\exists^\omega t P(t, n)) \Leftrightarrow \exists k (\mathfrak{G}_n \cong H_k).$$

It is well known that the group A_ω is simple, the proof is actually the same as the proof of simplicity for the groups of kind A_m, $m \neq 4$ (see, for example [12]). We already noted that each H_n has a nontrivial congruence relation. Therefore, $\mathfrak{G}_n$ has no nontrivial congruences if and only if $\exists^\omega t P(t, n)$ if and only if $n \in B$. $\square$

The proof of the theorem above gives us the following corollary:

Corollary 1. *The property of grops "to have no nontrivial normal subgroups" is Π_2^0-complete.*

Theorem 2. *The property "to have a finite number of congruences" is Σ_3^0-complete.*

Proof. In order to prove the theorem we need the following simple fact.

Lemma 1. *For each congruence η, $\eta = \sup\{\theta(a,b) \mid a\eta b\}$ holds.*

It is trivial that if $a\eta b$ then $\theta(a,b) \subseteq \eta$. Now assume that a congruence η' contains all congruences of the type $\theta(a,b)$ for $a\eta b$. One notices that $\eta \subseteq \eta'$ is true because of the following implications:

$$a\eta b \Rightarrow \theta(a,b) \subseteq \eta' \Rightarrow a\eta' b.$$

This proves the lemma.

We prove p. 1 of the main definition. By Lemma 1, an algebra has a finite number of congruences if and only if it has a finite number of principal congruences $\theta(a,b)$. This can be expressed by the following condition:

$$\exists k \, \exists a_0 \exists b_0 \ldots \exists a_k \exists b_k \, \forall a \forall b \left(\bigvee_{i=0,\ldots,k} \theta(a_i,b_i) = \theta(a,b) \right),$$

which is Σ_3^0.

Now we prove p. 2 of the main definition. Let B be a Σ_3^0-complete set. The set B can be presented as

$$n \in B \Leftrightarrow \exists k \exists^\omega t P(k,t,n),$$

for some computable predicate P (see [7]).

Fix an $n \in \omega$. The following construction will be uniform in n. Define intervals I_k^t, where $k, t < \omega$, on ω by induction as follows:

1. For all $k \in \omega$ we set: $I_k^0 = [k,k]$.
2. Assume that the intervals I_k^t have already been defined.
   (a) Set $I_k^{t+1} = I_k^t$, for all k such that $\neg \exists k' \leqslant k P(k',t,n)$.
   (b) Let k_0 be the minimal $k < \omega$ with the property $P(k,t,n)$. Set $I_{k_0}^{t+1} = I_{k_0}^t \cup I_{k_0+1}^t$, $I_{k_0+1}^{t+1} = I_{k_0+2}^t$, $I_{k_0+2}^{t+1} = I_{k_0+3}^t$, etc. (i.e., we join intervals $I_{k_0}^t$ and $I_{k_0+1}^t$ and enumerate again all intervals in the order they follow).

The proof of the following lemma is by induction and follows from the definition of the intervals.

Lemma 2. *The intervals I_k^t satisfy the following properties:*

1. *Each interval I_k^t is convex.*
2. *Each I_i^t precedes I_{i+1}^t, for all $i, t < \omega$.*
3. *For all $t \in \omega$ we have: $\omega = \bigcup_{i<\omega} I_i^t$.* $\qquad\square$

Denote the left and the right ends of the interval I_k^t by $\ell(I_k^t)$ and $r(I_k^t)$, respectively. Clearly, the functions $\ell(I_k^t)$ and $r(I_k^t)$ are computable on parameters t and k. The next lemma gives properties of these functions. The proof is a routine check that follows the definition of the intervals.

Lemma 3. *The functions $\ell(I_k^t)$ and $r(I_k^t)$ satisfy the following properties:*

1. *Assume that $k_0 < \omega$ is the minimal among all $k < \omega$ such that $\exists^\omega t P(k,t,n)$. Then*
   (a) *for each $k < k_0$, the limit $\lim_{t\to\infty} \ell(I_k^t) < \infty$ exists.*
   (b) *for each $k < k_0$, the limit $\lim_{t\to\infty} r(I_k^t) < \infty$ exists.*
   (c) *$\lim_{t\to\infty} r(I_{k_0}^t) = \infty$.*
   (d) *for each $k > k_0$ $\lim_{t\to\infty} \ell(I_k^t) = \lim_{t\to\infty} r(I_k^t) = \infty$.*
2. *Assume that no $k < \omega$ exists such that $\exists^\omega t P(k,t,n)$. Then for each $k < \omega$ the limits $\lim_{t\to\infty} \ell(I_k^t)$ and $\lim_{t\to\infty} r(I_k^t)$ exist and are finite.* $\qquad\square$

We now define the group G_n^t as follows: The elements of the group are all the permutations τ on the set $\bigcup_{k \leqslant t} I_j^t \times \{0,\ldots,t\}$ such that for all $k \leqslant t$ we have $\tau(I_k^t \times \{0,\ldots,t\}) = I_k^t \times \{0,\ldots,t\}$ and $\tau \upharpoonright I_k^t \times \{0,\ldots,t\}$ is an even permutation. From the definitions one checks that $G_n^t \subseteq G_n^{t+1}$, for all $t < \omega$.

For each $n \in \omega$, we now set $\mathfrak{G}_n = \bigcup_{t<\omega} G_n^t$. The following lemma follows from the constructions above:

Lemma 4. *The groups $\mathfrak{G}_n$ satisfy the following properties:*

1. *If $n \notin B$ then $\mathfrak{G}_n$ is isomorphic to a direct product of infinitely many groups of kind A_ω.*
2. *If $n \in B$ then $\mathfrak{G}_n$ is isomorphic to a direct product of a finite number of simple groups of kind A_ω.* $\qquad\square$

Maybe it is worth to note that in our proof we could use a similar construction without taking products with $\{0,\ldots,t\}$. In this case we should take extra care of the situation when one of the groups in a decomposition might be isomorphic to the group A_4, which is not simple. We could also take products with ω instead of $\{0,\ldots,t\}$, which is a matter of a taste.

The next lemma shows that our coding of B into the sequence $\{\mathfrak{G}_n\}_{n\in\omega}$ is a desired coding.

Lemma 5

1. *A direct product of an infinite number of nontrivial simple groups has infinite number of normal subgroups.*
2. *A direct product of a finite number of groups isomorphic to A_ω has a finite number of normal subgroups only.*

The first part of the lemma is obvious. We prove the second part. Assume G is a normal subgroup of $S_1 \times S_2 \times \ldots \times S_k$, where each S_i, $i = 1,\ldots,k$ is isomorphic to A_ω.

We prove that G coincides with a product of groups S_j such that there exists an element $g \in G$ such that in the decomposition $g = g_1 g_2 \ldots g_k$, where $g_i \in S_i$, $i = 1, \ldots, k$, its jth component $g_j \neq 1$.

Assume $g = g_1 g_2 \ldots g_k \in G$ and $g_j \neq 1$. Take some $h \in S_j$ such that $h g_j \neq g_j h$. Such h exists since $S_j \cong A_\omega$. We have:

$$1 \neq g^{-1} h^{-1} g h = g_j^{-1} h^{-1} g_j h \in G \cap S_j.$$

Since G is a normal subgroup in G, $S_j \subseteq G$. Thus, G contains $S_{j_1} \times \ldots S_{j_m}$, where $j_1 \ldots j_m$ is the list of all indices j such that G contains an element g whose decomposition has a nontrivial multiplier from S_j. On the other hand if $g = g_1 g_2 \ldots g_k \in G$ then obviously $g \in S_{j_1} \times \ldots S_{j_m}$. We proved $G = S_{j_1} \times \ldots S_{j_m}$. By this, there exists a finite number of normal subgroups in $S_1 \times S_2 \times \ldots S_k$. We have proved the lemma.

Thus, if $n \in B$ then $\mathfrak{G}_n$ has a finite number of congruences and if $n \notin B$ then $\mathfrak{G}_n$ has infinite number of congruences. Note that the family $(\mathfrak{G}_n)_{n<\omega}$ is a computable family of groups. This proves the theorem.

Since the algebras we constructed to prove the theorem are groups, we also have the following corollary:

Corollary 2. *The property of groups "to have finitely many normal subgroups" is Σ_3^0-complete.*

Finally, we prove the last theorem of this paper. Recall that a partially ordered set has a *decreasing chain property* if every sequence $a_0 \geqslant a_1 \geqslant \cdots$ stabilizes, i.e., there exists an $n \in \omega$ such that for all $m \geqslant n$ holds $a_m = a_n$. The definition of the *increasing chain property* is obtained from the definition above by replacing of $a_0 \geqslant a_1 \geqslant \cdots$ with $a_0 \leqslant a_1 \leqslant \cdots$.

Theorem 3

1. The property "to have congruence lattice satisfying the increasing chain property" is Π_1^1-complete.
2. The property "to have congruence lattice satisfying the decreasing chain property" is Π_1^1-complete.

Proof. We first verify that p. 1 of the main definition is satisfied by both properties. For this, we code sequences of subsets of ω through the following definition. We say that the set X codes the sequence $(Y_i)_{i<\omega}$ of subsets of ω if for all $i < \omega$ $Y_i = \{(x)_0 \mid (x)_1 = i \ \& \ x \in X\}$. Clearly, each $X \subseteq \omega$ codes some sequence of sets and each sequence of sets is coded by some set. For an arbitrary set $X \subseteq \omega$, we let $[X]_i = \{(x)_0 \mid (x)_1 = i \ \& \ x \in X\}$.

As above, subsets of ω can code binary relations. Namely, each set $X \subseteq \omega$ codes the following binary relation $R_X = \{\langle (x)_0, (x)_1 \rangle \mid x \in X\}$.

Using these codings, it is possible to express the decreasing chain property in the following manner:

$$\forall X (X \text{ codes a decreasing sequence of congruences } (R_{[X]_j})_{j<\omega} \text{ of the algebra } \mathcal{A}_i$$
$$\text{implies that } \exists m \forall n \geq m (R_{[X]_n} = R_{[X]_{n+1}}).$$

The increasing chain property is coded in the same way. Thus, the increasing chain and decreasing chain properties both are always Π_1^1, and hence they both satisfy p. 1 of the main definition.

We now verify p. 2 of the main definition. Recall that a nonempty subset $T \subseteq \omega^{<\omega}$ is called a *tree* if for each $s \in T$, each initial segment of s belongs to T as well. It is known that there exists a computable family of trees $(T_n)_{n<\omega}$ such that the property "T_n is well–founded" is Π_1^1–complete (see [7]). Given a tree T consider the following Kleene–Brouwer ordering on the tree. For nodes $s, t \in T$

$s < t$ if and only if *(t is an initial segment of s) or (t and s are not initial segments of each other and s lexicographically precedes t).*

It can be easily checked that $\langle T, < \rangle$ is a well–ordering if and only if T is well–founded. Moreover, if T is computable then the linear order $(T, <)$ is computable. This actually proves that there exists a computable sequence $L_n = \langle T_n, < \rangle$ of linear orderings such that the set $\{i \mid L_i$ is well–ordered $\}$ is Π_1^1–complete. We now need the following algebraic lemma that connects linear orders with computable algebras.

Lemma 6. *Let $\langle L, < \rangle$ be an ordered set. Consider the algebra*

$$\mathcal{A}(L) = \left\langle A, f_{x_0,x_1,y}^2 \right\rangle_{x_0,x_1,y \in A, x_0 < x_1 \leqslant y}$$

for which $A = L$ and the operations $f_{x_0,x_1,y}$ are defined as follows:

$$f_{x_0,x_1,y}(a,b) = \begin{cases} x_1, & \text{if } a = b = x_0 \\ y, & \text{if } a = x_0 \text{ and } b = x_1 \\ \max(a,b) & \text{otherwise.} \end{cases}$$

Then the congruence lattice of $\mathcal{A}(L)$ is isomorphic to the lattice of all upwards closed subsets of $\langle L, < \rangle$ by inclusion.

Prove the lemma. First note that if for some $x_0 < x_1$ and for some congruence θ the condition $\theta(x_0, x_1)$ holds then for all $y > x_0$ holds $\theta(x_0, y)$. Indeed, in case $x_0 < y < x_1$ we have

$$y = \max(x_0, y) = f_{x_0,x_1,x_1}(x_0, y) \sim_\theta f_{x_0,x_1,x_1}(x_1, y) = \max(x_1, y) = x_1 \sim_\theta x_0.$$

If $x_1 < y$ then

$$x_0 \sim_\theta x_1 = f_{x_0,x_1,y}(x_0, x_0) \sim_\theta f_{x_0,x_1,y}(x_0, x_1) = y.$$

Thus, each congruence of $\mathcal{A}(L)$ has the property that each of its equivalence classes either consists of one element or it is a union of sets of the type $\{x \in L \mid x > a\}$. It follows that it contains only one upwards closed class and the other classes contain exactly one element each.

Let U be an upwards closed subset of L. Consider the equivalence relation θ_U defined as follows:

$$\theta_U(x,y) \iff (x, y \in U) \vee (x, y \notin U \ \& \ x = y).$$

To prove that this is a congruence relation of the algebra, we first claim the following:

Claim. Assume $a \in U$ or $b \in U$. Then for all $x_0, x_1, y \in A$ such that $x_0 < x_1 \leqslant y$, $f_{x_0,x_1,y}(a, b) \in U$.

To prove this claim, consider the following cases:

Case 1: $a = b = x_0$. In this case we have: $f_{x_0,x_1,y}(a, b) = x_1$ and since $a = b = x_0 < x_1$, this implies $x_1 \in U$.

Case 2: $a = x_0$ and $b = x_1$. In this case we have: $f_{x_0,x_1,y}(a, b) = y$ and since $a = x_0 < x_1 = b < y$, this implies $y \in U$.

Case 3: None of the cases above. For this case we have: $f_{x_0,x_1,y}(a, b) = \max(a, b) \in U$. Thus the claim is proved.

Assume now that $a_0 \sim_{\theta_U} a_0'$ and $a_1 \sim_{\theta_U} a_1'$. We would like to check that $f_{x_0,x_1,y}(a_0, a_1) \sim_{\theta_U} f_{x_0,x_1,y}(a_0', a_1')$. If at least one pair a_0, a_0', a_1, a_1' is in U then by the claim above $f_{x_0,x_1,y}(a_0, a_1)$, $f_{x_0,x_1,y}(a_0', a_1')$ are in U and thus are θ_U–equivalent. If both pairs a_0, a_0', a_1, a_1' are not in U then $a_0 = a_0'$ and $a_1 = a_1'$ and the equality we need to check is trivial.

Finally, one can easily check that $\theta_U \subseteq \theta_V$ if and only if $U \subseteq V$.

Thus, the congruence lattice of $\mathcal{A}(\mathcal{U})$ is isomorphic to the the lattice of all upwards closed subsets of $\langle A, < \rangle$ by inclusion. Lemma is complete.

Now one notices that the congruence lattice of the algebra $\mathcal{A}(L)$ has increasing chain property if and only if $\langle L, < \rangle$ is well ordered, and it has the decreasing chain property if and only if the set L^* with the reversed order is well ordered.

Thus, for all $i < \omega$, the congruence lattice of the algebra $\mathcal{A}_i = \mathcal{A}(L_i)$ has the increasing chain property if and only if L_i is well–ordered, for all $i < \omega$; and the congruence lattice of the algebra $\mathcal{A}_i^* = \mathcal{A}(L_i^*)$ has decreasing chain property if and only if L_i is well–ordered.

To complete the theorem, it remains to note that the families $(\mathcal{A}_i)_{i<\omega}$ and $(\mathcal{A}_i^*)_{i<\omega}$ are computable.   □

The authors are grateful to anonymous referees for their useful remarks.

References

1. Grätzer, G., Schmidt, E.T.: Characterizations of congruence lattices of abstract algebras. Acta Sci. Math. (Szeged) 24 (1963)
2. Lampe, W.: The independence of certain related structures of a universal algebra. IV. The triple is independent. Algebra Universalis 2, 296–302 (1972)
3. Lampe, W.: Results and problems on congruence lattice representations. Special issue dedicated to Walter Taylor. Algebra Universalis 55(2-3), 127–135 (2006)
4. Kjos-Hanssen, B., Brodhead, P.: The strength of the Grätzer-Schmidt theorem. In: Ambos-Spies, K., Löwe, B., Merkle, W. (eds.) CiE 2009. LNCS, vol. 5635, pp. 59–67. Springer, Heidelberg (2009)
5. Morozov, A.S.: Groups of computable automorphisms. In: Handbook of recursive mathematics, Part 1. Stud. Logic Found. Math., vol. 138, pp. 311–345. North-Holland, Amsterdam (1998)
6. Soare, R.I.: Recursively enumerable sets and degrees. In: A Study of Computable Functions and Computably Generated Sets. Perspectives in Mathematical Logic. Springer, Berlin (1987)

7. Rogers, H.: Theory of recursive functions and effective computability. McGraw-Hill, New York (1967)

8. Handbook of recursive mathematics. In: Ershov, Y.L., Goncharov, S.S., Nerode, A., Remmel, J.B., Marek, V.W. (eds.) Recursive model theory, Part 1. Studies in Logic and the Foundations of Mathematics, vol. 138. North-Holland, Amsterdam (1998)

9. Handbook of recursive mathematics. In: Ershov, Y.L., Goncharov, S.S., Nerode, A., Remmel, J.B., Marek, V.W. (eds.) Recursive algebra, analysis and combinatorics. Studies in Logic and the Foundations of Mathematics, vol. 2, 139. North-Holland, Amsterdam (1998)

10. Ershov, Y.L., Goncharov, S.S.: Constructive models. In: Siberian School of Algebra and Logic. Consultants Bureau, New York (2000)

11. Grätzer, G.: Universal algebra, 2nd edn. Springer, New York (1979)

12. Kargapolov, M.I., Merzlyakov, Y.I.: Fundamentals of the theory of groups. Springer, Heidelberg (1979)

The Strength of the Besicovitch-Davies Theorem

Bjørn Kjos-Hanssen[1] and Jan Reimann[2]

[1] Department of Mathematics, University of Hawai'i at Mānoa, Honolulu HI 96822
bjoern@math.hawaii.edu
http://www.math.hawaii.edu/~bjoern
[2] Department of Mathematics, University of California, Berkeley CA 94720
reimann@math.berkeley.edu
http://www.math.berkeley.edu/~reimann

Abstract. A theorem of Besicovitch and Davies implies for Cantor space 2^ω that each Σ_1^1 (analytic) class of positive Hausdorff dimension contains a Π_1^0 (closed) subclass of positive dimension. We consider the weak (Muchnik) reducibility $\leq_w$ in connection with the mass problem $S(U)$ of computing a set $X \subseteq \omega$ such that the Σ_1^1 class U of positive dimension has a $\Pi_1^0(X)$ subclass of positive dimension.

We determine the difficulty of the mass problems $S(U)$ through the following results:

(1) Y is hyperarithmetic if and only if $\{Y\} \leq_w S(U)$ for some U;
(2) there is a U such that if Y is hyperarithmetic, then $\{Y\} \leq_w S(U)$;
(3) if Y is Π_1^1-complete then $S(U) \leq_w \{Y\}$ for all U.

Keywords: geometric measure theory, algorithmic randomness, computability theory.

1   Introduction

One of the most useful properties of Lebesgue measure λ is its *regularity*: For any measurable set E,

$$\lambda(E) = \sup\{\lambda(K) \colon K \subseteq E, K \text{ compact}\} \tag{1}$$
$$= \inf\{\lambda(U) \colon U \supseteq E, U \text{ open}\}. \tag{2}$$

This implies that (using the appropriate convergence theorems), for measure theoretic considerations, E can be replaced by a G_δ or F_σ set of the same measure, simplifying the complicated topological structure of arbitrary Borel sets.

The regularity properties (1) and (2) hold more generally for any positive Borel measure on a σ-compact Hausdorff space in which any compact set has finite measure, but fails to be true in general. It is one of the major results in geometric measure theory that the s-dimensional *Hausdorff measures* $\mathcal{H}^s$ are still *inner regular*[1].

---

[1] While *outer regularity* (2) fails for Hausdorff measures in general (open sets have infinite $\mathcal{H}^s$-measure for $s < 1$), one can still find, for any set E, a G_δ set of the same Hausdorff measure. This is often referred to as G_δ-*regularity* (see Rogers[19]).

F. Ferreira et al. (Eds.): CiE 2010, LNCS 6158, pp. 229–238, 2010.
© Springer-Verlag Berlin Heidelberg 2010

Theorem 1. *For any analytic ($\mathbf{\Sigma}_1^1$) set $E \subseteq \mathbb{R}^n$ and for any $s \geq 0$,*

$$\mathcal{H}^s(E) = \sup\{\mathcal{H}^s(K) \colon K \subseteq E, K \text{ compact}\}.$$

Theorem 1 was shown by Besicovitch [1] for $\mathbf{\Sigma}_3^0$ sets and extended by Davies [2] to $\mathbf{\Sigma}_1^1$ sets. It was subsequently generalized to various non-Euclidean settings. In 1995, Howroyd [10] showed that inner regularity holds for $\mathcal{H}^s$ on any compact metric space, in particular for *Cantor space* 2^ω with the standard metric

$$d(X, Y) = \begin{cases} 2^{-\min\{n \colon X(n) \neq Y(n)\}} & X \neq Y \\ 0 & X = Y. \end{cases}$$

In the following, we refer to the inner regularity property of Hausdorff measure in Euclidean and compact metric spaces simply as the *Besicovitch-Davies Theorem*.

The hierarchies of effective descriptive set theory allow for a further ramification of regularity properties. Any (boldface) Borel set is effectively (lightface) Borel relative to a parameter. Hence we can, for instance, given a (lightface) Σ_α^0 set, measure how hard it is to find a $\Sigma_2^0(Y)$ subset of the same measure, by proving lower bounds on the parameter $Y \in 2^\omega$.

Dobrinen and Simpson [4] investigated this question for Σ_3^0 sets in Lebesgue measure and discovered an interesting connection with measure-theoretic *domination properties*. Subsequently, measure-theoretic domination properties were linked to LR-reducibility, a reducibility concept from algorithmic randomness. Recently, Simpson [22] gave a complete characterization of the regularity problem for Borel sets with respect to Lebesgue measure. One of his results states that the property that every $\Sigma_{\alpha+2}^0$ (α a recursive ordinal) subset of 2^ω has a $\Sigma_2^0(Y)$ subset of the same Lebesgue measure holds if and only if $0^{(\alpha)} \leq_{LR} Y$. His paper [22] also contains a survey of previous results along with an extensive bibliography.

In this paper, we study the complexity of the corresponding inner regularity for Hausdorff measure on 2^ω. We will see that, in contrast to the case of Lebesgue measure, finding subsets of positive Hausdorff measure can generally not be done with the help of a *hyperarithmetical* oracle. The core observation is that determining whether a set of reals has *positive Hausdorff measure* is more similar to determining whether it is *non-empty* than to determining whether it has *positive Lebesgue measure*.

Determining the exact strength of the Besicovitch-Davies Theorem is not only of intrinsic interest. A family of important problems in theoretical computer science ask some version of the question to what extent randomness (which is a useful computational tool) can be extracted from a weakly random source (which is often all that is available). Such questions can also be expressed in computability theory. The advantage, and simultaneously the disadvantage, of doing so is that one abstracts away from considering any particular model of efficient computation. One way to conceive of weak randomness is in terms of *effective Hausdorff dimension*. Miller [16] and Greenberg and Miller [8] obtained a negative result for randomness extraction: there is a real of effective Hausdorff dimension 1, that does not Turing compute any Martin-Löf random real. Despite this negative result, effective Hausdorff dimension, which is a "lightface" form of Hausdorff dimension, has independent interest, as it seems to offer a way to redevelop much of geometric measure theory (for example Frostman's Lemma [18]) in a more effective way.

Another conception of weak randomness comes from considering sets that differ from Martin-Löf random sets only on a sparse set of bits [11], or sets that are subsets of Martin-Löf sets [9,12,13]. Actually, these conceptions are related, as we will try to illustrate with the help of the set MIN of all reals that have minimal Turing degree.

The following result seems rather surprising.

Theorem 2. *The set* MIN *has Hausdorff dimension 1.*

Proof. This is merely a relativization of the theorem of Greenberg and Miller [8] that there is a minimal Turing degree of effective Hausdorff dimension 1. [2]   □

Theorem 2 says that high effective Hausdorff dimension is not sufficient to be able to extract randomness. It can also be used to deduce that infinite subsets of random sets are not sufficiently close to being random, either.

The set MIN is Π_4^0, so by the Besicovitch-Davies Theorem, MIN has a closed subset C that still has Hausdorff dimension as close to 1 as desired. Each closed set C in Cantor space is $\Pi_1^0(X)$ for some oracle X. By a reasoning similar to [3]*Theorem 4.3, each X-random closed set contains a member of C. It follows by reasoning as in [12] that each X-random set has an infinite subset of minimal Turing degree - in particular an infinite subset that Turing computes no 1-random (Martin-Löf random) set. Thus, if X could be chosen recursive, we would have a positive answer to the following question.

Question 1. Does each 1-random subset of ω have an infinite subset that computes no 1-random sets?

A partial answer to this question is known, using other methods:

Theorem 3 ([13]). *Each 2-random set has an infinite subset that computes no 1-random sets.*

But it is easy to see that the set X just referred to cannot be chosen recursive. To wit, by the computably enumerable degree basis theorem there is no nonempty Π_1^0 class consisting entirely of sets of minimal Turing degree. In the present article we show that X can be taken recursive in Kleene's O, but in general, for arbitrary Σ_1^1 classes (or even just arbitrary Π_2^0 classes) in place of MIN, X cannot be taken hyperarithmetical.

We expect the reader to be familiar with basic descriptive set theory and the effective part on hyperarithmetic sets and Kleene's $\mathcal{O}$. Standard references are [20] and [21]. We also assume basic knowledge of Hausdorff measures and dimension, as can be found in [19]. The proofs are rather succinct, but details can easily be filled in using basic results and methods of the above theories.

---

[2] Their work in turn builds on the construction of a diagonally non-recursive function of minimal Turing degree in Kumabe and Lewis [14].

2   Index Set Complexity

Recall that the *Hausdorff dimension* of a set E in a metric space X is defined as

$$\dim_{\mathrm{H}}(E) = \inf\{s \geq 0 \colon \mathcal{H}^s(E) = 0\}.$$

To keep the presentation simple, we concentrate on the problem of finding, given a set $E \subseteq 2^\omega$, a closed subset of positive Hausdorff dimension. Note that if $\dim_{\mathrm{H}}(E) > 0$ and E is Σ^1_1, then by the Besicovitch-Davies Theorem there exists a closed subset $C \subseteq E$ such that $\dim_{\mathrm{H}}(C) > 0$.

Davies' argument is based on the representation of analytic sets via *Souslin schemes*. A Souslin scheme in a metric space X is a family $(P_s \colon s \in \omega^{<\omega})$ of closed sets. A set A is analytic if and only if it can be represented as

$$A = \bigcup_{f \in \omega^\omega} \bigcap_{n \in \omega} P_{f \restriction n}$$

for some Souslin scheme $(P_s \colon s \in \omega^{<\omega})$. Using a technique now known as the *increasing sets lemma*, Davies constructs a function $f : \omega \to \omega$ such that for each n, the closed set

$$F_n := \bigcup_{s \leq f \restriction n} \bigcap_{i < |s|} P_{s \restriction i}$$

(where $s \leq f \restriction n$ means $|s| = n$ and $s(i) \leq f(i)$ for $i < n$) has sufficiently large $\mathcal{H}^s_{\varepsilon_n}$-measure, for some $\varepsilon_n > 0$. The intersection of the F_n is then the desired closed subset of positive measure. If we translate this to the canonical representation of Σ^1_1 classes in Cantor space, we obtain the following version of the Besicovitch-Davies Theorem.

Theorem 4. *For each Σ^1_1 class C of dimension $d > 0$ and for $\varepsilon > 0$, written in canonical form*

$$C = \{X \mid \exists Y \, \forall a \, \exists b \; R(X, Y, a, b)\}$$

where R is a recursive predicate, there exists a function $g \in \omega^\omega$ such that for each f majorizing g, the class

$$C_f := \{X \mid \exists Y \, \forall a \, \exists b < f(a) \; R(X, Y, a, b)\}$$

is a closed subclass of dimension at least $d - \varepsilon$.

Initially, one may think that the computational difficulty in determining whether a set of reals has *positive Hausdorff dimension* could be similar to the difficulty in determining whether it has *positive Lebesgue measure*, but we find that it is more similar to the determining whether it is *non-empty* – and this is more difficult than the measure question. While questions about Lebesgue measure can often be answered using an arithmetical oracle, for non-emptiness we often have to go beyond even the hyperarithmetical. As we shall see this level of difficulty first arises at the G_δ ($\mathbf{\Pi}^0_2$) level; we start by going over the simpler cases of open ($\mathbf{\Sigma}^0_1$), closed ($\mathbf{\Pi}^0_1$), and F_σ ($\mathbf{\Sigma}^0_2$) sets.

Theorem 5. *The following families are identical, and have Σ_1^0-complete index sets.*

1. *Σ_1^0 classes that are nonempty;*
2. *Σ_1^0 classes that have positive Hausdorff dimension;*
3. *Σ_1^0 classes that have positive measure.*

Proof. Given an c.e. set $W_e \subseteq 2^{<\omega}$, let $\mathcal{W}_e = \bigcup_{\sigma \in 2^{<\omega}} N_\sigma$, where $N_\sigma = \{X \in 2^\omega : \sigma \subset X\}$. Since any non-empty open set has positive Lebesgue measure, and having positive Lebesgue measure implies having positive dimension, the three statements are equivalent. The corresponding index sets are c.e. since $W_e \neq \emptyset$ if and only if $\exists s, \sigma(\varphi_{e,s}(\sigma) \downarrow$, and they are complete by Rice's Theorem. $\square$

The case of Π_1^0 classes is only slightly more complicated.

Theorem 6. *The set of indices of Π_1^0 classes that are nonempty is Π_1^0-complete.*

Proof. A tree T does not have an infinite path if and only if for some level n, no string of length n is in T. If T is co-c.e. the latter event is c.e. and hence the set $\{e : [T_e] \neq \emptyset\}$ is Π_1^0. It is Π_1^0-hard by Rice's Theorem. $\square$

Theorem 7. *The set of indices of Π_1^0 classes that have positive Lebesgue measure is Σ_2^0-complete.*

Proof (Sketch). Given a tree T_e, $[T_e]$ has positive Lebesgue measure if and only if $\exists n \forall m(|T_e \cap \{0,1\}^m| \geq 2^{m-n}|)$. Hence the corresponding index set is Σ_2^0. One can reduce the Σ_2^0 complete set Fin $= \{e : W_e$ finite$\}$ to it by effectively building, for each e, a tree T_e such that if and only if a given W_e is finite, the measure is positive. This is achieved by cutting the measure in half (i.e. terminating an appropriate number of nodes) whenever another number enters W_e. $\square$

Theorem 8. *The set of indices of Π_1^0 classes of Hausdorff dimension zero is Π_2^0-complete.*

Proof (Sketch). A Π_1^0 class $\mathcal{C}$ has Hausdorff dimension zero if and only if for each $d > 0$ and n, there is a clopen set U_n, induced by finitely many strings $\sigma_1, \ldots, \sigma_k$, so that $\sum 2^{-d|\sigma_i|} \leq 2^{-n}$, and such that the Σ_1^0 statement $\mathcal{C} \subseteq U_n$ holds. Thus the set of indices of Π_1^0 classes of Hausdorff dimension zero is Π_2^0. To see that this set is in fact Π_2^0-complete, we reduce the Π_2^0 complete set Inf $= \{e : W_e$ infinite$\}$ to it. This is done by controlling the branching rate: Given e, we construct a co-r.e. tree T_e. Each time a new number enters the c.e. set, the branching rate of T_e is reduced: When we see the n-th number enter W_e at stage s, we thin out T_e by delaying, above level s, the level at which the next splitting occurs by one. $\square$

Theorem 9. *The set of indices of Σ_2^0 classes that are nonempty is Σ_2^0-complete.*

Proof. A Σ_2^0 class is nonempty if and only if some of the Π_1^0 classes in the effective union are nonempty. $\square$

Theorem 10. *The set of indices of Σ_2^0 classes of positive Hausdorff dimension is Σ_2^0-complete.*

Proof. Hausdorff dimension is *countably stable*, that is, if $E = \bigcup_n E_n$, then $\dim_H(E) = \sup_n \dim_H(E_n)$. Hence a Σ_2^0 class has positive Hausdorff dimension if and only if some of the Π_1^0 classes in the effective union have positive dimension. $\square$

Theorem 11. *The set of indices of Π_2^0 classes that have positive Hausdorff dimension is Σ_1^1-complete.*

Proof. For a given Π_2^0 class $\mathcal{G}$, we can consider the "Cartesian product"

$$\mathcal{P} = \mathcal{G} \times 2^\omega := \{G \oplus H \mid G \in \mathcal{G}, \ H \in 2^\omega\} \tag{1}$$

where $\oplus$ denotes the usual recursion-theoretic join. By the product formula for Hausdorff dimension (adapted to Cantor space, see [15,17]), if $\mathcal{G} \neq \emptyset$,

$$\dim_H(\mathcal{G} \times 2^\omega) \geq \frac{\dim_H(\mathcal{G}) + \dim_H(2^\omega)}{2} = \frac{\dim_H(\mathcal{G})}{2} + \frac{1}{2},$$

Hence the set $\mathcal{P}$ has positive Hausdorff dimension if and only if it has dimension at least $1/2$, if and only if $\mathcal{G} \neq \emptyset$. Since the set of indices of Π_2^0 classes in 2^ω that are nonempty is Σ_1^1-hard, so is the set of indices of Π_2^0 classes that have positive Hausdorff dimension. By the Besicovitch-Davies Theorem 4, the set of indices of Σ_1^1 classes that are of positive dimension is Σ_1^1, since

$$\dim\{X \mid \exists Y \forall a \exists b \, R(X,Y,a,b)\} > 0 \Leftrightarrow \exists f \, \dim C_f > 0. \qquad \square$$

Thus, the Besicovitch-Davies Theorem (Theorem 4) turns out to be enough information to completely classify the index set complexity of classes that have positive Hausdorff dimension from arithmetical pointclasses and up to Σ_1^1; see Figure 1.

Question 2. What is the complexity of the set of indices of Π_1^1 classes that have positive Hausdorff dimension?

At the level Π_2^0 it is far more complicated to determine whether a class has positive dimension than whether it has positive Lebesgue measure (this is Σ_3^0).

Question 3. What is the complexity of the set of indices of Σ_1^1 classes that have positive Lebesgue measure?

3   Closed Subsets of Positive Dimension

Recall that $A \leq_T B$ if A is Turing reducible to B; $A \leq_h B$ if A is hyperarithmetical in B; and, for sets of reals $\mathcal{A}$, $\mathcal{B}$, $\mathcal{A} \leq_w \mathcal{B}$ if $\mathcal{A}$ is weakly (Muchnik) reducible to $\mathcal{B}$, i.e., for each $B \in \mathcal{B}$ there is some $A \in \mathcal{A}$ such that $A \leq_T B$.

Family	Nonempty?	Positive Hausdorff dimension?
Σ_1^0	Σ_1^0-complete (5)	Σ_1^0-complete (5)
Π_1^0	Π_1^0-complete (6)	Σ_2^0-complete (8, 10)
Σ_2^0	Σ_2^0-complete (9)	
Π_2^0	Σ_1^1-complete	Σ_1^1-complete (11)
Σ_1^1		

Fig. 1. Index set complexity of some classes of reals. For example, the set of indices of Π_2^0 classes that are of positive Hausdorff dimension is Σ_1^1-complete, and this is shown in Theorem 11.

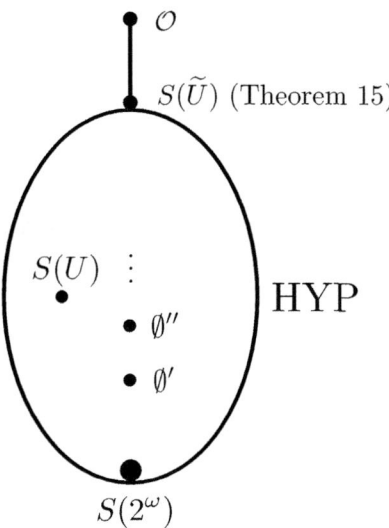

Fig. 2. The relative position in the Muchnik lattice of the various mass problems $S(U)$. At the top is Kleene's $\mathcal{O}$, according to Theorems 12(1) and 13. The ellipse represents the hyperarithmetical sets HYP with their cofinal sequence $\{0^{(\alpha)} : \alpha < \omega_1^{CK}\}$. The top $S(\widetilde{U})$ class is located as indicated in Theorem 16. Each $S(U)$ bounds only sets in HYP, per Theorem 15. It is not known which of the classes U here displayed might represent the set of minimal Turing degrees MIN.

Definition 1. *Let U be a Σ_1^1 class of positive Hausdorff dimension. The mass problem $S(U)$ is defined to be the collection of sets $X \subseteq \omega$ such that U has a $\Pi_1^0(X)$ subclass of positive dimension.*

We determine the difficulty of the mass problems $S(U)$ in Theorems 13, 15, and 16 below; the situation is summarized in Figure 2.

In the following definition, we are interested in the case $\Gamma = \Sigma_1^1$.

Definition 2. *A subset $\mathcal{B}$ of ω^ω is called a* basis *for a pointclass Γ if each nonempty collection of reals that belongs to Γ has a member in $\mathcal{B}$.*

Theorem 12 (Basis theorems for Σ_1^1). *Each of the following classes are bases for Σ_1^1:*

(1) $\{X \mid X \leq_T \mathcal{O}\}$, the sets recursive in some Π_1^1 set (see Rogers [20]*XLII(b));
(2) $\{X \mid X <_h \mathcal{O}\}$, the sets of hyperdegree strictly below $\mathcal{O}$ (Gandy [6]; see also Rogers [20]*XLIII(a));
(3) $\{X \mid X \not\leq_h A \ \& \ A \not\leq_h X\}$ (where A is any given non-hyperarithmetical set) (Gandy, Kreisel, and Tait [7]).

Theorem 13. *For each set $\mathcal{B}$ that is a basis for Σ_1^1 and each Σ_1^1 class U of positive dimension, there is some $X \in \mathcal{B}$ such that U has a $\Pi_1^0(X)$ subclass of positive dimension.*

Proof. Consider a Σ_1^1 class of the form

$$\{X \mid (\exists Y)(\forall a)(\exists b) R(a, b, Y, X)\}$$

and the closed subclass from Theorem 4,

$$C_f = \{X \mid (\exists Y)(\forall a)(\exists b < f(a)) R(a, b, Y, X)\}$$

which is a $\Pi_1^0(f)$ class in 2^ω. Now consider

$$\{f \in \omega^\omega \mid \dim C_f > 0\}$$

This is a Σ_2^0 class in ω^ω, in particular it is Σ_1^1, hence it has a member f in $\mathcal{B}$; and C_f for such an f is a $\Pi_1^0(G_f)$ class, where G_f is the graph of f. □

In particular, U always has a $\Pi_1^0(\mathcal{O})$ subclass of positive dimension.

Definition 3 (Solovay [23]). *A family F of infinite sets of natural numbers is said to be dense if each infinite set of natural numbers has a subset in F. A set A of natural numbers is said to be recursively encodable if the family of infinite sets in which A is recursive is dense.*

Theorem 14 (Solovay [23]). *The recursively encodable sets coincide with the hyperarithmetic sets.*

Theorem 15. *For each Y, if $\{Y\} \leq_w S(U)$ for some U then Y is hyperarithmetical.*

Proof. Suppose Y is recursive in each tree defining a closed subset of positive dimension of some Π_2^0 class U. By Theorem 4, Y is recursively encodable. So by Theorem 14, Y is hyperarithmetic. □

Theorem 16. *There is a Π_2^0 class $\widetilde{U}$ such that for each hyperarithmetical set Y, $\{Y\} \leq_w S(\widetilde{U})$.*

Proof. Let HYP denote the collection of all hyperarithmetical sets. Note that the class

$$U = \{X \mid \forall H \in \text{HYP} \, H \leq_T X\}$$

is Σ_1^1 (an observation made by Enderton and Putnam [5]). This class U already has positive Hausdorff dimension, but the more involved proof of this fact can be avoided by again passing to the product set $\widetilde{U} = U \times 2^\omega$.

Suppose X is such that there is a $\Pi_1^0(X)$ subclass of U that is of positive dimension and hence nonempty. Then by the relativized low and hyperimmune-free basis theorems, respectively, each H in HYP is recursive in a set A that is low relative to X, and recursive in a set B that is hyperimmune-free relative to X. But A and B form a minimal pair over X, so $H \leq_T X$.

Now, U, being Σ_1^1, is the projection of a Π_2^0 class. Of course, if every member of the projection computes H then so does every member of the original Π_2^0 class (since a pair $A \oplus B$ computes both A and B). So we can replace U with such a Π_2^0 class; $\widetilde{U}$ is then still Π_2^0. □

Acknowledgments. Kjos-Hanssen was partially supported by NSF grants DMS-0901020 and DMS-0652669, the latter being part of a Focused Research Group in algorithmic randomness. Reimann was partially supported by NSF grant DMS-0801270 and Templeton Foundation Grant 13404 "Randomness and the Infinite".

References

1. Besicovitch, A.S.: On existence of subsets of finite measure of sets of infinite measure. Nederl. Akad. Wetensch. Proc. Ser. A. **55** = Indagationes Math. 14, 339–344
2. Davies, R.O.: Subsets of finite measure in analytic sets. Nederl. Akad. Wetensch. Proc. Ser. A. **55** = Indagationes Math. 14, 488–489
3. Diamondstone, D.: Martin-löf randomness and galton-watson processes
4. Dobrinen, N.L.: Almost everywhere domination. J. Symbolic Logic 69(3), 914–922
5. Enderton, H.B.: A note on the hyperarithmetical hierarchy. J. Symbolic Logic 35, 429–430
6. Gandy, R.O.: On a problem of kleene's. Bull. Amer. Math. Soc. 66, 501–502
7. Gandy, R.O.: Set existence. Bull. Acad. Polon. Sci. Sér. Sci. Math. Astronom. Phys. 8, 577–582
8. Greenberg, N.: Diagonally non-recursive functions and effective hausdorff dimension
9. Greenberg, N.: Lowness for kurtz randomness. J. Symbolic Logic 74(2), 665–678
10. Howroyd, J.D.: On dimension and on the existence of sets of finite positive Hausdorff measure. Proc. London Math. Soc. 70(3), 581–604
11. Kjos-Hanssen, B.: Extracting algorithmic randomness from adaptive bit-fixing sources
12. Kjos-Hanssen, B.: Infinite subsets of random sets of integers. Math. Res. Lett. 16(1), 103–110
13. Kjos-Hanssen, B.: A law of weak subsets
14. Kumabe, M.: A fixed-point-free minimal degree. J. London Mathematical Society 80(3), 785–797 (2009)

15. Mattila, P.: Geometry of sets and measures in Euclidean spaces. In: Cambridge Studies in Advanced Mathematics, vol. 44. Cambridge University Press, Cambridge
16. Miller, J.S.: Extracting information is hard: a turing degree of non-integral effective hausdorff dimension. Advances in Mathematics
17. Reimann, J.: Computability and fractal dimension
18. Reimann, J.: Effectively closed sets of measures and randomness. Ann. Pure Appl. Logic 156(1), 170–182
19. Rogers, C.A.: Hausdorff Measures. Cambridge University Press, Cambridge
20. Rogers, H.: Theory of recursive functions and effective computability, 2nd edn. MIT Press, Cambridge
21. Sacks, G.E.: Higher recursion theory. In: Perspectives in Mathematical Logic. Springer, Berlin
22. Simpson, S.G.: Mass problems and measure-theoretic regularity
23. Solovay, R.M.: Hyperarithmetically encodable sets. Trans. Amer. Math. Soc. 239, 99–122

Circuit Complexity and Multiplicative Complexity of Boolean Functions[*]

Arist Kojevnikov and Alexander S. Kulikov

St. Petersburg Department of Steklov Institute of Mathematics
{arist,kulikov}@logic.pdmi.ras.ru

Abstract. In this note, we use lower bounds on Boolean multiplicative complexity to prove lower bounds on Boolean circuit complexity. We give a very simple proof of a $7n/3 - c$ lower bound on the circuit complexity of a large class of functions representable by high degree polynomials over GF(2). The key idea of the proof is a circuit complexity measure assigning different weights to XOR and AND gates.

1 Introduction

Proving lower bounds on the circuit complexity of explicitly defined Boolean functions is one of the most famous and difficult problems in Theoretical Computer Science. Already in 1949 Shannon [1] showed by a counting argument that almost all Boolean functions have circuits of size $\Omega(2^n/n)$ only. Still, we have no example of an explicit function requiring super linear circuit size. Moreover, only a few proofs of linear lower bounds are known. Namely, Schnorr [2] proved a $2n - c$ lower bound for a class of functions with the property that by fixing the values of any two variables one gets at least three different subfunctions. Then Paul [3] proved a $2.5n - c$ lower bound for a modification of the storage access function. Stockmeyer [4] obtained the same lower bound for a class of symmetric functions satisfying a certain simple property. Finally, Blum slightly modified Paul's function and proved a $3n - o(n)$ bound on it. This bound was published in 1984 and is still the best result for circuits over the full binary basis B_2. The current record lower bound $5n - o(n)$ for the basis $U_2 = B_2 \setminus \{\oplus, \equiv\}$ was given in 2002 by Iwama and Morizumi [5].

All bounds mentioned above are proved by the gate elimination method. The main idea of this method is the following. One considers a Boolean function on n variables from a certain class of functions and shows (usually by a long case analysis) that for any circuit computing this function setting some variables to constants one obtains a subfunction of the same type and eliminates several gates. Usually, a gate is eliminated just because one of its inputs becomes a

---

[*] Research is partially supported by Federal Target Programme "Scientific and scientific-pedagogical personnel of the innovative Russia" 2009–2013, RFBR (08-01-00640 and 09-01-12137), RAS Program for Fundamental Research, Grant of the President of Russian Federation (MK-3912.2009.1 and NSh-5282.2010.1), and ANR, France (NAFIT ANR-08-EMER-008-01).

F. Ferreira et al. (Eds.): CiE 2010, LNCS 6158, pp. 239–245, 2010.
© Springer-Verlag Berlin Heidelberg 2010

constant. By induction, one concludes that the original circuit must have many gates. Though this method is essentially the only known method for proving nontrivial lower bounds for general circuit complexity, as many authors note it is unlikely that it will allow to prove nonlinear bounds.

The multiplicative complexity of a Boolean function f is defined as the minimal number of AND gates in a circuit over $\{\wedge, \oplus, 1\}$ computing f. Again, by a counting argument one can show that almost all Boolean functions of n variables have multiplicative complexity $2^{n/2} - O(n)$ [6]. As for the constructive lower bounds, the situation is even worse than with circuit complexity. Namely, the best known lower bound is $n - 1$ [7]. This bound holds for any function representable by a polynomial over $GF(2)$ of degree n. It is quite easy to see that, e.g., the conjunction of n variables has multiplicative complexity $n - 1$. Even a function with multiplicative complexity at least n is not known.

In this note, we prove a lower bound $7n/3 - c$ on the circuit complexity of a large class of functions representable by high degree polynomials over $GF(2)$. The key idea of the proof is a circuit complexity measure assigning different weights to XOR and AND gates. Note that while the proven lower bound is weaker than Stockmeyer's $2.5n - c$ bound and Blum's $3n - o(n)$ bound, it is applicable for a much wider class of functions (Stockmeyer's proof works for symmetric functions only, Blum's bound works for a particular single function) and its proof is much simpler (in particular, contains almost no case analysis).

2   General Setting

2.1   Circuit Complexity and Gate Elimination

By B_n we denote the set of all Boolean functions $f\colon \{0,1\}^n \to \{0,1\}$. A circuit over a basis $\Omega \subseteq B_2$ is a directed acyclic graph with nodes of in-degree 0 or 2. Nodes of in-degree 0 are marked by variables from $\{x_1, \ldots, x_n\}$ and are called inputs. Nodes of in-degree 2 are marked by functions from Ω and are called gates. There is also a special output gate where the result is computed. The size of a circuit is its number of gates. By $C_\Omega(f)$ we denote the minimum size of a circuit over Ω computing f. The two commonly studied bases are B_2 and $U_2 = B_2 \setminus \{\oplus, \equiv\}$. In this note we consider circuits over B_2 and denote $C_{B_2}(f)$ by just $C(f)$.

We call a function $f \in B^n$ degenerated if it does not depend essentially on some of its variables, i.e., there is a variable x_i such that the subfunctions $f|_{x_i=0}$ and $f|_{x_i=1}$ are equal. It is easy to see that a gate computing a degenerated function from B_2 can be easily eliminated from a circuit without increasing its size (when eliminating this gate one may need to change the functions computed at its successors). The set B_2 contains the following sixteen functions $f(x, y)$:

- six degenerate functions: 0, 1, x, $x \oplus 1$, y, $y \oplus 1$;
- eight functions of the form $((x \oplus a) \wedge (y \oplus b)) \oplus c$, where $a, b, c \in \{0, 1\}$; we call them type-$\wedge$ functions;
- two functions of the form $x \oplus y \oplus a$, where $a \in \{0, 1\}$; these are called type-$\oplus$ functions;

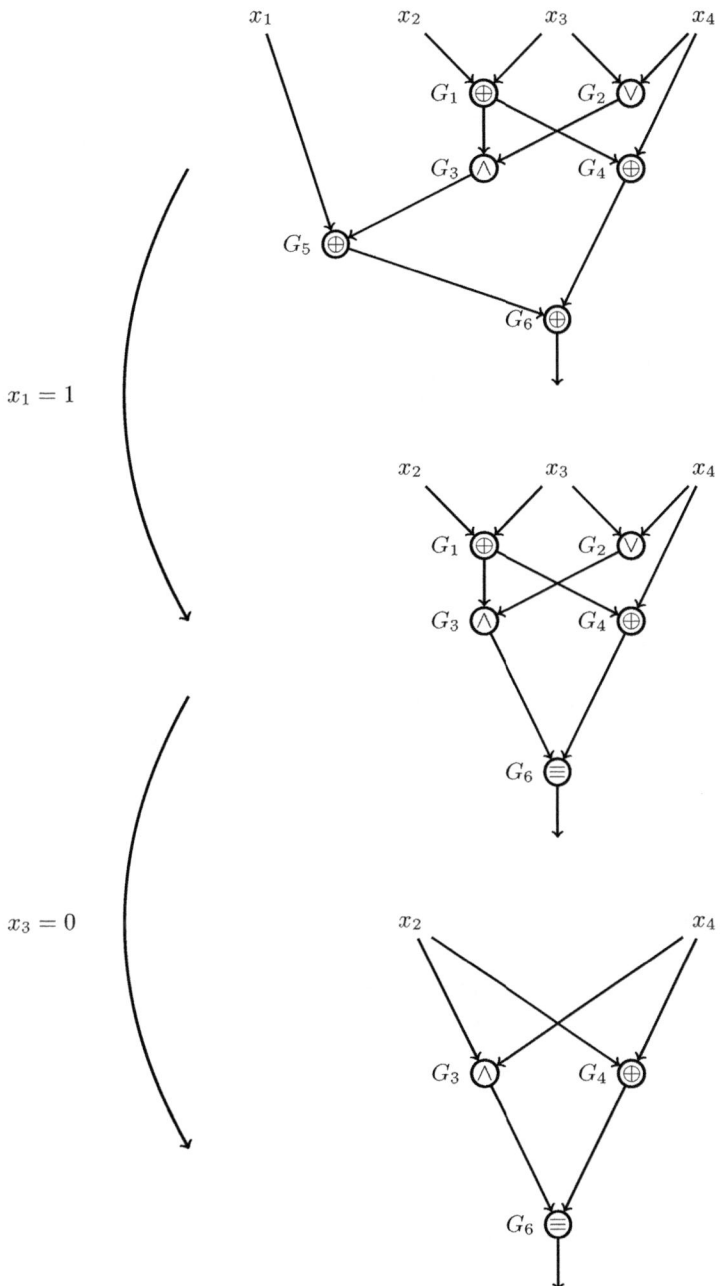

Fig. 1. Example of gate elimination

The main difference between type-$\wedge$ and type-$\oplus$ functions is that it is always possible to make a type-$\wedge$ binary function constant by assigning a Boolean value to any of its inputs (e.g., after replacing x by a one gets the constant c; in this case we say that a gate is trivialized), while this is not possible for a type-$\oplus$ function. Thus, if there is a variable feeding two type-$\oplus$ gates, then by assigning any value to it one eliminates only these two gates. However, if at least one of the two successors of a variable is of type-$\wedge$, then by assigning an appropriate value to it one eliminates both these gates and also all successors of this type-$\wedge$ gate. This is the reason why currently best lower bounds for circuits over U_2 are stronger than those for circuits over B_2.

Fig. 1 shows an example on how assigning values to input variables affects a circuit. We first replace x_1 by 1. Then G_5 computes $G_3 \oplus 1$ and is not needed anymore, while G_6 computes $G_3 \oplus G_4 \oplus 1$, i.e., $G_3 \equiv G_4$. We now assign the value 0 to x_3. Then both G_1 and G_2 can be eliminated as they compute x_2.

2.2 Multiplicative Complexity and Polynomials over GF(2)

The multiplicative complexity of a Boolean function f is defined as the minimal number of AND gates in a circuit over $\{\wedge, \oplus, 1\}$ computing f. It is easy to see that the minimal possible number of type-$\wedge$ gates in a circuit over B_2 computing f is exactly the multiplicative complexity of f.

Any function $f \in B_n$ can be easily represented as a multilinear polynomial over GF(2), i.e., a XOR (sum) of conjunctions (monomials). To obtain this representation one can take a sum of all "literal monomials" (i.e., conjunctions of literals, where a literal is either a variable or its negation) corresponding to elements of $f^{-1}(1)$. E.g., the polynomial over GF(2) for the majority function of three bits is

$$x_1 x_2 x_3 + (1 + x_1)x_2 x_3 + x_1(1 + x_2)x_3 + x_1 x_2(1 + x_3) = x_1 x_2 + x_2 x_3 + x_1 x_3 \, .$$

It is well known that such a representation is unique. Indeed, each function has a representation by a polynomial and the number of functions of n variables is equal to the number of multilinear polynomials over GF(2) (and is equal to 2^{2^n}). We denote this polynomial by $\chi(f)$. The important characteristics of a function f is the degree of $\chi(f)$ denoted by $\deg(f)$. Clearly, if a circuit contains only a few type-$\wedge$ gates, then it cannot compute a function of high degree. For example, to compute the conjunction of n variables $n - 1$ type-$\wedge$ gates are necessary and sufficient.

Lemma 1 ([7]). *Any circuit computing a Boolean function $f \in B_n$ contains at least $(\deg(f) - 1)$ type-$\wedge$ gates.*

3 $7n/3$ Lower Bound

In this section, we give a proof of a $7n/3 - c$ lower bound. It is as simple as Schnorr's proof of a $2n - c$ lower bound [2]. We first define a class of functions

S_n^k similar to the one used by Schnorr. We then define a circuit complexity measure by putting different weights to type-$\oplus$ and type-$\wedge$ gates and prove a lower bound for it. This lower bound is then used to prove a lower bound on the circuit complexity of a class of functions obtained by taking the sum of functions from S_n^k and a function of full degree.

Definition 1. *For a constant k, let S_n^k be the class of all functions $f \in B_n$, $n \geq k$, with the following properties:*

1. *for any two variables x_i and x_j one obtains at least three different subfunctions of f by fixing the values of x_i and x_j;*
2. *for any variable x_i and any constant $c \in \{0,1\}$, $f|_{x_i=c} \in S_{n-1}^k$.*

A natural function satisfying both these properties is a modular function $\mathrm{MOD}_{m,r}^n$ defined as follows:

$$\mathrm{MOD}_{m,r}^n(x_1,\ldots,x_n) = 1 \text{ iff } \sum_{i=1}^n x_i \equiv r \pmod{m}.$$

It is easy to see that, for any $m \geq 3$ and r, $\mathrm{MOD}_{m,r}^n \in S_n^{m+2}$. Indeed,

$$\mathrm{MOD}_{m,r}^n|_{x_i=0,x_j=0} = \mathrm{MOD}_{m,r}^{n-2}, \ \mathrm{MOD}_{m,r}^n|_{x_i=1,x_j=1} = \mathrm{MOD}_{m,r-2}^{n-2},$$

$$\mathrm{MOD}_{m,r}^n|_{x_i=0,x_j=1} = \mathrm{MOD}_{m,r}^n|_{x_i=1,x_j=0} = \mathrm{MOD}_{m,r-1}^{n-2}.$$

For $m \geq 3$ and $n \geq m+2$, $\mathrm{MOD}_{m,r}^{n-2}$, $\mathrm{MOD}_{m,r-1}^{n-2}$, $\mathrm{MOD}_{m,r-2}^{n-2}$ are not constant and pairwise different (note that for $m = 2$ this is not true as $\mathrm{MOD}_{2,r-2}^{n-2} = \mathrm{MOD}_{2,r}^{n-2}$). Also, $\mathrm{MOD}_{m,r}^n|_{x_i=c} = \mathrm{MOD}_{m,r-c}^{n-1}$.

In order to prove the stated lower bound we use the following circuit complexity measure: $\mu(D) = 3X(D) + 2A(D)$, where $X(D)$ and $A(D)$ denote, respectively, the number of type-$\oplus$ and type-$\wedge$ gates in a circuit D.

Lemma 2. *For any circuit D computing a function $f \in S_n^k$,*

$$\mu(D) \geq 6(n - k - 1).$$

Proof. We prove the statement by induction on n. The case $n \leq k+1$ is obvious. Assume that $n > k+1$ and let D be an optimal (w.r.t. μ) circuit computing f. We show that it is possible to assign a value to one of the variables such that μ is reduced by at least 6. Since the resulting subfunction belongs to S_{n-1}^k, the required inequality follows by induction. Note that the resulting subfunction is not a constant (otherwise it would not have three different subfunctions w.r.t. any two variables). Thus, if a gate is replaced by a constant during such a substitution, then this gate is not an output and hence has at least one successor that is also eliminated.

All degenerate gates can be eliminated from D without increasing $\mu(D)$. Let Q be a top-gate of D and x_i and x_j be its input variables. These variables are different as Q is non-degenerate. Since there are at least three different subfunctions of f w.r.t. x_i and x_j, one of them must feed at least one other gate. W.l.o.g. we assume that x_i feeds also a gate $P \neq Q$. There exist two cases.

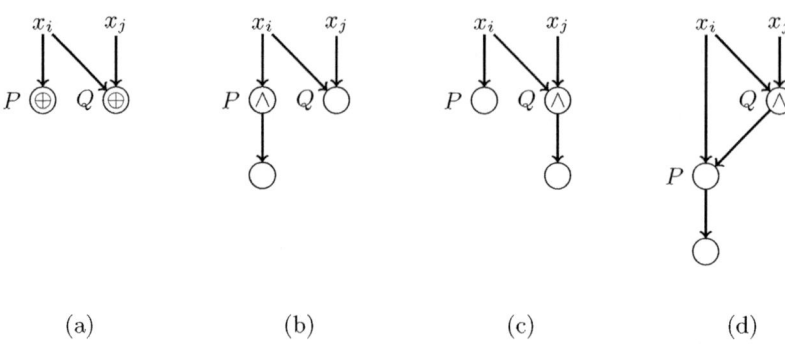

1. If both P and Q are type-$\oplus$ gates (see Fig. 2(a); note that only the types of gates are shown, but not the exact functions computed at them), we just assign a value to x_i. Clearly μ is reduced by at least 6.
2. If one of P and Q is a type-$\wedge$ gate (Fig. 2(b–d)), we assign to x_i the value which trivializes this gate. This eliminates both P and Q and at least one of their successors. Note that P can be a successor of Q (Fig. 2(d)). In this case after assigning x_i the right constant, not only Q becomes a constant, but also the gate P, as both its inputs are constants. Hence all successors of P are eliminated. Since at least three gates are eliminated, μ is again reduced by at least 6. □

Lemma 3. *Let* $f \in S_n^k$ *and* $\deg(f) = n$. *Then* $C(f) \geq 7n/3 - c(k)$.

Proof. Let D be a circuit computing f. Recall that the number of gates in a circuit D is at least $X(D) + A(D)$. The required inequality then follows from the following two inequalities:

$$3X(D) + 2A(D) \geq 6n - 6(k+1),$$
$$A(D) \geq n - 1.$$

□

The lemma above provides a $7n/3 - c$ lower bound for functions from S_k^n of degree n. In fact, it can be used to prove the same lower bound for a much wider class of functions. Assume that $f \in S_n^k$ and $\deg(f) < n$. Consider now *any* function $g \in B_{n-1}$ of degree $n - 1$ (note that the number of such functions is $2^{2^{n-1}-1}$, as we only need that the monomial $x_1 x_2 \ldots x_{n-1}$ is present in $\chi(g)$) and define $h \in B_n$ as follows:

$$h(x_1, \ldots, x_n) = f(x_1, \ldots, x_n) \oplus x_n g(x_1, \ldots, x_{n-1}).$$

Consider a circuit D computing h. Since $\deg(h) = n$, $A(D) \geq n - 1$. Note also that $h|_{x_n=0} = f|_{x_n=0} \in S_{n-1}^k$. Hence $\mu(D) \geq \mu(D|_{x_n=0}) \geq 6(n-1) - 6(k+1)$ and a lower bound $7n/3 - c(k)$ for the size of D follows.

4   Further Directions

It would be interesting to give an example of an explicit function with multiplicative complexity at least an, for $a > 1$. Such a function would immediately give a stronger than $7n/3$ lower bound for circuit complexity. As said above, the best known lower bound $n-1$ follows from the fact that a function is represented by a polynomial of high degree over GF(2). Such is, e.g., the function $\mathrm{MOD}_{3,r}^n$: $\deg(\mathrm{MOD}_{3,r}^n) \geq n-1$, for any r and $n \geq 3$. Thus, at least $n-2$ type-$\wedge$ gates are needed just to compute a high degree monomial in $\chi(\mathrm{MOD}_{3,r}^n)$. It is natural to expect that to compute the $\mathrm{MOD}_{3,r}^n$ function more type-$\wedge$ gates are needed in the worst case. Interestingly, this is not true: Boyar et al. [6] showed that the multiplicative complexity of any symmetric function is at most $n + o(n)$.

As to the lower bounds on circuit complexity, it would be interesting to improve lower bounds $3n - o(n)$ by Blum [8] and $5n - o(n)$ by Iwama and Morizumi [5] on the circuit complexity over B_2 and U_2, respectively. An (apparently) easier problem is to close one of the following gaps (see [4], [9], [10]):

$$2.5n - c \leq C_{B_2}(\mathrm{MOD}_3^n) \leq 3n + c, \ 4n - c \leq C_{U_2}(\mathrm{MOD}_4^n) \leq 5n + c.$$

Acknowledgments

We would like to thank Edward A. Hirsch for helpful comments.

References

1. Shannon, C.E.: The synthesis of two-terminal switching circuits. Bell System Technical Journal 28, 59–98 (1949)
2. Schnorr, C.: Zwei lineare untere Schranken für die Komplexität Boolescher Funktionen. Computing 13, 155–171 (1974)
3. Paul, W.J.: A 2.5n-lower bound on the combinational complexity of Boolean functions. SIAM Journal of Computing 6(3), 427–433 (1977)
4. Stockmeyer, L.J.: On the combinational complexity of certain symmetric Boolean functions. Mathematical Systems Theory 10, 323–336 (1977)
5. Iwama, K., Morizumi, H.: An explicit lower bound of $5n - o(n)$ for boolean circuits. In: Diks, K., Rytter, W. (eds.) MFCS 2002. LNCS, vol. 2420, pp. 353–364. Springer, Heidelberg (2002)
6. Boyar, J., Peralta, R., Pochuev, D.: On The Multiplicative Complexity of Boolean Functions over the Basis $(\wedge, \oplus, 1)$. Theoretical Computer Science 235(1), 1–16 (2000)
7. Schnorr, C.P.: The Multiplicative Complexity of Boolean Functions. In: Mora, T. (ed.) AAECC 1988. LNCS, vol. 357, pp. 45–58. Springer, Heidelberg (1989)
8. Blum, N.: A Boolean function requiring $3n$ network size. Theoretical Computer Science (28), 337–345 (1984)
9. Demenkov, E., Kojevnikov, A., Kulikov, A.S., Yaroslavtsev, G.: New upper bounds on the Boolean circuit complexity of symmetric functions. Information Processing Letters 110(7), 264–267 (2010)
10. Zwick, U.: A $4n$ lower bound on the combinational complexity of certain symmetric boolean functions over the basis of unate dyadic Boolean functions. SIAM Journal on Computing 20, 499–505 (1991)

Definability in the Subword Order

Oleg V. Kudinov[1,*], Victor L. Selivanov[2,**], and Lyudmila V. Yartseva[2,***]

[1] S.L. Sobolev Institute of Mathematics
Siberian Division Russian Academy of Sciences
kud@math.nsc.ru
[2] A.P. Ershov Institute of Informatics Systems
Siberian Division Russian Academy of Sciences
vseliv@iis.nsk.su, kotofejnik@gmail.com

Abstract. We develop a theory of (first-order) definability in the subword partial order in parallel with similar theories for the h-quasiorder of finite k-labeled forests and for the infix order. In particular, any element is definable (provided that words of length 1 or 2 are taken as parameters), the first-order theory of the structure is atomic and computably isomorphic to the first-order arithmetic. We also characterize the automorphism group of the structure and show that any arithmetical predicate invariant under the automorphisms of the structure is definable in the structure.

Keywords: Subword order, infix order, definability, automorphism, least fixed point, first-order theory, biinterpretability.

1 Introduction

The study of definability in natural structures is a central issue of logic and computation theory. For computation theory, the study of structures on words and trees is the most relevant. In particular, many deep facts on definability and (un)decidability are known for finitely presented semigroups and groups. The study of natural quasiorders on words and trees is also a traditional subject (see e.g. [10,9] and references therein) with several deep and interesting definability and (un)decidability results.

In this paper we develop a complete definability theory for the structure $(A^*; \preceq)$, where A^* is the set of words over a finite alphabet A with at least 2 letters and $\preceq$ is the subword partial order on A^* defined as follows: $u \preceq v$ iff u is a subword of v, i.e., u is obtained from v by deleting some letters. W.l.o.g. we may assume that $A = A_k = \{0, \ldots, k-1\}$ for some $k \geq 2$. As is well-known, $u \preceq v$ iff there is an embedding $f : u \to v$, i.e. an increasing function $f : A_{|u|} \to A_{|v|}$ such that $u(i) = v(f(i))$ for all $i < |u|$ where $u(i)$ is the i-th letter of u and $|u|$ is the length

* Supported by DFG-RFBR (grant No 436 RUS 113/1002/01, grant No 09-01-91334) and by RFBR (grants 08-01-00336, 07-01-00543a).
** Supported by DFG-RFBR (grant No 436 RUS 113/1002/01, grant No 09-01-91334) and by RFBR grant 07-01-00543a.
*** Supported by RFBR grant 07-01-00543a.

F. Ferreira et al. (Eds.): CiE 2010, LNCS 6158, pp. 246–255, 2010.
© Springer-Verlag Berlin Heidelberg 2010

of u. The structure $(A^*; \preceq)$ is interesting for several fields including the theory of well partial orders [4,6], combinatorics on words [10] and automata theory [3,13].

This paper makes a heavy use of [7,8] where a similar theory was developed for the h-quasiorder of finite labeled forests and for the structure $(A^*; \leq)$, where $\leq$ is the infix partial order ($u \leq v$ iff $v = xuy$ for some $x, y \in A^*$). We use some notation and terminology of those papers.

We also use some standard notation on words, in particular ε denotes the empty word, $xu = x \cdot u$ denotes the concatenation of words x and u, $|u|_a$ denotes the number of occurrences of the letter $a \in A$ in the word u, $A^{[m,n]}$ denotes the set $\{u \in A^* \mid m \leq |u| \leq n\}$. For $i \leq j \leq |u|$, let $u[i,j) = u(i) \cdots u(j-1)$; notation $u[i,j]$ is defined similarly. We identify letters with the corresponding one-letter words.

For a given structure $\mathbf{A}$ of signature σ, a predicate on A is *definable* if it is defined by a first-order formula of signature σ (in fact, this definition is not completely precise; to get the well-known precise definition, one has to fix also a suitable list of variables, as in the definition of the operator Γ in the next section). A function on $\mathbf{A}$ is definable if its graph is definable. An element is definable if the corresponding singleton set is definable. A structure is definable if its universe and all signature predicates are definable.

Section 2 recalls some necessary facts from [7]. In Section 3 we establish some useful definability results. In particular, we show that any element of A^* is definable in the $A^{[1,2]}$-expansion of $(A^*; \preceq)$ (i.e. in the expansion obtained by adding to the signature $\{\preceq\}$ the constant symbols denoting words of lengths 1 or 2) and that the infix order is definable in the $A^{[1,2]}$-expansion of $(A^*; \preceq)$. In Section 4 we characterize the automorphism group of $(A^*; \preceq)$. In Section 5 we establish our main definability results for the subword order, in particular we show that any arithmetical predicate on A^* which is invariant under the automorphisms of $(A^*; \preceq)$ is definable in $(A^*; \preceq)$.

In computability theory, people actively discuss several versions of the so called biinterpretability conjecture stating that some structures of degrees of unsolvability are biinterpretable (in parameters) with $(\omega; +, \cdot)$ (see e.g. [12] and references therein). The conjecture (which seems still open for the most important cases) is considered as in a sense the best possible definability result about degree structures. This paper and the papers [7,8] show that some natural structures on words and forests are biinterpretable (even without parameters) with $(\omega; +, \cdot)$. We believe that our methods could be applied for proving such kind of results for many other structures including those considered in [9].

2   Preliminaries on Gandy Theorem and Definability

For convenience of the reader we recall an approach to the study of first-order definability developed in [7] which applies well to the subword order.

Along with first-order definability, we are interested in definability by formulas of special kind related to admissible set theory and inductive definability. We briefly recall some relevant notions from [11,1,2]. Let σ be a finite signature containing a binary relation symbol $\leq$ and possibly some other relational or constant symbols. *RQ-Formulas* of σ are constructed from the atomic formulas according to the usual rules of first-order logic concerning $\wedge, \vee, \neg, \rightarrow$, the usual rules for the (unbounded) quantification with $\forall, \exists$ and the following additional formation rules for the bounded quantification: if φ is an *RQ*-formula and x, y are variables then the expressions $\forall x \leq y \varphi$ and $\exists x \leq y \varphi$ are *RQ*-formulas. As for the usual first-order formulas, any *RQ*-formula is equivalent to a *special RQ*-formula (i.e., a formula without implications that has negations only on the atomic formulas).

Δ_0-*Formulas* of signature σ are constructed inductively according to the following rules: any atomic formula of signature σ is a Δ_0-formula; if φ and ψ are Δ_0-formulas then $\neg\varphi, (\varphi \wedge \psi), (\varphi \vee \psi)$ and $(\varphi \rightarrow \psi)$ are Δ_0-formulas; if x, y are variables and φ is a Δ_0-formula then $\forall x \leq y \varphi$ and $\exists x \leq y \varphi$ are Δ_0-formulas. Σ-*Formulas* of signature σ are constructed inductively according to the following rules: any Δ_0-formula is a Σ-formula; if φ and ψ are Σ-formulas then $(\varphi \wedge \psi)$ and $(\varphi \vee \psi)$ are Σ-formulas; if x, y are variables and φ is a Σ-formula then $\forall x \leq y \varphi$, $\exists x \leq y \varphi$ and $\exists x \varphi$ are Σ-formulas.

A predicate on a σ-structure A is Δ_0-*definable* (resp. Σ-*definable*) if it is defined by a Δ_0-formula (resp. by a Σ-formula). A predicate on A is Δ-*definable* if both the predicate and its negation are Σ-definable.

Let P be an n-ary predicate symbol not in σ, φ an *RQ*-formula of signature $\sigma \cup \{P\}$, and $\bar{x} = x_1, \ldots, x_n$ a list of variables that includes all free variables of φ. We say that P *occurs positively in* φ if φ is in negation normal form and has no subformulas $\neg P(y_1, \ldots, y_n)$. For any n-ary predicate Q on a σ-structure $\mathbf{A}$, we denote by $(\mathbf{A}, Q)$ the expansion of $\mathbf{A}$ to the $\sigma \cup \{P\}$-structure where P is interpreted as Q. Then we can define the operator $\Gamma = \Gamma_{\varphi, \bar{x}}$ on the n-ary predicates on A that sends any Q to the predicate

$$\Gamma_{\varphi, \bar{x}}(Q) = \{(a_1, \ldots, a_n) \mid (\mathbf{A}, Q) \models \varphi(a_1, \ldots, a_n)\}.$$

A σ-structure $\mathbf{A}$ is *bounded* iff $\leq$ is a transitive directed relation (directed means that for all $x, y \in A$ there is $z \in A$ with $x, y \leq z$) and satisfies the Σ-collection principle [1,2] (which means that, for any Δ_0-formula φ, $\mathbf{A} \models \forall x \leq t \exists y \varphi$ implies $\mathbf{A} \models \exists v \forall x \leq t \exists y \leq v \varphi$.) As is well-known [11,2], if P is an n-ary predicate symbol that occurs positively in a Σ-formula φ with the free variables among $\bar{x}$ then the operator Γ is monotone in any bounded σ-structure. By the Tarski-Knaster fixed-point theorem, the operator Γ has the least fixed point denoted by $LFP(\Gamma)$. In general, the least fixed points defined in this way may be complicated. But for some structures $\mathbf{A}$ it turns out that the least fixed point of any Σ-formula φ as above is a Σ-predicate (in this case we say that $\mathbf{A}$ has the *Gandy property*). Let us recall a sufficient condition for $\mathbf{A}$ to have the Gandy property established in [7].

We say that a σ-structure $\mathbf{A}$ *admits a* Δ-*coding of finite sets* if there is a binary Δ-predicate $E(x, y)$ on $\mathbf{A}$ such that $E(x, y)$ implies $x \leq y$ for all $x, y \in A$ and, for all $n < \omega$ and $x_1, \ldots, x_n \in A$, there is $y \in A$ with

$$\mathbf{A} \models \forall x (E(x,y) \leftrightarrow x = x_1 \vee \cdots \vee x = x_n).$$

Note that our coding of finite sets is related to coding considered much earlier by R.A. Vaught [14].[1]

We call a σ-structure $\mathbf{A}$ *locally finite* if $\{x \mid x \leq y\}$ is finite for each $y \in A$. The next result is useful for understanding definability in some structures.

Theorem 1 ([7]). *Let $\mathbf{A}$ be a bounded locally finite σ-structure that admits a Δ-coding of finite sets and a Δ-definable copy of $(\omega; \leq)$. Then $\mathbf{A}$ has the Gandy property.*

Recall that a structure $\mathbf{A}$ equipped with a numbering α (i.e., a surjection from ω onto A) is *arithmetical*, if the equality predicate and all signature predicates are arithmetical modulo α. Obviously, any definable predicate on an arithmetical structure $(\mathbf{A}; \alpha)$ is arithmetical (w.r.t. α) and invariant under the automorphisms of $\mathbf{A}$; we say that $(\mathbf{A}; \alpha)$ has the *maximal definability property* if the converse is also true, i.e., any arithmetical predicate invariant under the automorphisms of $\mathbf{A}$ is definable.

Let again $\mathbf{A}$ be a countable σ-structure and let α be a numbering of A. We say that *the elements of $(\mathbf{A}; \alpha)$ are uniformly Σ-definable* if there is an arithmetical sequence of unary Σ-formulas $\{\psi_n(v_0)\}$ such that ψ_n defines the element $\alpha(n)$ in $\mathbf{A}$ for each $n < \omega$.

Recall (cf. [5,12]) that a structure $\mathbf{B}$ of a finite relational signature τ is *biinterpretable* with structure $\mathbf{C}$ of a finite relational signature ρ if $\mathbf{B}$ is definable in $\mathbf{C}$ (in particular, there is a bijection $f : B \to B_1$ on a definable set $B_1 \subseteq C^m$ for some $m \geq 1$ which induces an isomorphism on the τ-structure $\mathbf{B}_1$ definable in $\mathbf{C}$), $\mathbf{C}$ is definable in $\mathbf{B}$ (in particular, there is a similar bijection $g : C \to C_1$ on a definable set $C_1 \subseteq B^n$ for some $n \geq 1$), the function $g^m \circ f : B \to B^{nm}$ is definable in $\mathbf{B}$ and the function $f^n \circ g : C \to C^{mn}$ is definable in $\mathbf{C}$.

Theorem 2 ([7]). *Let $(\mathbf{A}; \alpha)$ be an arithmetical σ-structure with the uniformly Σ-definable elements such that $\mathbf{A}$ is bounded, locally finite and admits a Δ-coding of finite sets and a Δ-definable copy of $(\omega; \leq)$. Then $\mathbf{A}$ has the maximal definability property and is biinterpretable with $(\omega; +, \cdot)$.*

3   Some Definability Results

In this section we establish some facts on definability in the $A^{[1,2]}$-expansion of $(A^*; \preceq)$.

For $x, y \in A^*$, let $S(x,y)$ mean that y is an immediate successor of x w.r.t. $\preceq$. Since $S(x,y)$ is equivalent to $x \prec y \wedge \neg \exists z \preceq y (x \prec z \prec y)$, S is a Δ_0-predicate in $(A^*; \preceq)$.

For any $a \in A^*$, the sets a^* and a^+ are Δ_0-definable in the $A^{[1]}$-expansion of $(A^*; \preceq)$ (we use notation in the style of regular expressions). E.g., $x \in a^+$ iff $a \preceq x$ and $\neg \bigvee \{b \preceq x \mid b \in A, b \neq a\}$.

---

[1] We are grateful to an anonymous referee for this reference.

For all distinct $a, b \in A$ the sets a^*b^*, a^*b^+, a^+b^* and a^+b^+, are Δ_0-definable in the $A^{[1,2]}$-expansion of $(A^*; \preceq)$. E.g., $x \in a^*b^+$ iff $b \preceq x$, $ba \not\preceq x$ and $\neg \bigvee\{c \preceq x \mid c \in A \setminus \{a, b\}\}$.

For all $a \in A$ and $u, v \in A^*$, let $P_a(u, v)$ mean that $a \preceq v$ and u is obtained from v by deleting exactly one occurrence of a in v, let $p_a(u) = ua$ and $p^a(u) = au$, and let $Q_a(u, v) \leftrightarrow v \in ua^*$, $Q^a(u, v) \leftrightarrow v \in a^*u$.

Lemma 1. *For any $a \in A^*$, the predicates P_a, Q_a, Q^a and the functions p_a, p^a are Δ_0-definable in the $A^{[1,2]}$-expansion of $(A^*; \preceq)$.*

Proof. Since $P_a(u, v)$ is equivalent to

$$S(u, v) \wedge \exists x \preceq u \exists y \preceq v(y \in a^+ \wedge y \not\preceq u \wedge S(x, y)),$$

P_a is Δ_0-definable. For p_a, it suffices to check that

$$v = ua \leftrightarrow P_a(u, v) \wedge \bigwedge_{b \in A \setminus \{a\}} \forall x \preceq v(x \in a^*b^+ \rightarrow x \preceq u).$$

From left to right, this is obvious. Conversely, let the right condition be true. If $u \in a^*$ then clearly $v = ua$. Otherwise, $v = v_1 b^m a^n$ for unique $b \in A \setminus \{a\}, m \geq 1, n \geq 0$ and $v_1 \notin A^*b$. Let $x = a^k b^m$ where $k = |v_1|_a$, then $x \preceq v$, hence also $x \preceq u$. Since $P_a(u, v)$, $v = ua$.

For Q_a, it suffices to check that

$$v \in ua^* \leftrightarrow u \preceq v \wedge \bigwedge_{b \in A \setminus \{a\}} \forall x \preceq v(p_b(x) \preceq v \rightarrow p_b(x) \preceq u).$$

From left to right, this is obvious. Conversely, let the right condition be true. If $v \in a^*$ then clearly $v \in ua^*$. Otherwise, $v = xba^n$ for unique $b \in A \setminus \{a\}, n \geq 0$ and $x \in A^*$. Since $xb = p_b(x) \preceq v$, $xb \preceq u$. Since $u \preceq v$, u is obtained from v by deleting some letters. Since there is only one embedding of xb into v, the deleted letters can not be in xb. Therefore, u is of the form xba^m for some $m \leq n$, hence $v \in ua^*$.

For p^a and Q^a the arguments are symmetric to those for p_a and Q_a, respectively. □

Corollary 1. *The elements of A^* are uniformly Σ-definable in the $A^{[1,2]}$-expansion of $(A^*; \preceq)$.*

Proof. By induction on $|w|$, $w \in A^*$, we find a Σ-formula $\phi_w(x)$ with the unique free variable x that defines w in the $A^{[1,2]}$-expansion of $(A^*; \preceq)$. For $w = \varepsilon$ we can take the formula $\neg \exists y \preceq x(y \prec x)$ as ϕ_ε. For $|w| \geq 1$, represent w as va, $a \in A$, and take the Σ-formula $\exists y(\phi_v(y) \wedge \psi(y, x))$ as ϕ_w where ϕ_v is the Σ-formula that defines v (existing by induction) and ψ is a Δ_0-formula that defines p_a (existing by Lemma 1).

Let α be the natural numbering of $A^* = A_k^*$ induced by the length-first lexicographic ordering of the words in A^* (i.e., $\alpha(0) = \varepsilon$, $\alpha(1) = 0, \ldots, \alpha(k) =$

$k - 1$, $\alpha(k + 1) = 00, \alpha(k + 2) = 01, \ldots)$. Note that the numbered structure $(A^*; \preceq, \alpha)$ is computable.

Writing down (recursively) the explicit Σ-definitions of elements in the constructive proof above, we obtain a computable (hence arithmetical) sequence $\{\theta_n(x)\}$ of Σ-formulas such that θ_n defines $\alpha(n)$ for any $n \geq 0$. □

For all $u, v \in A^*$, let $u \leq_p v$ (resp. $u \leq_s v$) denote that u is a prefix (resp. a suffix) of v, i.e. $v = ux$ (resp. $v = xu$) for some $x \in A^*$.

Proposition 1. *The relations $\leq_p, \leq_s, \leq$ are Δ_0-definable in the $A^{[1,2]}$-expansion of $(A^*; \preceq)$.*

Proof. For $\leq_p$, by Lemma 1 it suffices to check that

$$u \leq_p v \leftrightarrow u \preceq v \wedge \bigwedge_{a \in A} \exists y \preceq v (Q_a(u, y) \wedge \neg \exists z \preceq v P_a(y, z)).$$

Let $u \leq_p v$, so $v = ux$ for some $x \in A^*$, in particular $u \preceq v$. For a given $a \in A$, let $y = ua^n$ where $n = |x|_a$. Then $y \in ua^*$ and $\neg \exists z \preceq v P_a(y, z)$.

Conversely, let the right condition be true, in particular $u \preceq v$ and $|u| \leq |v|$. Toward a contradiction, suppose $u \not\leq_p v$ and take the smallest $i < |u|$ with $u(i) \neq v(i)$. Let $a = v(i)$ and choose $y \preceq v$ such that $y \in ua^*$ and $\neg \exists z \preceq v P_a(y, z)$. Let f be an embedding of u into v and let $z = v[0, i)y[i, |y|] = u[0, i)au[i, |y|)a^n$ where n satisfies $y = ua^n$. Then $P_a(y, z)$ and $z \preceq v$ via the embedding $g : A_{|z|} \to A_{|v|}$ defined by $g(j) = j$ for $j \leq i$ and $g(j) = f(j - 1)$ for $i < j < |y|$. A contradiction.

For $\leq_s$ the proof is symmetric to that for $\leq_p$. Since

$$u \leq v \leftrightarrow \exists w \preceq v (u \leq_p w \wedge w \leq_s v),$$

$\leq$ is Δ_0-definable. □

Corollary 2. *The $A^{[1,2]}$-expansion of $(A^*; \preceq)$ admits a Δ-coding of finite sets.*

Proof. By Theorem 5 in [8], the $A^{[1,2]}$-expansion of $(A^*; \leq)$ admits a Δ-coding $E(x, y)$ of finite sets, so there are Σ-formulas $\phi(x, y)$ and $\psi(x, y)$ that define, respectively, $E(x, y)$ and $\neg E(x, y)$ in the $A^{[1,2]}$-expansion of $(A^*; \leq)$. By Proposition 1, there is a Σ-formula $\theta(u, v)$ that defines the relation $u \leq v$ in the $A^{[1,2]}$-expansion of $(A^*; \preceq)$. Replacing any occurrence of $u \leq v$ in ϕ and ψ by $\theta(u, v)$ (and using the standard renaming procedure for the bounded variables when required) we obtain Σ-formulas $\phi'(x, y)$ and $\psi'(x, y)$ that define, respectively, $E(x, y)$ and $\neg E(x, y)$ in the $A^{[1,2]}$-expansion of $(A^*; \preceq)$. □

4   Characterizing Automorphisms

Next we characterize the automorphism group $Aut(A^*; \preceq)$ of $(A^*; \preceq)$. We start with the following immediate corollary of Corollary 1.

Corollary 3. *Any automorphism of $(A^*; \preceq)$ identical on $A^{[1,2]}$ is the identity automorphism.*

Let $\mathbf{S}_k$ denote the symmetric group on k elements $\{0, \ldots, k-1\}$ and let $\mathbf{A} \simeq \mathbf{B}$ denote that structures $\mathbf{A}$ and $\mathbf{B}$ are isomorphic. Since for $k = 1$ we have $(A_k^*; \preceq) \simeq (\omega; \leq)$, $Aut(A_k^*; \preceq)$ is the trivial one-element group. For $k \geq 2$, along with the identity automorphism e the group $Aut(A_k^*; \preceq)$ has also some other elements, in particular the reverse automorphism r defined by $r(i_1 \cdots i_n) = i_n \cdots i_1$ for all $n \geq 0$ and $i_1, \ldots, i_n < k$ (note that $r \circ r = e$).

Lemma 2. *Let $k \geq 2$ and let f be an automorphism of $(A_k^*; \preceq)$ such that $f(i) = i$ and $f(ij) = ji$ for all $i, j < k$. Then $f = r$.*

Proof. The function $f \circ r$ is an automorphism of $(A_k^*; \preceq)$ identical on $A_k^{[1,2]}$. By Corollary 3, $f \circ r = e$, hence $f = f \circ r \circ r = r$.   □

The next theorem and the corresponding theorem in [8] imply that the automorphism groups of $(A_k^*; \preceq)$ and $(A_k^*; \leq)$ are isomorphic (though the structures themselves are far from being isomorphic). The proof here follows closely the corresponding proof in [8] where the sets of constants and of automorphisms are exactly the same as here. Nevertheless, some details are of course different (cf. case 1 below and the corresponding case in [8]).

Theorem 3. *For any $k \geq 2$, $Aut(A_k^*; \preceq) \simeq \mathbf{S}_k \times \mathbf{S}_2$.*

Proof. The restriction map $f \mapsto f|_{A_k}$ is easily seen to be a group homomorphism from $Aut(A_k^*; \preceq)$ onto $\mathbf{S}_k$. We check that the kernel K of this homomorphism coincides with $\{e, r\}$. Obviously, $e, r \in K$. Conversely, let $f \in K$, then $f(i) = i$ for all $i < k$. Since for every distinct $i, j < k$ the set of minimal elements in $(\{u \in A_k^* \mid i, j \prec u\}; \preceq)$ coincides with $\{ij, ji\}$, $f(01) \in \{01, 10\}$. We distinguish two cases.

Case 1. $f(01) = 01$. We show that in fact $f(ij) = ij$ for all $i, j < k$. For $i = j$ this is obvious, so let $i \neq j$. Assume first that $0 \in \{i, j\}$, say $0 = i$ (the case $0 = j$ is treated similarly). As above, $f(0j) \in \{0j, j0\}$. Toward a contradiction, suppose that $f(0j) = j0$. For the word $w = 10j$ we have $10, 0j \preceq w$, hence $10 = f(10) \preceq f(w)$ and $j0 = f(0j) \preceq f(w)$. But $|f(w)| = 3$ (because any automorphism of $(A_k^*; \preceq)$ obviously preserves the length of words), hence $f(w) \in \{1j0, j10\}$. In the case $f(w) = 1j0$ we have $f(1j) = 1j$ and $10, 0j, j1 \preceq f(j10)$ which is a contradiction. A similar contradiction is obtained in the remaining case $f(w) = j10$ (consider $f(1j0)$).

The case $1 \in \{i, j\}$ is symmetric to the case $0 \in \{i, j\}$, so it remains to consider the case $0, 1 \notin \{i, j\}$. Since $f(0i) = 0i$ and $i \in \{i, j\}$, $f(ij) = ij$ by taking i in place of 1 in the above argument. We have shown that f is identical on $A_k^{[1,2]}$. By Corollary 3, $f = e$.

Case 2. $f(01) = 10$. By the argument of case 1 one can show that in fact $f(ij) = ji$ for all $i, j < k$. By Lemma 2, $f = r$. This completes the proof of equality $K = \{e, r\}$.

Let $g \mapsto \tilde{g}$ be the embedding of $\mathbf{S}_k$ into $Aut(A_k^*; \preceq)$ defined by $\tilde{g}(i_1 \cdots i_n) = g(i_1) \cdots g(i_n)$ for all $n \geq 0$ and $i_1, \ldots, i_n < k$ (note that $\tilde{g}|_{A_k} = g$ for each $g \in \mathbf{S}_k$). Then $\tilde{\mathbf{S}}_k = \{\tilde{g} \mid g \in \mathbf{S}_k\}$ and K are subgroups of $Aut(A_k^*; \preceq)$,

$\tilde{\mathbf{S}}_k \cap K = \{e\}$, and r commutes with any element of $\tilde{\mathbf{S}}_k$. Then each element of $Aut(A_k^*; \preceq)$ is uniquely representable in the form $\tilde{g} \circ h$ where $g \in \mathbf{S}_k$ and $h \in K$. Therefore, $Aut(A_k^*; \preceq) \simeq \tilde{\mathbf{S}}_k \times K$. This completes the proof because $\tilde{\mathbf{S}}_k \simeq \mathbf{S}_k$ and $K \simeq \mathbf{S}_2$. □

5 Main Results

Now we are ready to establish the following main result of this paper.

Theorem 4. *The $A^{[1,2]}$-expansion of $(A^*; \preceq)$ has the Gandy property, the maximal definability property and is biinterpretable with $(\omega; +, \cdot)$.*

Proof. The expansion is clearly bounded and locally finite. By Corollary 1, the elements of A^* are uniformly Σ-definable. The structure $(a^*; \preceq)$, $a \in A$, is a Δ-definable copy of $(\omega; \leq)$. By Corollary 2, the expansion admits a Δ-coding of finite sets. Thus all conditions of Theorems 1 and 2 are satisfied. Conclusions of these theorems give the desired properties. □

We formulate an immediate consequence of the main theorem (cf. [9]):

Corollary 4. *The structure $(\omega; +, \cdot)$ is definable in the $A^{[1,2]}$-expansion of the structure $(A^*; \preceq)$. Therefore, the first-order theory of the $A^{[1,2]}$-expansion of $(A^*; \preceq)$ is computably isomorphic to $FO(\omega; +, \cdot)$.*

We conclude the paper with the complete characterization of definable predicates on $(A^*; \preceq)$.

Theorem 5. *For any $k \geq 2$, the structure $(A_k^*; \preceq)$ has the maximal definability property.*

Proof. Let $S(x, y)$ be the predicate from Section 3. Let $\{v_a\}_{a \in A_k^{[1,2]}}$ be distinct variables, let $\bar{v}$ denote the tuple

$$\left(v_0, \ldots, v_{k-1}, v_{00}, \ldots, v_{0(k-1)}, \ldots, v_{(k-1)0}, \ldots, v_{(k-1)(k-1)}\right)$$

of variables, and let $\rho = \rho(\bar{v})$ be a formula of signature $\{\preceq\}$ equivalent to the conjunction of the following formulas:

  - $S(\varepsilon, v_i)$ for all $i < k$,
  - $\neg v_i = v_j$ for all distinct $i, j < k$,
  - $S(v_i, v_{ij}) \wedge S(v_j, v_{ij})$ for all $i, j < k$,
  - $\neg v_{ij} = v_{ji}$ for all distinct $i, j < k$,
  - $\exists x (S(v_{ij}, x) \wedge S(v_{jl}, x) \wedge S(v_{il}, x))$ for all pairwise distinct $i, j, l < k$.

We claim that ρ defines in $(A_k^*; \preceq)$ the orbit $Orb(\bar{b})$ of the tuple

$$\bar{b} = (0, \ldots, k-1, 00, \ldots, 0(k-1), \ldots, (k-1)0, \ldots, (k-1)(k-1))$$

(recall that $Orb(\bar{b}) = \{f(\bar{b}) \mid f \in Aut(A^*; \preceq)\}$ where $\bar{b} = (b_0, \ldots, b_n)$ and $f(\bar{b}) = (f(b_0), \ldots, f(b_n))$). Indeed, since $(A_k^*; \preceq) \models \rho(\bar{b})$, $(A_k^*; \preceq) \models \rho(f(\bar{b}))$ for each $f \in Aut(A^*; \preceq)$ and therefore $(A_k^*; \preceq) \models \rho(\bar{a})$ for each $\bar{a} \in Orb(\bar{b})$.

Conversely, let $(A_k^*; \preceq) \models \rho(\bar{v})$ for some values of $\bar{v}$ in A_k^*; we have to show that $\bar{v} \in Orb(\bar{b})$. Since $S(\varepsilon, v_i)$ and $v_i \neq v_j$ for all distinct $i, j < k$, $(v_0, \ldots, v_{k-1}) = (g(0), \ldots, g(k-1))$ for some $g \in \mathbf{S}_k$. Extend g to the automorphism of $(A_k^*; \preceq)$ (also denoted g) by $g(i_1 \cdots i_n) = g(i_1) \cdots g(i_n)$ where $n \geq 0$ and $i_1, \ldots, i_n < k$. Since $S(v_i, v_{ij}) \wedge S(v_j, v_{ij})$ for all $i, j < k$ and $v_{ij} \neq v_{ji}$ for all distinct $i, j < k$, $\{g(ij), g(ji)\} = \{v_{ij}, v_{ji}\}$ for all $i, j < k$, in particular $v_{01} \in \{g(01), g(10)\}$. Repeating the proof of Theorem 3 (cases 1,2) and using the condition $\exists x(S(v_{ij}, x) \wedge S(v_{jl}, x) \wedge S(v_{il}, x))$ for all pairwise distinct $i, j, l < k$ we obtain that $\bar{v} = g(\bar{b})$ in case $v_{01} = g(01)$ and $\bar{v} = r(g(\bar{b}))$ in case $v_{01} = g(10)$ where r is the reversing automorphism. In any case, $\bar{v} \in Orb(\bar{b})$.

Let now $P(\bar{x})$, $\bar{x} = (x_1, \ldots, x_n)$, be an n-ary arithmetical predicate on A_k^* which is invariant under the automorphisms of $(A_k^*; \preceq)$. We have to show that P is definable in $(A_k^*; \preceq)$. W.l.o.g. we assume that variables $x_1, \ldots, x_n$ are distinct from the variables v_a above. By Theorem 4, there is a formula $\phi(\bar{x})$ of signature $\{\preceq, a\}_{a \in A_k^{[1,2]}}$ that defines P. Let ϕ_1 be the formula of signature $\{\preceq\}$ obtained from ϕ by substituting the variable v_a in place of the constant symbol a for all $a \in A_k^{[1,2]}$. Finally, let θ be the formula obtained from $\rho \wedge \phi_1$ by existential quantification over the variables v_a for all $a \in A_k^{[1,2]}$. Then θ defines P in $(A_k^*; \preceq)$. $\qquad\square$

From the last theorem and finiteness of any orbit $Orb(\bar{b})$ we immediately obtain the following important model-theoretic and algorithmic properties of $(A^*; \preceq)$.

Corollary 5. *The structure $(A^*; \preceq)$ is atomic, minimal and $FO(A^*; \preceq)$ is computably isomorphic to $FO(\omega; +, \cdot)$.*

References

1. Barwise, J.: Admissible Sets and Structures. Springer, Berlin (1975)
2. Ershov, Y.L.: Definability and Computability. Plenum, New-York (1996)
3. Glaßer, C., Schmitz, H.: The Boolean Structure of Dot-Depth One. Journal of Automata, Languages and Combinatorics 6, 437–452 (2001)
4. Higman, G.: Ordering by divisibility in abstract algebras. Proceegings of London Mathematical Society 3, 326–336 (1952)
5. Hodges, W.: Model Theory. Cambridge Univ. Press, Cambidge (1993)
6. Kruskal, J.B.: The theory of well-quasi-ordering: a frequently discovered concept. J. Combinatorics Th. (A) 13, 297–305 (1972)
7. Kudinov, O.V., Selivanov, V.L.: A Gandy theorem for abstract structures and applications to first-order definability. In: Ambos-Spies, K., Löwe, B., Merke, W. (eds.) CiE 2009. LNCS, vol. 5635, pp. 290–299. Springer, Heidelberg (2009)
8. Kudinov, O.V., Selivanov, V.L.: Definability in the infix order on words. In: Diekert, V., Nowotka, D. (eds.) DLT 2009. LNCS, vol. 5583, pp. 454–465. Springer, Heidelberg (2009)
9. Kuske, D.: Theories of orders on the set of words. RAIRO Theoretical Informatics and Applications 40, 53–74 (2006)
10. Lothaire, M.: Combinatorics on words, Cambridge Mathematical Library. Cambridge University Press, Cambridge (1997)

11. Moschovakis, Y.: Elementatry induction on abstract structures. North-Holland, Amsterdam (1974)
12. Nies, A.: Definability in the c.e. degrees: questions and results. Contemporary Mathematics 257, 207–213 (2000)
13. Selivanov, V.L.: A logical approach to decidability of hierarchies of regular star-free languages. In: Ferreira, A., Reichel, H. (eds.) STACS 2001. LNCS, vol. 2010, pp. 539–550. Springer, Heidelberg (2001)
14. Vaught, R.A.: Axiomatizability by a schema. Journal of Symbolic Logic 32(4), 473–479 (1967)

Undecidability in Weihrauch Degrees

Oleg V. Kudinov[1,*], Victor L. Selivanov[2,**], and Anton V. Zhukov[3,***]

[1] S.L. Sobolev Institute of Mathematics, 4 Acad. Koptyug avenue,
630090 Novosibirsk, Russia
kud@math.nsc.ru

[2] A.P. Ershov Institute of Informatics Systems, 6 Acad. Lavrentjev pr.,
630090 Novosibirsk, Russia
vseliv@iis.nsk.su

[3] Novosibirsk State Pedagogical University, 28 Vilyuiskaya ul.,
630126 Novosibirsk, Russia
zhukan@ngs.ru

Abstract. We prove that the 1-quasiorder and the 2-quasiorder of finite k-labeled forests and trees have hereditarily undecidable first-order theories for $k \geq 3$. Together with an earlier result of P. Hertling, this implies some undecidability results for Weihrauch degrees.

Keywords: Weihrauch reducibility, labeled forest, 1-quasiorder, 2-quasiorder, undecidability, theory.

1 Introduction

As is well-known, different notions of hierarchies and reducibilities serve as useful tools for understanding the complexity (or non-computability) of decision problems on discrete structures. In computable analysis, many problems of interest turn out to be non-computable (even non-continuous), so there is a need for tools to measure their non-computability. Accordingly, also in this context of decision problems on continuous structures, people employed several hierarchies and reducibilities closely related to descriptive set theory.

In [18,19] K. Weihrauch introduced some notions of reducibility for functions on topological spaces which turned out useful for understanding the non-computability and non-continuity of interesting decision problems in computable analysis [8,7,2] and constructive mathematics [1]. In particular, the following three notions of reducibility between functions $f, g : X \to Y$ on topological spaces were introduced: $f \leq_0 g$ (resp. $f \leq_1 g$, $f \leq_2 g$) iff $f = g \circ H$ for some continuous function $H : X \to X$ (resp. $f = F \circ g \circ H$ for some continuous functions $H : X \to X$ and $F : Y \to Y$, $f(x) = F(x, g(H(x)))$ for some continuous

---

* Supported by DFG-RFBR (Grant 436 RUS 113/1002/01, 09-01-91334), and by RFBR Grant 07-01-00543a.

** Supported by DFG-RFBR (Grant 436 RUS 113/1002/01, 09-01-91334) and by RFBR Grant 07-01-00543a.

*** Supported by RFBR Grant 07-01-00543a.

F. Ferreira et al. (Eds.): CiE 2010, LNCS 6158, pp. 256–265, 2010.

functions $H : X \to X$ and $F : X \times Y \to Y$). In this way we obtain preorders $(Y^X; \leq_i)$, $i \leq 2$, on the set Y^X of all functions from X to Y.

The notions are nontrivial even for the case of discrete spaces $Y = k = \{0, \ldots, k-1\}$ with $k < \omega$ points (we call functions $f : X \to k$ k-partitions of X because they are in a natural bijective correspondence with the partitions $(A_0, \ldots, A_{k-1})$ of X where $A_i = f^{-1}(i)$). E.g., for $k = 2$ the relation $\leq_0$ coincides with the classical Wadge reducibility [9].

In [6,7] P. Hertling gave a useful "combinatorial" characterization of important segments of the degree structures under Weihrauch reducibilities on k-partitions of the Baire space $\mathbb{B} = \omega^\omega$. In [16] the characterization from [6,7] for $\leq_0$ was extended to a larger segment of $(k^{\mathbb{B}}; \leq_0)$. Note that the Baire space is important because it is commonly used in computable analysis [19] for representing many other spaces of interest.

From this characterization and results in [11] it follows that for any $k \geq 3$ the first-order theory of this segment for $\leq_0$ is hereditary undecidable (and it is even computably isomorphic to the first-order arithmetic [12]). In [17] it was shown that in fact many natural initial segments of $(k^{\mathbb{B}}; \leq_0)$ for any $k \geq 3$ have hereditary undecidable first-order theories. The main result of this paper states similar undecidability properties for the reducibilities $\leq_1$ and $\leq_2$. Recall that a first-order theory of a finite signature is called hereditary undecidable if any its subtheory of the same signature is undecidable. In particular, any hereditary undecidable theory is undecidable.

In this paper we consider only the simplest versions of Weihrauch reducibilities defined above. In the literature one could find some generalized versions, in particular reducibilities on partial multi-valued functions [1,2] and on sets of functions [6]. It would be interesting to investigate definability and (un)decidability issues for such generalized degree structures.

In Section 2 we recall some relevant definitions and precise formulation of the result in [6,7]. In Section 3 we establish auxiliary properties of some relevant preorders on labeled forests and trees. Section 4 recalls the interpretation scheme from [11] which is used to obtain our undecidability results. Finally, in Sections 5 and 6 we prove the undecidability results.

2   Preliminaries

We use some standard notation and terminology on posets which may be found e.g. in [4]. Sometimes we apply notions concerning posets also to quasiorders (known also as preorders); in such cases we mean the corresponding quotient-poset of the preordered set. Throughout this paper, k denotes an arbitrary integer, $k \geq 2$, which is identified with the set $\{0, \ldots, k-1\}$. By a *forest* we mean a finite poset in which every lower cone $\hat{x}$ is a chain. A *tree* is a forest that has a least element (called the *root* of the tree). W.l.o.g. we assume that all finite posets under consideration are subsets of ω.

A k-*labeled poset* (or just a k-poset) is an object $(P; \leq, c)$ consisting of a poset $(P; \leq)$ and a labeling $c : P \to k$. A k-poset P can be seen as a node-weighted directed graph where every $x \in P$ carries the label $c(x)$. We call a k-poset

$(P; \leq, c)$ *repetition-free* iff $c(x) \neq c(y)$ whenever x is an immediate predecessor of y in P. We usually simplify the notation of a k-poset to (P, c) or even P. We are mainly interested in the set $\mathcal{F}_k$ of finite k-labeled forests, but also in the set $\mathcal{T}_k$ of finite k-labeled trees. For any k-forest $F \in \mathcal{F}_k$ and $i < k$, let $p_i(F)$ denote a k-tree obtained from F by adding a new smallest element with the label i.

A *homomorphism* (or 0-morphism) $f : (P; \leq_P, c_P) \rightarrow (Q; \leq_Q, c_Q)$ between k-posets is a monotone function $f : (P; \leq_P) \rightarrow (Q; \leq_Q)$ respecting the labelings, i.e. satisfying $c_P = c_Q \circ f$.

A *1-morphism* $f : (P; \leq_P, c_P) \rightarrow (Q; \leq_Q, c_Q)$ between k-posets is a monotone function $f : (P; \leq_P) \rightarrow (Q; \leq_Q)$ for which there exists a mapping $g : k \rightarrow k$ such that $c_P = g \circ c_Q \circ f$.

A *2-morphism* $f : (P; \leq_P, c_P) \rightarrow (Q; \leq_Q, c_Q)$ between k-posets is a monotone function $f : (P; \leq_P) \rightarrow (Q; \leq_Q)$ which maps comparable elements (nodes) with different labels to elements with different labels, i.e.

$$\forall x, y \in P((x \leq_P y \wedge c_P(x) \neq c_P(y)) \rightarrow c_Q(f(x)) \neq c_Q(f(y))).$$

We say that a k-poset P is *0-morphic* (resp. *1-morphic*, *2-morphic*) to a k-poset Q, denoted $P \leq_0 Q$ (respectively $P \leq_1 Q$, $P \leq_2 Q$), iff there exists a 0-morphism (resp. 1-morphism, 2-morphism) $f : P \rightarrow Q$. It is easy to see that any 0-morphism is a 1-morphism and any 1-morphism is a 2-morphism. Therefore, $\leq_0$ implies $\leq_1$ and $\leq_1$ implies $\leq_2$.

Let $\equiv_0$ (0-equivalence, h-equivalence or just equivalence), $\equiv_1$ (1-equivalence) and $\equiv_2$ (2-equivalence) denote the equivalence relations induced by the preorders $\leq_0$, $\leq_1$ and $\leq_2$, respectively. The quotient set of $\mathcal{F}_k$ ($\mathcal{T}_k$) under $\equiv_i$, $i \leq 2$, is denoted by $\mathfrak{F}_k^i$ (respectively $\mathfrak{T}_k^i$). We use the same symbols $\leq_0$, $\leq_1$ and $\leq_2$ to denote the partial orders induced by the corresponding preorders on the corresponding quotient sets.

We assume the reader to be familiar with the Borel and projective hierarchies in the Baire space [9]. Levels of the Borel (resp. projective) hierarchies are denoted $\mathbf{\Sigma}_n^0, \mathbf{\Pi}_n^0, \mathbf{\Delta}_n^0$ for $1 \leq n < \omega_1$ (resp. $\mathbf{\Sigma}_n^1, \mathbf{\Pi}_n^1, \mathbf{\Delta}_n^1$ for $1 \leq n < \omega$).

For any class $\mathcal{C}$ of subsets of a set M, $BC(\mathcal{C})$ denotes the class of finite Boolean combinations of sets in $\mathcal{C}$, and for any $k \geq 2$ $\mathcal{C}_k$ denotes the set of $\mathcal{C}$-partitions, i.e. partitions $\nu \in k^M$ such that $\nu^{-1}(i) \in \mathcal{C}$ for each $i < k$. Let $P(M)$ be the class of subsets of M.

The above-mentioned result of P. Hertling [6,7] states that the quotient-structure of the preorder $(BC(\mathbf{\Sigma}_1^0)_k; \leq_0)$ (resp. $(BC(\mathbf{\Sigma}_1^0)_k; \leq_1)$, $(BC(\mathbf{\Sigma}_1^0)_k; \leq_2)$) on the k-partitions of the Baire space is isomorphic to $(\mathfrak{F}_k^0; \leq_0)$ (resp. to $(\mathfrak{F}_k^1; \leq_1)$, $(\mathfrak{F}_k^2; \leq_2)$). Using this result, we may consider the last structures instead of the corresponding structures of Weihrauch degrees. This motivates the study of posets $(\mathfrak{F}_k^0; \leq_0)$, $(\mathfrak{F}_k^1; \leq_1)$, $(\mathfrak{F}_k^2; \leq_2)$. For additional motivation to study similar structures see [13,14]. The main technical result of this paper (to be proved in Section 5) states that the first-order theories of the last two structures are hereditary undecidable.

3   Properties of the Preorders on Forests and Trees

Here we observe only some necessary properties of $\mathfrak{F}_k^i$ and $\mathfrak{T}_k^i$, $i = 1, 2$, used further in this paper. We give short proofs or even only sketches of proofs because of the restrictions on the size of the paper.

Lemma 1. *Let $A \leq_2 B$ for $A, B \in \mathcal{T}_k$. Then there exists a 2-morphism from A to B which maps the root of A to the root of B.*

Proof. By induction on the height of A. □

It is a simple and well-known fact that $(\mathfrak{F}_k^0, \leq_0)$ is a distributive upper semilattice (and even distributive lattice, see [15]). For $a, b \in \mathfrak{F}_k$, their supremum $a \sqcup b$ is the 0-equivalence class of the join (i.e. disjoint union) $A \sqcup B$ of any k-forests $A \in a$ and $B \in b$.

Proposition 1. *The structures $(\mathfrak{F}_k^2, \leq_2)$ and $(\mathfrak{T}_k^2, \leq_2)$ are distributive upper semilattices.*

Proof. It is easy to verify that the suprema in $(\mathfrak{F}_k^2, \leq_2)$ are defined in the same way as for $(\mathfrak{F}_k^0, \leq_0)$, so $(\mathfrak{F}_k^2, \leq_2)$ is an upper semilattice. Now let $a, b \in \mathfrak{T}_k^2$. Consider arbitrary disjoint k-trees $A \in a$ and $B \in b$. We can assume w.l.o.g. that the roots of A and B are labeled with 0. Let C be the k-tree that is obtained from the union of A and B by identifying their roots, $C = p_0(A \sqcup B)$. By the previous lemma, it is straightforward to prove that $[C]_{\equiv_2}$ is the supremum of a and b in $(\mathfrak{T}_k^2, \leq_2)$.

To prove the distributivity of $(\mathfrak{F}_k^2, \leq_2)$, note that any k-forest X can be represented as the disjoint union of some k-trees $X_1, \ldots X_n$. Let $X \leq_2 Y \sqcup Z$ by some 2-morphism f. Then $X = Y_1 \sqcup Z_1$ for $Y_1 = \bigsqcup \{X_i \mid f(X_i) \subseteq Y\}$ and $Z_1 = \bigsqcup \{X_j \mid f(X_j) \subseteq Z\}$, and $Y_1 \leq_2 Y$, $Z_1 \leq_2 Z$. The case of $(\mathfrak{T}_k^2, \leq_2)$ is treated similarly (note that any k-tree T is 2-equivalent to $p_0(T_1 \sqcup \cdots \sqcup T_n)$ for some disjoint k-trees $T_1, \ldots T_2$ with non-zero labels on their roots). □

For an arbitrary distributive upper semilattice $(S; \sqcup, \leq)$ where the partial order $\leq$ corresponds to $\sqcup$, $x \in S$ is called *join-irreducible* iff

$$\forall y \leq x \forall z \leq x (x \leq y \sqcup z \rightarrow (x \leq y \lor x \leq z)).$$

Let $ir(x)$ be a formula of signature $\{\leq\}$ which defines exactly the non-zero join-irreducible elements in $(S; \sqcup, \leq)$, and let $ir(S)$ denote the set of all non-zero join-irreducible elements of S. Let $\mathcal{Y}_k$ be the set of all finite k-trees whose root has exactly one immediate successor and the labels of the root and its successor differ.

Proposition 2. *(i) $ir(\mathfrak{F}_k^2) = \mathfrak{T}_k^2$.*
(ii) For any $r(\mathfrak{T}_k^2)$, $t \in ir(\mathfrak{T}_k^2)$ iff t includes a k-tree from $\mathcal{Y}_k$.

Proof is a simple routine which uses the monotonicity of 2-morphisms. □

The structures $(\mathfrak{F}_k^1, \leq_1)$ and $(\mathfrak{T}_k^1, \leq_1)$ are in a sense more complicated. In particular, we will show that they are not semilattices. Let $(P, \leq)$ be an arbitrary poset or preordered set and $a, b \in P$. As usual, $c \in P$ is called a *minimal upper bound* of a and b iff

$$a \leq c \wedge b \leq c \wedge \forall x \leq c((a \leq x \wedge b \leq x) \rightarrow c \leq x),$$

we call c also a *quasijoin* of a and b in P.

Let S_k denote the group of all permutations of the elements $0, \ldots, k - 1$. For any $\alpha \in S_k$ and for any k-forest $F = (F; \leq_F, c_F) \in \mathcal{F}_k$, let $\alpha(F)$ denote the k-forest $(F; \leq_F, \alpha \circ c_F)$ (i.e. all labels in F are replaced by their images via α).

By $\mathcal{F}'_k$ and $\mathcal{T}'_k$ we denote the sets of all finite k-forests and k-tress, respectively, which carry all labels from the set $k = \{0, 1, \ldots, k - 1\}$. We call a k-forest 1-minimal iff it is not 1-equivalent to any k-forest of lesser cardinality.

Proposition 3. *(i) For any k-forests $F_1, \ldots, F_n \in \mathcal{F}'_k$, if a k-forest F is a minimal upper bound of $F_1, \ldots, F_n$ in $(\mathcal{F}_k, \leq_1)$ then $F \equiv_1 \alpha_1(F_1) \sqcup \ldots \sqcup \alpha_n(F_n)$ for some $\alpha_1, \ldots \alpha_n \in S_k$.*

(ii) For any k-trees $T_1, \ldots, T_n \in \mathcal{T}'_k$, if a k-tree T is a minimal upper bound of $T_1, \ldots, T_n$ in $(\mathcal{T}_k, \leq_1)$ then $T \equiv_1 p_0(\alpha_1(T_1) \sqcup \ldots \sqcup \alpha_n(T_n))$ for some $\alpha_1, \ldots \alpha_n \in S_k$.

Proof. (i) Obviously, $F_i \leq_1 \alpha_i(F_i) \leq_1 \alpha_1(F_1) \sqcup \ldots \sqcup \alpha_n(F_n)$. Let $F_i \leq_1 G$ by some 1-morphism f_i and $G \in \mathcal{F}_k$. Since each F_i carries all labels from k, there exists $\alpha_i \in S_k$ such that f_i is a 0-morphism from $\alpha_i(F_i)$ to G. Thus $\bigcup f_i$ is a 0-morphism from $\alpha_1(F_1) \sqcup \ldots \sqcup \alpha_n(F_n)$ to G. Item (ii) is considered in a similar fashion. □

Proposition 4. *Let $T_1, \ldots, T_n \in \mathcal{T}'_k$ be 1-minimal pairwise 1-incomparable k-trees.*

(i) A k-forest F is a minimal upper bound of $T_1, \ldots, T_n$ in $(\mathcal{F}_k, \leq_1)$ iff $F \equiv_1 \alpha_1(T_1) \sqcup \ldots \sqcup \alpha_n(T_n)$ for some $\alpha_1, \ldots \alpha_n \in S_k$.

(ii) If $T_1, \ldots, T_n \in \mathcal{Y}_k$ then a k-tree T is a minimal upper bound of $T_1, \ldots, T_n$ in $(\mathcal{T}_k, \leq_1)$ iff $T \equiv_1 p_0(\alpha_1(T_1) \sqcup \ldots \sqcup \alpha_n(T_n))$ for some $\alpha_1, \ldots \alpha_n \in S_k$.

Proof. The direct implications of the both items follow from the previous propositions. For converse implication in item (i), suppose that $\alpha_1(T_1) \sqcup \ldots \sqcup \alpha_n(T_n) \leq_1 \beta_1(T_1) \sqcup \ldots \sqcup \beta_n(T_n)$ by some 1-morphism f. Then $f(\alpha_i(T_i)) \subseteq \beta_i(T_i)$ for all $i < k$ since T_i are pairwise 1-incomparable, and $f(\alpha_i(T_i)) = \beta_i(T_i)$ for all $i < k$ since T_i are 1-minimal. Therefore $\beta_1(T_1) \sqcup \ldots \sqcup \beta_n(T_n) \leq_1 \alpha_1(T_1) \sqcup \ldots \sqcup \alpha_n(T_n)$ by the 1-morphism f^{-1}. Item (ii) is considered in a similar fashion. □

For an arbitrary preordered set $(P; \leq)$ and $x, y \in P$, let $mub_P(x, y)$ denote the set of all minimal upper bounds of x and y in P. For an arbitrary $u \in P$, we call $x \in P$ *u-quasijoin-irreducible* iff

$$\forall y, z, v \leq u((v \in mub_P(y, z) \wedge x \leq v) \rightarrow (x \leq y \vee x \leq z)).$$

Since the relation $x \in mub_P(y, z)$ can be defined by a formula of signature $\{\leq\}$, there exists a formula $ir(x, u)$ of signature $\{\leq\}$ which defines (in every preordered set) exactly the non-zero u-quasijoin-irreducible elements. We call

$x \in P$ *u-quasijoin-irreducible* if x is u-quasijoin-irreducible for any $u \in P$. If P is an upper semilattice then $mub_P(x, y) = \{x \sqcup y\}$ for all $x, y \in P$. One can easily show that, in any distributive upper semilattice, the quasijoin-irreducible elements are exactly the join-irreducible elements. Let $ir(P, u)$ denote the set of all non-zero u-quasijoin-irreducible elements in P.

Proposition 5. *(i) If $t \in \mathfrak{F}'_k$ then $t \in ir(\mathfrak{F}^{1'}_k, u)$ for any $u \in F^1_k$ with $t \leq_1 u$.*
(ii) If $t \in \mathcal{Y}'_k$ then $t \in ir(\mathfrak{T}^{1'}_k, u)$ for any $u \in \mathfrak{T}^1_k$ with $t \leq_1 u$.

Proof is a simple routine. $\qquad\square$

We will need also the following technical lemma which shows that the posets $(\mathfrak{T}^1_k, \leq_1)$ and $(\mathfrak{T}^2_k, \leq_2)$ have antichains with any finite number of elements (though they have neither infinite antichains nor infinite descending chains [10]).

Lemma 2. *For all $k > 2$ and $n \in \omega$ there exist pairwise 2-incomparable (i.e. incomparable in $\leq_2$) repetition-free k-chains $C_0, \dots, C_n$.*

Proof. For every repetition-free k-word α whose first symbol is not $k - 1$, let D_α denote the repetition-free k-chain presented by the k-word $0 \dots (k - 1)\alpha$. It is straightforward to check that D_α and D_β are 2-incomparable if $|\alpha| = |\beta|$ and $\alpha \neq \beta$ (consider non-equal labels on the same place in α and β). It remains to note that $|\{D_\alpha : |\alpha| = n\}| = (k - 1)^n$ for any $n \in \omega$. $\qquad\square$

4   Interpretation Scheme

For proving our undecidability results we will use the same interpretation scheme as in [11]. As is well known [5], for establishing hereditary undecidability of the first-order theory of a structure, say of a partial order $(P; \leq)$, it suffices to show that the class of finite models of the theory of two equivalence relations is relatively elementarily definable in $(P; \leq)$ with parameters [5]. It turns out that the following very particular case of the last notion is sufficient for our proof.

It suffices to find first-order formulas $\phi_0(x, \bar{p})$, $\phi_1(x, y, \bar{p})$ and $\phi_2(x, y, \bar{p})$ of signature $\{\leq\}$ (where x, y are variables and $\bar{p}$ is a string of variables called parameters) with the following property:

(*) for every $n < \omega$ and for all equivalence relations ξ, η on $\{0, \dots, n\}$ there are values of parameters $\bar{p} \in P$ such that the structure $(\{0, \dots, n\}; \xi, \eta)$ is isomorphic to the structure $(\phi_0(P, \bar{p}); \phi_1(P, \bar{p}), \phi_2(P, \bar{p}))$.

Here

$$\phi_0(P, \bar{p}) = \{a \in P | (P; \leq) \models \phi_0(a, \bar{p})\},$$

$$\phi_1(P, \bar{p}) = \{(a, b) \in P | (P; \leq) \models \phi_1(a, b, \bar{p})\}$$

and similarly for ϕ_2. In other words, for all n, ξ, η as above there are parameter values $\bar{p} \in P$ such that the set $\{0, \dots, n\}$ and the relations ξ, η are first-order definable in $(P; \leq)$ with parameters $\bar{p}$.

5 Main Technical Fact

The main technical result of this paper is formulated as follows.

Theorem 1. *For any $k \geq 3$, the first-order theories of the structures $(\mathfrak{F}_k^1, \leq_1)$, $(\mathfrak{F}_k^2, \leq_2)$, $(\mathfrak{T}_k^1, \leq_1)$, $(\mathfrak{T}_k^2, \leq_2)$ are hereditarily undecidable.*

Sketch of proof. By the preceding section, for each structure it suffices to find suitable formulas ϕ_0, ϕ_1, ϕ_2 and to specify parameter values $\bar{p}$ as described in (*). Assume that $\xi, \eta \subseteq (n+1)^2$ are arbitrary equivalence relations on $n + 1 = \{0, \ldots, n\}$ and $\{\xi_0, \ldots, \xi_m\}$, $\{\eta_0, \ldots, \eta_l\}$ are the partitions of $n + 1$ induced by ξ and η, respectively. Note that the parameters $\bar{p}$ will depend on ξ and η.

For interpreting the elements $0, \ldots, n$ of the set $n + 1$, we use 2-incomparable k-chains $C_0, \ldots, C_n$ from Lemma 2. By the proof of Lemma 2, we can assume w.l.o.g. that all k-chains $C_0, \ldots, C_n$ are disjoint, their roots are labeled with 0 and the successors of the roots are labeled with 1.

We construct the parameters $\bar{p} = (u, v, w, a)$ from $C_0, \ldots, C_n$ in the following way. For the case of $(\mathfrak{F}_k^1, \leq_1)$ and $(\mathfrak{F}_k^2, \leq_2)$, let $U = C_0 \sqcup \ldots \sqcup C_n$ and $V = V_1 \sqcup \ldots \sqcup V_m$ where V_i is obtained from $\{C_j \mid j \in \xi_i\}$ by identifying the roots, $V_i \equiv_0 p_0(\bigsqcup\{C_j \mid j \in \xi_i\})$ (see Picture 1). Define W as V with η instead of ξ. Namely, $W = W_1 \sqcup \ldots \sqcup W_l$ where W_i is obtained from $\{C_j \mid j \in \eta_i\}$ by identifying the roots, $W_i \equiv_0 p_0(\bigsqcup\{C_j \mid j \in \eta_i\})$. For $\mathfrak{F}_k^1, \mathfrak{F}_k^2$ we set $u = [U]_{\equiv_i}$, $v = [V]_{\equiv_i}$ and $w = [W]_{\equiv_i}$.

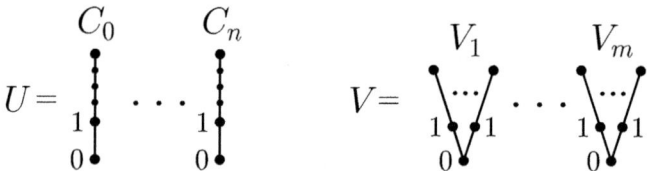

Picture 1. Parameters U and V for $\mathfrak{F}_k^1$ and $\mathfrak{F}_k^2$.

For the case of $(\mathfrak{T}_k^1, \leq_1)$ and $(\mathfrak{T}_k^2, \leq_2)$, the parameters u, v, w are a bit more complicated than in the previous case, see Picture 2.

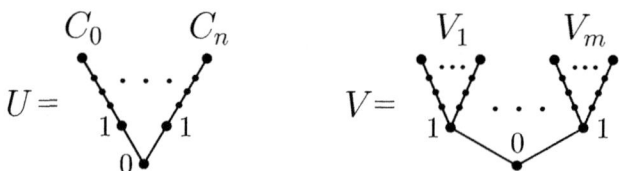

Picture 2. Parameters U and V for $\mathfrak{T}_k^1$ and $\mathfrak{T}_k^2$.

Namely, let U be obtained from $C_0, \ldots, C_n$ by identifying the roots, so we have $U \equiv_0 p_0(C_0 \sqcup \cdots \sqcup C_n)$. For each i from 1 to l, let V_i be obtained from $\{C_j \mid j \in \xi_i\}$ by identifying the roots and their successors. Now V is constructed by identifying the roots of all V_i, $1 \leq i \leq l$. Define W as V with η instead of ξ. For $\mathfrak{T}_k^1$ and $\mathfrak{T}_k^2$ we set $u = [U]_{\equiv_i}$, $v = [V]_{\equiv_i}$ and $w = [W]_{\equiv_i}$.

Finally, we define the parameter a as $[A]_{\equiv_1}$ where $A = 01 \ldots (k-1)$. This parameter is in fact essential only for $\mathfrak{F}_k^1$ and $\mathfrak{T}_k^1$.

Next we construct the formulas ϕ_0, ϕ_1, ϕ_2 and prove the property (*) from the previous section. Let $\tau_1(x, u, a)$ be obtained from the formula

$$x \leq u \wedge ir(x, u) \wedge \neg \exists y > x(y \leq u \wedge ir(y, u))$$

by adding the requirements that every variable (free or bounded) is greater or equal than a. The formula $\tau_1(x, u)$ means that x is a maximal non-zero u-quasijoin-irreducible element in the upper cone $\breve{a}$. Let $\tau_2(x, u)$ be the formula

$$x \leq u \wedge ir(x) \wedge \neg \exists y > x(y \leq u \wedge ir(y)),$$

which means that x is a maximal non-zero join-irreducible element below u.

The case $\underline{\mathfrak{F}_k^1}$. In this case we take

$$\phi_0(x, \bar{p}) \Leftrightarrow \tau_1(x, u, a),$$

$$\phi_1(x, y, \bar{p}) \Leftrightarrow \tau_1(x, u, a) \wedge \tau_1(y, u, a) \wedge \exists t \leq v(\tau_1(t, v, a) \wedge x \leq t \wedge y \leq t),$$

and let $\phi_2(x, y, \bar{p})$ be obtained from $\phi_1(x, y, \bar{p})$ by substituting w in place of v.

Let $c_0 = [C_0]_{\equiv_1}, ..., c_n = [C_n]_{\equiv_1}$. Note that $C_0, \ldots, C_n \in T_k'$ are u-quasijoin-irreducible in $\breve{A}$ by Proposition 5. Let f be an arbitrary u-quasijoin-irreducible from the set $\{x \mid x \in \mathfrak{F}_k^1, a \leq_1 x \leq_1 u\}$. By Proposition 4, u is a minimal upper bound of $c_0, \ldots, c_n$, since $C_0, \ldots, C_n$ are 1-minimal and pairwise 1-incomparable. Therefore $f \leq_1 c_i$ for some $i \leq n$. Thus we proved that $\phi_0(\mathfrak{F}_k^1, \bar{p}) = \{c_0, \ldots, c_n\}$. One can show similarly that $\phi_1(c_i, c_j, \bar{p})$ holds iff $(i, j) \in \xi$.

The case $\underline{\mathfrak{T}_k^1}$. The formulas ϕ_0, ϕ_1, ϕ_2 are the same as in the previous case, but the parameters are different (see Pic.2 and the corresponding description). The proof is similar to the previous case, we use the second items of Propositions 4 and 5.

The case $\underline{\mathfrak{F}_k^2}$. Let

$$\phi_0(x, \bar{p}) \Leftrightarrow \tau_2(x, u),$$

$$\phi_1(x, y, \bar{p}) \Leftrightarrow \tau_2(x, u) \wedge \tau_2(y, u) \wedge \exists t \leq v(\tau_2(t, v) \wedge x \leq t \wedge y \leq t),$$

and let $\phi_2(x, y, \bar{p})$ be obtained from $\phi_1(x, y, \bar{p})$ by substituting w in place of v. The desired properties can be easily deduced using Proposition 1.

The case $\underline{\mathfrak{T}_k^2}$. Here we use the same formulas as in the previous case, but the parameters are different (see Pic.2 and the corresponding description). Note that for $\mathfrak{F}_k^2$ and $\mathfrak{T}_k^2$ we do not need the parameter a. $\qquad \square$

6   Consequences for Weihrauch Degrees

Here we explain how the results of the last section imply undecidability of many initial segments of Weihrauch degrees. We start with formulating two auxiliary facts.

Recall that Δ_0-*formulas* of signature $\{\leq\}$ are constructed inductively according to the following rules: any atomic formula of signature $\{\leq\}$ is a Δ_0-formula; if φ and ψ are Δ_0-formulas then $\neg\varphi$, $(\varphi \wedge \psi)$, $(\varphi \vee \psi)$ and $(\varphi \to \psi)$ are Δ_0-formulas; if x, y are variables and φ is a Δ_0-formula then $\forall x \leq y \varphi$ and $\exists x \leq y \varphi$ are Δ_0-formulas.

The first fact is a particular case of a well-known property of Δ_0-formulas (see e.g. [3]).

Lemma 3. *Let* $\mathbf{A} = (A; \leq)$ *and* $\mathbf{B} = (B; \leq)$ *be two posets such that A is an initial segment of B. Let some evaluation of the free variables of a Δ_0-formula φ in $\mathbf{A}$ be fixed. Then* $\mathbf{A} \models \varphi$ *iff* $\mathbf{B} \models \varphi$.

The second fact follows in a straightforward way from the definition of Weihrauch's reducibility $\leq_2$.

Lemma 4. *Let $k \geq 2$ and let $\mathcal{C}$ be one of the classes $P(\omega^\omega)$, $\boldsymbol{\Delta}_n^1$, $BC(\boldsymbol{\Sigma}_n^1)$, $BC(\boldsymbol{\Sigma}_n^0)$, $\boldsymbol{\Delta}_{n+1}^0$ in the Baire space, where $n \geq 1$. Then $BC(\boldsymbol{\Sigma}_1^0)_k$ is an initial segment of $(\mathcal{C}_k; \leq_2)$.*

Now we are able to deduce undecidability results for Weihrauch degrees.

Theorem 2. *Let $k \geq 3$ and let $\mathcal{C}$ be one of the classes $P(\omega^\omega)$, $\boldsymbol{\Delta}_n^1$, $BC(\boldsymbol{\Sigma}_n^1)$, $BC(\boldsymbol{\Sigma}_n^0)$, $\boldsymbol{\Delta}_{n+1}^0$ in the Baire space, where $n \geq 1$. Then the first-order theories of the quotient-posets of $(\mathcal{C}_k; \leq_1)$ and $(\mathcal{C}_k; \leq_2)$ are hereditary undecidable.*

Proof. By Lemma 4, the quotient-poset of $(BC(\boldsymbol{\Sigma}_1^0)_k; \leq_i)$ is an initial segment of the quotient-posets of $(\mathcal{C}_k; \leq_i)$ for any $i = 1, 2$. By the result of P. Hertling, the quotient-poset of $(BC(\boldsymbol{\Sigma}_1^0)_k; \leq_i)$ is isomorphic to $(\mathfrak{F}_k^i, \leq_i)$. Therefore, it suffices to show that if $\mathfrak{F}_k^i$ is an initial segment of a poset $(P; \leq)$ then the first-order theory of $(P; \leq)$ is hereditary undecidable.

Observe that the formulas ϕ_0, ϕ_1 and ϕ_2 (see the previous section) that interpret the theory of two equivalence relations on finite sets in poset $(\mathfrak{F}_k^i, \leq_i)$ are Δ_0-formulas. By Lemma 3, the formulas ϕ_0, ϕ_1 and ϕ_2 interpret the theory of two equivalence relations on finite sets in $(P; \leq)$ (with the same parameter values as in $\mathfrak{F}_k^i$). By Section 4, the first-order theory of $(P; \leq)$ is hereditary undecidable. $\square$

References

1. Brattka, V., Gherardi, G.: Weihrauch degrees, omniscience principles and weak computability (2009), http://arxive.org/abs0905.4679
2. Brattka, V., Gherardi, G.: Effective choice and boundedness principles in computable analysis (2009), http://arxive.org/abs0905.4685

3. Ershov, Y.L.: Definability and Computability. Plenum, New-York (1996)
4. Davey, B.A., Priestley, H.A.: Introduction to Lattices and Order, 2nd edn. Cambridge University Press, Cambridge (2002)
5. Ershov, Y.L., Lavrov, T.A., Taimanov, A.D., Taitslin, M.A.: Elementary theories. Uspechi Mat. Nauk 20(4), 37–108 (1965) (in Russian)
6. Hertling, P.: Topologische Komplexitätsgrade von Funktionen mit endlichem Bild. In: Informatik-Berichte, vol. 152, Fernuniversität Hagen (1993)
7. Hertling, P.: Unstetigkeitsgrade von Funktionen in der effektiven Analysis, Informatik Berichte 208-11/1996, Fernuniversität Hagen (1996)
8. Hertling, P., Weihrauch, K.: Levels of degeneracy and exact lowercomplexity bounds for geometric algorithms. In: Proc. of the 6th Canadian Conf. on Computational Geometry Saskatoon, pp. 237–242 (1994)
9. Kechris, A.S.: Classical Descriptive Set Theory. Springer, New York (1994)
10. Kruskal, J.B.: The theory of well-quasi-ordering: a frequently discovered concept. J. Combinatorics Theory (A) 13, 297–305 (1972)
11. Kudinov, O.V., Selivanov, V.L.: Undecidability in the homomorphic quasiorder of finite labeled forests. In: Beckmann, A., Berger, U., Löwe, B., Tucker, J.V. (eds.) CiE 2006. LNCS, vol. 3988, pp. 289–296. Springer, Heidelberg (2006)
12. Kudinov, O.V., Selivanov, V.L.: Undecidability in the homomorphic quasiorder of finite labelled forests. Journal of Logic and Computation 17, 1135–1151 (2007)
13. Kosub, S., Wagner, K.: The Boolean hierarchy of NP-partitions. In: Reichel, H., Tison, S. (eds.) STACS 2000. LNCS, vol. 1770, pp. 157–168. Springer, Heidelberg (2000)
14. Lehtonen, E.: Labeled posets are universal. European J. Combin. 29, 493–506 (2008)
15. Selivanov, V.L.: Boolean hierarchy of partitions over reducible bases. Algebra and Logic 43(1), 44–61 (2004)
16. Selivanov, V.L.: Hierarchies of Δ_2^0-measurable k-partitions. Math. Logic Quarterly 53, 446–461 (2007)
17. Selivanov, V.L.: Undecidability in Some Structures Related to Computation Theory. Journal of Logic and Computation 19(1), 177–197 (2009)
18. Weihrauch, K.: The degrees of discontinuity of some translators between representations of the real numbers. Technical Report TR-92-050, International Computer Science Institute, Berkeley (1992)
19. Weihrauch, K.: Computable Analysis. Springer, Berlin (2000)

Degrees with Almost Universal Cupping Property

Jiang Liu and Guohua Wu

Division of Mathematical Sciences
School of Physical and Mathematical Sciences
Nanyang Technological University
Singapore 637371, Singapore

1 Introduction

The notion of cupping/noncupping has played an essential role in the study of various degree structures in the Ershov hierarchy. As an approach to refute Shoenfield conjecture, Yates (see [2]) proved the existence of a nonzero noncuppable r.e. degree, a degree cupping no incomplete r.e. degree to $\mathbf{0}'$. In contrast to this, Arslanov proved in [1] that nonzero noncuppable degrees do not exist in the structure of d.r.e. degrees, which shows that the structures of r.e. degrees and d.r.e. degrees are not elementary equivalent.

On the other hand, Posner and Robinson proved in [7] and [8] that any degree below $\mathbf{0}'$ has a complement in the Δ_2^0 degrees. Slaman and Steel [9] improved this result by showing that such complements can be 1-generic degrees. This implies the existence of a Δ_2^0 degree such that $\mathbf{0}$ and $\mathbf{0}'$ are the only r.e. degrees comparable with it — such degrees are called *Yates degrees*, as introduced by Wu in [11].

Say that an incomplete degree has universal cupping property if it cups every nonzero r.e. degree to $\mathbf{0}'$. By Lachlan's observation that every nonzero n-r.e. degree bounds a nonzero r.e. degree, no universal cupping degree can be n-r.e. In [5], Li, Song and Wu proved that in terms of the Ershov hierarchy, universal cupping degrees can be ω-r.e.

In this paper, we consider those degrees with *almost universal cupping property*. Here an incomplete degree $\mathbf{d}$ has almost universal cupping property if it cups every nonzero r.e. degree to $\mathbf{0}'$, except for those degrees below $\mathbf{d}$. In [4], Cooper, Harrington, Lachlan, Lempp and Soare showed the existence of an incomplete maximal d.r.e. degree. Obviously, such maximal d.r.e. degrees have the almost universal cupping property.

The construction in Cooper et al.'s proof involves an extremely complicated $\mathbf{0}'''$-priority argument, and it is fairly hard to combine their construction with other strategies. For example, we do not know whether there are two maximal d.r.e. degrees forming a minimal pair, even in the d.r.e. degrees. If it is true, then it can be regarded as a strong unification of many well-known results on d.r.e. degrees, including Arslanov's cupping theorem, Downey's diamond embedding theorem, and of course, Cooper et al.'s nondensity theorem.

F. Ferreira et al. (Eds.): CiE 2010, LNCS 6158, pp. 266–275, 2010.

In this paper, we provide a direct construction of almost universal cupping degrees.

Theorem 1. *There is an almost universal cupping d.r.e. degree* **d** *such that all the r.e. degrees below it are bounded by an r.e. degree* **b**.

We remark here that according to Cooper and Yi [3], **d** is said to be isolated by r.e. degree **b**.

Our proof of Theorem 1 involves a $0'''$ argument, which could be combined with other strategies to get more structural properties of the d.r.e. degrees.

Our terminology and notation are quite standard and follows Soare [10] and Odifreddi [6].

2   Requirements

To prove Theorem 1, we will construct a d.r.e. set D, an r.e. set B and an auxiliary r.e. set A, partial recursive (p.r. for short) functionals $\{\Gamma_e\}_{e\in\omega}$ and $\{\Delta_e\}_{e\in\omega}$ satisfying the following requirements:

$\mathcal{P}_e\colon A \neq \Phi_e^B$,
$\mathcal{I}_e\colon W_e = \Phi_e^{B,D} \Rightarrow \exists\Theta_e(W_e = \Theta_e^B)$,
$\mathcal{R}_e\colon K = \Gamma_e^{B,D,W_e} \vee W_e = \Delta_e^B$.

where K is the halting problem set and $\{W_e\}_{e\in\omega}$ is an effective list of all r.e. sets. Let $\mathbf{b} = deg_T(B)$ and $\mathbf{d} = deg_T(B \oplus D)$. Then the $\mathcal{P}$-requirements ensure that **b** is incomplete and the $\mathcal{R}$-requirements ensure that every r.e. degree **w** either cups **d** to $\mathbf{0}'$ or $\mathbf{w} \leq \mathbf{b}$. The $\mathcal{I}_e$-requirements ensure that no r.e. degree is between **b** and **d**, and hence Theorem 1 is proved, if all the requirements are satisfied.

3   Strategies

In this section, we describe how each strategy works and consider the interactions between various strategies. Here, we assume that each strategy is located on a node of the priority tree.

3.1   A $\mathcal{P}$-Strategy

A $\mathcal{P}$-strategy is simply a Friedberg-Muchnik's diagonalization argument. Let α be a $\mathcal{P}_e$-strategy. For simplicity, we use $\Phi_\alpha^B(x)$ to denote $\Phi_{e(\alpha)}^B(x)$. α works as follows:

1. Choose a witness x, and wait for $\Phi_\alpha^B(x) \downarrow = 0$.
2. Put x into A and restrain numbers $\leq \varphi_\alpha(B;x)$ to enter B.

There are two possible cases.

$w\colon$ α waits forever at step 1. Then $\Phi_\alpha^B(x)$ never converges to 0, and hence $A(x) = 0 \neq \Phi_\alpha^B(x)$, $\mathcal{P}_e$ is satisfied. Let w denote this outcome.
$d\colon$ α reaches step 2 at some stage. Then $\Phi_\alpha^B(x) \downarrow = 0 \neq 1 = A(x)$, and again, $\mathcal{P}_e$ is satisfied. Let d denote this outcome.

3.2   An $\mathcal{R}$-Strategy

Let β be an $\mathcal{R}_e$-strategy. For the given W_e, β will construct a p.r. functional Γ_β (as before, we use Γ_β to denote $\Gamma_{e(\beta)}$ and W_β to denote $W_{e(\beta)}$) such that $K = \Gamma_\beta^{B,D,W_\beta}$. Γ_β is constructed as follows:

1. At a stage s, for all $z < s$ with $\Gamma_\beta^{B,D,W_\beta}(z)[s] \uparrow$, define $\Gamma_\beta^{B,D,W_\beta}(z)[s] = K_s(z)$ with use $\gamma_\beta(z)[s]$ a fresh number.
2. If $\Gamma_\beta^{B,D,W_\beta}(z)[s] \downarrow \neq K_s(z)$, then put $\gamma_\beta(z)[s]$ into D to undefine $\Gamma_\beta^{B,D,W_\beta}(z)$.

In the construction, β may enumerate infinitely many numbers into D to ensure that Γ_β^{B,D,W_e} computes K correctly. This will cause an obvious conflict between β and those $\mathcal{I}$-strategies η below β, since η attempts to restrain D from changes to preserve a computation. This kind of interactions is the major concern of the whole construction, as we will see soon after introducing the $\mathcal{I}$-strategies.

3.3   An $\mathcal{I}$-Strategy

Let η be an $\mathcal{I}_e$-strategy for some $e \in \omega$. We write W_η for $W_{e(\eta)}$ and $\Phi_\eta^{B,D}$ for $\Phi_{e(\eta)}^{B,D}$ as we did above. Given a stage s, define the length of agreement function $l(\eta, s)$ over η as

$$l(\eta, s) = \max\{x \mid \forall y < x(W_\eta \lceil y[s] = \Phi_\eta^{B,D} \lceil y[s])\},$$

and the maximal length function

$$m(\eta, s) = \max\{l(e, t), 0 \mid t < s \text{ and } t \text{ is an } \eta\text{-stage}\},$$

where t is called an η-stage if η is visited at stage t. We say s is an η-expansionary stage if s is an η-stage and $l(\eta, s) > m(\eta, s)$.

The basic strategy for η is to extend the definition of Θ_η at each η-expansionary stage s such that $\Theta_\eta^B \lceil l(\eta, s) = W_{\eta,s} \lceil l(\eta, s)$. However, in the construction, after stage s, numbers entering/exiting D or B may injure the current computations $\Phi_\eta^{B \oplus D} \lceil l(\eta, s)[s]$. In the case that a smaller number is enumerated into B first, Θ_η^B will be undefined correspondingly. A problem arises when numbers, z say, are enumerated into or removed from D, as some $\Phi_\eta^{B \oplus D}(y)$, $y < l(\eta, s)$, is destroyed, leading to a change of both $\Phi_\eta^{B \oplus D}(y)$ and $W_\eta(y)$, from 0 to 1. In this case, as no number is enumerated into B, $\Theta_\eta^B(y)$ is kept the same, and equals to 0, which shows that Θ_η^B does not compute W_η correctly.

We get around this problem as follows. Assume that $\Theta_\eta^B(y)[s] = 0$ is defined at η-expansionary stage s. That is, $\Phi_\eta^{B \oplus D}(y) = W_{\eta,s}(y) = 0$. Suppose that $s_1 > s$ is the least η-expansionary stage at which $W_{\eta,s_1}(y) = 1$. This means that between stages s and s_1, there is a stage s_0 at which a number less than $\varphi_\eta(y)[s]$ is enumerated into D. In this case, η removes all numbers enumerated into D after stage s_0 to recover the computation $\Phi_\eta^{B \oplus D}(y)$ to $\Phi_\eta^{B \oplus D}(y)[s]$, which produces an inequality between $W_\eta(y)$ and $\Phi_\eta^{B \oplus D}(y)$ as

$$\Phi_\eta^{B \oplus D}(y)[s_1] = \Phi_\eta^{B \oplus D}(y)[s] = 0 \neq 1 = W_{\eta,s_1}(y) = W_\eta(y).$$

Such a y (the least one) is called *a witness or a potential disagreement point* for the isolation strategy η. To preserve this inequality, it is sufficient to protect the computation $\Phi_\eta^{B \oplus D}(y)$ as $\Phi_\eta^{B \oplus D}(y)[s_1]$ since W_η is an r.e. set, and $W_\eta(y)$ cannot be changed back to 0.

η has three outcomes f, d and ∞ with $\infty <_L f <_L d$, where f means that there are only finitely many η-expansionary stages, ∞ means that there are infinitely many η-expansionary stages and d means that η succeeds in creating a disagreement between $\Phi_\eta^{B \oplus D}$ and W_η as above.

3.4   The Interaction among $\mathcal{I}$-Strategies

The crucial point of the construction is that whenever numbers are extracted from D, some small numbers will be enumerated into B, to ensure the consistency between different strategies.

Recall that when $\Theta_\eta^B(y)$ is defined at stage s, we have $\Phi_\eta^{B \oplus D}(y)[s] = W_{\eta,s}(y)$ and we define $\Theta_\eta^B(y)[s] = W_{\eta,s}(y)$. This computation $\Phi_\eta^{B \oplus D}(y)$ can be destroyed by the changes of D or B, at a bigger stage $s' > s$. As seen above, a computation $\Phi_\eta^{B \oplus D}(y)$ can be destroyed by enumeration of small numbers into D, and we can extract such numbers from D to recover D to a previous stage, and hence to recover computation $\Phi_\eta^{B \oplus D}(y)$ to $\Phi_\eta^{B \oplus D}(y)[s]$, and hence η is satisfied. But, on the other hand, such an action also causes problems to those $\mathcal{I}$-strategies with higher priority. To see this, let $\eta' < \eta$ be an $\mathcal{I}$-strategy with $\eta'^\frown \infty \subset \eta$.

1. At an η-expansionary stage s_0 (of course it's also η'-expansionary), a number z is enumerated into D.
2. At an η'-expansionary stage $s_1 > s_0$, η' defines $\Theta_{\eta'}^B(y')[s_1] = W_{\eta',s_1}(y')$, as 0 (for a general purpose), with use $\theta_{\eta'}(B; y')[s_1] = s_1$ for some y'. We assume that the computation $\Phi_{\eta'}^{B \oplus D}(y')[s_1]$ uses the fact that z is in D_{s_1}.
3. At an η-expansionary stage $s_2 > s_1$, η sees that $W_{\eta,s_2}(y) \neq \Theta_\eta^B(y)[s_2]$, and hence η recovers D to D_{s_0} *to recover the* $\Phi_\eta^{B \oplus D}(y)$ *to* $\Phi_\eta^{B \oplus D}(y)[s_0]$ *to create a disagreement between* $\Phi_\eta^{B \oplus D}(y)$ *and* $W_\eta(y)$. In particular, z is extracted from D.
4. At an η'-expansionary stage $s_3 > s_2$, η' sees that $W_{\eta',s_3}(y') \neq \Theta_{\eta'}^B(y')[s_3]$. This is so because z is removed from D and $\Phi_{\eta'}^{B \oplus D}(y')[s_3]$ is a new computation, converging to 1.

As we are constructing D as a d.r.e. set, z cannot be enumerated into D again to recover $\Phi_{\eta'}^{B \oplus D}(y')$ to $\Phi_{\eta'}^{B \oplus D}(y')[s_2]$. To avoid such a situation, at stage s_2, when z is extracted from D, we also enumerate s_0 into B, where s_0 is the stage at which z enters D. This enumeration undefines $\Theta_{\eta'}^B(y')$ as s_0 is less than s_1, and hence the inequality in step 4 can never occur. This ensures that η's action is consistent with those $\mathcal{I}$-strategies with higher priority. Also note that extra enumeration of s_0 into B is also consistent with η itself, as s_0 is bigger than the use of $\Phi_\eta^{B \oplus D}(y)[s_0]$.

Another point is that there are perhaps infinitely many $\mathcal{I}$-strategies below $\eta^\frown \infty$, and we need to make sure that for a given $y \in \omega$, $\Theta_\eta^B(y)$ can be undefined in this way by only finitely many such $\mathcal{I}$-strategies. Note that after η sees that $\Phi_\eta^{B\oplus D}(y)$ converges, and when $\Theta_\eta^B(y)$ is first defined at stage s, only finitely many $\mathcal{I}$-strategies are visited below η's outcome ∞, and only these strategies can undefine $\Theta_\eta^B(y)$ later, as the use $\theta_\eta(B;y)$ is always defined as s, and hence after all these $\mathcal{I}$-strategies stop to act, $\Theta_\eta^B(y)$ cannot be undefined anymore.

3.5   One $\mathcal{I}$-Strategy Below One $\mathcal{R}$-Strategy

Now we consider how an $\mathcal{I}$-strategy η works below an $\mathcal{R}$-strategy β. Let s be an η-expansionary stage at which $\Theta_\eta^B(y)[s]$ is defined. Assume that $s_1 > s_0$ is a stage at which β would put some small number $\gamma_\beta(k)[t] < \varphi_\eta(y)[s_1]$ into D as k enters K currently, which may change $\Phi_\eta^{B\oplus D}(y)$ and injure η's strategy. Such a process can happen infinitely many times, and hence η cannot satisfy the requirement $\mathcal{I}_e$. To avoid this, we need to make the computation $\Phi_\eta^{B\oplus D}(y)$ be clear of the γ_β-uses, and we apply the threshold strategy as follows.

When η is first visited, it fixes a parameter p_η as a threshold for the enumeration of γ_β-uses. Whenever a number $k \leq p_\eta$ enters K, we enumerate the current $\gamma_\beta(p_\eta)$-use into D and *reset the strategy* η by cancelling all the parameters defined by η, except for p_η, and initializing all the strategies with lower priority. Once the parameter p_η is settled down, this resetting process can happen at most finitely often. Suppose that after stage s', η cannot be reset. Then at stage s_1, η puts $\gamma_\beta(p_\eta)[s_1]$ into D to *start an attack* on β with outcome g_β by defining

$$\Delta_{\eta\beta}^B \lceil \gamma_\beta(p_\eta)[s_1] = W_{\beta,s_1} \lceil \gamma_\beta(p_\eta)[s_1]$$

with use s_1 and waiting for a W_β-change below $\gamma_\beta(p_\eta)[s_1]$. We say that the attack associated with $\gamma_\beta(p_\eta)[s_1]$ is *activated* at stage s_1.

There are two possible cases.

1. If $W_\beta \lceil \gamma_\beta(p_\eta)[s_1]$ never changes since stage s_1, then the attack associated with $\gamma_\beta(p_\eta)[s_1]$ keeps active and in this case, η fails to pass the threshold p_η for β but succeeds in defining $\Delta_{\eta\beta}^B$ up to $\gamma_\beta(p_\eta)[s_1]$.

2. Otherwise, there is an η-expansionary stage $t > s_1$, and

$$W_{\beta,s_1} \lceil \gamma_\beta(p_\eta)[s_1] \neq W_{\beta,t} \lceil \gamma_\beta(p_\eta)[s_1].$$

   Then η uses this W_β-change, *instead of enumerating $\gamma_\beta(p_\eta)[s_1]$ into D*, to lift the value of $\gamma_\beta(p_\eta)$. η's action is to remove all numbers being enumerated into D after stage s to recover the computation $\Phi_\eta^{B\oplus D}(y)$ to $\Phi_\eta^{B\oplus D}(y)[s]$, where s is the stage indicated in the last section. *Note that after stage t, the enumeration of γ_β-uses will not affect the computation $\Phi_\eta^{B\oplus D}(y)[s]$, as by the choice of s', no $k \leq p_\eta$ can enter K after stage s'.* We say that the attack is completed at stage t, and that η passes threshold p_η for β.

Accordingly, η has one more outcome:

g_β: η starts infinitely many attacks but none of these cycles can pass the threshold p_η for β. Then $\Delta_{\eta\beta}^B$ is totally defined and computes W_β correctly — $\mathcal{R}_e$ is satisfied, even though $\Gamma_\beta^{B,D,W_\beta}(p_\eta)$ diverges.

η can now have four outcomes with the priority:

$$g_\beta <_L \infty <_L f <_L d.$$

If η has outcome g_β, then $\mathcal{I}_e$ is not satisfied at η, and a backup strategy for the requirement $\mathcal{I}_e$ is given below the outcome g_β. Let η' be such a backup strategy below $\eta^\frown g_\beta$. Then η' knows that eventually, (almost) all γ_β-uses will go to infinity — in particular, $\gamma_\beta(p_\eta)$ goes to infinity. η' only believes a computation $\Phi_{\eta'}^{B \oplus D}(y)$ at stage s if $\gamma_\beta(p_\eta)[s]$ is bigger than $\varphi_{\eta'}(y)[s]$ — we will call such a computation η'-believable in the remainder of this paper. By only using η'-believable computations, η' satisfies the requirement $\mathcal{I}_e$ in a usual way, as β's further enumeration will not affect such computations anymore.

3.6   One $\mathcal{I}$-Strategy Below Many $\mathcal{R}$-Strategies

We now consider the case when an $\mathcal{I}$-strategy η is below several $\mathcal{R}$-strategies $\beta_0 \subset \ldots \subset \beta_n$. Then, when η sees that y is a potential witness for a disagreement between $\Phi_\eta^{B,D}$ and W_η, η needs to ensure that the computation $\Phi_\eta^{B \oplus D}(y)$ is clear of all the γ_{β_j}-uses for $0 \leq j \leq n$, so that after this disagreement is established, it will not be injured by these β-strategies. As before, η first sets a threshold p_η (*again, we reset η by cancelling all papameters and cycles of η, except for p_η, when a number $k \leq p_\eta$ enters K*), and to ensure that no such a β_j can enumerate a number into D later, destroying the wanted computation $\Phi_\eta^{B \oplus D}(y)$, we need to force a change of W_{β_j}, for each j, to have a small change to lift these γ_{β_j}-uses. η's actions consist of several cycles (perhaps infinitely many) with cycle 0 started first, and each cycle i can starts cycle $i + 1$, and can also cancel all cycles $i' > i$ — when the latter happens, a (small) number will be enumerated into B, for the sake of consistency between the $\mathcal{I}$-strategies. The numbers being enumerated into B are relatively large and will not affect η's purpose of creating an inequality.

Fix $i \geq 0$. Cycle i works as follows:

$i1$. At an η-expansionary stages s, extend the definition of Θ_η^B up to $l(\eta, s)$ if $\Theta_\eta^B[s]$ and $W_{\eta,s}$ agree on all the numbers in the domain of $\Theta_\eta^B[s]$. In this case, β has outcome ∞.

$i2$. At an η-expansionary stage s_1, η sees a (least) y with $\Theta_\eta^B(y)$ defined but not equal to $W_{\eta,s_1}(y)$, then η starts to lift the γ_{β_j}-uses by running the following steps, from $j = n$. Each step can have two phases, attacking phase $i2 - j1$ (at stage s_{n-j}^1) and completing phase $i2 - j2$ (at stage s_{n-j}^2):

$i2 - j1$. *Attack on β_j.*

Enumerate $\gamma_{\beta_j}(p_\eta)[s_{n-j}^1]$ into D and extend the definition of $\Delta_{\eta \beta_j}^B$ by letting

$$\Delta_{\eta \beta_j}^B(z)[s_{n-j}^1] = W_{\beta_j, s_{n-j}^1}(z)$$

with use s_{n-j}^1 for those $z < \gamma_{\beta_j}(p_\eta)[s_{n-j}^1]$ and $\Delta_{\eta \beta_j}^B(z)$ is not defined yet. β has outcome g_j. Wait for W_{β_j} to change below $\gamma_{\beta_j}(p_\eta)[s_{n-j}^1]$, and at the same time start cycle $i + 1$.

$i2 - j2$. *Completion on* β_j.

At an η-expansionary stage $s_{n-j}^2 > s_{n-j}^1$, η sees that

$$W_{\beta_j, s_{n-j}^1}\lceil \gamma_{\beta_j}(p_\eta)[s_{n-j}^1] \neq W_{\beta_j, s_{n-j}^2}\lceil \gamma_{\beta_j}(p_\eta)[s_{n-j}^1].$$

Extract from D all the numbers, including $\gamma_{\beta_j}(p_\eta)[s_{n-j}^1]$, which are enumerated into D after stage s_{n-j}^1, *to recover the computation* $\Phi_\eta^{B \oplus D}(y)$ *to* $\Phi_\eta^{B \oplus D}(y)[s]$, *which is now clear of* γ_{β_j}-*uses*. Also enumerate s_{n-j}^1 into B for the sake of consistency between the $\mathcal{I}$-strategies. Cancel all the cycles $i' > i$, and we say that cycle i passes the threshold p_η for β_j.

If $j > 0$, then run step $j - 1$. In this case, β has outcome g_{j-1}.

If $j = 0$, then extract from D the numbers enumerated into D between stages s (the last η-expansionary stage at which η sees that Θ_η^B is correct) and s_n^1 (when we started attack on β_n). Enumerate s into B and declare that η is satisfied. From now on, η will always have outcome d.

Now η has outcomes

$$g_0 <_L g_1 <_L \dots <_L g_n,$$

all of which are to the left of $\infty <_L f <_L d$ (we write g_j for g_{β_j}, for convenience).

If η runs only finitely many cycles, then either a cycle i passes the threshold p_η for all β_j, $0 \leq j \leq n$, in which case $i2 - 02$ is reached for cycle i and a disagreement is created, η is satisfied, or a cycle i nevers goes to at $i2$, which means that there are only finitely many η-expansionary stages or Θ_η^B defined by this cycle i agrees with W_η on all arguments, and we can argue as usual that η is also satisfied.

If η runs infinitely many cycles, then as each cycle has only finitely many possible outcomes, one of them (the least one), g_j say, will be the true outcome for infinitely many cycles, which means $\Delta_{\eta\beta_j}^B$ will be defined infinitely often. Let s' be the least stage after which no cycle can have outcome $g_{j'}$ with $j' < j$. It is easy to see that $\Delta_{\eta\beta_j}^B$ computes W_{β_j} correctly, by the choice of s' and also our assumption on j. In this case, η shows that W_{β_j} is reducible to B, and hence β_j is satisfied at η.

At the end of this section, we clarify when a cycle is considered to be active. Let $i_1 < i_2$ be two cycles and suppose that cycle i_1 is waiting for a W_{β_j}-change. Cycle i_1 becomes inactive automatically when cycle i_2 starts its attack on β_j. This means that when cycle i_1 becomes inactive, then we give up the potential disagreement of cycle i_1, which is acceptable in the sense that we now turn to use the potential disagreement of cycle i_2.

3.7   Two $\mathcal{I}$-Strategies Below Two $\mathcal{R}$-Strategies

We now consider how two $\mathcal{I}$-strategies η_1, η_2 work below two $\mathcal{R}$-strategies β_1, β_2, or even more. A nontrivial case is when $\beta_1 \subset \beta_2 \subset \eta_1^\frown g_2 \subset \eta_2$. Let p_{η_1}, p_{η_2} be the thresholds on η_1 and η_2 respectively. Also let $y_{1,s}$ and $y_{2,s}$ be potential disagreements seen by cycles, i_{η_1}, i_{η_2}, of η_1 and η_2 at stage s, respectively.

The first problem is as follows.

1. At an η_2-expansionary stage s_1, η_k sees $\Theta^B_{\eta_k}(y_{k,s_1})[s_1] \neq W_{\eta_k,s_1}(y_{k,s_1})$ for $k = 1, 2$.

   At this stage, for $k = 1, 2$, both cycles i_{η_k} start their attacks — cycle i_{η_k} enumerates $\gamma_{\beta_{3-k}}(p_{\eta_k})[s_1]$ into D, and also extend the definition of $\Delta^B_{\eta_k\beta_{3-k}}$ for those $z < \gamma_{\beta_{3-k}}(p_{\eta_k})[s_1]$ with $\Delta^B_{\eta_k\beta_{3-k}}(z)$ is currently not defined, and wait for a change of $W_{\beta_{3-k}}$ below $\gamma_{\beta_{3-k}}(p_{\eta_k})[s_1]$.

2. At an η_1-expansionary stage $s_2 > s_1$, cycle i_{η_1} sees a change of W_{β_2}, i.e.

$$W_{\beta_2,s_1} \lceil \gamma_{\beta_2}(p_{\eta_1})[s_1] \neq W_{\beta_2,s_2} \lceil \gamma_{\beta_2}(p_{\eta_1})[s_1].$$

So $\Delta^B_{\eta_1}$ becomes incorrect, and η_1 wants to recover the computation $\Phi^{B \oplus D}_{\eta_1}(y_{1,s_1})$ to the one at stage s_0 (where s_0 is the stage at which $\Theta^B_{\eta_1}(y_{1,s_1})[s_1]$ was defined). η_1's extraction of only $\gamma_{\beta_2}(p_{\eta_1})[s_1]$ from D is not enough for this purpose, as $\gamma_{\beta_1}(p_{\eta_2})[s_1]$ is still in D.

To avoid this, at stage s_2, η_1 also extracts $\gamma_{\beta_1}(p_{\eta_2})[s_1]$ out of D. η_1 can take such an extraction because η_2 is initialized at stage s_2.

The second problem occurs when more strategies are involved, which shows the necessity of enumerating numbers into B, when numbers are removed from D. We already had such discussions in the previous sections.

1. At an η_2-expansionary stage s_0, η_2 sees that $\Theta^B_{\eta_2}(y_{2,s_0})[s_0] \neq W_{\eta_2,s_0}(y_{2,s_0})$. η_2 enumerates $\gamma_{\beta_1}(p_{\eta_2})[s_0]$ into D to start an attack for cycle i_{η_2}, extends the definition of $\Delta^B_{\eta_2\beta_1}$ and waits for a change of W_{β_1} below $\gamma_{\beta_1}(p_{\eta_2})[s_0]$.

2. At an η_1-expansionary stage $s_1 > s_0$, η_1 sees that $\Theta^B_{\eta_1}(y_{1,s_1})[s_1] \neq W_{\eta_1,s_1}(y_{1,s_1})$. η_1 enumerates $\gamma_{\beta_2}(p_{\eta_1})[s_1]$ into D to start an attack for cycle i_{η_1}, extends the definition of $\Delta^B_{\eta_1\beta_2}$ and waits for a change of W_{β_2} below $\gamma_{\beta_2}(p_{\eta_1})[s_1]$.

3. At an η_2-expansionary stage s_2, η_2 sees that

$$W_{\beta_1,s_0} \lceil \gamma_{\beta_1}(p_{\eta_2})[s_0] \neq W_{\beta_1,s_2} \lceil \gamma_{\beta_1}(p_{\eta_2})[s_0].$$

Then η_2 extracts $\gamma_{\beta_1}(p_{\eta_2})[s_0]$ from D to recover the computation $\Phi^{B \oplus D}_{\eta_2}(y_{2,s_0})$ to $\Phi^{B \oplus D}_{\eta_2}(y_{2,s_0})[s_0]$, and hence complete cycle i_{η_2}. η_2 creates a disagreement between $\Phi^{B,D}_{\eta_2}$ and W_{η_2} and η_2 is satisfied. However, it may be true that $\varphi_{\eta_1}(B \oplus D; y_{1,s_1})[s_1] > \gamma_{\beta_1}(p_{\eta_2})[s_0]$, and such an extraction may injure η_1.

4. At an η_1-expansionary stage s_3, η_1 sees that

$$W_{\beta_2,s_1} \lceil \gamma_{\beta_2}(p_{\eta_1})[s_1] \neq W_{\beta_2,s_3} \lceil \gamma_{\beta_2}(p_{\eta_1})[s_1].$$

Then according to the $\mathcal{I}$-strategy, η_1 extracts $\gamma_{\beta_2}(p_{\eta_1})[s_1]$ from D to recover the computation $\Phi^{B \oplus D}_{\eta_1}(y_{1,s_1})$ to $\Phi^{B \oplus D}_{\eta_1}(y_{1,s_1})[s_1]$ to complete cycle i_{η_1}. To recover such a computation, we also need to reenumerate the number $\gamma_{\beta_1}(p_{\eta_2})[s_0]$ into D, which are not allowed as we are constructing D as a d.r.e. set.

To get round this problem, when η_2 extracts $\gamma_{\beta_1}(p_{\eta_2})[s_0]$ from D, it also enumerates s_0, the stage at which cycle i_{η_2} starts an attack on β_1, into B. This enumeration undefines $\Delta^B_{\eta_1\beta_2}(z)$, which is defined at stage s_1. Note that the enumeration of s_0 into B allows us to cancel all the cycles which are started after stage s_0.

So, generally speaking, η acts by running several cycles as before, but with a few of modifications added. Especially, $i2 - j2$ works as follows:

$i2 - j2.$ (Completion on β_j)

    At an η-expansionary stage $s_2 > s_1$, η sees a W_{β_j}-change below $\gamma_{\beta_j}(p_\eta)[s_1]$, then η extracts all the numbers, including $\gamma_\beta(p_\eta)[s_1]$, and all numbers being enumerated into D by strategies below outcome g_j, out of D to recover the computation $\Phi^{B\oplus D}_\eta(y)$ to the one at stage s, where $s < s'$ is the stage at which $\Theta^B_\eta(y)[s']$ is defined (y is the witness chosen by cycle i at stage s'). At the same time, η puts s^1_{n-j} into B to undefine all $\Delta^B_{\eta'\beta'}(z)$ being defined after stage s^1_{n-j} with $\eta' \subset \eta$. Note that this enumeration into B cancels all cycles of η' started after stage s^1_{n-j}. Cycle i passes the threshold p_η for β_j.

As pointed out before, for any z, $\Delta^B_{\eta\beta_j}(z)$ can also be undefined by those η-strategies below the outcome g_j. It will not be a problem, as when $\Delta^B_{\eta\beta_j}(z)$ is defined, there are only finitely many $\mathcal{I}$-strategies below the outcome g_j having been visited, and only these strategies can undefine $\Delta^B_{\eta\beta_j}(z)$. As discussed in the part of the interactions between $\mathcal{I}$-strategies, $\Delta^B_{\eta\beta_j}(z)$ will be eventually defined. Therefore, if η has outcome g_j, then $\Delta^B_{\eta\beta_j}$ is defined as total, and computes W_{β_j} correctly. β_j is satisfied at η.

The whole construction proceeds on a priority tree in a standard way. A full construction will be given in a forthcoming paper.

Acknowledgement

Wu is partially supported by AcRF grants RG58/06 (M52110023) and RG37/09 (M52110101).

References

1. Arslanov, M.M.: Structural properties of the degrees below $\mathbf{0}'$. Dokl. Akad. Nauk SSSR(N. S.) 283, 270–273 (1985)
2. Cooper, S.B.: On a theorem of C. E. M. Yates. Handwritten Notes (1974)
3. Cooper, S.B., Yi, X.: Isolated d.r.e. degrees. Preprint series, vol. 17, 25 p. University of Leeds, Dept. of Pure Math. (1995)
4. Cooper, S.B., Harrington, L., Lachlan, A.H., Lempp, S., Soare, R.I.: The d.r.e. degrees are not dense. Ann. Pure Appl. Logic 55, 125–151 (1991)
5. Li, A., Song, Y., Wu, G.: Universal cupping degrees. In: Cai, J.-Y., Cooper, S.B., Li, A. (eds.) TAMC 2006. LNCS, vol. 3959, pp. 721–730. Springer, Heidelberg (2006)

6. Odifreddi, P.: Classical recursion theory. In: Studies in Logic and the Foundations of Mathematics, vol. 125. North-Holland, Amsterdam (1989)

7. Posner, D.: The upper semilattice of degree $\leq \mathbf{0}'$ is complemented. J. Symbolic Logic 46, 705–713 (1981)

8. Posner, D., Robinson, W.: Degrees joining to $\mathbf{0}'$. J. Symbolic Logic 46, 714–722 (1981)

9. Slaman, T.A., Steel, J.R.: Complementation in the Turing degrees. Journal of Symbolic Logic 54, 160–176 (1989)

10. Soare, R.I.: Recursively Enumerable Sets and Degrees. Springer, Berlin (1987)

11. Wu, G.: Jump operator and Yates degrees. Journal of Symbolic Logic 71, 252–264 (2006)

12. Yates, C.E.M.: Recursively enumerable degrees and the degrees less than $0^{(1)}$, Sets, Models and Recursion Theory. In: Proc. Summer School Math. Logic and Tenth Logic Colloq., Leicester, 1965, pp. 264–271. North-Holland, Amsterdam (1967)

Incomputability in Physics

Giuseppe Longo

Informatique, CNRS - Ecole Normale Supérieure et CREA, Paris, France
http://www.di.ens.fr/users/longo

Abstract. Computability originated from Logic and followed the original path proposed by the founding fathers of the modern foundational analysis of Mathematics (Frege, Hilbert). This theoretical path departed in principle from the contemporary renewed relations between Geometry and Physics. In particular, the key issue of physical measure, as our only access to "reality", is not part of its theoretical frame, in contrast to Physics, since Poincaré, Planck and Einstein. Computability though, by its fine analysis of undecidability, provides a very useful tool for the investigation of "unpredictability" in Physics. Unpredictability coincides with physical randomness, in classical and quantum frames. And an understanding of randomness turns out to be a key component of intelligibility in Physics.

1 The Issue of Physical Measure

Before the crisis of the foundations in Mathematics, that is before the invention of non-euclidean Geometries, Mathematics was secured by the direct link between Euclid's geometry and physical space. Certainty was in the relation between our space of human action and Euclid's constructions, by ruler and compass, in particular once these were embedded by Descartes in abstract spaces and turned into an absolute by Newton. And a theorem proved at our scale could be transferred to the Stars or to Democritus atoms: the sum of internal angles of a triangle, say, would always be 180^0. But Riemann claimed that the relevant space manifolds, where "the cohesive forces among physical bodies could be related to the metrics" [17], were not closed under homotheties (this is Klein's version, to be precise). Thus, the invariance of scale (by homotheties), extending our proofs by ruler and compass to Stars and Atoms, was lost. Finally, with Einstein, Euclid's spaces turned o! ut to be an irrelevant singularity, corresponding to curvature 0, or, at most, a local approximation in the novel geometrization of Physics. In short, the physically crucial meaning of curved spaces demolished the presumed absolute correspondence between intuition as action in human space and physical space-time.

Frege's and Hilbert's response to this major crisis, though different as regards to to meaning and existence, agreed in giving a central role to Arithmetics, with its "absolute laws", away from the "delirium" of the non-Euclidean turn as regards the intuitive relation to space (in Frege's words). Logic or formal systems should categorically (for Frege, if we put in modern terms) or completely (for Hilbert) allow mathematics to be reconstructed.

F. Ferreira et al. (Eds.): CiE 2010, LNCS 6158, pp. 276–285, 2010.

This "royal way out" in the foundation of Mathematics lead it to depart from the relation to physical space and, thus to Physics. In particular, it proposed a foundational culture, in Mathematics, programmatically disregarding the relation to the physical space and, thus, neglecting our forms of "access to the world", by measure, in space and time. Mathematics had to be founded away from our forms of life, from action and space. As a matter of fact, we "access" to physical processes only by measure, from cognitive-perceptive measures as relations to our environment to the very refined tools of Quantum Measure.

And an entire arithmetizing community passed by a key aspect of the revolution that happened in physics at the turn of the XXth century: the novel relevance of physical measure in understanding Nature. When turning back to natural phenomena, some project on them these arithmetic-computational views and their discrete structures of determination, as I will explain below.

Poincaré first understood, by his Three Body Theorem (1890), that the intrinsically approximated measure of the initial conditions, jointly with the nonlinearity of the mathematical description (and gravitational "resonances"), led to unpredictable, though deterministic, continuous dynamics. The Geometry of Dynamical Systems was born and classical randomness since then became understood as deterministic unpredictability (see below). Quantum Mechanics, though differently, also brought the process of measure to the limelight. It introduced an intrinsic randomness as soon as Schrödinger's deterministic evolutions, in Hilbert spaces, are made accessible (are measured in our space and time) and by Planck's h as a lower bound for joint measure of conjugated variables.

Shortly later, the arithmetizing culture, explicitly born against Riemann's and Poincaré's physicalization of Geometry (and geometrization of physics), produced remarkable arithmetic machines. Computability was invented within Logic, for purely foundational purposes, by Herbrand, Gödel, Church, Kleene . . . . And Turing's Logical Computing Machine introduced a further – metaphysical – split from the world: the perfect (Cartesian) dualism of the distinction between hardware and software, based on a revolutionary theoretical (and later practical) separation between the "logic" of a process and its physical realization. Turing's idea was the description of a human in the "least act of computing", or actually of thought. Wasn't Frege's and even Hilbert's project also an analysis of general human reasoning? Weren't spikes, discovered by the new electroencephalography in the '30s, some sort of 0, 1 in the brain?

The physical realization of the Machine preserved Turing's fundamental split: its soft "soul" could act on discrete data types, with no problem as for access, independently from the hardware. No need of physically approximated measure nor random information: the data types are exactly given. Digital processes are exact, evolve away from the world, over a very artificial device, a most remarkable human alphanumeric invention: a discrete state physical process, capable of iterating identically whatever program it is given and, by this, reliable. Nothing like this existed before in the world. So rarely Nature iterates exactly (perhaps a few chemical processes, *in vitro* – we will go back to this).

I insist that the mathematical foundation of the Machine is in its exactness over discrete data types and, thus, in the reliable and identically iteratable interaction between hardware and software. It is in the arithmetical certainty postulated by Frege and Hilbert within Logic, away from the fuzzy measures of classical/relativistic dynamics, from the randomness of Quantum Mechanics. Networks of concurrent computers – distributed in physical space/time – are challenging today this original view.

2   Preliminaries: From Equational Determination to Incompleteness

In a short note of 2001, I suggested that Poincaré's three-body theorem may be considered an epistemological predecessor of Gödel's undecidability result[1], by understanding Hilbert's completeness conjecture as a meta-mathematical revival of Laplace's idea of the predictability of formally (equationally) determined systems. For Laplace, once the equations are given, one can completely derive the future states of affairs (with some – preserved – approximation). Or, more precisely, in "Le système du monde", he claims that the mathematical mechanics of moving particles, one by one, two by two, three by three ... compositionally and *completely* "covers", or makes understandable, the entire Universe. And, as for celestial bodies, by this progressive mathematical integration ... "We should be able to deduce all facts of astronomy", sayd he.

The challenge, for a closer comparison, is that Hilbert was speaking about complete deducibility or *decidability* of purely mathematical *"yes or no"* questions, while unpredictability shows up in the relation between a physical system and a mathematical set of equations (or evolution function).

In order to consistently relate unpredictability to undecidability, one needs to effectivize the dynamical spaces and measure theory (typically, by a computable Lebesgue measure), the loci for dynamic randomness. This allows one to have a sound and purely mathematical treatment of the epistemological issue (and obtain a convincing correspondence between unpredictability and undecidability, see the next section).

A methodological remark. Gödel's and Poincaré's theorems share also a methodological aspect: they both destroy the conjecture of predictability/decidability from inside. Poincaré does not need to refer concretely to a physical process that would not be predictable, by measuring it "before and after". He shows, by a pure analysis of the equations, that the resulting bifurcations and "homoclinic intersections" (intersection points of stable and unstable manifolds or trajectories) lead to deterministic unpredictability (of course, the equations are derived in reference to three bodies in their gravitational fields, similarly as Peano Axioms are invented in reference to the ordered structure of numbers). Similarly, in the *statements* and *proofs* in his 1931 paper, Gödel formally constructs an undecidable sentence, by playing the purely syntactic game,

---

[1] This is more extensively discussed in [12], [13].

with no reference whatsoever to "semantics", "truth" or suchlike, that is to the underlying mathematical structure.

Modern "concrete incompleteness" theorems (that is, Girard's normalization, Paris-Harrington or Friedman-Kruskal theorems, see [10] and [13] for references and discussions) resemble instead Laskar's results of the '90s [9], where "concrete unpredictability" is shown for the solar system, in reference to the *best possible astronomical measures*. Similarly, concrete incompleteness was given by proving (unprovability and) truth over the (standard) model, thus comparing formal syntax and the intended structure.

Philosophically, the incompleteness of our formal (and equational) approaches to knowledge is a general and fundamental epistemological issue. It motivates our permanent need for new science: by inventing new *contentual* – i. e. meaningful – principles and conceptual constructions, we change directions, propose new intelligibilities and meaningfully grasp or organize new fragments of the world. There no such thing as "the final solution to the foundational problem" in mathematics (as Hilbert dreamed – a true nightmare), nor in other sciences.

3   Randomness vs. Undecidability

As mentioned above, classical (physical) randomness is unpredictability of deterministic systems in finite time (dice trajectories are perfectly determined: they follow the Hamiltonian, thus a unique geodetics; yet, they are very sensitive to initial and contour conditions). Now, algorithmic randomness, that is Martin-Löf's (and Chaitin's) number-theoretic randomness, is defined for infinite sequences, [5]. How may this then yield a connection between Poincaré's unpredictability and Gödel's undecidability?

Classical physical randomness is deterministic unpredictability. It is, thus, a matter at the interface "equations/process" and shows up at finite time. Yet, also physical randomness may be expressed as a limit or asymptotic notion and, by this, it may be soundly turned into a purely mathematical issue: this is Birkhoff's ergodicity (for any observable, limit time averages coincide with space averages: an equality of two infinite sums or integrals, see [14]). And this sense applies in (weakly chaotic) dynamical systems, within the frame of Poincaré's geometry of dynamical systems.

As for algorithmic randomness, Martin-Löf randomness is a "Gödelian" notion of randomness, since it is based on recursion theory and yields a strong form of undecidability for infinite 0-1 sequences (in short, a sequence is random if it passes all *effective statistical tests*; as a consequence, it contains no infinite r.e. subsequences). Recently, under Galatolo's and my supervision, M. Hoyrup and C. Rojas proved that dynamic randomness (a la Poincaré, thus, but at the purely mathematical limit, in the ergodic sense), in suitable *effectively given* measurable dynamical systems, is equivalent to (a generalization of) Martin-Löf randomness (Schnorr's randomness). This is a non-obvious result, based also on a collaboration with P. Gacs, developed in those two parallel doctoral dissertations (defended in June 2008) [8].

As for quantum randomness now, note that, because of entanglement, it mathematically differs from classical: if two classical dice interact and then separate, the probabilistic analysis of their values are independent. When two quanta interact and form a "system", they can no longer be separated: measures on them give correlated probabilities of the results (mathematically, they violate Bell's inequalities), see [2] for a comparative introduction.

Algorithmic randomness provides a close analysis of classical randomness: how can this mathematics of discrete structures tell us more about randomness, in general?

4   Discrete vs. Continua

One of the ideas developed in a book, ([1], and in several papers with Francis Bailly and Thierry Paul, two physicists, is that the mathematical structures, constructed for the intelligibility of physical phenomena, according to their continuous (mostly in physics) or discrete nature (generally in computing), may propose different understandings of Nature[2]. In particular, the "causal relations", as structures of intelligibility (we "understand Nature" by them), are mathematically related to the use of the continuum or the discrete and may deeply differ (in modern terms: they induce different symmetries and symmetry-breakings).

But what discrete (mathematical) structures are we referring to? There is one clear mathematical definition of "discrete": a structure is *discrete* when the discrete topology on it is "natural". Of course, this is not a formal definition, but in mathematics we all know what "natural" means. For example, one can endow Cantor's real line with the discrete topology, but this is not "natural" (you do not do much with it); on the other hand, the integer numbers or a digital data base are naturally endowed with the discrete topology (even though one may have good reasons to work with them also under a different structuring).

Church's thesis, introduced in the 1930s after the functional equivalence proofs of various formal systems for computability, concerns only computability over integers or discrete data types. As such, it is an extremely robust thesis: it ensures that any sufficiently expressive *finitistic formal system* over integers (a Hilbertian-type logico-formal system) computes exactly the recursive functions, as defined by Herbrand, Gödel, Kleene, Church, Turing .... This thesis therefore emerged within the context of mathematical logic, as grounded on formal systems for arithmetic computations, on discrete data types.

The very first question to ask is the following: what happens if we extend the formal framework? If we want to refer to continuous (differentiable) mathematical structures, the extension to consider is to the computable real numbers, see [16]. Are the various formalisms for computability over real numbers equivalent, when they are maximal? An affirmative answer could suggest an extension of Church thesis to computability on "continua". Of course, the computable reals

---

[2] The enlightening collaboration with Francis, has been fundamental. Francis recently passed away: a recorded Colloquium in his memory may be accessed from my web page.

are countably many, but they are dense in the "natural" topology over Cantor's reals. As we shall see below, there is a crucial difference.

Posing this question, we get closer to current physics, since it is within spatial and often also temporal continuity that we represent dynamical systems, that is, most mathematical models for (classical) physics. This does not imply that the World is continuous, but only that, since Newton and Leibniz, we have said many things by continuous tools as very well specified by Cantor (but his continuum is not the only possible one: Lawvere and Bell, say, proposed a topos-theoretic one, without points, [4]).

Now, as for this equivalence of formalisms, which is at the heart of Church's thesis, there remains very little when passing to computability over real numbers: the theories proposed are demonstrably different in terms of computational expressiveness (the classes of defined functions). The various systems (recursive analysis, first developed by Lacombe and Grezgorzcyk, in 1955-57; the Blum, Shub and Smale, BSS, system; the Moore-type recursive real functions; different forms of "analog" systems, such as threshold neurones, the GPAC ...) yield different classes of "continuous" computable functions. Some recent work established links, reductions between the various systems (more precisely: pairwise relations between subsystems and/or extensions). Yet, the full equivalence as in the discrete case is lost. Moreover, and this is crucial, these systems have no "universal function" in Turing's sense. This function is constructed by an isomorphism between spaces of different dimension. More generally, in computability on the discrete, the work spaces may be of any finite dimension: they are all effectively isomorphic or the "Cartesian dimension" does not matter!

This is highly unsuitable for physics. First, dimensional analysis is a fundamental tool (one cannot confuse energy with force, nor with the square of energy ...). Second, dimension is a topological invariant, in all space manifolds for classical and relativistic physics. That is, take physical measure, an interval, as a "natural" starting point for the metric (thus the interval or "real" topology), then you can *prove* that, if two such spaces have isomorphic open subsets, they have the same dimension. The topological invariance of dimension, on physically meaningful topologies, is a very simple, yet beautiful correspondence between mathematics and physics.

These facts weaken the computational approaches to the analysis of *physical* invariants, over continua, as two fundamental *computational* invariants are lost: equivalence (that is, the grounds for Church Thesis) and universality.

In summary, in discrete computability, a cloud of isolated points has no dimension, per se, and, for all theoretical purposes, one may encode them on a line. When you have dimension back, in computability over continua, where the trace of the interval topology maintains good physical properties, you lose the universal function and the equivalence of systems. Between the theoretical world of discrete computability and physico-mathematical continua there is a huge gap. While I believe that one should do better than Cantor as for continua, I would not give a penny for a physical theory where dynamics takes place only on discrete spaces, departing from physical measure, dimensional analysis and the general relevance of dimensions in physics (again, from heat propagation to mean field theory, to relativity theory ... space dimension is crucial). The analysis over "computable continua" provides a more interesting frame for physics, adds relevant information, but loses two key invariants of computations over the discrete.

By the way, are the main physical constants, G, c, h, *computable* (real numbers)? Of course, it depends on the choice of the reference system and the metrics. So, fix $h = 1$. Then, you have to renormalize all metrics and re-calculate, by equations, dimensional analyses *and* physical measure, G and c. But physical measure will always give an interval, as we said, or, in quantum frame, the probability of a value. If one interprets the classical measure interval as a Cantorian continuum (the best way, so far, to grasp fluctuations), then ... where are G and c? Computable reals form a dense subset of Lebesgues measure 0, with plenty of gaps and no jumps. Why should (the mathematical understanding of) fluctuations avoid falling in gaps and jump from computable real to computable real? Cristian Calude and I conjecture instead that random (thus highly incomputable) reals are a better structure of intelligibility for non-observable events.

Yet, the most striking mistake of many "computationalists" is to say: but, if "physics is not fully computable", then, some physical processes would super-compute (that is, "compute" non-computable functions)! No, this is not the point. Most physical processes, simply do not *define* a mathematical function. And the issue, again, is our only form of access to the (physical) "world": *measure*. In order to force a classical process to define a function, you have to fix a time for input, associate a (rational) number to the interval of measure and ... let the process go. Then you wait for the output time and measure again. In order, for the process, to define $f(x) = y$, at a rational input x it must always associate a rational output y. But if you restart, say, your physical double pendulum on x, that is within the interval of the measure which gave you x, a minor (thermal, say) fluctuation, *below that interval* defined by x, will yield a different observable result y' after very short time. So, a good question would be, instead: consider a physical process that *defines* a function, is this function computable?

The idea then is that the process should be sufficiently insensitive to initial conditions (some say: robust) as to actually define a function. But, then one should be able to partition the World in little cubes of the smallest size, according to the best measure as for insensitivity (fluctuations below that measure do not affect the dynamics). If the Accessible World is considered finite (but ... is it? What about Riemann's sphere?), then one can make a finite list out of the input-output relation established by the given process. This is a "program": is it compressible?

As for the relevance of the discrete, quantum mechanics started exactly by the discovery of a key (and unexpected) discretization of energy absorption or emission spectra of atoms. Then, a few dared to propose a discrete lower bound to measure of *action*, that is of the product (energy × time). It is this physical dimension that bares a discrete structure. Clearly, one can then compute, by assuming the relativistic maximum for the speed of light, a Planck's length and time. But in no way space and time are thus discretized in small "quantum boxes or cubes". And this is the most striking and crucial feature of quantum mechanics: the "systemic" or entanglement effects, which yield inseparability of observables. No discrete space topology is natural. That is, these quantum effects are the opposite of a discrete, separated organization of space, while being at the core of its scientific originality. In particular, they motivate quantum computing (as well as our analysis of quantum randomness above). As a matter of fact, Thierry Paul and I claim that the belief in an absolutely separable topology of space continua is Einstein's mistake in EPR [7], where entanglement is first examined (and considered impossible).

In summary, continua, Cantorian or topos theoretic, take care rather well (but they are not an absolute) of the approximated nature of physical measure, which is represented as an interval: the unknowable fluctuation is within the interval. And (physical) measure is our only form to access "reality". The arithmetizing foundation of mathematics went along another (and very fruitful) direction, based on perfectly accessible data types.

5   The Originality of the Discrete State Machine

As I mentioned above, the Discrete State Alphanumeric Machine that compute is a remarkable and very original human invention, based on a long history. As hinted in [11], this story begins with the invention of the alphabet, probably the oldest experience of discretization. The continuous song of speech, instead of being captured by the design of concepts and ideas (by recalling "meaning", like in ideograms), is discretized by annotating phonetic pitches, an amazing idea (the people of Altham, in Mesopotamia, 3300 B.C.). Meaning is reconstructed by the sound, which acts as a compiler, either loud or in silence (but only after the IV century A.D. we learned to read "within the head"!).

The other key passage towards alphanumeric discretization is the invention of a discrete coding structure. This originated with Gödel-numbering, an obvious practice now, but another as remarkable as artificial idea. Turing's work followed: the Logical Computing Machine (LCM), as he first called it, is at the core of computing (right/left, $0, 1 \ldots$). Of course, between the alphabet and Turing, you also have Descartes "discretization' of thought (stepwise reasoning, along a discrete chain of intuitive certitudes ...) and much more.

When, in the late '40s, Turing works again in physics, he changes the name to his LCM: in [19] and [20], he refers to it as ! !
!
Discrete State. Machine (this is what matters as for its physical behavior). And twice in his 1950 paper (the "imitation game"), he calls it "Laplacian". Its evolution is theoretically predictable, even if there may be practical unpredictability (too long programs to be grasped, says he).

So, we invented an incredibly stable processor, which, by working on discrete data types, does what it is expected to do. And it iterates, very faithfully, I insist, this is its key feature. Primitive recursion and portability of software are forms of iterability: iteration and update of a register, do what you are supposed to do, respectively, even in slightly different contexts, over and over again. For example, program the evolution function of the most chaotic strange attractor you know. Push "restart": the digital evolution, by starting on the same initial digits, will follow exactly the same trajectory. This makes no physical sense, but it is very useful (also in meteorology: you may restart your turbulence, exactly, and try to better understand how it evolves ...). Of course, one may imitate unpredictability by some pseudo-random generator or by ... true physical randomness, added ad hoc. But this is cheating the observer, in the same way Turing's *imitation* of a woman's brain is meant to cheat the observer, not to "model" the brain. He says this explicitly, all the while working, in his 1952 paper, at a *model* of morphogenesis, as (non-)linear dynamics. The brain activity, says he, may depend on fluctuations below measure, not his DSM (see [11] for a closer analysis and, of course, Turing's two papers, which should always be read simultaneously). Only a mathematical model tries to propose a "structure of determination" (equations for action, reaction, diffusion in the 1952 paper). Observe, finally, that our colleagues in networks and concurrency are so good that also programming in concurrent networks is reliable: programs do what they are supposed to do, they iterate and ... give you the web page you want, identically, one thousands time, one million times. And this is hard, as physical space-time, which we better understand by continua and continuous approximations, steps in, yet still on discrete data types, which allow iteration. Of course, identical iteration is the opposite of randomness (many define a process to be random when, iterated on the "same" – physical – initial conditions, it follows a different evolution).

Those who claim that the Universe is a big digital computer, miss the originality of this machine of ours, its history, from the alphabet to Hilbert's formal systems, to the current work in concurrency in networks and its reliability (as iteratability) as key objective. When we construct computers, we make a remarkable achievement, by producing a reliable, thus programmable, physical, but artificial device, far away from the natural world, iterating as we wish and any time we wish, even in networks. One should not miss the principles that guided this invention, as well as the principles by which we understand physical dynamics, since Poincaré.

6   Incomputability in Biology

For lack of space in the proceedings, the final sections of this paper are missing here: they may be found in the complete version (*downloadable*: http://www.di.ens.fr/users/longo).

References

1. Bailly, F., Longo, G.: Mathematics and Natural Sciences. In: The Physical Singularity of Life Phenomena. Imperial College Press/World Scientific (2010) (In French: Hermann, Paris, 2006)

2. Bailly, F., Longo, G.: Randomness and Determination in the interplay between the Continuum and the Discrete. Mathematical Structures in Computer Science 17(2), 289–307 (2007)
3. Braverman, M., Yampolski, M.: Non-computable Julia sets. Journ. Amer. Math. Soc. 19, 551–578 (2006)
4. Bell, J.: A Primer in Infinitesimal Analysis. Cambridge U.P., Cambridge (1998)
5. Calude, C.: Information and randomness. Springer, Berlin (2002)
6. Cooper, B.S., Odifreddi, P.: Incomputability in Nature. In: Cooper, et al. (eds.) Computability and Models - Perspectives East and West, pp. 137–160. Kluwer Academic/Plenum, New York (2003)
7. Einstein, A., Podolsky, B., Rosen, N.: Can Quantum-Mechanical Description of Physical Reality be Considered complete? Phys. Rev. 41, 777 (1935)
8. Gacs, P., Hoyrup, M., Rojas, C.: Randomness on Computable Metric Spaces: A dynamical point of view. In: 26th International Symposium on Theoretical Aspects of Computer Science, STACS 2009 (2009)
9. Laskar, J.: Large scale chaos in the Solar System. Astron. and Astrophys 287, 9–12 (1994)
10. Longo, G.: Reflections on Incompleteness. In: Callaghan, P., Luo, Z., McKinna, J., Pollack, R. (eds.) TYPES 2000. LNCS, vol. 2277, pp. 160–180. Springer, Heidelberg (2002)
11. Longo, G.: Critique of Computational Reason in the Natural Sciences. In: Gelenbe, E., Kahane, J.-P. (eds.) Fundamental Concepts in Computer Science. Imperial College Press/World Sci. (2008)
12. Longo, G.: From exact sciences to life phenomena: following Schroedinger and Turing on Programs, Life and Causality. Special issue of Information and Computation 207(5), 543–670 (2009)
13. Longo, G.: Interfaces of Incompleteness, downloadable, Italian version. La Matematica, Einaudi (2010); French version, Les Mathématiques, Editions du CNRS (2011)
14. Petersen, K.: Ergodic Theory. Cambridge Univ. Press, Cambridge (1983)
15. Pilyugin, S.Y.: Shadowing in Dynamical Systems. Springer, Heidelberg (1999)
16. Pour-El, M.B., Richards, J.I.: Computability in analysis and physics. In: Perspectives in Mathematical Logic. Springer, Berlin (1989)
17. Riemann, B.: On the hypothesis which lie at the basis of Geometry (1854); English Transl. by Clifford, W., Nature (1873)
18. Svozil, K.: Randomness and undecidability in Physics. World Sci., Singapore (1993)
19. Turing, A.M.: Computing Machines and Intelligence. Mind LIX(236), 433–460 (1950)
20. Turing, A.M.: The Chemical Basis of Morphogenesis. Philo. Trans. Royal Soc. B237, 37–72 (1952)

Approximate Self-assembly
of the Sierpinski Triangle*

Jack H. Lutz and Brad Shutters

Department of Computer Science, Iowa State University
lutz@cs.iastate.edu, shutters@iastate.edu

Abstract. Winfree introduced the Tile Assembly Model in order to
study the nanoscale self-assembly of DNA crystals. Lathrop, Lutz, and
Summers proved that the Sierpinski triangle **S** cannot self-assemble in
the "strict" sense in which tiles are not allowed to appear at positions
outside the target structure. Here we investigate the strict self-assembly
of sets that approximate **S**. We show that every set that does strictly self-
assemble disagrees with **S** on a set with fractal dimension at least that of
S (≈ 1.585), and that no subset of **S** with fractal dimension greater than
1 strictly self-assembles. We show that our bounds are tight by presenting a strict self-assembly that adds communication fibers to the fractal
structure without disturbing it. To verify this strict self-assembly we develop a generalization of the local determinism method of Soloveichik
and Winfree to tile assembly systems that use a blocking technique.

1 Introduction

Self-assembly is a process in which simple objects autonomously combine to
form complex structures as a consequence of specific, local interactions among
the objects themselves. It occurs spontaneously in nature as well as in engineered
systems and is a fundamental principle of structural organization at all scales.
Since the pioneering work of Seemen [9], the self-assembly of DNA molecules has
developed into a field with rich interactions between the theory of computing
(the information processing properties of DNA) and geometry (the structural
properties of DNA), and with many applications to nanotechnology [10].

Winfree [12] introduced the Tile Assembly Model (TAM) as a mathematical
model of self-assembly in order to study the growth of complex (typically aperi-
odic) DNA crystals. The TAM models the self-assembly of unit square *tiles* that
can be translated, but not rotated. A tile has a *glue* on each side that is made up
of a *color* and an integer *strength* (usually 0, 1, or 2). Two tiles with the same
glue on each side are of the same *tile type*. Two tiles placed next to each other
interact if the glues on their abutting sides match in both color and strength.
A *tile assembly system* (TAS) is a finite set of tile types, a single tile for the
seed, and a specified integer *temperature* (usually 2). The process starts with the
seed tile placed at the origin and growth occurs by single tiles attaching one at

---

* This research was supported in part by NSF grants 0652569 and 0728806.

F. Ferreira et al. (Eds.): CiE 2010, LNCS 6158, pp. 286–295, 2010.

a time. A tile can attach at a site where the summed strength of the glues on sides that interact with the existing structure is at least the temperature.

This paper is concerned with the self-assembly of fractals. Structures that self-assemble in naturally occurring biological systems are often fractals of low dimension, which have advantages for materials transport, heat exchange, information processing, and robustness [5]. Fractals are normally bounded and have the same detail at arbitrarily small scales. But, the TAM models the bottom-up self-assembly of tiles which are discrete objects. So, structures that self-assemble in the TAM are fundamentally discrete. Thus, we consider the self-assembly of discrete fractals which are unbounded and have the same detail at arbitrarily large scales. There are two main notions of the self-assembly of a fractal. In *weak self-assembly*, one typically causes a two-dimensional surface to self-assemble with the desired fractal structure appearing as a labeled subset of the surface. In contrast, *strict self-assembly* requires only the fractal structure, and nothing else, to self-assemble. For many purposes, strict self-assembly is needed in order to achieve the above mentioned advantages of fractal structures.

The Sierpinski triangle **S** is a canonical "toy" problem for self-assembly. Winfree [12] showed that **S** weakly self-assembles, and Rothemund, Papadakis, and Winfree [7] achieved a molecular implementation of this self-assembly. Lathrop, Lutz, and Summers [5] proved that **S** cannot strictly self-assemble. It is an open question whether any self-similar fractal strictly self-assembles . Thus, techniques are needed to approximate self-similar fractals with strict-self-assembly. The only previously known technique, introduced by Lathrop, Lutz, and Summers [5], and generalized by Patitz and Summers [6], enables strict self-assembly by adding communication fibers that shift successive stages of the fractal causing the result to only visually resemble, but not contain, the intended fractal structure.

In this paper we address a quantitative question: given that **S** cannot strictly self-assemble, how closely can strict self-assembly approximate **S**? That is, if X is a set that *does* strictly self-assemble, how small can the fractal dimension of the symmetric difference $X \triangle S$ be? Our first main theorem says that the fractal dimension of $X \triangle S$ is at least the fractal dimension of **S**. To gain further insight, we restrict our attention to subsets of **S** and show that here the limitation is even more severe. Any subset of the Sierpinski triangle that strictly self-assembles must have fractal dimension 0 or 1. Roughly speaking, the axes that bound **S** form the largest subset of **S** that strictly self-assembles.

Our second main theorem shows that our first main theorem is tight, even when restricted to supersets of **S**. To prove this we demonstrate the existence of a set X such that $S \subseteq X$, the fractal dimension of $X \triangle S$ is the fractal dimension of **S**, and X strictly self-assembles. What we have achieved here is a means of *fibering* **S** *in place*, i.e., adding the needed communication fibers (the set $X \setminus S$) without disturbing the set **S**.

The local determinism method of Soloveichik and Winfree [11] is a common technique for proving correctness of a TAS. However, the TAS in the proof of our second main theorem is not locally deterministic. We thus introduce *conditional determinism*, a generalization of local determinism, to verify this TAS.

2   Preliminaries

We work in the discrete Euclidean plane $\mathbb{Z}^2$. We write U_2 for the set of all *unit vectors* in $\mathbb{Z}^2$. We often refer to the elements of U_2 as the cardinal directions, and write $\vec{u}_N$ for $(0,1)$, $\vec{u}_S$ for $(0,-1)$, $\vec{u}_E$ for $(1,0)$, and $\vec{u}_W$ for $(-1,0)$.

Let X and Y be sets. We write $[X]^2$ for the set of all 2-element subsets of X. For a partial function $f : X \dashrightarrow Y$, we write $f(x)\downarrow$ if $x \in \operatorname{dom} f$ and $f(x)\uparrow$ otherwise. For a *Boolean* expression ϕ, $[\![\phi]\!] = 1$ if ϕ is true, and 0 if ϕ is false.

All graphs here are undirected graphs of the form $G = (V, E)$, where $V \subseteq \mathbb{Z}^2$ is a set of *vertices* and $E \subseteq [V]^2$ is a set of *edges*. A *grid graph* is a graph where each $\{\vec{m}, \vec{n}\} \in E$ satisfies $\vec{m} - \vec{n} \in U_2$. If E contains every $\{\vec{m}, \vec{n}\} \in [V]^2$ such that $\vec{m} - \vec{n} \in U_2$, we say it is the *full grid graph* on V, written $G_V^{\#}$. A *cut* of a graph is a partition of V into two subsets. A *binding function* on a graph is a function $\beta : E \to \mathbb{N}$. If β is a binding function on G and C is a cut of G, then the *binding strength of β on C* is $\beta_C = \sum \{\beta(e) \mid e \in E, e \cap C_0 \neq \emptyset, \text{ and } e \cap C_1 \neq \emptyset\}$, and the *binding strength of β on G* is $\beta_G = \min \{\beta_C \mid C \text{ is a cut of } G\}$. A *binding graph* is an ordered triple (V, E, β), where β is a binding function on (V, E). For $\tau \in \mathbb{N}$, a binding graph (V, E, β) is *τ-stable* when $\beta_{(V,E)} \geq \tau$.

The most commonly used fractal dimension for discrete fractals is ζ-dimension [2]. For $A \subseteq \mathbb{Z}^2$ and $I \subseteq [0,\infty)$, let $A_I = \{\vec{m} \in A \mid \|\vec{m}\| \in I\}$, where $\|\vec{m}\|$ is the Euclidean distance from the origin to $\vec{m}$. The *ζ-dimension* of a set $A \subseteq \mathbb{Z}^2$ is

$$\operatorname{Dim}_\zeta(A) = \limsup_{n\to\infty} \frac{\log_2 |A_{[0,n]}|}{\log_2 n}. \tag{2.1}$$

Note that ζ-dimension has the following properties of a fractal dimension [2].

Observation 2.1. *Let $A, B \subseteq \mathbb{Z}^2$. Then,*

(1) $A \subseteq B \implies \operatorname{Dim}_\zeta(A) \leq \operatorname{Dim}_\zeta(B)$ (monotonicity), and

(2) $\operatorname{Dim}_\zeta(A \cup B) = \max \{\operatorname{Dim}_\zeta(A), \operatorname{Dim}_\zeta(B)\}$ (stability).

Formally, the discrete Sierpinski triangle $\mathbf{S}$ is a set of points in $\mathbb{Z}^2$. Let $V = \{(1,0), (0,1)\}$ and define the sets $\mathbf{S}_0, \mathbf{S}_1, \ldots$ by the recursion

$$\mathbf{S}_0 = \{(0,0)\} \quad \text{and} \quad \mathbf{S}_{i+1} = \mathbf{S}_i \cup (\mathbf{S}_i + 2^i V) \tag{2.2}$$

where $A + cB = \{\vec{m} + c\vec{n} \mid \vec{m} \in A \text{ and } \vec{n} \in B\}$. Then, $\mathbf{S} = \bigcup_{i=0}^\infty \mathbf{S}_i$. We often refer to $\mathbf{S}_i$ as the ith stage of $\mathbf{S}$. Note that $\mathbf{S}$ can also be defined as the nonzero residues modulo 2 of Pascal's triangle [1]. It is also a numerically self-similar fractal [4]. It is easy to see that $|\mathbf{S}_n| = 3^n$.

Observation 2.2. $\operatorname{Dim}_\zeta(\mathbf{S}) = \log_2 3 \approx 1.585$.

Theorem 2.3 (Lathrop, Lutz, and Summers [5]). $\mathbf{S}$ *cannot strictly self-assemble in the Tile Assembly Model.*

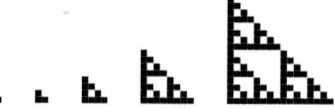

Fig. 1. Stages 0 through 4 of the discrete Sierpinski triangle

2.1   The Tile Assembly Model

We now review the Tile Assembly Model [12,8]. Our notation follows that of [5].

A *tile* t is a unit square that can be translated, but not rotated, so it has a well defined "side $\vec{u}$" for each $\vec{u} \in U_2$. Each side $\vec{u}$ of t has a *glue* $t(\vec{u}) = (\mathrm{col}_t(\vec{u}), \mathrm{str}_t(\vec{u}))$ where $\mathrm{col}_t(\vec{u})$ is the glue *color*, and $\mathrm{str}_t(\vec{u}) \in \mathbb{N}$ is the glue *strength*. Two tiles with the same glue on each side are of the same *tile type*.

Let T be a finite set of tile types. A T-configuration is a partial function $\alpha : \mathbb{Z}^2 \dashrightarrow T$. For $\vec{m}, \vec{n} \in \mathrm{dom}\,\alpha$, the tiles at locations $\vec{m}$ and $\vec{n}$ *interact* with *strength*

$$\mathrm{str}_\alpha(\vec{m}, \vec{n}) = [\![\vec{n} - \vec{m} \in U_2]\!] \cdot \mathrm{str}_{\alpha(\vec{m})}(\vec{n} - \vec{m}) \cdot [\![\alpha(\vec{m})(\vec{n} - \vec{m}) = \alpha(\vec{n})(\vec{m} - \vec{n})]\!].$$

The *binding graph* of α is $G_\alpha = (\mathrm{dom}\,\alpha, E, \beta)$, where $E = \{\{\vec{m}, \vec{n}\} \in [V]^2 \mid \mathrm{str}_\alpha(\vec{m}, \vec{n}) > 0\}$ and $\beta(\{\vec{m}, \vec{n}\}) = \mathrm{str}_\alpha(\vec{m}, \vec{n})$ for all $\{\vec{m}, \vec{n}\} \in E$. For $\tau \in \mathbb{N}$, α is τ-*stable* if G_α is τ-stable. We write $\mathcal{A}_T^\tau$ for the set of all τ-stable T-configurations. Let $\alpha, \alpha' \in \mathcal{A}_T^\tau$. If $\mathrm{dom}\,\alpha \subseteq \mathrm{dom}\,\alpha'$ and $\alpha(\vec{m}) = \alpha'(\vec{m})$ for all $\vec{m} \in \mathrm{dom}\,\alpha$, then α is a *subconfiguration* of α' and we write $\alpha \sqsubseteq \alpha'$. If $|\mathrm{dom}\,\alpha' \setminus \mathrm{dom}\,\alpha| = 1$, then α' is a *single-tile-extension* of α and we write $\alpha' = \alpha + (\vec{m} \mapsto t)$ where $\{\vec{m}\} = \mathrm{dom}\,\alpha' \setminus \mathrm{dom}\,\alpha$ and $t = \alpha'(\vec{m})$. For each $t \in T$, the τ-t-*frontier* of α is

$$\partial_t^\tau \alpha = \left\{ \vec{m} \in \mathbb{Z}^2 \setminus \mathrm{dom}\,\alpha \mid \left(\Sigma_{\vec{u} \in U_2}\,\mathrm{str}_{\alpha + (\vec{m} \mapsto t)}(\vec{m}, \vec{m} + \vec{u}) \right) \geq \tau \right\}, \text{ and}$$

the τ-*frontier* of α is $\partial^\tau \alpha = \bigcup_{t \in T} \partial_t^\tau \alpha$. We say α is *terminal* when $\partial^\tau \alpha = \emptyset$.

A *tile assembly system* (TAS) is an ordered triple $\mathcal{T} = (T, \sigma, \tau)$ where T is a finite set of tile types, the *seed assembly* $\sigma \in \mathcal{A}_T^\tau$ is such that $\mathrm{dom}\,\sigma = \vec{0}$, and $\tau \in \mathbb{N}$ is the *temperature*. An *assembly sequence* in $\mathcal{T}$ is a sequence $\vec{\alpha} = (\alpha_i \mid 0 \leq i < k)$ where $\alpha_0 \in \mathcal{A}_T^\tau$, $k \in \mathbb{Z}^+ \cup \{\infty\}$ and for each $0 \leq i < k$, $\alpha_{i+1} = \alpha_i + (\vec{m} \mapsto t)$ for some $t \in T$ and $\vec{m} \in \partial_t^\tau \alpha_i$. The *result* of $\vec{\alpha}$, written $\mathrm{res}\,\vec{\alpha}$, is the unique $\alpha \in \mathcal{A}_T^\tau$ satisfying $\mathrm{dom}\,\alpha = \bigcup_{0 \leq i < k} \mathrm{dom}\,\alpha_i$ and for each $0 \leq i < k$, $\alpha_i \sqsubseteq \alpha$. We write $\alpha_0 \longrightarrow \mathrm{res}\,\vec{\alpha}$ if an assembly sequence from α_0 to $\mathrm{res}\,\vec{\alpha}$ exists. The set of *producible assemblies* is $\mathcal{A}[\mathcal{T}] = \{\alpha \in \mathcal{A}_T^\tau \mid \sigma \longrightarrow \alpha\}$ and the set of *terminal assemblies* is $\mathcal{A}_\square[\mathcal{T}] = \{\alpha \in \mathcal{A}[\mathcal{T}] \mid \partial^\tau \alpha = \emptyset\}$. $\mathcal{T}$ is *directed* if $|\mathcal{A}_\square[\mathcal{T}]| = 1$. A set X *strictly self-assembles in* $\mathcal{T}$ if every $\alpha \in \mathcal{A}_\square[\mathcal{T}]$ satisfies $\mathrm{dom}\,\alpha = X$. We say X *strictly self-assembles* if X strictly self-assembles in some TAS.

Let $\mathcal{T} = (T, \sigma, \tau)$ be a TAS, $\vec{\alpha} = (\alpha_i \mid 0 \leq i < k)$ be an assembly sequence in $\mathcal{T}$, and $\alpha = \mathrm{res}\,\vec{\alpha}$. For each $\vec{m} \in \mathbb{Z}^2$, the $\vec{\alpha}$-*index* of $\vec{m}$ is $i_{\vec{\alpha}}(\vec{m}) = \min\{i \in \mathbb{N} \mid \vec{m} \in \mathrm{dom}\,\alpha_i\}$. If $\vec{m}, \vec{n} \in \mathrm{dom}\,\alpha$ and $i_{\vec{\alpha}}(\vec{m}) < i_{\vec{\alpha}}(\vec{n})$, we say $\vec{m}$ *precedes* $\vec{n}$ *in* $\vec{\alpha}$, and write $\vec{m} \prec_{\vec{\alpha}} \vec{n}$. For each $\vec{m} \in \mathrm{dom}\,\alpha$, define [11] the sets $\mathrm{IN}^{\vec{\alpha}}(\vec{m}) = \{\vec{u} \in U_2 \mid \vec{m} + \vec{u} \prec_{\vec{\alpha}} \vec{m} \text{ and } \mathrm{str}_{\alpha_{i_{\vec{\alpha}}(\vec{m})}}(\vec{m}, \vec{m} + \vec{u}) > 0\}$, and $\mathrm{OUT}^{\vec{\alpha}}(\vec{m}) = \{\vec{u} \in U_2 \mid -\vec{u} \in \mathrm{IN}^{\vec{\alpha}}(\vec{m} + \vec{u})\}$. For $X \subseteq \mathrm{dom}\,\alpha$, $\alpha \restriction X$ is the unique T-configuration satisfying $(\alpha \restriction X) \sqsubseteq \alpha$ and $\mathrm{dom}(\alpha \restriction X) = X$. Then, $\vec{\alpha}$ is *locally deterministic* [11] if

(1) for all $\vec{m} \in \mathrm{dom}\,\alpha \setminus \mathrm{dom}\,\alpha_0$, $\sum_{\vec{u} \in \mathrm{IN}^{\vec{\alpha}}(\vec{m})} \mathrm{str}_{\alpha_{i_{\vec{\alpha}}(\vec{m})}}(\vec{m}, \vec{m} + \vec{u}) = \tau$,

(2) for all $\vec{m} \in \mathrm{dom}\,\alpha \setminus \mathrm{dom}\,\alpha_0$ and $t \in T \setminus \{\alpha(\vec{m})\}$,

$$\vec{m} \notin \partial_t^\tau(\alpha \restriction (\mathrm{dom}\,\alpha \setminus (\{\vec{m}\} \cup (\vec{m} + \mathrm{OUT}^{\vec{\alpha}}(\vec{m}))))),$$

(3) and $\partial^\tau \alpha = \emptyset$.

A TAS is *locally deterministic* if it has a locally deterministic assembly sequence. Soloveichik and Winfree [11] proved every locally deterministic TAS is directed.

3   Limitations on Approximating the Sierpinski Triangle

In this section we present our first main theorem. We show that every set that strictly self-assembles disagrees with $\mathbf{S}$ on a set with ζ-dimension at least that of $\mathbf{S}$. We then show that for subsets of $\mathbf{S}$, the limitation is even more severe.

Lemma 3.1. *Let T be a set of tile types and $\tau \in \mathbb{N}$. If $\alpha \in \mathcal{A}_T^\tau$ and $\operatorname{dom}\alpha = \mathbf{S}_n$ for some $n \in \mathbb{N}$, then for each $\vec{m} \in \operatorname{dom}\alpha$ and $\vec{u} \in U_2$,*

$$\vec{m} + \vec{u} \in \operatorname{dom}\alpha \implies \alpha(\vec{m})(\vec{u}) = \alpha(\vec{m} + \vec{u})(-\vec{u}) \text{ and } \operatorname{str}_{\alpha(\vec{m})}(\vec{u}) \geq \tau.$$

Lemma 3.2. *If $\mathbf{S}_n$ strictly self-assembles in a TAS (T, σ, τ), then $|T| \geq 2^{n+1}-1$.*

Proof. Assume the hypothesis with $n \in \mathbb{N}$ and TAS $\mathcal{T} = (T, \sigma, \tau)$ as witness. Let $\alpha \in \mathcal{A}_\square[\mathcal{T}]$. If $n = 0$ the lemma is trivially true, so assume $n > 1$. Let $A = A_h \cup A_v$, where $A_h = \{(i,0) \mid 0 \leq i < 2^n\}$ and $A_v = \{(0,i) \mid 0 \leq i < 2^n\}$. Let $T_A = \{\alpha(\vec{m}) \mid \vec{m} \in A\}$. It suffices to show that $|T_A| \geq 2^{n+1} - 1$. By equation (2.2) and $\alpha \in \mathcal{A}_\square[\mathcal{T}]$, $\vec{m} \in \operatorname{dom}\alpha$ for all $\vec{m} \in A$. Then, there exists $\vec{m}, \vec{n} \in A$ such that $\vec{m} \neq \vec{n}$ and $\alpha(\vec{m}) = \alpha(\vec{n})$. Either $\vec{m}, \vec{n} \in A_h$, $\vec{m}, \vec{n} \in A_v$, or $\vec{m} \in A_h \setminus \{\vec{0}\}$ and $\vec{n} \in A_v \setminus \{\vec{0}\}$. It suffices to show that in every case, $\mathbf{S}_n$ does not strictly self-assemble in $\mathcal{T}$.

Case 1. Suppose $\vec{m}, \vec{n} \in A_h$. Without loss of generality, let $\vec{m} = (i, 0)$ and $\vec{n} = (j, 0)$ where $0 \leq i < j < 2^n$. Let β be the unique T-configuration such that for all $\vec{k} = (k_1, k_2) \in \mathbb{N}$,

$$\beta(\vec{k}) = \begin{cases} \uparrow & \text{if } k_1 > 2^n \text{ or } k_2 \neq 0 \\ \alpha(\vec{k}) & \text{if } k_1 < j \\ \alpha(\vec{m} + ((k_1 - j) \bmod (j - i), 0)) & \text{otherwise.} \end{cases}$$

By (2.2) and $j > 0$, $\vec{n} + \vec{u}_W \in \operatorname{dom}\alpha$. So, by Lemma 3.1, $\alpha(\vec{n})(\vec{u}_W) = \alpha(\vec{n} + \vec{u}_W)(\vec{u}_E)$ and $\operatorname{str}_{\alpha(\vec{n})}(\vec{u}_W) \geq \tau$. But since $\alpha(\vec{n}) = \alpha(\vec{m})$, $\alpha(\vec{m})(\vec{u}_W) = \alpha(\vec{n} + \vec{u}_W)(\vec{u}_E)$ and $\operatorname{str}_{\alpha(\vec{m})}(\vec{u}_W) \geq \tau$. So, $\beta \in \mathcal{A}[\mathcal{T}]$. Then, there exists a $\gamma \in \mathcal{A}_\square[\mathcal{T}]$ such that $\beta \sqsubseteq \gamma$ and since $\beta((2^n, 0)) \downarrow$, $\gamma((2^n, 0)) \downarrow$. But $(2^n, 0) \notin \mathbf{S}_n$. So, $\mathbf{S}_n$ does not strictly self-assemble in $\mathcal{T}$.

Case 2. The case for $\vec{m}, \vec{n} \in A_v$ is similar to Case 1.

Case 3. Suppose $\vec{m} \in A_h \setminus \{\vec{0}\}$ and $\vec{n} \in A_v \setminus \{\vec{0}\}$. Let $\vec{m} = (i, 0)$ and $\vec{n} = (0, j)$, where $i, j \in \{1, \ldots, 2^n - 1\}$. Let β be the unique T-configuration such that for all $\vec{k} = (k_1, k_2) \in \mathbb{N}$,

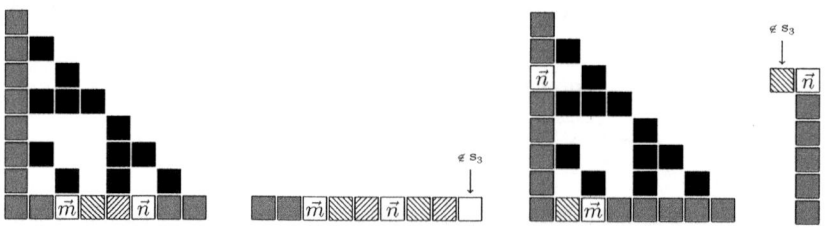

Fig. 2. Illustrating the proof of Lemma 3.2

$$\beta(\vec{k}) = \begin{cases} \alpha(\vec{k}) & \text{if } k_1 = 0 \text{ and } k_2 \leq j \text{ or } k_2 = 0 \text{ and } k_1 \leq i \\ \alpha(\vec{m} + \vec{u}_W) & \text{if } k_1 = -1 \text{ and } k_2 = j \\ \uparrow & \text{otherwise.} \end{cases}$$

By (2.2) and $i > 0$, $\vec{m} + \vec{u}_W \in \operatorname{dom}\alpha$. So, by Lemma 3.1, $\alpha(\vec{m})(\vec{u}_W) = \alpha(\vec{m} + \vec{u}_W)(\vec{u}_E)$ and $\operatorname{str}_{\alpha(\vec{m})}(\vec{u}_W) \geq \tau$. But since $\alpha(\vec{m}) = \alpha(\vec{n})$, $\alpha(\vec{n})(\vec{u}_W) = \alpha(\vec{m} + \vec{u}_W)(\vec{u}_E)$ and $\operatorname{str}_{\alpha(\vec{n})}(\vec{u}_W) \geq \tau$. So, $\beta \in \mathcal{A}[\mathcal{T}]$. Then, there exists a $\gamma \in \mathcal{A}_\square[\mathcal{T}]$ such that $\beta \sqsubseteq \gamma$ and since $\beta((-1,j)) \downarrow$, $\gamma((-1,j)) \downarrow$. But $(-1,j) \notin \mathbf{S}_n$. So, $\mathbf{S}_n$ does not strictly self-assemble in $\mathcal{T}$. $\square$

Even if we only require that $\mathbf{S}_n$ appear in the terminal assembly (not necessarily at the origin), we still have an exponential lower bound on the minimum number of tile types needed. To show this we use the *ruler function* $\rho : \mathbb{Z}^+ \to \mathbb{N}$ defined by the recurrence $\rho(2k + 1) = 0$ and $\rho(2k) = \rho(k) + 1$ for all $k \in \mathbb{N}$. The value of $\rho(n)$ is the exponent of the largest power of 2 that divides n [3].

Lemma 3.3. *Let* $n \in \mathbb{N}$ *and* $\vec{m} \in \mathbb{Z}^2$. *If* $\mathcal{T} = (T, \sigma, \tau)$ *is a TAS such that for every* $\alpha \in \mathcal{A}_\square[\mathcal{T}]$, $\operatorname{dom}\alpha \cap (\vec{m} + \{0, \dots, 2^n - 1\}^2) = \vec{m} + \mathbf{S}_n$, *then* $|T| \geq 2^n - 2$.

Proof. Assume the hypothesis with $n \in \mathbb{N}$, $\vec{m} = (m_1, m_2) \in \mathbb{Z}^2$, and TAS $\mathcal{T} = (T, \sigma, \tau)$ as witness. Let $\alpha \in \mathcal{A}_\square[\mathcal{T}]$. Without loss of generality, assume all of the glue colors on tiles in T do not contain numbers. Suppose $|T| < 2^n - 2$. Then we can construct the TAS $\mathcal{T}' = (T', \sigma', \tau)$ as follows.

(1) For every $\vec{n} \in \vec{m} + \{1, 2, \dots, 2^{n-1} - 1\}^2$, if $\alpha(\vec{n}) \downarrow$, then $\alpha(\vec{n}) \in T'$.

(2) Let $i = 2^{n-1}$. There exists tiles $h_i, v_i \in T'$ such that
$$h_i(\vec{u}_N) = v_i(\vec{u}_N) = \alpha(\vec{m} + (2^{n-1}, 1))(\vec{u}_S), \quad v_i(\vec{u}_S) = h_i(\vec{u}_W) = (i - 1, \tau),$$
$$h_i(\vec{u}_E) = v_i(\vec{u}_E) = \alpha(\vec{m} + (1, 2^{n-1}))(\vec{u}_W), \quad \text{and } v_i(\vec{u}_W) = h_i(\vec{u}_S) = (0, 0).$$

(3) For each $0 < i < 2^{n-1}$, there exists tiles $v_i, h_i \in T'$ such that
$$v_i(\vec{u}_W) = h_i(\vec{u}_S) = (0, 0),$$
$$v_i(\vec{u}_E) = \begin{cases} \alpha(\vec{m} + (2^{n-1} - 2^{\rho(i)} + 1, 2^{n-1})) & \rho(i) > 0 \\ (0, 0) & \rho(i) = 0, \end{cases}$$
$$h_i(\vec{u}_N) = \begin{cases} \alpha(\vec{m} + (2^{n-1}, 2^{n-1} - 2^{\rho(i)} + 1)) & \rho(i) > 0 \\ (0, 0) & \rho(i) = 0, \end{cases}$$
$$v_i(\vec{u}_N) = h_i(\vec{u}_E) = (i, \tau),$$
$$\text{and } v_i(\vec{u}_S) = h_i(\vec{u}_W) = (i - 1, \tau).$$

(4) There exists a tile $\sigma' \in T'$ such that
$$\sigma'(\vec{u}_N) = \sigma'(\vec{u}_E) = (0, \tau) \text{ and } \sigma'(\vec{u}_S) = \sigma'(\vec{u}_W) = (0, 0).$$

Each tile type added to T' in step (1) is also a tile type in T, so, by assumption, we add at most $2^n - 3$ tile types to T' in step (1). Then, 2 tile types are added to T' in step (2), $2^n - 2$ tile types are added to T' in step (3). and 1 tile type is added to T' in step (4). Thus, $|T'| \leq 2^{n+1} - 2$. But, by using equation (2.2) and the ruler function properties, it is easy to verify that $\mathbf{S}_n$ strictly self-assembles in $\mathcal{T}'$. This contradicts Lemma 3.2. Thus, no such TAS exists. $\square$

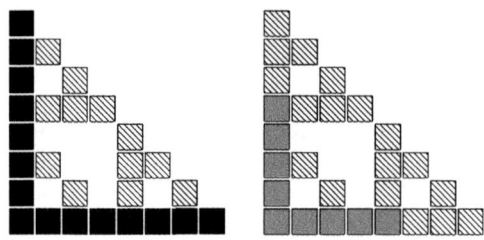

The left figure shows an $\alpha \in \mathcal{A}_\square[T]$, with tiles at locations used to construct T' drawn with dots. The right figure shows the unique $\beta \in \mathcal{A}_\square[T']$, with tiles added in step 1 drawn with diagonal lines, and tiles added in steps 2 through 4 drawn in gray.

Fig. 3. Illustrating of the proof of Lemma 3.3

Theorem 3.4. *If $X \subseteq \mathbb{Z}^2$ strictly self-assembles, then $\mathrm{Dim}_\zeta(X \triangle \mathbf{S}) \geq \mathrm{Dim}_\zeta(\mathbf{S})$.*

Proof. Assume the hypothesis with $X \subseteq \mathbb{Z}^2$ and TAS $\mathcal{T} = (T, \sigma, \tau)$ as witness. Let $V = \{(0,0), (0,1), (1,0)\}$ and $n = \lceil \log_2(|T| + 2) + 1 \rceil$. Since X strictly self-assembles in $\mathcal{T}$, for every $\alpha \in \mathcal{A}_\square[T]$, $X = \mathrm{dom}\,\alpha$. Let $d : \mathbb{Z}^2 \times \mathbb{N} \to \mathbb{N}$ where

$$d(\vec{m}, k) = \left| (X \cap (\vec{m} + \{0, \ldots, 2^{n+k} - 1\}^2)) \triangle (\vec{m} + \mathbf{S}_{n+k}) \right|$$

for all $\vec{m} \in \mathbb{Z}^2$ and $k \in \mathbb{N}$. Then, by (2.2), $d(\vec{m}, k) \geq \sum_{\vec{v} \in V} d(\vec{m} + 2^{n+k-1}\vec{v}, k-1)$. Since $|T| < 2^n - 2$, by Lemma 3.3, for all $\vec{m} \in \mathbb{Z}^2$, $X \cap (\vec{m} + \{0, \ldots, 2^n - 1\}^2) \neq \vec{m} + \mathbf{S}_n$. So, for all $\vec{m} \in \mathbb{Z}^2$, $d(\vec{m}, 0) \geq 1$. Then, the recurrence solves to $d(\vec{m}, k) \geq 3^k$ for all $\vec{m} \in \mathbb{Z}^2$. So,

$$\mathrm{Dim}_\zeta(X \triangle \mathbf{S}) \overset{(2.1)}{=} \limsup_{k \to \infty} \frac{\log_2 d(\vec{0}, k)}{n + k} \geq \mathrm{Dim}_\zeta(\mathbf{S}) \text{ by Observation 2.2.} \quad \square$$

To gain further insight, we consider the strict self-assembly of subsets of $\mathbf{S}$.

Lemma 3.5. *If $X \subseteq \mathbf{S}$ strictly self-assembles in a TAS (T, σ, τ), then for all $\vec{m} = (m_1, m_2) \in \mathbb{Z}^2$ such that $m_1 \geq |T|$ and $m_2 \geq |T|$, $\vec{m} \notin X$.*

Lemma 3.6. *If $X \subseteq \mathbf{S}$ strictly self-assembles in a TAS (T, σ, τ), then for every $n \in \mathbb{N}$, $|X_{[0,n]}| \leq 2|T|(n+1)$.*

Proof. Assume the hypothesis with $X \subseteq \mathbb{Z}^2$ and $\mathcal{T} = (T, \sigma, \tau)$ as witness. Let $\alpha \in \mathcal{A}_\square[T]$ and let $n \in \mathbb{N}$. If $n \leq |T|$ the theorem is trivially true, so assume $n > |T|$. Let $A = \{0, \ldots, |T| - 1\}^2$, $B = \{0, \ldots, |T| - 1\} \times \{|T|, \ldots, n\}$, $C = \{|T|, \ldots, n\} \times \{0, \ldots, |T| - 1\}$, and $D = \{|T|, \ldots, n\}^2$. It is clear that A, B, C, D is a partition of $\{0, \ldots, n\}^2$. Then,

$$\begin{aligned}
|X_{[0,n]}| &= |\{0, \ldots, n\}^2 \cap \mathrm{dom}\,\alpha| && \text{since } X \subseteq \mathbb{N}^2 \\
&= |A \cap \mathrm{dom}\,\alpha| + |B \cap \mathrm{dom}\,\alpha| + |C \cap \mathrm{dom}\,\alpha| && \text{by Lemma 3.5} \\
&\leq 2|T|(n+1). && \square
\end{aligned}$$

Theorem 3.7. *If $X \subseteq \mathbf{S}$ strictly self-assembles, then $\mathrm{Dim}_\zeta(X) \in \{0, 1\}$.*

Proof. Assume the hypothesis with $X \subseteq \mathbf{S}$ and TAS (T, σ, τ) as witness. By Lemma 3.6, $|X_{[0,n]}| \leq 2|T|(n+1)$. Then, $\mathrm{Dim}_\zeta(X) \leq 1$. But, the binding graph of any $\alpha \in \mathcal{A}_T^\tau$ must be connected and any infinite connected structure has ζ-dimension at least 1, it follows that either $\mathrm{Dim}_\zeta(X) = 1$ or X is finite, in which case X has ζ-dimension 0. So, $\mathrm{Dim}_\zeta(X) \in \{0, 1\}$. $\square$

Note that boundary of $\mathbf{S}$ is a subset of $\mathbf{S}$ that strictly self-assembles and has ζ-dimension 1. A single tile placed at the origin is a subset of $\mathbf{S}$ that strictly self-assembles and has ζ-dimension 0. Hence, Theorem 3.7 is trivially tight.

4   Conditional Determinism

The method of local determinism introduced by Soloveichik and Winfree [11] is a common technique for showing that a TAS is directed. However, there exists very natural constructions that are directed but not locally deterministic. Consider the TAS $\mathcal{T}$ of Figure 4. Clearly, there is only one assembly sequence $\vec{\alpha}$ in $\mathcal{T}$ such that res $\vec{\alpha}$ is terminal. Hence, $\mathcal{T}$ is directed. However, $\vec{\alpha}$ fails condition (2) of local determinism at the location $(0, 1)$. The culprit is the blocking technique used by this TAS which is marked by a red X in the figure. After introducing some new notation, we give sufficient conditions for proving such a TAS is directed.

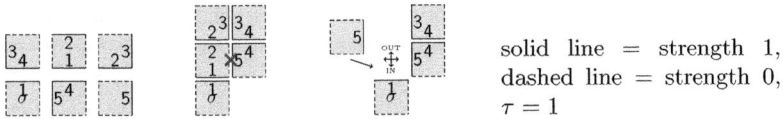

solid line = strength 1,
dashed line = strength 0,
$\tau = 1$

Fig. 4. An example of a blocking tile assembly system

For $\vec{m}, \vec{n} \in \mathbb{Z}^2$, if $\vec{m} \prec_{\vec{\alpha}} \vec{n}$ for every assembly sequence $\vec{\alpha}$ in a TAS $\mathcal{T}$, then $\vec{m}$ *precedes* $\vec{n}$ *in* $\mathcal{T}$, written $\vec{m} \prec_{\mathcal{T}} \vec{n}$. Now, for each $\vec{m} \in \mathbb{Z}^2$, we define the set

$$\mathrm{DEP}^{\mathcal{T}}(\vec{m}) = \{\vec{u} \in U \mid \vec{m} \prec_{\mathcal{T}} \vec{m} + \vec{u}\}.$$

Now, let $\mathcal{T}$ be a TAS, $\vec{\alpha}$ an assembly sequence in $\mathcal{T}$, and $\alpha = \mathrm{res}\,\vec{\alpha}$. Then, $\vec{\alpha}$ is *conditionally deterministic* if the following three conditions hold.

(1) For all $\vec{m} \in \mathrm{dom}\,\alpha \setminus \mathrm{dom}\,\alpha_0$, $\sum_{\vec{u} \in \mathrm{IN}^{\vec{\alpha}}(\vec{m})} \mathrm{str}_{\alpha_{i_{\vec{\alpha}}(\vec{m})}}(\vec{m}, \vec{m} + \vec{u}) = \tau$.

(2) For all $\vec{m} \in \mathrm{dom}\,\alpha \setminus \mathrm{dom}\,\alpha_0$ and all $t \in T \setminus \{\alpha(\vec{m})\}$,

$$\vec{m} \notin \partial_t^{\mathcal{T}}(\alpha \upharpoonright (\mathrm{dom}\,\alpha \setminus (\{\vec{m}\} \cup (\vec{m} + (\mathrm{OUT}^{\vec{\alpha}}(\vec{m}) \cup \mathrm{DEP}^{\mathcal{T}}(\vec{m})))))).$$

(3) $\partial^\tau \alpha = \emptyset$.

A TAS is *conditionally deterministic* if it has a conditionally deterministic assembly sequence. Although conditional determinism is weaker than local determinism, it is sufficient for showing a TAS is directed.

Theorem 4.1. *Every locally deterministic TAS is conditionally deterministic.*

Theorem 4.2. *Every conditionally deterministic TAS is directed.*

5   Fibering the Sierpinski Triangle in Place

In this section we present our second main theorem. We construct a TAS in which a superset of $\mathbf{S}$ with the same ζ-dimension strictly self-assembles. Thus, our first main theorem is tight, even when restricted to supersets of $\mathbf{S}$.

Conceptually, the communication fibers added to $\mathbf{S}$ enable a superset of $\mathbf{S}_{i+1}$ to strictly self-assemble when given a superset of $\mathbf{S}_i$ as input. By (2.2), $\mathbf{S}_{i+1}$ can be constructed by placing a copy of $\mathbf{S}_i$ on top and to the right of itself. This is achieved by copying the left boundary of $\mathbf{S}_i$ to the right of $\mathbf{S}_i$, and the bottom boundary of $\mathbf{S}_i$ to the top of $\mathbf{S}_i$. To ensure that the newly added bars are of the proper length, *counter* fibers control their attachment. The counter fibers increment until they have reached same height as the middle point of the largest diagonal in $\mathbf{S}_i$, and then decrement to zero. To know where the middle point is, the counter fibers initiate the attachment of *test fibers* which grow back to $\mathbf{S}_i$, test whether the middle point is reached, and return the result to the counters. Thus, $\mathbf{S}_i$ must be fully attached before the test fibers for $\mathbf{S}_{i+1}$ can attach. This is achieved by the *diagonal cap* fibers that attach along the largest diagonal in $\mathbf{S}$ on the side opposite the seed. They force the necessary part of $\mathbf{S}_i$ to complete attaching before the counters for $\mathbf{S}_{i+1}$ can begin to attach. Then, a blocking technique is used for the test fibers. The bottom row of the test fibers runs from the counters until blocked by the cap fibers. This attachment forms a path on which information can propagate from the diagonals back to the counters in a controlled manner. We now describe how the self-assembly determines the center of $\mathbf{S}_i$. A location is at the center of $\mathbf{S}_i$ when it sits directly above the left boundary of the $\mathbf{S}_{i-1}$ structure on the right part of $\mathbf{S}_i$ and directly to the right of the bottom boundary of the $\mathbf{S}_{i-1}$ structure on the top part of $\mathbf{S}_i$. This is computed in our construction by assigning to each bar of $\mathbf{S}$ a boolean value that is true (represented in Figure 5 by orange) only if it meets the criteria above. Every new bar that attaches to an existing bar will carry a true value unless it is the unique bar that attaches at the halfway point. Then, when two true bars meet,

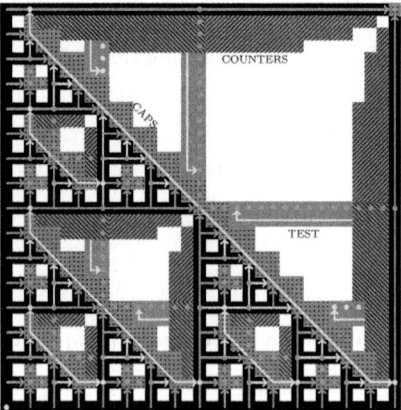

Fig. 5. Fibering the discrete Sierpinski triangle in place

it is always at a location in the middle of the largest diagonal of some stage of **S**. Note that every bar that attaches on the boundary has a true value.

We have constructed a TAS $\mathcal{T}$ that implements the techniques described above using a tile set of ninety-five tile types. We use the method of conditional determinism introduced in Section 4 to prove that $\mathcal{T}$ is directed. From Figure 5 it is clear that the terminal assembly of $\mathcal{T}$ is a superset of **S**. A rigorous analysis of this structure shows the set of added communication fibers (and hence the structure itself) has the same ζ-dimension as **S**. It is also interesting to note that **S** weakly self-assembles in $\mathcal{T}$. Instructions for simulating $\mathcal{T}$ are at http://www.cs.iastate.edu/~shutters/asast

Theorem 5.1. *There exists a set* $X \subseteq \mathbb{Z}^2$ *with the following properties.*
(1) $\mathbf{S} \subseteq X$.
(2) $\mathrm{Dim}_\zeta(X \triangle \mathbf{S}) = \mathrm{Dim}_\zeta(\mathbf{S})$.
(3) X *strictly self-assembles in the Tile Assembly Model.*

Acknowledgments

We thank Jim Lathrop, Xiaoyang Gu, Scott Summers, Dave Doty, Matt Patitz, and Brian Patterson for useful discussions.

References

1. Bondarenko, B.A.: Generalized Pascal Triangles and Pyramids, Their Fractals, Graphs and Applications. The Fibonacci Association (1993)
2. Doty, D., Gu, X., Lutz, J.H., Mayordomo, E., Moser, P.: Zeta-dimension. In: Proceedings of the 30th International Symposium on Mathematical Foundations of Computer Science (2005)
3. Graham, R.L., Knuth, D.E., Patashnik, O.: Concrete Mathematics. Addison-Wesley, Reading (1994)
4. Kautz, S.M., Lathrop, J.I.: Self-assembly of the Sierpinski carpet and related fractals. In: Proceedings of the 15th International Meeting on DNA Computing and Molecular Programming (2009)
5. Lathrop, J.I., Lutz, J.H., Summers, S.M.: Strict self-assembly of discrete Sierpinski triangles. Theoretical Computer Science 410, 384–405 (2009)
6. Patitz, M.J., Summers, S.M.: Self-assembly of discrete self-similar fractals. Natural Computing (to appear)
7. Rothemund, P.W.K., Papadakis, N., Winfree, E.: Algorithmic self-assembly of DNA Sierpinski triangles. PLoS Biology 2(12) (2004)
8. Rothemund, P.W.K.: Theory and Experiments in Algorithmic Self-Assembly. Ph.D. thesis, University of Southern California, Los Angeles, California (2001)
9. Seeman, N.C.: Nucleic-acid junctions and lattices. Journal of Theoretical Biology 99, 237–247 (1982)
10. Seeman, N.C.: DNA in a material world. Nature 421, 427–431 (2003)
11. Soloveichik, D., Winfree, E.: Complexity of self-assembled shapes. SIAM Journal on Computing 36, 1544–1569 (2007)
12. Winfree, E.: Algorithmic Self-Assembly of DNA. Ph.D. thesis, California Institute of Technology, Pasadena, California (1998)

Hairpin Lengthening*

Florin Manea[1,3], Carlos Martín-Vide[2], and Victor Mitrana[3,4]

[1] Otto-von-Guericke-University Magdeburg, Faculty of Computer Science
PSF 4120, D-39016 Magdeburg, Germany
flmanea@fmi.unibuc.ro
[2] Research Group in Mathematical Linguistics, Rovira i Virgili University
Avinguda Catalunya, 35, 43002 Tarragona, Spain
carlos.martin@urv.cat
[3] Faculty of Mathematics and Computer Science, University of Bucharest
Str. Academiei 14, 010014 Bucharest, Romania
mitrana@fmi.unibuc.ro
[4] Departamento de Organización y Estructura de la Información
Escuela Universitaria de Informática, Universidad Politécnica de Madrid
Crta. de Valencia Km. 7, 28031, Madrid, Spain

Abstract. The hairpin completion is a natural operation of formal languages which has been inspired by molecular phenomena in biology and by DNA-computing. We consider here a new variant of the hairpin completion, called hairpin lengthening, which seems more appropriate for practical implementation. The variant considered here concerns the lengthening of the word that forms a hairpin structure, such that this structure is preserved, without necessarily completing the hairpin. Although our motivation is based on biological phenomena, the present paper is more about some algorithmic properties of this operation. We prove that the iterated hairpin lengthening of a language recognizable in $\mathcal{O}(f(n))$ time is recognizable in $\mathcal{O}(n^2 f(n))$ time, while the one-step hairpin lengthening of such a language is recognizable in $\mathcal{O}(nf(n))$ time. Finally, we propose an algorithm for computing the hairpin lengthening distance between two words in quadratic time.

Keywords: DNA computing, hairpin structure, hairpin completion, hairpin lengthening, formal languages.

1 Introduction

This paper is a continuation of a series of works started with [2] (based on some ideas from [1]), where, inspired by the DNA manipulation, a new formal operation on words, called hairpin completion, was introduced. The initial work was followed by a series of related papers ([8,9,10,11]), where both the hairpin

* Florin Manea's work was supported by the *Alexander von Humboldt Foundation*.
Victor Mitrana's work was supported by the PIV Program of the *Agency for Management of University and Research Grants*.

F. Ferreira et al. (Eds.): CiE 2010, LNCS 6158, pp. 296–306, 2010.

completion, as well as its inverse operation, the hairpin reduction, were further investigated.

Single-stranded DNA molecules (ssDNA) are composed by nucleotides which differ from each other by their bases: A (adenine), G (guanine), C (cytosine), and T (thymine). Therefore each ssDNA may be viewed as a finite string over the four-letter alphabet $\{A, C, G, T\}$. Two single strands can bind to each other, forming the secondary structure of DNA, if they are pairwise Watson-Crick complementary: A is complementary to T, and C to G. The binding of two strands is also called *annealing*. Similarly, RNA molecules are chains of nucleotides having the bases A, G, C and U (uracil), with A complementary to U, and C to G. An intramolecular base pairing, known as *hairpin*, is a pattern that can occur in single-stranded DNA or RNA molecules. Hairpin or hairpin-free structures have numerous applications to DNA-computing and molecular genetics. In many DNA-based algorithms, these DNA molecules cannot be used in the subsequent computations. Therefore, it is important to design methods for constructing sets of DNA sequences which are unlikely to lead to such "bad" hybridizations. This problem was considered in a series of papers, see e.g. [12,4,5,7] and the references therein.

In [2] a new formal operation on words is introduced, namely the *hairpin completion*. It consists of three biological principles. Besides the Watson-Crick complementarity and annealing, the third biological phenomenon is that of *lengthening DNA by polymerases*. In our case the phenomenon produces a new molecule as follows: one starts with a hairpin - which is, here, a single-stranded molecule, such that one of its ends (a prefix or, respectively, a suffix) is annealed to another part of itself by Watson-Crick complementarity -, and a *polymerization buffer* with many copies of the four basic nucleotides. Then, the initial hairpin is prolonged by polymerases (thus adding a suffix or, respectively, a prefix), until a complete hairpin structure is obtained (the beginning of the strand is annealed to the end of the strand). Of course, all these phenomena are considered here in an idealized way. For instance, we allow polymerase to extend the strand at either end (usually denoted in biology with 3' and 5') despite that, due to the greater stability of 3' when attaching new nucleotides, DNA polymerase can act continuously only in the 5' $\longrightarrow$ 3' direction. However, polymerase can also act in the opposite direction, but in short "spurts" (Okazaki fragments).

In this paper we consider a new variant of the hairpin completion, called hairpin lengthening, which seems more appropriate for practical implementation. This variant concerns the prolongation of a strand which forms a hairpin, similarly to the process described above, but not necessarily until a complete hairpin structure is obtained. The main motivation in introducing this operation is that, in practice, it may be a difficult task to control the completion of a hairpin structure, and it seems easier to model only the case when such a structure is extended.

Nevertheless, it seems interesting to consider the iterated versions of the hairpin completion or lengthening. Since these operations can be seen as phenomena by which a single-stranded molecule evolves into a new single-stranded molecule, it is natural to consider the situation when multiple evolution steps occur, thus the initial word is transformed by multiple hairpin completion/lengthening steps. In this context a natural algorithmic question occurs: "given two words, can we decide if the smaller one evolved (in one-step or by iterated hairpin completion/ lengthening) into the longer one?". Moreover, one can be also interested in finding what is the minimum number of steps needed to transform a word into another by iterated application of hairpin completion/lengthening. In the case of the hairpin completion, these problems were approached in [8,11]. In this paper, we prove that the iterated hairpin lengthening of a language recognizable in $\mathcal{O}(f(n))$ time is recognizable, in its turn, in $\mathcal{O}(n^2 f(n))$ time; by customizing the proof of this result, one can show that the one-step hairpin lengthening of a language recognizable in $\mathcal{O}(f(n))$ time is recognizable in $\mathcal{O}(nf(n))$ time. Then we define the hairpin lengthening distance between two words and propose an algorithm for computing it in quadratic time. Note that all the time complexity bounds we show here hold on the unit cost RAM model.

2   Preliminaries

Given a word w over an alphabet V, we denote by $|w|$ its length, while $w[i..j]$ denotes the subword of w starting at position i and ending at position j, $1 \leq i \leq j \leq |w|$. If $i = j$, then $w[i..j]$ is the i-th letter of w, which is simply denoted by $w[i]$.

Let Ω be a "superalphabet", that is an infinite set such that any alphabet considered in this paper is a subset of Ω. In other words, Ω is the *universe* of the languages in this paper, i.e., all words and languages are over alphabets that are subsets of Ω. An *involution* over a set S is a bijective mapping $\sigma : S \longrightarrow S$ such that $\sigma = \sigma^{-1}$. Any involution σ on Ω such that $\sigma(a) \neq a$ for all $a \in \Omega$ is said to be, in this paper's context, a *Watson-Crick involution*. Despite that this is nothing more than a fixed point-free involution, we prefer this terminology since the hairpin lengthening defined later is inspired by the DNA lengthening by polymerases, where the Watson-Crick complementarity plays an important role. Let $\bar{}$ be a Watson-Crick involution fixed for the rest of the paper. The Watson-Crick involution is extended to a morphism from Ω^* to Ω^* in the usual way. We say that the letters a and $\bar{a}$ are complementary to each other. For an alphabet V, we set $\overline{V} = \{\bar{a} \mid a \in V\}$. Note that V and $\overline{V}$ could be disjoint or intersect or be equal. We denote by $(\cdot)^R$ the mapping defined by $^R : V^* \longrightarrow V^*$, $(a_1 a_2 \ldots a_n)^R = a_n \ldots a_2 a_1$. Note that ^{R} is an involution and an *anti-morphism* $((xy)^R = y^R x^R$ for all $x, y \in V^*)$. Note also that the two mappings $\bar{}$ and $\cdot^R$ commute, namely, for any word x, $(\overline{x})^R = \overline{x^R}$ holds.

Let V be an alphabet, for any $w \in V^+$ we define the *k-hairpin lengthening* of w, denoted by $HL_k(w)$, for some $k \geq 1$, as follows:

- $HLP_k(w) = \{\overline{\delta^R} w \mid w = \alpha\beta\overline{\alpha^R}\gamma, |\alpha| = k, \alpha, \beta, \gamma \in V^+$ and δ is a prefix of $\gamma\}$,
- $HLS_k(w) = \{w\overline{\delta^R} \mid w = \gamma\alpha\beta\overline{\alpha^R}, |\alpha| = k, \alpha, \beta, \gamma \in V^+$ and δ is a suffix of $\gamma\}$,
- $HL_k(w) = HLP_k(w) \cup HLS_k(w)$.

The *hairpin lengthening* of w is defined by $HL(w) = \bigcup_{k \geq 1} HL_k(w)$. Clearly, $HL_{k+1}(w) \subseteq HL_k(w)$ for any $w \in V^+$ and $k \geq 1$. The hairpin lengthening is naturally extended to languages by $HL_k(L) = \bigcup_{w \in L} HL_k(w)$.

This operation is schematically illustrated in Figure 1.

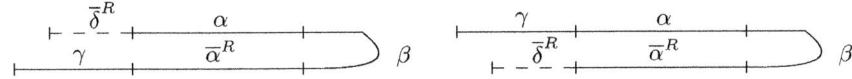

Fig. 1. Hairpin lengthening

The iterated version of the hairpin lengthening is defined as usual by:

$$HL_k^0(w) = \{w\}, \; HL_k^{n+1}(w) = HL_k(HL_k^n(w)), \; HL_k^*(w) = \bigcup_{n \geq 0} HL_k^n(w),$$

and $HL_k^*(L) = \bigcup_{w \in L} HL_k^*(w)$.

3   Complexity of the Hairpin Lengthening

A key means in this section is the rather well-known Knuth-Morris-Pratt algorithm (KMP for short, [6]), a classical algorithm used to locate all the occurrences of a given word x, usually called pattern, in another given word w, usually called text, with linear time-complexity $\mathcal{O}(|x| + |w|)$ and linear working-space. Note that while running the main procedure of this algorithm (see [3] for a detailed presentation) we can compute, without changing the overall time or space complexity, an array of $|x|$ natural numbers $LO_{x,w}$ (Leftmost Occurrence) defined by

$$LO_{x,w}[i] = \begin{cases} t, & \text{if the leftmost occurrence of } x[1..i] \text{ in } w \text{ is } w[t-i+1..t], \\ 0, & \text{if } x[1..i] \text{ does not appear in } w. \end{cases}$$

Clearly, if $LO_{x,w}[i] = 0$, then $LO_{x,w}[j] = 0$ for all $j > i$; moreover, for all $i \geq 1$ $LO_{x,w}[i] \leq LO_{x,w}[i+1]$, provided that $LO_{x,w}[i+1] \neq 0$.

The main result of this section is:

Theorem 1. *For every $k \geq 1$ and every language L recognizable in $\mathcal{O}(f(n))$ time, the iterated k-hairpin lengthening of L is recognizable in $\mathcal{O}(n^2 f(n))$ time.*

Proof. It is worth mentioning that the argument used for proving a similar complexity result in [8] for hairpin completion does not work here. The algorithm proposed in [8] is based on the fact that every word obtained by iterated hairpin completion in more than one step has a non-trivial suffix that equals the reverse of the complement of its prefix, and this property can be efficiently tested. In the case of the hairpin lengthening this property does not hold, thus we have to develop another approach. However, in both cases we rely on a dynamic programming strategy (which may make the two algorithms seem similar).

Let w be a word of length n. We define a function, $Member_{L,k}(w)$, which decides whether or not $w \in HL_k^*(L)$. The algorithm implemented by this function computes, as the main data structure, a $n \times n$ matrix M with binary entries defined by $M[i][j] = (w[i..j] \in HL_k^*(L))$, that is $M[i][j]$ has the same truth value as $w[i..j] \in HL_k^*(L)$. The computation of M is based on a dynamic programming approach. Initially, for all $i, j \in \{1, \ldots, n\}$, we set $M[i][j] = 1$, provided that $w[i..j] \in L$, and $M[i][j] = 0$, otherwise. Further on, we analyze all the subwords of w, in increasing order of their length; in order to decide whether $w[i..j]$ can be obtained by iterated k-hairpin lengthening from a word in L we simply have to check whether one of the following conditions is satisfied:

- there exists an index $i + 2k + 2 \leq s < j$ such that $w[i..s] \in HL_k^*(L)$ and $w[i..j] \in HLS_k(w[i..s])$,
- there exists an index $i < t \leq j - 2k - 2$ such that $w[t..j] \in HL_k^*(L)$ and $w[i..j] \in HLP_k(w[t..j])$.

Note that the search for the indices s and t can be carried out because of the dynamic programming strategy.

However, the approach described above cannot be implemented efficiently in a direct way. To this aim we need some additional data structures. We define two $n \times n$ upper triangular matrices P_s and P_p, having natural number entries, with the following meaning:

- $P_s[i][j]$ stores the position on which the rightmost occurrence of $\overline{w[i..j]}^R$ starts in $w[1..i-1]$. By default, we set $P_s[1][j] = 0$ for all $j \leq n$ and $P_s[i][j] = 0$ for all $j < i \leq n$.
- $P_p[i][j]$ stores the position on which the leftmost occurrence of $\overline{w[i..j]}^R$ ends in $w[j+1..n]$. By default, we set $P_p[i][n] = 0$ for all $i \leq n$ and $P_p[i][j] = 0$ for all $j < i \leq n$.

We claim that the nontrivial elements of the two matrices can be computed as follows:

- $P_s[i][j] = i - LO_{w[i..n], \overline{w[1..i-1]}^R}[j-i+1]$ for all i and j such that $n \geq j \geq i > 1$.
- $P_p[i][j] = j + LO_{\overline{w[1..j]}^R, w[j+1..n]}[j-i+1]$ for all i and j such that $n > j \geq i \geq 1$.

Indeed, $P_s[i][j] = t$, $t < i$, implies $i - LO_{w[i..n], \overline{w[1..i-1]}^R}[j - i + 1] = t$, hence $w[i..j] = \overline{w[t..t+j-i]}^R$ and t is the greatest number, with $t + j - i < i$, that verifies this relation. On the other hand, $P_s[i][j] = i$ implies that $w[i..j]$ does not occur as a factor of $\overline{w[1..i-1]}^R$.

Analogously, $P_p[i][j] = t$, $t > j$, implies $\overline{w[i..j]}^R = w[t - j + i..t]$ and t is the smallest number, with $t - j + i > j$, that verifies this relation. On the other hand, $P_p[i][j] = j$ implies that $\overline{w[i..j]}^R$ does not occur as a factor of $w[j+1..n]$.

By these considerations, we may easily conclude that the two matrices can be computed in quadratic time and space by Algorithm 1.

Algorithm 1. $ComputeMat(w)$: returns the values of the two matrices

---

1: **for** $i = 2$ to n **do**
2:    Compute $LO_{w[i..n], \overline{w[1..i-1]}^R}$;
3: **end for**
4: **for** $j = 1$ to $n - 1$ **do**
5:    Compute $LO_{\overline{w[1..j]}^R, w[j+1..n]}$;
6: **end for**
7: **for** $i = 1$ to n **do**
8:    **for** $j = 1$ to n **do**
9:       **if** $i = 1$ or $j < i$ **then**
10:          $P_s[i][j] = 0$
11:       **else**
12:          $P_s[i][j] = i - LO_{w[i..n], \overline{w[1..i-1]}^R}[j - i + 1]$
13:       **end if**
14:       **if** $j = n$ or $j < i$ **then**
15:          $P_p[i][j] = 0$
16:       **else**
17:          $P_p[i][j] = j + LO_{\overline{w[1..j]}^R, w[j+1..n]}[j - i + 1]$
18:       **end if**
19:    **end for**
20: **end for**
21: Return the pair of matrices (P_s, P_p)

As far as the computation of M defined in the beginning of this proof is concerned, we conclude that $M[i][j] = 1$ if and only if one of the following conditions holds:

- $w[i..j] \in L$.
- There exists an index s such that $i \le s \le j$, $M[i][s] = 1$ and $w[s - k + 1..j]$ is a subword of $\overline{w[i..s - k]}^R$, hence $P_s[s - k + 1][j] \ge i$.
- There exists an index t such that $i \le t \le j$, $M[t][j] = 1$ and $w[i..t + k - 1]$ is a subword of $\overline{w[i..t + k]}^R$, hence $P_p[i][t + k - 1] \le j$.

On the other hand, it is rather plain that $P_s[i][j] > P_s[i'][j]$ for $i' < i$, and $P_p[i][j] < P_p[i][j']$ for $j < j'$. From these considerations we deduce:

- If there exists the index s such that $w[i..j] \in HLS_k(w[i..s])$, then $w[i..j] \in HLS_k(w[i..s'])$ for every index s' with $j > s' > s$. Indeed, from $w[i..j] \in HLS_k(w[i..s])$ it follows that $P_s[s - k + 1][j] \ge i$, thus $P_s[s' - k + 1][j] \ge i$, for all $s' > s$, which is equivalent to $w[i..j] \in HLS_k(w[i..s'])$.
- If there exists the index t such that $w[i..j] \in HLP_k(w[t..j])$, then $w[i..j] \in HLP_k(w[t'..j])$ for every index t' with $i < t' < t$. Indeed, from $w[i..j] \in HLP_k(w[t..j])$ it follows that $P_s[i][t + k - 1] \le j$, thus $P_s[i][t' + k - 1] \le j$, for all $t' < t$, which is equivalent to $w[i..j] \in HLP_k(w[t'..j])$.

This shows that $M[i][j] = 1$ if and only if one of the following holds:

- $w[i..j] \in HLS_k(w[i..s])$ where s is the greatest index such that $M[i][s] = 1$ and $s < j$;
- $w[i..j] \in HLP_k(w[t..j])$ where t is the smallest index such that $M[t][j] = 1$ and $i < t$.

Algorithm 2. $Member_{L,k}(w)$: returns the truth value of $w \in HL_k^*(L)$

1: Initialize matrix M: if $w[i..j] \in L$ set $M[i][j] = 1$, otherwise set $M[i][j] = 0$
2: $(P_s, P_p) = ComputeMat(w)$;
3: Initialize arrays r and l;
4: **for** $len = 1$ to n **do**
5:     **for** $i = 1$ to $n - len + 1$ **do**
6:         $j = i + len - 1$;
7:         **if** $M[i][j] = 0$ and $r[i] \neq 0$ and $P_s[r[i] - k + 1][i] \geq i$ **then**
8:             $M[i][j] = 1$;
9:         **end if**
10:        **if** $M[i][j] = 0$ and $l[i] \neq 0$ and $P_p[i][l[j] + k - 1] \leq j$ **then**
11:            $M[i][j] = 1$;
12:        **end if**
13:        **if** $M[i][j] = 1$ **then**
14:            $r[i] = j$ and $l[j] = i$;
15:        **end if**
16:     **end for**
17: **end for**
18: Return **true** if $M[1][n] = 1$ or **false** otherwise.

In conclusion, M can be computed as shown in Algorithm 2 which makes use of two further arrays, l and r with n positions each, where $r[i]$ is the greatest s found so far such that $M[i][s] = 1$, and $l[j]$ is the smallest t found so far such that $M[t][j] = 1$.

The soundness of the algorithm follows from the aforementioned consideration. It is easy to note that the most time consuming part of the algorithm is that formed by the step 1 which requires $\mathcal{O}(n^2 f(n))$ time. All the other parts require quadratic time. $\square$

A closer look to the proof reveals that the overall complexity of the algorithm is $\mathcal{O}(\max(f(n), n^2))$, provided that all the subwords of a word w (with $|w| = n$) contained in L can be found also in $\mathcal{O}(f(n))$ time. This is the case of context-free languages ($f(n) = n^3$) and regular languages ($f(n) = n^2$).

Using techniques inspired by the above algorithms we can prove that:

Theorem 2. *For every $k \geq 1$ and every language L recognizable in $\mathcal{O}(f(n))$ time, the k-hairpin lengthening of L is recognizable in $\mathcal{O}(nf(n))$ time. If L is regular (context-free), $HL_k(L)$ is recognizable in $\mathcal{O}(n)$ (respectively, $\mathcal{O}(n^3)$) time.*

4   Hairpin Lengthening Distance

The *k-hairpin lengthening distance* between two words x and y is defined as the minimal number of hairpin lengthening operations which can be applied either to x in order to obtain y or to y in order to obtain x. If none of them can be obtained from the other by iterated hairpin lengthening, then the distance is ∞. Formally, the k-hairpin lengthening distance between x and y, denoted by $HLD_k(x, y)$, is defined by: $HLD_k(x, y) = \begin{cases} \min\{p \mid x \in HL_k^p(y) \text{ or } y \in HL_k^p(x)\}, \\ \infty, \text{ if neither } x \in HL_k^p(y) \text{ nor } y \in HL_k^p(x) \end{cases}$

We stress from the very beginning that the function HLD_k defined above is not a distance function in the mathematical sense, since it does not necessarily verify the triangle inequality. However, we call it distance as similar measures (based on different operations on words) are called distances in the literature.

In our view, it is rather surprising that the hairpin lengthening distance can be computed in quadratic time, using a greedy strategy. We recall from [11] that the best known algorithm for computing the hairpin completion distance requires $\mathcal{O}(n^2 \log n)$ time, where n is the length of the longest input word.

Theorem 3. *The k-hairpin lengthening distance between two words x and w can be computed in $\mathcal{O}(max(|x|, |w|)^2)$.*

Proof. First, let us define the notion of derivation, in the context of hairpin lengthening. We say that the word x derives the word w, and denote it $x \to w$, if and only if $w \in HL_k(x)$. If x is a subword of w, $w \in HL_k(x)$, and w has length n, we define the *maximal derivation of w from x* as the sequence of p derivation steps $w_0 \to w_1 \to \ldots \to w_p$, where:

- $w_0 = x$ and $w_p = w$;
- for any i, with $p > i \geq 0$, we either have $w_i = w[s_i..t_i]$ and $w_{i+1} = w[s_i..t_{i+1}]$, where t_{i+1} is the maximum value t such that $w[s_i..t] \in HLS_k(w[s_i..t_i])$, or we have $w_i = w[s_i..t_i]$ and $w_{i+1} = w[s_{i+1}..t_i]$, where s_{i+1} is the minimum value s such that $w[s..t_i] \in HLS_k(w[s_i..t_i])$.

In the following we show that if $w \in HL_k^p(x)$ then there exists a maximal derivation of w from x consisting of at most p derivation steps. Clearly, if $p = 1$ then the derivation $x \to w$ is already a maximal derivation of w from x. Let us assume that $p > 1$.

Since $w \in HL_k^p(x)$ there exists a sequence of p derivation steps $x = w_0 \to w_1 \to \ldots \to w_p = w$. Assume by contradiction that this derivation is not maximal. Therefore in this derivation we have, for some $p - 1 > i \geq 0$, one of the following cases:

- $w_i \to w_{i+1}$, $w_i = w[s_i..t_i]$, $w_{i+1} = w[s_{i+1}..t_{i+1}]$ and there exists $t'_{i+1} > t_{i+1}$ such that $w[s_i..t'_{i+1}] \in HLS_k(w_i)$.
- $w_i \to w_{i+1}$, $w_i = w[s_i..t_i]$, $w_{i+1} = w[s_{i+1}..t_{i+1}]$ and there exists $s'_{i+1} < s_{i+1}$ such that $w[s'_{i+1}..t_i] \in HLS_k(w_i)$.

We show how this derivation can be transformed into a maximal derivation of w from x. We analyze only the first case, since the other can be treated in a similar fashion.

Let $w'_{i+1} = w[s_i..t'_{i+1}]$. If $w_{i+2} = w[s_{i+2}..t_{i+1}]$ with $s_{i+2} < s_i$ (i.e. it was obtained from w_{i+1} by hairpin lengthening with a prefix), then we can derive $w'_{i+2} = w[s_{i+2}..t'_{i+1}]$ from w'_{i+1}. This process can continue until we reach a derivation step where a suffix is added to the derived string. Without loss of generality, we may assume that w_{i+2} is actually obtained in this manner from w_{i+1}. That is $w_{i+2} = w[s_i..t_{i+2}]$ with $t_{i+2} > t_{i+1}$. There are two cases to be discussed: if $t_{i+2} < t'_{i+1}$, then we simply skip this derivation step; if $t_{i+2} > t'_{i+1}$, then we still can obtain w_{i+2} from w'_{i+1} (by arguments similar to those used in the proof of Theorem 1). It follows that we can still continue the derivation and obtain w in at most as many steps as in the original derivation by replacing the derivation $w_i \rightarrow w_{i+1}$ with the derivation $w_i \rightarrow w'_{i+1}$.

Consequently, any sequence of p derivation steps leading from x to w can be transformed into a maximal derivation of w from x, with at most p steps (note that such a property does not hold for the hairpin completion operation). Now we can deduce that $HLD_k(x, w)$ equals the minimum number of derivation steps performed in a maximal derivation of w from x. Moreover, one can easily note that if $x = w_0 \rightarrow w_1 \rightarrow \ldots \rightarrow w_p = w$ is a maximal derivation of w from x, then $w_0 \rightarrow \ldots \rightarrow w_i$ is a maximal derivation of w_i from x, for all $i \leq p$.

Now, let us return to the algorithm for computing the hairpin lengthening distance. Assume that x and w are two words of length m and n, respectively. We are interested in computing the distance $HLD_k(x, w)$. We can assume, without loss of generality, that $m < n$. As we have seen, $HLD_k(x, w)$ equals the minimum number of derivation steps performed in a maximal derivation of w from x, provided that such a derivation exists, or ∞, otherwise; it is clear that w can not be transformed into x.

The first step in computing the minimum number of derivation steps performed in a maximal derivation of w from x is to compute the $n \times n$ matrices C_s and C_p, defined by:

– $C_s[i][j] = t$ if and only if $w[i..t] \in HLS_k(w[i..j])$ and $w[i..t'] \notin HLS_k(w[i..j])$ for all the indices t' such that $n \geq t' > t$.
– $C_p[i][j] = s$ if and only if $w[s..j] \in HLP_k(w[i..j])$ and $w[s'..j] \notin HLP_k(w[i..j])$ for all the indices s' such that $1 \leq s' < s$.

To compute these matrices we will need the auxiliary $n \times n$ matrices P'_s and P'_p:

$$P'_s[i][j] = \max\{t \mid j \leq t \leq n, P_s[j][t] = i\} \text{ and } P'_p[i][j] = \min\{s \mid 1 \leq s \leq i, P_p[s][i] = j\}.$$

Clearly, matrices P'_s and P'_p can be computed using the $ComputeMat(w)$ function, within the same time: basically we initialize all the elements of these matrices with 0, and, then, we update an element ($P'_s[i][j]$, for instance) each time we need to, according to their definition (in our example, when we identify a new s such that $P_s[j][s] = i$ we set $P'_s[i][j]$ to the maximum value from its former value and s).

Now we can show how the matrices C_s and C_p are computed. It is not hard to see that the following recurrence relations hold:

– $C_s[i][j] = 0$ for $i > j$ or $j = n$, and $C_s[i][j] = \max\{P'_s[i][j], C_s[i + 1][j]\}$ otherwise.
– $C_p[i][j] = 0$ for $i > j$ or $i = 1$, and $C_p[i][j] = \min\{P'_p[i][j], C_p[i][j - 1]\}$ otherwise.

Finally we can compute the hairpin lengthening distance between the two words x and w. The strategy that we use is a mixture of dynamic programming and greedy: we analyze, in increasing order of their length (by dynamic programming),

---

Algorithm 3. $HLD_k(x, w)$: returns $HLD_k(x, w)$ $(2k + 1 < |x| < |w|)$

1: Initialize array H;
2: Compute matrices C_s and C_p;
3: **for** $l = 2k$ to n **do**
4:   **for** $i = 1$ to $n - l + 1$ **do**
5:     $j = i + l - 1$;
6:     **if** $C_s[i][j - k + 1] \neq 0$ **then**
7:       $H[i][C_s[i][j - k + 1]] = \min\{H[i][C_s[i][j - k + 1]], 1 + H[i][j]\}$
8:     **end if**
9:     **if** $C_p[i + k - 1][j] \neq 0$ **then**
10:       $H[C_p[i + k - 1][j]][j] = \min\{H[C_p[i + k - 1][j]][j], 1 + H[i][j]\}$
11:     **end if**
12:   **end for**
13: **end for**
14: Return $H[1][n]$

---

all the subwords of w and construct for each of them the subwords of w that can be derived from it by extending it as much as possible using hairpin lengthening - as in each step of a maximal derivation of w from x (greedy strategy); at the same time we count for each of the constructed words the minimum number of derivation steps needed to obtain that subword from x in a maximal derivation of w from x, and store these values in a $n \times n$ matrix H. In this manner $H[i][j]$ will store, at the end of the computation, the value $HLD_k(x, w[i..j])$.

In more detail, when we analyze a subword $w[i..j]$ we proceed as follows:

– Initially we set $H[i][j] = \infty$ for all $i, j \in \{1, \ldots, n\}$; then we set $H[i][j] = 0$ if $w[i..j] = x$.
– Further, for a pair of indices i, j, with $i < j$ and $j - i + 1 > 2k$, we set:

$H[i][C_s[i][j-k+1]] = \min\{H[i][C_s[i][j-k+1]], 1+H[i][j]\}$, if $C_s[i][j-k+1] \neq 0$,
$H[C_p[i+k-1][j]][j] = \min\{H[C_p[i+k-1][j]][j], 1+H[i][j]\}$, if $C_p[i+k-1][j] \neq 0$.

It is not hard to see that for each subword $w[s..t]$ of w we verify all the possible ways in which it can be derived from another subword of w using the rules of a maximal derivation of w from x. When we find such a derivation $w[i..j] \to w[s..t]$, we have already computed $H[i][j]$, thus we can update, if necessary, the value $H[s][t]$. Therefore, the relations given above will lead to the correct computation of the elements of the matrix H. To find the distance $HLD_k(x, w)$ we simply have to return $H[1][n]$.

The implementation of this strategy is given in the Algorithm 3. The time complexity of the above algorithm is $\mathcal{O}(n^2)$, where n is length of the longest input word. Indeed, steps 1 and 2 can be executed in quadratic time each. The part formed by steps $3 - 13$ requires quadratic time, as well. In conclusion, the overall running time of the algorithm is $\mathcal{O}(\max(|x|, |w|)^2)$. □

References

1. Bottoni, P., Labella, A., Manca, V., Mitrana, V.: Superposition based on Watson-Crick-like complementarity. Theory of Computing Systems 39(4), 503–524 (2006)
2. Cheptea, D., Martin-Vide, C., Mitrana, V.: A new operation on words suggested by DNA biochemistry: hairpin completion. In: Proc. Transgressive Computing, pp. 216–228 (2006)
3. Cormen, T.H., Leiserson, C.E., Rivest, R.R.: Introduction to Algorithms. MIT Press, Cambridge (1990)
4. Deaton, R., Murphy, R., Garzon, M., Franceschetti, D.R., Stevens, S.E.: Good encodings for DNA-based solutions to combinatorial problems. In: Proc. of DNA-based computers II. DIMACS Series, vol. 44, pp. 247–258 (1998)
5. Garzon, M., Deaton, R., Nino, L.F., Stevens Jr., S.E., Wittner, M.: Genome encoding for DNA computing. In: Proc. Third Genetic Programming Conference, Madison, MI, pp. 684–690 (1998)
6. Knuth, D.E., Morris, J.H., Pratt, V.R.: Fast pattern matching in strings. SIAM Journal of Computing 6, 323–350 (1977)
7. Kari, L., Konstantinidis, S., Sosik, P., Thierrin, G.: On hairpin-free words and languages. In: De Felice, C., Restivo, A. (eds.) DLT 2005. LNCS, vol. 3572, pp. 296–307. Springer, Heidelberg (2005)
8. Manea, F., Martín-Vide, C., Mitrana, V.: On some algorithmic problems regarding the hairpin completion. Discr. App. Math. 157(9), 2143–2152 (2009)
9. Manea, F., Mitrana, V.: Hairpin completion versus hairpin reduction. In: Cooper, S.B., Löwe, B., Sorbi, A. (eds.) CiE 2007. LNCS, vol. 4497, pp. 532–541. Springer, Heidelberg (2007)
10. Manea, F., Mitrana, V., Yokomori, T.: Two complementary operations inspired by the DNA hairpin formation: completion and reduction. Theor. Comput. Sci. 410(4-5), 417–425 (2009)
11. Manea, F.: A series of algorithmic results related to the iterated hairpin completion (submitted)
12. Sakamoto, K., Gouzu, H., Komiya, K., Kiga, D., Yokoyama, S., Yokomori, T., Hagiya, M.: Molecular computation by DNA hairpin formation. Science 288, 1223–1226 (2000)

Infinities in Quantum Field Theory and in Classical Computing: Renormalization Program

Yuri I. Manin

Max–Planck–Institut für Mathematik, Bonn, Germany, and
Northwestern University, Evanston, USA

1 Feynman Graphs and Perturbation Series: A Toy Model

Introduction. The main observable quantities in Quantum Field Theory, *correlation functions*, are expressed by the celebrated Feynman path integrals. A mathematical definition of them involving a measure and actual integration is still lacking. Instead, it is replaced by a series of *ad hoc* but highly efficient and suggestive heuristic formulas such as *perturbation formalism*. The latter interprets such an integral as a formal series of finite–dimensional but *divergent integrals*, indexed by Feynman graphs, the list of which is determined by the Lagrangian of the theory. *Renormalization* is a prescription that allows one to systematically "subtract infinities" from these divergent terms producing an asymptotic series for quantum correlation functions. On the other hand, graphs treated as *"flowcharts"*, also form a combinatorial skeleton of the abstract computation theory. Partial recursive functions that according to Church's thesis exhaust the universe of (semi)computable maps are generally not everywhere defined due to potentially infinite searches and loops. In this paper I argue that such infinities can be addressed in the same way as Feynman divergences. More details can be found in [9,10].

We start with a toy model in which a series over Feynman diagrams arises.

Notation. *Feynman path integral* is an heuristic expression of the form

$$\frac{\int_{\mathcal{P}} e^{S(\varphi)} D(\varphi)}{\int_{\mathcal{P}} e^{S_0(\varphi)} D(\varphi)} \tag{1.1}$$

or, more generally, a similar heuristic expression for *correlation functions*.

In the expression (1.1), $\mathcal{P}$ is imagined as a functional space of *classical fields* φ on a *space–time manifold* M. $S : \mathcal{P} \to \mathbf{C}$ is a functional of *classical action* measured in Planck's units. Usually $S(\varphi)$ itself is an integral over M of a local density on M which is called *Lagrangian*. In our notation $S(\varphi) = -\int_M L(\varphi(x))dx$. Lagrangian density may depend on derivatives, include distributions etc.

Finally, the integration measure $D(\varphi)$ and the integral itself $\int_{\mathcal{P}}$ should be considered as symbolic constituents of the total expression (1.1) expressing the idea of "summing over trajectories".

F. Ferreira et al. (Eds.): CiE 2010, LNCS 6158, pp. 307–316, 2010.
© Springer-Verlag Berlin Heidelberg 2010

In our toy model, we will replace $\mathcal{P}$ by a finite–dimensional real space. We endow it with a basis indexed by a finite set of "colors" A, and an Euclidean metric g encoded by the symmetric tensor (g^{ab}), $a, b \in A$. We put $(g^{ab}) = (g_{ab})^{-1}$.

The action functional $S(\varphi)$ will be a formal series in linear coordinates on $\mathcal{P}$, (φ^a), of the form

$$S(\varphi) = S_0(\varphi) + S_1(\varphi), \quad S_0(\varphi) := -\frac{1}{2} \sum_{a,b} g_{ab} \varphi^a \varphi^b,$$

$$S_1(\varphi) := \sum_{k=1}^{\infty} \frac{1}{k!} \sum_{a_1,\dots,a_k \in A} C_{a_1,\dots,a_k} \varphi^{a_1} \dots \varphi^{a_k} \tag{1.2}$$

where $(C_{a_1,\dots,a_n})$ are certain symmetric tensors. Below we will mostly consider (g_{ab}) and $(C_{a_1,\dots,a_n})$ as "formal coordinates on the space of theories".

We will express the toy version of (1.1) as a series over (isomorphism classes of) graphs. A graph τ for us consists of two finite sets, edges E_τ and vertices V_τ, and the incidence map sending E_τ to the set of unordered pairs of vertices. Halves of edges form flags F_τ. For more details, see sec. 2.

Theorem 1. *We have, for a formal parameter* λ

$$\frac{\int_\mathcal{P} e^{\lambda^{-1} S(\varphi)} D(\varphi)}{\int_\mathcal{P} e^{\lambda^{-1} S_0(\varphi)} D(\varphi)} = \sum_{\tau \in \Gamma} \frac{\lambda^{-\chi(\tau)}}{|\mathrm{Aut}\, \tau|} w(\tau) \tag{1.3}$$

where τ *runs over isomorphism classes of all finite graphs* τ. *The weight* $w(\tau)$ *of such a graph is determined by the action functional (1.2) as follows:*

$$w(\tau) := \sum_{u:\, F_\tau \to A} \prod_{e \in E_\tau} g^{u(\partial e)} \prod_{v \in V_\tau} C_{u(F_\tau(v))}. \tag{1.4}$$

Each edge e *consists of a pair of flags denoted* ∂e, *and each vertex* v *determines the set of flags incident to it denoted* $F_\tau(v)$, *and* $\chi(\tau)$ *is the Euler characteristic of* τ.

2   Graphs, Flowcharts, and Hopf Algebras

Combinatorial graphs and geometric graphs. *A combinatorial graph* τ consists of two sets F_τ (flags), V_τ (vertices) and two incidence relations. Each flag $f \in F_\tau$ is incident to exactly one vertex $v \in V_\tau$, its *boundary* which is $v = \partial_\tau(f)$, and the map $\partial_\tau : F_\tau \to V_\tau$ is a part of the data. Finally, some pairs of flags form "halves of an edge": this incidence relation is represented by an involution $j_\tau : F_\tau \to F_\tau$, $j_\tau^2 = \mathrm{id}$.

A geometric graph, geometric realization of τ, is the topological space $|\tau|$ which is obtained from the family of segments $\{[0, 1/2]_f\}$ indexed by $f \in F_\tau$. We glue together 0's of all segments, whose respective flags have one and the same boundary vertex v. Finally, we glue the end–points $1/2$ of each pair of segments indexed by two flags $f \neq f'$ such that $j_\tau(f) = f'$.

Pairs of flags $f \neq f'$ with $j_\tau(f) = f'$ are elements of the set of *edges* E_τ. Fixed points of j_τ are called *tails*, their set is denoted T_τ. Any morphism of combinatorial graphs $h : \tau \to \sigma$ that we will consider, will be *uniquely defined by two maps*:

$$h_V : V_\tau \to V_\sigma, \quad h^F : F_\sigma \to F_\tau$$

However, conditions, restricting allowed maps, will depend on the class of morphisms, and eventually on the decorations. Composition is composition of maps.

In particular, h is *an isomorphism*, iff h_V, h^F *are bijections, identifying the incidence maps*.

Graphs often are decorated. Let $L = (L_F, L_V)$ be two sets: *labels of flags and vertices*, respectively.

An *L–decoration* of the combinatorial graph τ consists of two maps $F_\tau \to L_F$, $V_\tau \to L_V$. Usually these maps are restricted by certain *compatibility with incidence relations*. L–decorated graphs for various specific sets of labels will also form a category. Any morphism h will be, as above, determined by h_V, h^F, but this time these maps will be restricted by certain *compatibility with decorations*. *An isomorphism* of decorated graphs is simply an isomorphism of underlying combinatorial graphs, preserving the decorations.

Orientation and flowcharts. Consider the set of labels $L_F := \{in, out\}$. A decoration $F_\tau \to \{in, out\}$ such that halves of any edge are decorated by different labels, is called *an orientation of τ*. On the geometric realization, a flag marked by *in* (resp. *out*) is decorated towards (resp. outwards) its vertex.

Tails of τ oriented *in* (resp. *out*) are called *(global) inputs* T_τ^{in} (resp. *(global) outputs* T_τ^{out}) of τ. Similarly, $F_\tau(v)$ is partitioned into inputs and outputs of the vertex v.

Imagine now that we have a set of operations Op that can be performed on certain objects from a set Ob and produce other objects of this kind. Take (names of) this set of operations as the set of labels of vertices L_V. Then an oriented graph τ with vertices decorated by L_V can sometimes be considered as a flowchart describing a set of computations.

To avoid loops, we will consider only a subclass of graphs.

An oriented graph τ is called *directed* if it satisfies the following condition:

On each connected component of the geometric realization $|\tau|$, one can define a continuous real valued function ("height") in such a way that moving in the direction of orientation along each flag decreases the value of this function.

In particular, oriented trees and forests are always directed.

Example: Dana Scott's flow diagrams. In [11], a class of (eventually infinite) decorated graphs is introduced called *flow diagrams*. The set L_V (labels of vertices) has the following structure:

$$L_V = \mathcal{F} \coprod \mathcal{S} \coprod \{\bot, \top\}.$$

Here $\mathcal{F}$ are functional labels: names of *operators* transforming its input to its output. This set includes the name I of identical operator. Furthermore, $\mathcal{S}$ are names of *switches*: a switch tests its input and redirects it to *one of its outputs,* depending on the results. Finally, symbols $\bot$ and $\top$ are used to generate some "improper" diagrams. In particular, a vertex labeled by $\bot$ describes "the smallest" flow diagram, whereas $\top$ corresponds to the "largest" one.

The first subclass of decorated graphs qualifying as Scott's flow diagrams are *oriented trees.* The orientation (describing the direction of the information flow) goes from one tail (input) to many tails (outputs). Each vertex has one input and either one output, or two outputs. If it has one output, it must be labeled by a symbol from $\mathcal{F}$ or else $\top, \bot$. If it has two outputs, it must be labeled by a switch from $\mathcal{S}$. Clearly, when such a finite tree processes a concrete input, the output will show *only at one of the output tails,* because of the semantics of switching. Hence we might imagine a special vertex accepting many inputs and producing the one which is "not silent". A price for it will be that our graphs will not be trees anymore. They will still remain *directed graphs.*

The following example gives a version of Kreimer–Connes Hopf algebra in quantum field theory.

Definition 1. *Let τ be an oriented graph. Call a proper cut C of τ any partition of V_τ into a disjoint union of two non–empty subsets V_τ^C (upper vertices) and $V_{\tau,C}$ (lower vertices) satisfying the following conditions:*

(i) For each oriented wheel in τ, all its vertices belong either to V_τ^C, or to $V_{\tau,C}$.

(ii) If an edge e connects a vertex $v_1 \in V_\tau^C$ to $v_2 \in V_{\tau,C}$, then it is oriented from v_1 to v_2 ("information flows down").

(iii) There are also two improper cuts: the upper improper cut is the partition $V_\tau^C = \emptyset$, $V_{\tau,C} = V_\tau$, whereas the lower one is the partition $V_\tau^C = V_\tau$, $V_{\tau,C} = \emptyset$.

A cut C defines two graphs: τ^C (upper part of τ wrt C) and τ_C (lower part wrt C). Vertices of τ^C (resp. τ_C) are V_τ^C (resp. $V_{\tau,C}$). Flags of τ^C (resp. τ_C) are all flags of τ incident a vertex of τ^C (resp. τ_C). Edges of τ^C (resp. τ_C) are all edges of τ whose both boundaries belong to τ^C (resp. τ_C). Orientations and labels remain the same. We have $\sigma = \tau^C \coprod \tau_C$ where $\coprod$ is the disjoint union.

Bialgebras of decorated graphs. Fix a set of labels $L = (L_F, L_V)$. Assume that $L_F = L_F^0 \times \{in, out\}$, The isomorphism class of a decorated graph τ is denoted $[\tau]$.

Definition 2. *A set Fl ("flowcharts") of L–decorated graphs is called admissible, if the following conditions are satisfied:*

(i) Each connected component of a graph in Fl belongs to Fl. Each disjoint union of a family of graphs from Fl belongs to Fl. Empty graph $\emptyset$ is in Fl.

(ii) For each $\tau \in Fl$ and each cut C of τ, τ^C and τ_C belong to Fl.

Let now Fl be an admissible set of graphs, and k a commutative ring. Denote by $H = H_{Fl}$ the k–linear span of isomorphism classes of graphs in Fl: the k–module of formal finite linear combinations $\{ \sum_{\tau \in Fl} a_{[\tau]}[\tau] \}$. Define two linear maps

$$m : H \otimes H \to H, \quad \Delta : H \to H \otimes H$$

by the following formulas extended by k–linearity:

$$m([\sigma] \otimes [\tau]) := [\sigma \coprod \tau], \quad \Delta([\tau]) := \sum_C [\tau^C] \otimes [\tau_C],$$

where the sum is taken over all cuts of τ.

Proposition 1. *(i)* m *defines on* H *the structure of a commutative* k–*algebra with unit* $[\emptyset]$*. Denote the respective ring homomorphism* $\eta : k \to H, 1_k \mapsto [\emptyset]$ *.*
(ii) Δ *is a coassociative comultiplication on* H*, with counit*

$$\varepsilon : H \to k, \quad \sum_{\tau \in Fl} a_{[\tau]}[\tau] \mapsto a_{[\emptyset]}$$

(iii) $(H, m, \Delta, \varepsilon, \eta)$ *is a commutative bialgebra with unit and counit.*

Hopf algebra of decorated graphs. In order to construct an antipode on the bialgebra $H = H_{Fl}$ we will show that one can introduce on H_{Fl} a grading by $\mathbf{N}$ turning it into a *connected graded bialgebra* in the sense of [5], 2.1. There are two kinds of natural gradings. One can simply define

$$H_n := \text{ the } k - \text{submodule of } H \text{ spanned by } [\tau] \text{ in } Fl \text{ with } |F_\tau| = n.$$

One can also introduce a weight function on the set of labels in 2.4: $|\cdot| : L \to \mathbf{N}$ and put

$$H_n := \text{ the } k - \text{submodule of } H$$
$$\text{spanned by } [\tau] \text{ in } Fl \text{ with } n = \sum_{f \in F_\tau} (|l(f)| + 1) + \sum_{v \in V_\tau} |l(v)|,$$

where $l : V_\tau \coprod F_\tau \to L$ is the structure decoration of τ.

From the definitions, it follows that for either choice we have

$$m(H_p \otimes H_q) \subset H_{p+q}, \quad \Delta(H_n) \subset \oplus_{p+q=n} H_p \otimes H_q,$$

and moreover, $H_0 = k[\emptyset]$ is one–dimensional, so that H is connected.

Hopf algebras from quantum computation: a review. One standard model of quantum computation starts with a classical Boolean circuit B which computes a map $f : X \to X$, where X is a finite set of Boolean words, say, of length n. After replacing bits with qubits, and X with the 2^n–dimensional Hilbert space H spanned by the ortho–basis of binary words, we have to calculate the linear operator $U_f : H \to H$, linearization of f. Only unitary operators U_f can be physically implemented. Clearly, U_f are unitary only for bijective f; moreover, they must be calculated by "reversible" Boolean circuits.

On the other hand, interesting f are only rarely bijective. There is a well–known trick, allowing one to transform any Boolean circuit B_f calculating f into another Boolean circuit B_F of comparable length consisting only of reversible gates and calculating a bijection F of another finite set. If we now focus on permutations of X, there naturally arise two Hopf algebras related to them: *group algebra of permutations* and a dual Hopf algebra. For infinite X, there are several versions of Hopf algebras associated to the group of unitary operators $H_X \to H_X$.

Below in 5.2 we will use this trick for renormalization of halting problem.

3   Regularization and Renormalization

"Minimal subtraction" algebras. A "minimal subtraction" scheme, that merges well with Hopf algebra renormalization techniques (cf. [5]), formally is based upon a commutative associative algebra $\mathcal{A}$ over a field K, together with two subalgebras $\mathcal{A}_-$ ("polar part") and $\mathcal{A}_+$ (regular part), such that $\mathcal{A} = \mathcal{A}_- \oplus \mathcal{A}_+$ as a linear space. One assumes $\mathcal{A}$ unital, and $1_{\mathcal{A}} \in \mathcal{A}_+$. Moreover, an augmentation homomorphism $\varepsilon_{\mathcal{A}} : \mathcal{A}_+ \to K$ must be given. Then any element $a \in \mathcal{A}$ is the sum of its polar part a_- and regular part a_+. The "regularized value" of a is $\varepsilon_{\mathcal{A}}(a_+)$.

A typical example: $K = \mathbf{C}$, $\mathcal{A} :=$ the ring of germs of meromorphic functions of z, say, at $z = 0$; $\mathcal{A}_- := z^{-1}\mathbf{C}[z^{-1}]$, $\mathcal{A}_+$ consists of germs of regular functions at $z = 0$, $\varepsilon_{\mathcal{A}}(f) := f(0)$. This is a toy model of situations arising in cut–off regularization schemes.

Hopf renormalization scheme. Let $\mathcal{H}$ be a Hopf algebra as above, $\mathcal{A}_+, \mathcal{A}_- \subset \mathcal{A}$ a minimal subtraction unital algebra. Consider the set $G(\mathcal{A})$ of K–linear maps $\varphi : \mathcal{H} \to \mathcal{A}$ such that $\varphi(1_{\mathcal{H}}) = 1_{\mathcal{A}}$. Then $G(\mathcal{A})$ with the convolution product

$$\varphi * \psi(x) := m_{\mathcal{A}}(\varphi \otimes \psi)\Delta(x) = \varphi(x) + \psi(x) + \sum_{(x)} \varphi(x')\psi(x'')$$

is a group, with identity $e(x) := u_{\mathcal{A}} \circ \varepsilon(x)$ and inversion

$$\varphi^{*-1}(x) = e(x) + \sum_{m=1}^{\infty} (e - \varphi)^{*m}(x)$$

where for any $x \in \mathrm{rker}\,\varepsilon$ the latter sum contains only finitely many non–zero summands.

Theorem 2. (Theorem on the Birkhoff decomposition). *If $\mathcal{A}$ is a minimal subtraction algebra, each $\varphi \in G(\mathcal{A})$ admits a unique decomposition of the form*

$$\varphi = \varphi_-^{*-1} * \varphi_+; \quad \varphi_-(1) = 1_{\mathcal{A}}, \ \varphi_-(\mathrm{rker}\,\varepsilon) \subset \mathcal{A}_-, \ \varphi_+(\mathcal{H}) \subset \mathcal{A}_+.$$

Values of renormalized polar (resp. regular) parts φ_- (resp. φ_+) on $\mathrm{rker}\,\varepsilon$ are given by the inductive formulas

$$\varphi_-(x) = -\pi \left(\varphi(x) + \sum_{(x)} \varphi_-(x')\varphi(x'') \right),$$

$$\varphi_+(x) = (\mathrm{rid} - \pi)\left(\varphi(x) + \sum_{(x)} \varphi_-(x')\varphi(x'')\right).$$

Here $\pi : \mathcal{A} \to \mathcal{A}_-$ is the polar part projection in the algebra $\mathcal{A}$.

If φ is a character, φ_+ and φ_- are characters as well.

4   Cut–Off Regularization and Anytime Algorithms

Cut–off regularization. Feynman integrals, say, in momentum space, may diverge when momentum becomes large (resp. small). In this case, the formal integral in question I is replaced by the finite integral I_P taken over momenta $p \le P := p_{cutoff}$ (resp. $p \ge P := p_{cutoff}$).

In computer science the natural "divergence" occurs in space–time: a computation uses discrete memory (space) and runtime. Application–oriented computer scientists recognize the practical necessity of time cut–offs, accompanied by sober estimates of quality of outputs. Systematic work on this problem resulted in the notion of "Anytime Algorithms", cf. [6]. In a stimulating paper [2], Ch. Calude and M. Stay addressed the problem of cut–off of runtime theoretically, and designed quantitative characteristics of such a cut–off.

Let f be a partial recursive function and F its description, a program calculating it. *Computation time (or runtime)* is another partial recursive function $t(F)$, with the same domain $D(t(F)) = D(f) \subset X$ and target $\mathbf{Z}^+$, whose precise definition depends on the details of the implied choice of our programming method. One can similarly define " storage volume" $m(F) : X \to \mathbf{Z}^+$. Yet another partial recursive function, $s(F)$, for Turing computations, is the sum total of lengths of filled parts of the tape over all steps of computation. One can define natural analogs of functions $t(F)$, $m(F)$, and $s(F)$ for rather general programming methods F. Returning to [2], we will show that some of the basic results of that paper related to cut–offs, admit a straightforward reformulation in such a way that they become applicable and true for *any* partial recursive function, including $t(F)$, $m(F)$, and $s(F)$.

Complexity. We will use Kolmogorov's ("exponential", or "program") complexity $C = C_u : \mathbf{Z}^+ \to \mathbf{Z}^+$, cf. [7], VI.9. In [2] it is called *the natural complexity* and denoted ∇_U or simply ∇. Moreover, we will use it also for partial recursive functions of any fixed number of variables m, as in VI.9.1 of [7]. This requires a choice of Kolmogorov optimal recursive function of $m + 1$ variables.

Proposition 2. *For any partial recursive function $f : \mathbf{Z}^+ \to \mathbf{Z}^+$ and $x \in D(f)$ we have $C(f(x)) \le c_f C(x) \le c'_f x$, If f and $x \in D(f)$ are allowed to vary, we have $C(f(x)) \le c\, C(f)C(x) \operatorname{rlog}(C(f)C(x))$.*

Runtimes according to [2]. The central argument of [2] is based upon two statements:

a) The runtime of the Kolmogorov optimal program at a point x of its definition domain is either $\le cx^2$, or is not "algorithmically random" (Theorem 5 of [2]).

b) Not "algorithmically random" integers have density zero for a class of computable probability distributions.

This last statement justifies the time cut–off prescription which is the main result of [2]:

if the computation on the input x did not halt after cx^2 Turing steps, stop it, decide that the function is not determined at x, and proceed to $x + 1$.

Proposition 3 below somewhat generalizes the statement a).

Consider a pair of functions $\varphi, \psi : \mathbf{R}_{>0} \to \mathbf{R}_{>0}$ satisfying the following conditions:

a) $\varphi(x)$ and $\dfrac{x}{\varphi(x)}$ are strictly increasing starting with a certain x_0 and tend to infinity as $x \to \infty$.

b) $\psi(x)$ and $\dfrac{\psi(x)}{x\varphi(\psi(x))}$ are increasing and tend to infinity as $x \to \infty$.

The simplest examples are $\varphi(x) = r\log(x + 2)$, $\psi(x) = (x + 1)^{1+\varepsilon}$, $\varepsilon > 0$.

In our context, φ will play the role of a "randomness scale". Call $x \in \mathbf{Z}^+$ *algorithmically φ–random*, if $C(x) > x/\varphi(x)$. The second function ψ will then play the role of associated growth scale.

Proposition 3. *Let f be a partial recursive function. Then for all sufficiently large x exactly one of the following alternatives holds:*

(i) $x \in D(f)$, and $f(x) \leq \psi(x)$.

(ii) $x \notin D(f)$.

(iii) $x \in D(f)$, and $f(x)$ is not algorithmically φ–random.

5   Regularization and Renormalization of the Halting Problem

Introduction. In this section, we devise tentative regularization/renormalization schemes tailored to fit the halting problem.

(a) Deforming the Halting Problem. Recognizing, whether a number $k \in \mathbf{Z}^+$ belongs to the definition domain $D(f)$ of a partial recursive function f, is translated into the problem, whether an analytic function $\Phi(k, f; z)$ of a complex parameter z has a pole at $z = 1$.

(b) Choosing a minimal subtraction algebra. Let $\mathcal{A}_+$ be the algebra of analytic functions in $|z| < 1$, continuous at $|z| = 1$. It is a unital algebra; we endow it with augmentation $\varepsilon_{\mathcal{A}} : \Phi(z) \mapsto \Phi(1)$. Put $\mathcal{A}_- := (1 - z)^{-1}\mathbf{C}[(1 - z)^{-1}]$, $\mathcal{A} := \mathcal{A}_+ \oplus \mathcal{A}_-$.

(c) Hopf algebra of an enriched programming method. Basically, $\mathcal{H} = \mathcal{H}_P$ is the symmetric algebra, spanned by isomorphism classes $[p]$ of certain descriptions. Comultiplication in $\mathcal{H}_P$ is dual to the composition of descriptions.

In order to produce a Hopf filtration and antipode, we must postulate in addition existence of a "size function" on descriptions as above.

(d) Characters, corresponding to the halting problem. The character $\varphi_k : \mathcal{H}_P \to \mathcal{A}$ corresponding to the halting problem at a point $k \in \mathbf{Z}^+$ for the partial recursive function computable with the help of a description $p \in P(\mathbf{Z}^+, \mathbf{Z}^+)$, is defined as $\varphi_k([p]) := \Phi(k, f; z) \in \mathcal{A}$.

Reduction of the general halting problem to the recognition of fixed points of permutations. Start with a partial recursive function $f : X \to X$ where X is an infinite constructive world. Extend X by one point, i. e. form $X \coprod \{*_X\}$. Choose a total recursive structure of an additive group without torsion on $X \coprod \{*_X\}$ with zero $*_X$. Extend f to the everywhere defined (but generally uncomputable) function $g : X \coprod \{*_X\} \to X \coprod \{*_X\}$, by $g(y) := *_X$ rif $y \notin D(f)$. Define the map

$$\tau_f : (X \coprod \{*_X\})^2 \to (X \coprod \{*_X\})^2, \quad \tau_f(x, y) := (x + g(y), y).$$

Clearly, it is a permutation. Since $(X \coprod \{*_X\}, +)$ has no torsion, the only finite orbits of $\tau_f^{\mathbf{Z}}$ are fixed points. Moreover, the restriction of τ_f upon the recursive enumerable subset $D(\sigma_f) := (X \coprod \{*_X\}) \times D(f)$ of the constructive world $Y := (X \coprod \{*_X\})^2$ induces a partial recursive permutation σ_f of this subset. Since $g(y)$ never takes the zero value $*_X$ on $y \in D(f)$, but always is zero outside it, the complement to $D(\sigma_f)$ in Y consists entirely of fixed points of τ_f.

Thus, the halting problem for f reduces to the fixed point recognition for τ_f.

The Kolmogorov order. We can define a Kolmogorov numbering on a constructive world X as a bijection $\mathbf{K} = \mathbf{K}_u : X \to \mathbf{Z}^+$ arranging elements of X in the increasing order of their complexities C_u. Let now $\sigma : X \to X$ be a partial recursive map, such that σ maps $D(\sigma)$ to $D(\sigma)$ and induces a permutation of this set. Put $\sigma_{\mathbf{K}} := \mathbf{K} \circ \sigma \circ \mathbf{K}^{-1}$ and consider this as a permutation of the subset

$$D(\sigma_{\mathbf{K}}) := \mathbf{K}(D(\sigma)) \subset \mathbf{Z}^+$$

consisting of numbers of elements of $D(\sigma)$ in the Kolmogorov order. If $x \in D(\sigma)$ and if the orbit $\sigma^{\mathbf{Z}}(x)$ is infinite, then there exist such constants $c_1, c_2 > 0$ that for $k := \mathbf{K}(x)$ and all $n \in \mathbf{Z}$ we have $c_1 \cdot \mathbf{K}(n) \le \sigma_{\mathbf{K}}^n(k) \le c_2 \cdot \mathbf{K}(n)$.

Proposition 4. *Let* $X = \mathbf{Z}^+$ *and* σ *be a partial recursive map, inducing a permutation on its definition domain. Put*

$$\Phi(k, \sigma; z) := \frac{1}{k^2} + \sum_{n=1}^{\infty} \frac{z^{\mathbf{K}(n)}}{(\sigma_{\mathbf{K}}^n(k))^2}.$$

Then we have: (i) If σ-orbit of x is finite, then $\Phi(x, \sigma; z)$ is a rational function in z whose all poles are of the first order and lie at roots of unity.

(ii) If this orbit is infinite, then $\Phi(x, \sigma; z)$ is the Taylor series of a function analytic at $|z| < 1$ and continuous at the boundary $|z| = 1$.

References

1. Baez, J., Stay, M.: Physics, topology, logic and computation: a Rosetta stone. Preprint arxiv:0903.0340
2. Calude, C., Stay, M.: Most programs stop quickly or never halt. Adv. in Appl. Math. 40, 295–308 (2008)

3. Calude, C., Stay, M.: Natural halting probabilities, partial randomness, and zeta functions. Information and Computation 204, 1718–1739 (2006)
4. Connes, A., Kreimer, D.: Renormalization in quantum field theory and the Riemann–Hilbert problem. I. The Hopf algebra structure of graphs and the main theorem. Comm. Math. Phys. 210(1), 249–273 (2000)
5. Ebrahimi-Fard, K., Manchon, D.: The combinatorics of Bogolyubov's recursion in renormalization. math-ph/0710.3675
6. Grass, J.: Reasoning about Computational Resource Allocation. An introduction to anytime algorithms. Posted on the Crossroads website,
   http://www.acm.org/crossroads/xrds3-1/racra.html
7. Manin, Y.: A Course in Mathematical Logic for Mathematicians, XVII+384 pp. Springer, Heidelberg (2010) (The second, expanded Edition, with collaboration by Zilber, B.)
8. Manin, Y.: Classical computing, quantum computing, and Shor's factoring algorithm. Séminaire Bourbaki 266(862), 375–404 (1999), quant-ph/9903008
9. Manin, Y.: Renormalization and computation I. Motivation and background. math.QA/0904.492
10. Manin, Y.: Renormalization and computation II: Time cut–off and the Halting Problem. math.QA/0908.3430
11. Scott, D.: The lattice of flow diagrams. In: Symposium on Semantics of Algorithmic Languages. LNM, vol. 188, pp. 311–372. Springer, Heidelberg (1971)

Computational Complexity Aspects
in Membrane Computing

Giancarlo Mauri, Alberto Leporati, Antonio E. Porreca, and Claudio Zandron

Dipartimento di Informatica, Sistemistica e Comunicazione
Università degli Studi di Milano – Bicocca
Viale Sarca 336/14, 20126 Milano, Italy
{mauri,leporati,porreca,zandron}@disco.unimib.it

Abstract. Within the framework of membrane systems, distributed parallel computing models inspired by the functioning of the living cell, various computational complexity classes have been defined, which can be compared against the computational complexity classes defined for Turing machines. Here some issues and results concerning computational complexity of membrane systems are discussed. In particular, we focus our attention on the comparison among complexity classes for membrane systems with active membranes (where new membranes can be created by division of membranes which exist in the system in a given moment) and the classes **PSPACE**, **EXP**, and **EXPSPACE**.

Keywords: Membrane systems, Computational complexity.

1 Membrane Systems

Living cells can be seen as very sophisticated devices for information processing, and computer science can benefit trying to understand the basic mechanisms they use and to abstract from them formal computation models. In this view, in [7] Gheorghe Păun introduced membrane systems (also called *P systems*), a class of distributed and parallel computing devices, inspired by the structure and functioning of living cells, and the way the cells are organized in tissues or higher order structures. More specifically, the feature of the living cells that is abstracted by membrane computing is their compartmentalized internal structure, where the cell membrane contains different organelles (compartments or regions), in turn delimited by membranes.

In the original definition, a membrane system consists of a hierarchical (tree-like) structure composed by several membranes, embedded into a main membrane called the *skin*, that separates the system (the cell) from the environment. Membranes divide the Euclidean space into *regions* that contain multisets of *objects* of different kind (representing molecules), denoted by symbols from an alphabet Γ, and region-specific *developmental rules* (bio–chemical reactions taking place in the different compartment of the cell). A *configuration* of the system is a mapping that associates with every membrane the multiset of objects it contains. The application of a rule modifies the content of the membrane where the

F. Ferreira et al. (Eds.): CiE 2010, LNCS 6158, pp. 317–320, 2010.

rule resides, and hence the configuration of the system, taking a multiset of objects contained in the membrane and producing a new multiset; furthermore, it specifies whether the resulting objects will stay into the current region, or will be transferred one level up (to the external region) or one level down (to one of the inner regions) in the hierarchy. The transfer thus provides communication between regions. More formally, a rule has the form $u \to v$, where u is a string over Γ, and v is a string over $\Gamma \times \{here, out, in\}$. Thus, each element of v is of the form (a, tar), where a is an object from Γ and tar (target) is either *here* or *out* or *in*. The rule can be applied, i.e. is *enabled*, if and only if the multiset w of objects present in the membrane includes u; its application will subtract u from w, and will produce the objects indicated in v, that will be sent to the corresponding target. The rules are usually applied in a nondeterministic and maximally parallel manner: at each step, all the rules enabled in any membrane will be applied, modifying the system configuration; a nondeterministic choice is done in the case where two different rules have non disjoint enabling multisets. A computing device is obtained in this way: a *computation* starts from an initial configuration of the system and terminates when no further rules can be applied. Usually, the result of a computation is the multiset of objects contained in an *output membrane* or emitted from the skin of the system.

Many variants have been defined up to now, with mathematical, computer science or biological motivation (see, e.g., [9], [10]), including symport/antiport, catalytic, spiking neural, tissue, insertion–deletion, splicing, energy–based P-systems. Applications were reported especially in biology and medicine, but also in other directions, such as economics, approximate optimization, and computer graphics. A number of implementations are currently attempted. An overview of the state-of-the-art and an up-to-date bibliography concerning P systems can be found in [10].

2   Computational Complexity of Membrane Systems

From the very beginning, one of the central research topics in the theory of P systems has been to establish the computational power of the various proposed types of membrane systems, comparing them with conventional models such as finite state automata or Turing machines. Moreover, it has been and remains of interest to describe "universal" membrane systems with minimum number of membranes, or to describe the power of various classes of such systems depending on the number of membranes.

In particular, in [8] membrane systems are considered where membranes play an active role in the computation. In this model, the evolution rules involve both objects and membranes, and the communication of objects through a membrane is performed with the direct participation of the membrane itself; moreover, the membrane structure can be modified also by adding new membranes, which can be created by dividing existing membranes, through a process inspired from cellular *mitosis*. There are two different types of division rules: division rules for elementary membranes (i.e. membranes not containing other membranes) and division rules for non-elementary membranes.

This variant with active membranes allows to produce an exponential number of membranes in a polynomial time. As a consequence, in [8] it was shown that such systems are able to efficiently solve **NP**-complete problems [6] (by trading time for space), and similar results are given in [3].

Starting from these results, various complexity classes for P systems were defined [11]. Such classes were then compared against the usual complexity classes such as **P** and **NP** (see, e.g., [12] and [2]); in [15], [1] it was shown that complexity classes defined in terms of two variants of P systems with active membranes contain all problems in **PSPACE**. In [13] it was shown that there exists a complexity class for P systems with active membranes which includes the class **PSPACE** and which is included in the class **EXP**; in [16] this result was strengthened by proving equality to **PSPACE**.

The previous results were given considering either deterministic systems or *confluent* systems, i.e. systems which can operate nondeterministic choices, but in such a way that the result (acceptance or rejection) of every computation is the same (thus, the simulation of a single computation is enough to determine the result of all computations).

In [17] it was shown that, unless **P** = **NP**, a confluent P system working without using membrane division cannot solve an **NP**-complete problem in polynomial time. In fact, the class of problems solved by such system, usually denoted by $\mathbf{PMC}_{\mathcal{NAM}}$, is equal to the standard complexity class **P**. In [14] non-confluent P systems without membrane division and operating in polynomial time are considered. First a solution for the decision problem SAT (satisfiability for Boolean formulas in conjunctive normal form) with a polynomially semi-uniform family of non-confluent P systems without membrane division is proposed, thus proving that **NP** $\subseteq$ $\mathbf{NPMC}_{\mathcal{NAM}}$. Then the reverse inclusion is proved, showing that non-confluent P systems without membrane division can be simulated in polynomial time by non-deterministic Turing machines. As a consequence, we have that $\mathbf{NPMC}_{\mathcal{NAM}}$ equals the complexity class **NP**.

3   Conclusion and Future Directions

The classes of P systems with active membranes we have considered are only defined according to which kinds of membrane division rules are available (none, just elementary or both elementary and non-elementary). The same questions may be also worth posing about other restricted classes, such as P systems without object evolution or communication [12, 4], P systems with division but without dissolution, or even purely communicating P systems, with or without polarizations. Finally, we think that the differences between P systems and traditional computing devices deserve to be investigated for their own sake also from the point of view of space-bounded computations.

References

1. Alhazov, A., Martín-Vide, C., Pan, L.: Solving a PSPACE-complete problem by P systems with restricted active membranes. Fundamenta Informaticae 58(2), 67–77 (2003)
2. Gutiérrez-Naranjo, M.A., Pérez-Jiménez, M.J.: P systems with active membranes, without polarizations and without dissolution: a characterization of P. In: Calude, C.S., Dinneen, M.J., Păun, G., Pérez-Jímenez, M.J., Rozenberg, G. (eds.) UC 2005. LNCS, vol. 3699, pp. 105–116. Springer, Heidelberg (2005)
3. Krishna, S.N., Rama, R.: A variant of P systems with active membranes: Solving NP-complete problems. Romanian J. of Information Science and Technology 2(4), 357–367 (1999)
4. Leporati, A., Zandron, C., Ferretti, C., Mauri, G.: On the Computational Power of Spiking Neural P Systems. Intern. J. of Unconventional Computing 5(5), 459–473 (2009)
5. Leporati, A., Mauri, G., Zandron, C., Păun, G., Pérez-Jiménez, M.J.: Uniform Solutions to SAT and Subset Sum by Spiking Neural P Systems. Natural Computing 8(4), 681–702 (2009)
6. Papadimitriou, C.H.: Computational Complexity. Addison-Wesley, Reading (1993)
7. Păun, G.: Computing with membranes. J. of Computer and System Sciences 61(1), 108–143 (2000)
8. Păun, G.: P systems with active membranes: attacking NP complete problems. In: Antoniou, I., Calude, C.S., Dinneen, M.J. (eds.) Unconventional Models of Computation, pp. 94–115. Springer, London (2000)
9. Păun, G.: Membrane Computing. An Introduction. Springer, Berlin (2002)
10. Păun, G., Rozenberg, G., Salomaa, A. (eds.): The Oxford Handbook of Membrane Computing. Oxford University Press, Oxford (2009)
11. Pérez-Jiménez, M.J., Romero-Jiménez, A., Sancho-Caparrini, F.: Complexity Classes in Cellular Computing with Membranes. Natural Computing 2(3), 265–285 (2003)
12. Pérez-Jiménez, M.J., Romero-Jiménez, A., Sancho-Caparrini, F.: The P versus NP problem through cellular computing with membranes. In: Jonoska, N., Păun, G., Rozenberg, G. (eds.) Aspects of Molecular Computing. LNCS, vol. 2950, pp. 338–352. Springer, Heidelberg (2004)
13. Porreca, A.E., Mauri, G., Zandron, C.: Complexity classes for membrane systems. RAIRO Theoretical Informatics and Applications 40(2), 141–162 (2006)
14. Porreca, A.E., Mauri, G., Zandron, C.: Non-confluence in divisionless P systems with active membranes. Theoretical Computer Science 411(6), 878–887 (2010)
15. Sosík, P.: The computational power of cell division in P systems: Beating down parallel computers? Natural Computing 2(3), 287–298 (2003)
16. Sosík, P., Rodríguez-Patón, A.: Membrane computing and complexity theory: a characterization of PSPACE. Journal of Computer and System Sciences 73(1), 137–152 (2007)
17. Zandron, C., Ferretti, C., Mauri, G.: Solving NP-complete problems using P systems with active membranes. In: Antoniou, I., Calude, C.S., Dinneen, M.J. (eds.) Unconventional Models of Computation, pp. 289–301. Springer, London (2000)

Computable Ordered Abelian Groups and Fields

Alexander G. Melnikov

The University of Auckland, New Zealand
a.melnikov@cs.auckland.ac.nz

Abstract. We present transformations of linearly ordered sets into ordered abelian groups and ordered fields. We study effective properties of the transformations. In particular, we show that a linear order L has a Δ^0_2 copy if and only if the corresponding ordered group (ordered field) has a computable copy. We apply these codings to study the effective categoricity of linear ordered groups and fields.

Keywords: computable algebra, effective categoricity.

We study complexity of isomorphisms between computable copies of ordered abelian groups and fields[1]. Recall that an ordered abelian group is one in which the order is compatible with the additive group operation. Ordered fields are defined in a similar manner. We say that an ordered abelian group $\mathcal{A} = (A; +, \leq)$ is computable if its domain A, the operation $+$, and the relation $\leq$ are computable. Similarly, a field is computable if its domain and its basic operations are computable. If $\mathcal{A}$ is computable and isomorphic to $\mathcal{B}$, we say that $\mathcal{A}$ is a computable copy (or equivalently, computable presentation) of $\mathcal{B}$. We mention that Malcev started a systematic study of computable abelian groups [18], and Rabin initiated a systematic development of the theory of computable fields [22].

One of the main themes in the study of abstract mathematical structures such as groups and fields is to find their isomorphism invariants. For instance, in the case of countable abelian p-groups, Ulm invariants are well-studied (see [26]). However, in the study of computable structures the isomorphism invariants do not always reflect computability-theoretic properties of the underlying structures. For instance, Malcev in [18] constructed computable abelian groups G_1 and G_2 that are isomorphic but not computably isomorphic. In fact, in G_1 the dependency relation is decidable, while in G_2 it is not. Another example is Khisamiev's criterion obtained for computable abelian p-groups of small Ulm length [13]. It is not clear how one can extend this result of Khisamaev to arbitrary Ulm length.

In the study of properties of computable structures, one can use different types of transformation methods between the structures. The idea consists in transforming a given class of computable structures into another class of computable structures in such a way that certain desirable properties of structures in the

---

[1] The author would like to thank his advisors Bakhadyr Khoussainov and Sergey S. Goncharov for suggesting this topic. Many thanks to Andre Nies for his help in writing the final version of the paper.

F. Ferreira et al. (Eds.): CiE 2010, LNCS 6158, pp. 321–330, 2010.

first class are preserved under the transformation to the second class. Here are some examples of this type. Hirschfeldt, Khoussainov, Shore and Slinko in [6] interpret directed graphs in the classes of groups, integral domains, rings, and partial orders. These transformations preserve degree spectra of relations and computable isomorphisms types. Goncharov and Knight [12] provide a method of transforming trees into p-groups to show that the isomorphism problem for these groups is Σ_1^1-complete. Downey and Montalban [5] code trees into torsion-free abelian groups to obtain the analogous result for the torsion-free case. For more examples see [11]. In this paper we present transformations of linearly ordered sets into ordered abelian groups and ordered fields. We study effective properties of these transformations. In particular, we prove the following result:

Theorem 1. *There are transformations Φ and Ψ of the class of linear orders into the classes of ordered abelian groups and ordered fields, respectively, such that both Φ and Ψ preserve the isomorphism types. Moreover, a linear order L has a Δ_2^0 copy if and only if the corresponding ordered group $\Phi(L)$ (ordered field $\Psi(L)$) has a computable copy.*

As a consequence of this theorem and the results from [12], *the isomorphism problems for ordered abelian groups and ordered fields are Σ_1^1-complete.*

The transformations Φ and Ψ are applied to investigate the complexity of isomorphisms between computable copies of ordered abelian groups and fields. We recall known concepts. A structure (e.g. group or field) is *computably categorical* if there is a computable isomorphism between any two computable copies of the structure. Computably categorical structures have been studied extensively. There are results on computably categorical Boolean algebras [8] [16], linearly ordered sets [23] [10], abelian p-groups [24] [9], torsion-free abelian groups [21], trees [17], ordered abelian groups [7] and linearly ordered sets with function symbol [4]. If a structure $\mathcal{A}$ is not computably categorical, then a natural question is concerned with measuring the complexity of the isomorphisms between computable copies of $\mathcal{A}$. Here one uses computable ordinals. Let α be a computable ordinal. A structure $\mathcal{A}$ is Δ_α^0-categorical, if any two computable copies of $\mathcal{A}$ are isomorphic via a Δ_α^0 function. In [19] McCoy studies Δ_2^0-categorical linear orders and Boolean algebras. In [1] Ash characterizes hyperarithmetical categoricity of ordinals. In [17], for any given $n > 0$, a tree is constructed that is Δ_{n+1}^0-categorical but not Δ_n^0-categorical. Similar examples are built in the class of abelian p-groups [3]. There are also examples of Δ_n^0-categorical torsion-free abelian groups for small n [20]. In this paper we apply the transformations Φ and Ψ to construct Δ_α^0-categorical ordered abelian groups and fields. We prove the following theorem:

Theorem 2. *Suppose $\alpha = \delta + 2n + 1$ is a computable ordinal, where δ is either 0 or a limit ordinal, and $n \in \omega$. Then there is there is an ordered abelian group (ordered field) which is Δ_α^0-categorical but not Δ_β^0-categorical, for any $\beta < \alpha$.*

Here is a brief summary of the rest of the paper. Section 1 provides some necessary background and notation for ordered abelian groups. Sections 2 and 3 prove Theorem 1 and Theorem 2 for the case of ordered abelian groups. Section 4 outlines both theorems for the case of ordered fields.

1   Basic Concepts

We briefly introduce basic concepts of ordered abelian groups and computability. Standard references are [25] for computability, [14] for the theory of ordered groups, and [26] for abelian groups. We use $+$ to denote the group operation, 0 for the neutral element, and $-a$ for the inverse of a. We use notation na for $\underbrace{a + \ldots + a}_{n \text{ times}}$, where $n \in \omega$ and $a \in A$. Note if $n = 0$ then $na = 0$. Therefore, every abelian group is a Z-module. So one can define the notion of linear independence (over Z). For more background on general linear algebra see [15].

Definition 1. *Let X be a set. For $x \in X$, set $Zx = \langle \{nx : n \in Z\}, + \rangle$, where $nx + mx = (n + m)x$. The free abelian group over X is the group $\bigoplus_{x \in X} Zx$.*

Thus, the free abelian group over X is isomorphic to the direct sum of $card(X)$ copies of Z. In this paper, when we write $\sum_{b \in B} m_b b$, where B is an infinite subset of an abelian group and all m_b's are integers, we mean that the set $B_0 = \{b \in B : m_b \neq 0\}$ is finite. Thus, $\sum_{b \in B} m_b b$ is just our notation for $\sum_{b \in B_0} m_b b$.

Recall that an ordered abelian group is a triple $\langle G, +, \leq \rangle$, where $(G, +)$ is an abelian group and $\leq$ is a linear order on G such that for all $a, b, c \in G$ the condition $a \leq b$ implies $a + c \leq b + c$.

If $\varphi : A \to B$ is a homomorphism of structures A and B, then we write $a\varphi$ to denote the φ-image of $a \in A$ in B, and we write $A\varphi$ for the range of φ. Recall that $\varphi : A \to B$ is a homomorphism between ordered abelian groups A and B if (1) $(a + b)\varphi = a\varphi + b\varphi$, and (2) $a \leq b$ implies $a\varphi \leq b\varphi$ for all $a, b \in A$.

If $G_1 = (A; +, \leq)$ is a computable ordered abelian group isomorphic to G then we say that G_1 is a computable copy of G. As our groups are infinite and countable, the domain A of G_1 can always be assumed to be ω.

Recall that the arithmetical hierarchy [25] can be extended to the hyperarithmetical hierarchy [2] by iterating the Turing jump up to any computable ordinal. We follow [2] in our notations for the hyperarithmetical hierarchy.

Definition 2. *Let α be a computable ordinal. A structure $\mathcal{A}$ is Δ^0_α-categorical if there exists a Δ^0_α-computable isomorphism between any two computable copies of $\mathcal{A}$. The structure is $\Delta^0_\alpha(X)$-categorical if any two X-computable copies of $\mathcal{A}$ have a $\Delta^0_\alpha(X)$ isomorphism between them.*

2   Proof of Theorem 1 for Ordered Abelian Groups

We define a transformation Φ of the class of linear orders into the class of ordered abelian groups as follows.

Definition 3 (Definition of $\Phi(L)$ and $G(L)$). *Let $L = \langle \{l_i : i \in I\}, \leq \rangle$ be a linear order. Let $G(L)$ be the free abelian group defined over L and ordered lexicographically with respect to L. More formally,*

1. *The domain of $G(L)$ consists of formal finite sums $m_{i_1}l_{i_1} + m_{i_2}l_{i_2} + \ldots + m_{i_k}l_{i_k}$, where $l_{i_j} \in L$, $0 \neq m_{i_j} \in Z$ and $l_{i_1} > l_{i_2} > \ldots > l_{i_k}$ in L.*
2. *The operation $+$ is defined coordinate-wise to make $\langle G(L), + \rangle$ free over L.*
3. *The order is defined by the positive cone as follows: $m_{i_1}l_{i_1} + m_{i_2}l_{i_2} + \ldots + m_{i_k}l_{i_k} > 0$, where $L \models l_{i_1} > l_{i_2} > \ldots > l_{i_k}$, iff $m_{i_1} > 0$.*

It is easy to check that the structure $G(L)$ is an ordered abelian group. Now define $L^\star = L \cup \{\star\}$, where the new element $\star$ is the least element of order $L^\star$. We set $\Phi(L) = G(L^\star)$.

The linear order L can clearly be identified with a subset of $G(L)$ under the mapping $l \to 1 \cdot l$. If $i : L \to Li$ is an isomorphism of linear orders then we say that $i : L \to Li \subset G(Li)$ is an L-embedding of L into $G(Li)$.

Definition 4. *For $g \in G$, where G is an ordered abelian group, the absolute value of $|g|$ is g if $g \geq 0$, and $-g$ otherwise. Two elements $g, h \in G$ are Archimedean equivalent, written $g \sim h$, if there is an integer $m > 0$ such that $|mg| > |h|$ and $|mh| > |g|$. The equivalence classes of $\sim$ are called the Archimedean classes of G.*

The following lemma describes the Archimedean classes of $G(L)$.

Lemma 1. *Two nonzero elements $g, h \in G(L)$ are Archimedean equivalent if and only if they can be written as follows:*

$$g = m_j l_j + \sum_{l_i < l_j} m_i l_i, \quad h = n_j l_j + \sum_{l_i < l_j} n_i l_i,$$

where $m_j \neq 0$ and $n_j \neq 0$.

Proof. If there are such decompositions then g and h are Archimedean equivalent. Now suppose $g \sim h$. By the definition of $G(L)$, every nonzero element a of $G(L)$ can be presented uniquely as $a = m_{i_1}l_{i_1} + m_{i_2}l_{i_2} + \ldots + m_{i_k}l_{i_k}$, where $l_{i_1} > l_{i_2} > \ldots > l_{i_k}$ in L. Thus $g = m_j l_j + \sum_{l_i < l_j} m_i l_i$ and $h = n_k l_k + \sum_{l_t < l_k} n_t l_t$. Suppose $l_j \neq l_k$. Without loss of generality we may assume $l_j < l_k$. Then $|mg| < |h|$, for all $m \in Z^+$. This is a contradiction.

Definition 5. *Let $[a]_\sim$ be the Archimedean equivalence class of $a \in G = G(L)$. Let $\mathcal{L}(G) = \langle \{[a]_\sim : a \in G, a \neq 0\}, \preceq \rangle$, where $[a]_\sim \preceq [b]_\sim$ if $G \models |a| < |b|$ or $[a]_\sim = [b]_\sim$.*

Proposition 1. *The linear orders L_0 and L_1 are isomorphic if and only if the ordered abelian groups $G_0 = G(L_0)$ and $G_1 = G(L_1)$ are isomorphic.*

Proof. If $L_0 \cong L_1$ then clearly $G_0 \cong G_1$. Suppose $L_0 = \langle \{\nu_i : i \in I\}, \leq \rangle$. By Lemma 1, $\mathcal{L}(G_0) = \langle \{[\nu_i]_\sim : i \in I\}, \preceq \rangle \cong \langle \{\nu_i : i \in I\}, \leq \rangle = L_0 \cong \mathcal{L}(G_1) \cong L_1$. Thus $G_0 \cong G_1$ implies $L_0 \cong L_1$.

As a consequence of *Proposition 1, Theorem 1* and ([12], *Theorem 4.4 (d)*), we have the following:

Corollary 1. *The isomorphism problem for ordered abelian groups is Σ_1^1-complete.*

Proposition 2. *Let L be a $0'$-computable linear order with least element. Then there is a computable presentation H of $G(L)$ and a $0'$-computable L-embedding $i : L \to H$.*

Proof. Without loss of generality we assume that the domain of L is ω, and the least element of L has ω-number 0. The diagram $D_0(L)$ of L is $0'$-computable. Thus its characteristic function is the limit of a computable function. At *Step t* we define a finite approximation of $D_0(L)$ enumerating this computable function, and we denote this approximation by $D_0^t(L)$. We require that $D_0^t(L)$ is consistent with the axioms of finite linear order on elements $\{0, 1, \ldots, t\}$. We denote this linear order by L_t. At *Step t*, the procedure enumerates a finite part $D_0^t(H)$ of $D_0(H)$, the finite ordered semigroup H_t which is described by $D_0^t(H)$, and a finite map $i_t : L_t \to H_t$.

The idea of the construction can be illustrated by the following example. Suppose at *Step $t-1$* we have $L_{t-1} = \ldots l_k < l_{k+1} < \ldots$ and $i_{t-1} : L_{t-1} \to H_{t-1}$. At *Step t* we may have $l_{k+1} < l_k$. In this case we declare $i_{t-1}l_{k+1}$ to be equal to $M_t l_k$, where M_t is a big natural number. We will require the i_t-images of l_j to be positive in the group for every t. Therefore $i_{t-1}l_{k+1} = M_t l_k i_{t-1}$ will preserve the order, but will glue the Archimedean classes of these elements. Then we may add new free generators to our group and redefine i. The existence of least element makes the construction simpler.

Step 0. Set $H_0 = \{0, a_0\}$, $L_0 = \{0\}$, $D_0^0(H) = \{a_0 > 0, 0 + 0 = 0\}$, $D_0^0(L) = \emptyset$ and $0 i_0 = a_0$.

Suppose that $D_0^{t-1}(L)$, L_{t-1}, $D_0^{t-1}(G)$, H_{t-1} and i_{t-1} have been defined by the previous stages. Then *Step t* proceeds as follows:

Step t. Perform at most t stages in approximation of $D_0(L)$ on $\{0, 1, \ldots, t\} \subset \omega$ starting with $D_0^{t-1}(L)$, until getting an approximation $D_0^t(L)$ that respects the axioms of a linear order with the least element 0.

We say that $n \in \omega$, $n \leq t - 1$, *is at the wrong place* if there is $k \in \omega$ such that (1.) $\omega \models k < n$, and (2.) $D_0^{t-1}(L) \vdash n < k$ but $D_0^t(L) \vdash k < n$, or $D_0^{t-1}(L) \vdash k < n$ but $D_0^t(L) \vdash n < k$.

Let R^t be the set of all numbers that are declared to be at the wrong place at *Step t*. There is a disjoint partition $R^t = R_0^t \cup R_1^t \cup \ldots$ such that $R_j^t = [r_{j,1}^t, \ldots, r_{j,h(t)}^t]$ is an interval (with end-points, i.e. closed) in L_{t-1}, for every j. By our hypothesis, L has least element 0 (evidently $0 \notin R^t$ for all t). Therefore the partition $R^t = R_0^t \cup R_1^t \cup \ldots$ can be chosen in such a way that for every j there is a number $k_j^t \in \omega$ such that $k_j^t \notin R^t$ and $D_0^{t-1}(L)$ says that the successor of k_j^t is in R_j^t.

Let M_t be a natural number greater then any number we have ever used in our procedure so far. For every $R_j^t = [r_{j,1}^t, \ldots, r_{j,h(t)}^t]$ add to $D_0^{t-1}(H)$ formulas corresponding to the following new equations:

$$r_{j,1}^t i_{t-1} = M_t \cdot k_j^t i_{t-1}, \quad r_{j,2}^t i_{t-1} = M_t^2 \cdot k_j^t i_{t-1}, \ldots, \quad r_{j,h(t)}^t i_{t-1} = M_t^{h(t)} \cdot k_j^t i_{t-1}.$$

Let $D(t)$ be $D_0^{t-1}(H)$ extended by this set of formulas.

For every $k \in R^t$ and for $k = t$ add a new element $x_k > 0$ to H_{t-1}, and for every $k \leq t$ set

$$k i_t = \begin{cases} k i_{t-1}, & \text{if } n \notin R^t \text{ and } k \neq t, \\ x_k, & \text{otherwise.} \end{cases}$$

Extend the order on H_{t-1} to $(R^t \cup \{t\})i_t$. This new extended order must meet the following requirements: (1) $L_t i_t \cong L_t$ (as linear orders) and (2) for all $k, s \notin R^t$, $D_0^{t-1}(G) \vdash k i_{t-1} < s i_{t-1}$ iff $D(t) \vdash k i_t < s i_t$. These are the only restrictions on the extended order on new elements $\{x_k\}_{k \in R^t}$. Therefore we can make these new elements look Archimedean independent of the all previously defined elements. Thus, we can always find an extended order satisfying (1) and (2). We can put x_k between the previously defined Archimedean classes to make sure that $L_t i_t \cong L_t$. Requirement (2) will be satisfied automatically.

Set $H_t = \{\sum_{0 \le k \le t} n_k \cdot (k i_t) : |n_k| \le M_t^{t+1}\}$ and define $+$ on this set using the restriction: $a, \bar{b} \in H_t$ implies $a + b \in H_t$, for all a, b. Then extend the linear order on $L_t i_t$ to the whole of G_t lexicographically. Set $D_0^t(G) = D_0(H_t)$.

Lemma 2. *The map* $\lim_t i_t = i : L \longrightarrow Li$ *is a $0'$-computable isomorphism of linear orders.*

Proof. For every $k \in \omega$ and every t_0, $k i_{t_0} = \lim_t k i_t$ if and only if $k \notin R_t$, for all $t \ge t_0$. Indeed, $k i_t \ne k i_{t-1}$ if and only if k is declared to be at the wrong place at *Step t*. But the diagram of L is $0'$-computable, and the definition of R_t uses the natural order of ω. By a simple inductive argument, $\lim_t k i_t$ exists for all k. Thus i is $0'$-computable. By the construction, $L_t i_t \cong L_t$. By our assumption, for all $k, s \in L$ there is t_0 such that $k, s \notin R_t$, for all $t \ge t_0$. Thus $D_0(L) \vdash s < k \Leftrightarrow D_0^t(L) \vdash s < k \Leftrightarrow D_0^t(H) \vdash s i_t < k i_t \Leftrightarrow D_0(H) \vdash si < ki$.

The output $D_0 = \cup_t D_0^t(H)$ of the procedure is a computable diagram of ordered abelian group $H = \cup_t H_t$. Finally, by our construction and by *Lemma 2*, $H \cong G(Li)$. This finishes the proof.

Now *Theorem 1* follows from *Propositions 1* and *2*. Indeed, $\Phi(L) = G(L^*)$, L^* has least element $\star$, L is X-computable if and only if L^* is, and $L^* \cong L_0^*$ if and only if $L \cong L_0$.

3   Proof of Theorem 2 for Ordered Abelian Groups

In this section all groups are free lexicographically ordered abelian groups. So let L be a linear order, and let $i : L \to G = G(Li)$ be an L-embedding. Consider the *projection* map $\pi : G \setminus \{0\} \to L$ defined by the following rule. If $g \in G$, $g \ne 0$ and $g = m_j l_j + \sum_{l_i < l_j} m_i l_i$, then $g\pi = (m_j l_j + \sum_{l_i < l_j} m_i l_i)\pi = l_j \in L$. Note that in this particular representation of g we list the corresponding free generators in the decreasing order. This map has a number of useful properties.

(1) $li\pi = l$, for all $l \in L$.

(2) π induces a homomorphism of the linear order $(G^+, \le)$ onto L, where G^+ is the cone of (strictly) positive elements.

(3) For all nonzero $a, b \in G$, $a \sim b$ if and only if $a\pi = b\pi$ (see *Lemma 1*).

(4) Let $\tau : G \setminus \{0\} \to \mathcal{L}(G)$ be the canonical map with respect to $\sim$, i.e. $a\tau = [a]_\sim$ for all $a \in G$, $a \ne 0$. If $\gamma : \mathcal{L}(G) \to L$ is an isomorphism such that $i\tau = \gamma^{-1}$, then $\tau\gamma = \pi$ (by $(1) - (3)$ and the proof of *Proposition 1*).

Lemma 3. *Let L and L_0 be linear orders and let $i : L \to G$ and $i_0 : L_0 \to H$ be L- and L_0-embeddings respectively. Also let $\pi_0 : H \to L_0$ be the projection. If $\varphi : G \to H$ is an isomorphism (of ordered abelian groups) then $i\varphi\pi_0 : L \to L_0$ is an isomorphism (of linear orders).*

Proof. It is not hard to see that $i\varphi\pi_0$ is a homomorphism of linear orders. The isomorphism φ preserves Archimedean equivalence, and every Archimedean class in $\mathcal{L}(G)$ has exactly one representative xi for some $x \in L$. Thus every element of $\mathcal{L}(H)$ has exactly one representative $xi\varphi$ for some $x \in L$. Therefore $i\varphi\tau_0 : L \to \mathcal{L}(H)$ is an isomorphism, where $\tau_0 : G \setminus \{0\} \to \mathcal{L}(H)$ is the canonical homomorphism, see (4) above. Thus, $i\varphi\tau_0\gamma_0$ is an isomorphism of L onto L_0, where $\gamma_0 : \mathcal{L}(H) \to L_0$ is an isomorphism such that $i_0\tau_0 = \gamma_0^{-1}$. By property (4) of projection maps applied to π_0, $i\varphi\tau_0\gamma_0 = i\varphi\pi_0$.

Now we turn to computable properties of the embedding described above.

Let $G = G(L)$ be computable. By the proof of *Proposition 1*, L is isomorphic to $\langle G \setminus \{0\}, \leq \rangle / \sim$, where $\sim$ is the Archimedean equivalence. The relation $\sim$ is Σ^0_1, thus L has a Σ^0_1-presentation, and it is $0'$-computable. There is a $0'$-computable set of representatives of $\mathcal{L}(G)$ in G. Recall that we identify L with the set of free generators of $G(L)$.

Proposition 3. *Suppose $G = G(L)$ is computable. Let $S \subset G$ be a $0'$-computable set of representatives of $\mathcal{L}(G)$. There is a set of representatives $C \subset G$ of $\mathcal{L}(G)$ and a $0''$-computable bijective map $\beta : S \to C$ such that:*
(1) $s \sim s\beta$, for all $s \in S$;
(2) there is an automorphism ψ of G such that $L\psi = C$, and $l\psi \sim l$ for all $l \in L$.

Proof. Consider the following $0''$-computable procedure, which builds the set C. We will show later that this set satisfies the needed requirements. We will define β later as well.

At *Step 0* we set $C_0 = \{l_0\}$, for some $l_0 \in L$. Suppose that C_{t-1} has been defined at *Step (t-1)*. At *Step t* search for a finite family of elements $C(t)$ such that:
(1) $c > 0$, for all $c \in C(t)$;
(2) $c \nsim c'$, for all $c \neq c'$ such that $c, c' \in C_{t-1} \cup C(t)$;
(3) for every $h \in G$, every $c \in C(t)$ and every integer k s.t. $|k| > 1$, $h \sim c$ implies $kh + c \sim c$;
(4) if $g \in G$ is the element with ω-number t then $g = \sum_{c \in C_{t-1} \cup C(t)} m_c c$ (where m_c is an integer, for every c).
Set $C_t = C_{t-1} \cup C(t)$ and proceed to *Step t+1*.

Lemma 4. *For every t, Step t halts.*

Proof. If $C(t)$ satisfies (3) and $c \in C(t)$, then $c = l_j + \sum_{l_i < l_j} m_i l_i$, for some $l_j \in L$. Indeed, if $c = l_j + \sum_{l_i < l_j} m_i l_i$, then (3) holds, by *Lemma 1*. If $c = m_j l_j + \sum_{l_i < l_j} m_i l_i$ and $|m_j| > 1$, then (3) fails for $h = l_j$ and $k = -m_j$.

Thus, if C_{t-1} has been defined then $C_{t-1} = \{l_j + \sum_{i \in K_j} m_i l_i : j \in J_{t-1}\}$, where J_{t-1} and all the K_j's are finite. The element g with ω-number t can be written as $g = \sum_{k \in I(t)} n_k l_k$, for some finite set $I(t)$. If we set $C(t) = \{l_k : k \in [I(t) \cup \bigcup_{j \in J_{t-1}} K_j] \setminus J_{t-1}\}$ then the requirements $(1) - (4)$ will be satisfied. Thus, there is at least one extension of C_{t-1} which satisfies $(1) - (4)$.

Let $C = \bigcup_t C_t$. By *Lemma 4* and requirement (4) of *Step t*, every element of G can be written as $\sum_{c \in C} m_c c$. This decomposition is unique because C is linearly independent (by requirement (2)). Therefore, G (without order) is isomorphic to the free abelian group over C.

Now suppose $s \in S$, $s = \sum_{c \in C} m_c c$. By requirement (2) of the procedure and *Lemma 1*, there is the unique $c_s \in C$ such that $c_s \sim s$. Let $s\beta = c_s$. This map is $0''$-computable.

For every $l_j \in L$ let $l_j \psi = c$, where $c \in C$ and $c \sim l_j$ (this element is unique, as above). Then close ψ under $+$ and $-$ by setting $(g - h)\psi = g\psi - h\psi$. It is not hard to see that ψ is bijective and preserves $+$ and $-$. On the other hand, $\sum_{j \in J} n_j l_j = n_{j_0} l_{j_0} + \sum_{l_j < l_{j_0}} n_j l_j > 0$ iff $n_{j_0} > 0$. But $(\sum_{j \in J} n_j l_j)\psi = \sum_{j \in J} n_j (l_j + \sum_{l_i < l_j} m_i l_i) > 0$ iff $n_{j_0} > 0$. Thus ψ is an automorphism of G.

Proposition 4. *Let L be a (countable) linear order with the least element and let $\alpha > 2$. Then L is $\Delta_\alpha^0(0')$-categorical if and only if $G(L)$ is Δ_α^0-categorical.*

Proof. ($\Rightarrow$). Suppose $G \cong G_0 \cong G(L)$. Consider the $0'$-computable sets $\mathcal{L}$ and $\mathcal{L}_0$ of representatives of $\mathcal{L}(G)$ and $\mathcal{L}(G_0)$ respectively. Sets $\mathcal{L}$ and $\mathcal{L}_0$ with corresponding induced orders are isomorphic to L (see *Proposition 1*). By our assumption, there is a Δ_α^0 isomorphism $\varphi : \mathcal{L} \to \mathcal{L}_0$. By *Proposition 3* there are $0''$-computable bijective maps $\beta : \mathcal{L} \to G(L)$ and $\beta_0 : \mathcal{L}_0 \to G(L_0)$ such that $G = G(\mathcal{L}\beta)$ and $G_0 = G(\mathcal{L}_0\beta_0)$. Thus $\beta^{-1}\varphi\beta_0 : L \to L_0$ can be extended to a Δ_α^0 isomorphism of G onto G_0.

($\Leftarrow$). Suppose that $G(L)$ is Δ_α^0-categorical. Let L_0 and L_1 be $0'$-computable copies of L. By *Proposition 2*, there are computable ordered abelian groups $G_0 \cong G_1 \cong G(L)$ and $0'$-computable L_0- and L_1-embeddings $i_0 : L_0 \to G_0$ and $i_1 : L_1 \to G_1$. By our assumption there is a Δ_α^0 isomorphism $\varphi : G_0 \to G_1$. Let $\pi_1 : G_1 \to L_1$ be a $0'$-computable projection such that $i_1\pi_1 = id_{L_1}$. By *Lemma 3*, the map $i_0\varphi\pi_1 : G_0 \to G_1$ is a Δ_α^0 isomorphism.

The following result has already been mentioned:

Theorem 3 (Ash, [1]). *Let α be a computable ordinal. Suppose $\omega^{\delta+n} \leq \alpha < \omega^{\delta+n+1}$, where δ is either 0 or a limit ordinal, and $n \in \omega$. Then α is $\Delta_{\delta+2n}^0$-categorical and is not Δ_β^0-categorical, for $\beta < \delta + 2n$.*

By Proposition 4 and relativized Theorem 3, $G(\omega^{\delta+n})$ is $\Delta_{\delta+2n+1}^0$-categorical and is not Δ_β^0-categorical, for $\beta < \delta + 2n + 1$. This proves *Theorem 2*.

4   Ordered Fields

Let $L = \langle \{x_i : i \in I\}, \leq \rangle$ be a (nonempty) linear order. Consider the ring of polynomials $Z[L]$ and the field of rational fractions $Q(L)$ with the set L used as the set of indeterminates. Informally, we will define an order on $Q(L)$ such that x_i is infinitely large relative to x_j if $L \models x_i > x_j$.

First, we order the set of monomials $\prod_i x_i^{\epsilon_i}$ *lexicographically* relative to L. We illustrate this definition by the following examples. Suppose $L \models x_3 < x_2 < x_1$. Then $x_1 x_3 > x_2^4 x_3^3$ since the first monomial has indeterminate x_1 which is L-greater then any indeterminate in the second monomial. We have $x_1 x_2 > x_1 x_3^3$ since $L \models x_2 > x_3$. Finally, $x_1 x_2^3 > x_1 x_2^2 x_3$ because the L-greatest indeterminate x_2 in which these two monomials differ has power 3 in the first monomial, and 2 in the second. For any given $p \in Z[L]$ we can find the largest monomial $B(p)$ of p relative to the order defined above. We denote by $\mathcal{C}(p)$ the integer coefficient of $B(p)$ and let $p > 0$ if $\mathcal{C}(p) > 0$. Finally, $\frac{p}{q} > 0$ $(q > 0)$ if and only if $p > 0$. One can see that $Q(L)$ with this order respects all the axioms of ordered fields (see [15]).

Definition 6. *Suppose L is a linear order. Recall $L^\star = L \cup \{\star\}$ with least element $\star$. Set $\Psi(L) = Q(L^\star)$.*

Definition 7. *Let F be an ordered field. For any $g \in F$ we define its absolute value $|g|$ to be g if $g \geq 0$, and $-g$ if $g < 0$. Two elements a and b of F are Archimedean equivalent, written $a \sim b$, if there is integer $m > 0$ such that $|a^m| > |b|$ and $|b^m| > |a|$.*

Suppose $a \in Q(L)$. Let $\pi(a) = x_i$ where x_i is the L-greatest element of L that appears in the numerator of a. Say, $\pi(\frac{x_1^3 x_2 + x_1^2 x_3}{x_2 x_3 - x_3}) = x_1$, if $L \models x_1 > x_2 > x_3$. Note that for $a, b > 1$ and $a, b \nsim 1$, $a \sim b$ if and only if $\pi(a) = \pi(b)$. Denote by $Q_1(L)$ the set of all elements of $Q(L)$ which are greater than 1 and that are not Archimedean equivalent to 1. Then $(Q_1(L)/\sim) \cong L$. Therefore, we have established the following

Proposition 5. *Let L and L_0 be linear orders. Then $Q(L) \cong Q(L_0)$ if and only if $L \cong L_0$.*

The proofs of *Theorems 1* and *2* for fields are similar to the proofs for ordered groups. In the proof of *Theorem 1* we declare $x_i = x_j^M$ (for a large number M) to make $x_i \sim x_j$. In the proof of *Theorem 2* we need the same technical propositions. In the proof of *Proposition 3* for fields we use slightly modified requirements to the extensions which respect the new definition of Archimedean equivalence.

References

1. Ash, C.J.: Recursive labelling systems and stability of recursive structures in hyperarithmetical degrees. Trans. of Amer. Math. Soc. 298, 497–514 (1986)
2. Ash, C.J., Knight, J.F.: Computable structures and the hyperarithmetical hyerarchy. Elsevier, Amsterdam (2000)
3. Barker, E.J.: Back and forth relations for reduced abelian p-groups. Annals of Pure and Applied Logic 75, 223–249 (1995)

4. Cenzer, D., Csima, B.F., Khoussainov, B.: Linear orders with distinguished function symbol. Archive for Mathematical Logic, Special Issue: University of Florida Special Year in Logic 48(1), 63–76 (2009)
5. Downey, R., Montolban, A.: The isomorphism problem for torsion-free abelian groups is analytic complete. Journal of Algebra 320, 2291–2300 (2008)
6. Hirschfeldt, D., Khoussainov, B., Shore, R., Slinko, A.: Degree Spectra and Computable Dimensions in Algebraic Structures. Annals of Pure and Applied Logic 115, 71–113 (2002)
7. Goncharov, S., Lempp, S., Solomon, R.: The computable dimension of ordered abelian groups. Advances in Mathematics 175, 102–143 (2003)
8. Goncharov, S.: Countable Boolean algebras and decidability. In: Siberian School of Algebra and Logic, Novosibirsk, Nauchnaya Kniga (1996)
9. Goncharov, S.: Autostability of models and abelian groups. Algebra and Logic 19, 13–27 (1980) (English translation)
10. Goncharov, S., Dzgoev, V.: Autostability of models. Algebra and Logic 19(1), 45–58 (1980)
11. Goncharov, S., Ershov, Y.: Constructive models. Kluwer Academic Pub., Dordrecht (2000)
12. Goncharov, S., Knight, J.: Computable Structure and Non-Structure Theorems. Algebra and Logic 41(6), 351–373 (2002)
13. Khisamiev, N.: Constructive abelian groups. In: Handbook of recursive mathematics, part 2, vol. 2. Elsevier, Amsterdam (1998)
14. Kopytov, A., Kokorin, V.: Fully Ordered Groups. John Wiley and Sons, Chichester (1974)
15. Lang, S.: Algebra. Springer, Heidelberg (2002)
16. La Roche, P.: Recursively presented Boolean algebras. Notices AMS 24, 552–553 (1977)
17. Lempp, S., McCoy, C., Miller, R., Solomon, R.: Computable Categoricity of Trees of Finite Height. Journal of Symbolic Logic 70(1), 151–215 (2005)
18. Maltsev, A.I.: On Recursive Abelian Groups, transl. Soviet Math. Dokl. 3
19. McCoy, C.: Δ_2^0-Categoricity in Boolean algebras and linear orderings. Annals of Pure and Applied Logic 119, 85–120 (2003)
20. Melnikov, A.G.: $0''$-Categorical completely decomposable torsion-free abelian groups. In: Proceedings of CiE 2009. Springer, Heidelberg (2009)
21. Nurtazin, A.T.: Computable classes and algebraic criteria of autostability. Summary of Scientific Schools, Math. Inst., Novosibirsk (1974)
22. Rabin, M.: Computable algebra, general theory, and theory of com- putable elds. Transactions of the American Mathematical Society 95, 341–360 (1960)
23. Remmel, J.B.: Recursively categorical linear orderings. Proc. Amer. Math. Soc. 83, 387–391 (1981)
24. Smith, R.L.: Two theorems on autostability in p-groups. In: Logic Year 1979 - 80. LNM, vol. 859, pp. 302–311. Springer, University of Connecticut, Berlin, Storrs (1981)
25. Soare, R.I.: Recursively Enumerable Sets and Degrees. Springer, New York (1987)
26. Fuchs, L.: Infinite Abelian Groups, vol. I. Academic Press, London (1975)

Focusing in Asynchronous Games

Samuel Mimram[*]

CEA LIST / École Polytechnique
CEA LIST, Laboratory for the Modelling and Analysis of Interacting Systems,
Point Courrier 94, 91191 Gif-sur-Yvette, France
samuel.mimram@cea.fr

Abstract. Game semantics provides an interactive point of view on
proofs, which enables one to describe precisely their dynamical behavior during cut elimination, by considering formulas as games on which
proofs induce strategies. We are specifically interested here in relating
two such semantics of linear logic, of very different flavor, which both
take in account concurrent features of the proofs: asynchronous games
and concurrent games. Interestingly, we show that associating a concurrent strategy to an asynchronous strategy can be seen as a semantical
counterpart of the focusing property of linear logic.

A cut-free proof in sequent calculus, when read from bottom up, progressively
introduces the connectives of the formula that it proves, in the order specified
by the syntactic tree constituting the formula, following the conventions induced
by the logical rules. In this sense, a formula can be considered as a playground
that the proof will explore. The formula describes the rules that this exploration
should obey, it can thus be abstractly considered as a *game*, whose moves are
its connectives, and a proof as a *strategy* to play on this game. If we follow the
principle given by the Curry-Howard correspondence, and see a proof as some
sort of program, this way of considering proof theory is particularly interesting
because the strategies induced by proofs describe very precisely the interactive
behavior of the corresponding program in front of its environment.

This point of view is at the heart of *game semantics* and has proved to be very
successful in order to provide denotational semantics which is able to describe
precisely the dynamics of proofs and programs. In this interactive perspective,
two players are involved: the *Proponent*, which represents the proof, and the
Opponent, which represents its environment. A formula induces a game which is
to be played by the two players, consisting of a set of moves together with the
rules of the game, which are formalized by the polarity of the moves (the player
that should play a move) and the order in which the moves should be played. The
interaction between the two players is formalized by a *play*, which is a sequence
of moves corresponding to the part of the formula being explored during the
cut-elimination of the proof with another proof. A proof is thus described in this

---

[*] This work has been supported by the CHOCO (ANR-07-BLAN-0324) ANR project.

F. Ferreira et al. (Eds.): CiE 2010, LNCS 6158, pp. 331–341, 2010.

setting by a *strategy* which corresponds to the set of interactions that the proof is willing to have with its environment.

This approach has been fruitful for modeling a wide variety of logics and programming languages. By refining Joyal's category of Conway games and Blass' games, Abramsky and Jagadeesan were able to give the first fully complete game model of the multiplicative fragment of linear logic extended with the MIX rule [3], which was later refined into a fully abstract model of PCF [4], a prototypical programming language. Here, "fully complete" and "fully abstract" essentially mean that the model is very precise, in the sense that every strategy is *definable* (i.e. is the interpretation of a proof or a program). At exactly the same time, Hyland and Ong gave another fully abstract model of PCF based on a variant of game semantics called *pointer games* [10]. In this model, definable strategies are characterized by two conditions imposed on strategies (well-bracketing and innocence). This setting was shown to be extremely expressive: relaxing in various ways these conditions gave rise to fully abstract models of a wide variety of programming languages with diverse features such as references, control, etc. [8].

Game semantics is thus helpful to understand how logic and typing regulate computational processes. It also provides ways to analyze them (for example by doing model checking [2]) or to properly extend them with new features [8], and this methodology should be helpful to understand better concurrent programs. Namely, concurrency theory being relatively recent, there is no consensus about what a good process calculus should be (there are dozens of variants of the π-calculus and only one λ-calculus) and what a good typing system for process calculus should be: we believe that the study of denotational semantics of those languages is necessary in order to reveal their fundamental structures, with a view to possibly extending the Curry-Howard correspondence to programming languages with concurrent features. A few game models of concurrent programming languages have been constructed and studied. In particular, Ghica and Murawski have built a fully abstract model of Idealized Algol (an imperative programming language with references) extended with parallel composition and mutexes [9] and Laird a game semantics of a typed asynchronous π-calculus [11].

In this paper, we take a more logical point of view and are specifically interested in concurrent denotational models of linear logic. The idea that multiplicative connectives express concurrent behaviors is present since the beginnings of linear logic: it is namely very natural to see a proof of $A \,\mathring{\gamma}\, B$ or $A \otimes B$ as a proof of A in "parallel" with a proof of B, the corresponding introduction rules being

$$\frac{\vdash \Gamma, A, B}{\vdash \Gamma, A \,\mathring{\gamma}\, B}[\mathring{\gamma}] \qquad \text{and} \qquad \frac{\vdash \Gamma, A \quad \vdash \Delta, B}{\vdash \Gamma, \Delta, A \otimes B}[\otimes]$$

with the additional restriction that the two proofs should be "independent" in the case of the tensor, since the corresponding derivations in premise of the rule are two disjoint subproofs. Linear logic is inherently even more parallel: it has the *focusing* property [6] which implies that every proof can be reorganized into one in which all the connectives of the same polarity at the root of a formula are introduced at once (this is sometimes also formulated using *synthetic connectives*). This property, originally discovered in order to ease proof-search has later on revealed to be fundamental in semantics and type theory. Two game models of linear logic have been developed in order to capture this intuition. The first, by Abramsky and Melliès, called *concurrent games*, models strategies as closure operators [5] following the domain-theoretic principle that computations add information to the current state of the program (by playing moves). It can be considered as a big-step semantics because concurrency is modeled by the ability that strategies have to play multiple moves at once. The other one is the model of *asynchronous games* introduced by Melliès [12] where, in the spirit of "true concurrency", playing moves in parallel is modeled by the possibility for strategies to play any interleaving of those moves and these interleavings are considered to be equivalent. We recall here these two models and explain here that concurrent games can be related to asynchronous games using a semantical counterpart of focusing. A detailed presentation of these models together with the proofs of many properties evoked in this paper can be found in [14,15].

1 Asynchronous Games

Recall that a *graph* $G = (V, E, s, t)$ consists of a set V of vertices (or *positions*), a set E of edges (or *transitions*) and two functions $s, t : E \to V$ which to every transition associate a position which is called respectively its *source* and its *target*. We write $m : x \longrightarrow y$ to indicate that m is a transition with x as source and y as target. A *path* is a sequence of consecutive transitions and we write $t : x \longrightarrow\!\!\!\!\to y$ to indicate that t is a path whose source is x and target is y. The concatenation of two consecutive paths $s : x \longrightarrow\!\!\!\!\to y$ and $t : y \longrightarrow\!\!\!\!\to z$ is denoted $s \cdot t$. An *asynchronous graph* $G = (G, \diamond)$ is a graph G together with a *tiling* relation $\diamond$, which relates paths of length two with the same source and the same target. If $m : x \longrightarrow y_1$, $n : x \longrightarrow y_2$, $p : y_1 \longrightarrow z$ and $q : y_2 \longrightarrow z$ are four transitions, we write

$$
\begin{array}{ccc}
 & z & \\
{\scriptstyle p}\nearrow & \sim & \nwarrow{\scriptstyle q} \\
y_1 & & y_2 \\
{\scriptstyle m}\nwarrow & & \nearrow{\scriptstyle n} \\
 & x &
\end{array}
\tag{1}
$$

to diagrammatically indicate that $m \cdot p \diamond n \cdot q$. We write $\sim$ for the smallest congruence (wrt concatenation) containing the tiling relation. This relation is called *homotopy* because it should be thought as the possibility, when s and t are two homotopic paths, to deform "continuously" the path s into t. From the concurrency point of view, a homotopy between two paths indicates that these paths are the same up to reordering of independent events, as in Mazurkiewicz traces. In the diagram (1), the transition q is the *residual* (in the sense of rewriting theory) of the transition m after the transition n, and similarly p is the residual of n after m; the *event* (also called *move*) associated to a transition is therefore its equivalence class under the relation identifying a transition with its residuals. In the asynchronous graphs we consider, we suppose that given a path $m \cdot p$ there is at most one path $n \cdot q$ forming a tile (1). We moreover require that a transition should have at most one residual after another transition.

We consider formulas of the multiplicative and additive fragment of linear logic (MALL), which are generated by the grammar

$$A \quad ::= \quad A \,\invamp\, A \quad | \quad A \otimes A \quad | \quad A \,\&\, A \quad | \quad A \oplus A \quad | \quad X \quad | \quad X^*$$

where X is a variable (for brevity, we don't consider units). The $\invamp$ and $\&$ (resp. $\otimes$ and $\oplus$) connectives are sometimes called *negative* or *asynchronous* (resp. *positive* or *synchronous*). A *position* is a term generated by the following grammar

$$x \quad ::= \quad \dagger \quad | \quad x \,\invamp\, x \quad | \quad x \otimes x \quad | \quad \&_L x \quad | \quad \&_R x \quad | \quad \oplus_L x \quad | \quad \oplus_R x$$

The de Morgan dual A^* of a formula is defined as usual, for example $(A \otimes B)^* = A^* \,\invamp\, B^*$, and the dual of a position is defined similarly. Given a formula A, we write $\mathrm{pos}(A)$ for the set of valid positions of the formula which are defined inductively by $\dagger \in \mathrm{pos}(A)$ and if $x \in \mathrm{pos}(A)$ and $y \in \mathrm{pos}(B)$ then $x \,\invamp\, y \in \mathrm{pos}(A \,\invamp\, B)$, $x \otimes y \in \mathrm{pos}(A \otimes B)$, $\&_L x \in \mathrm{pos}(A \,\&\, B)$, $\&_R y \in \mathrm{pos}(A \,\&\, B)$, $\oplus_L x \in \mathrm{pos}(A \oplus B)$ and $\oplus_R y \in \mathrm{pos}(A \oplus B)$.

An *asynchronous game* $G = (G, *, \lambda)$ is an asynchronous graph $G = (V, E, s, t)$ together with a distinguished *initial position* $* \in V$ and a function $\lambda : E \to \{O, P\}$ which to every transition associates a *polarity*: either O for Opponent or P for Proponent. A transition is supposed to have the same polarity as its residuals, polarity is therefore well-defined on moves. We also suppose that every position x is *reachable* from the initial position, i.e. that there exists a path $* \longrightarrow\!\!\!\!\!\rightarrow x$. Given a game G, we write G^* for the game G with polarities inverted. Given two games G and H, we define their asynchronous product $G \| H$ as the game whose positions are $V_{G \| H} = V_G \times V_H$, whose transitions are $E_{G \| H} = E_G \times V_H + V_G \times E_H$ (by abuse of language we say that a transition is "in G" when it is in the first component of the sum or "in H" otherwise) with the source of $(m, x) \in E_G \times V_H$ being $(s_G(m), x)$ and its target being $(t_G(m), x)$, and similarly for transitions in $V_G \times E_H$, two transitions are related by a tile whenever they are all in G (resp. in H) and the corresponding transitions in G (resp. in H) are related by a tile or when two of them are an instance of a transition in G and the two other are instances of a transition in H, the initial position is $(*_G, *_H)$ and the polarities of transitions are those induced by G and H.

To every formula A, we associate an asynchronous game G_A whose vertices are the positions of A as follows. We suppose fixed the interpretation of the free variables of A. The game $G_{A \invamp B}$ is obtained from the game $G_A \| G_B$ by replacing every pair of positions (x, y) by $x \,\invamp\, y$, and adding a position $\dagger$ and an Opponent transition $\dagger \longrightarrow \dagger \,\invamp\, \dagger$. The game $G_{A \& B}$ is obtained from the disjoint union $G_A + G_B$ by replacing every position x of G_A (resp. G_B) by $\&_L x$ (resp. $\&_R x$), and adding a position $\dagger$ and two Opponent transitions $\dagger \longrightarrow \&_L \dagger$ and $\dagger \longrightarrow \&_R \dagger$. The games associated to the other formulas are deduced by de Morgan duality: $G_{A^*} = G_A^*$. This operation is very similar to the natural embedding of event structures into asynchronous transition systems [17]. For example, if we interpret the variable X (resp. Y) as the game with two positions $\dagger$ and x (resp. $\dagger$ and y) and one transition between them, the interpretation of the formula $(X \otimes X^*) \,\&\, Y$ is depicted in (2). We have made explicit the positions of the games in order to underline the fact that they correspond to partial explorations of formulas, but the naming of a position won't play any role in the definition of asynchronous strategies.

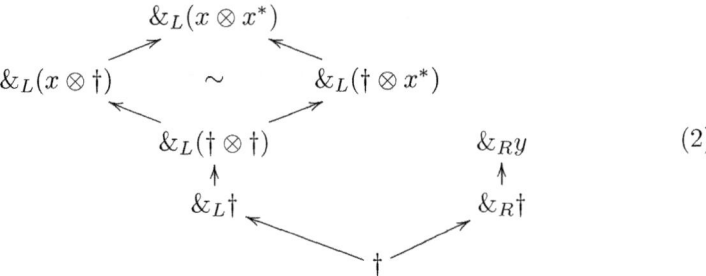

$$(2)$$

A *strategy* σ on a game G is a prefix-closed set of *plays*, which are paths whose source is the initial position of the game. To every proof π of a formula A, we associate a strategy, defined inductively on the structure of the proof. Intuitively, these plays are the explorations of formulas allowed by the proof. For example, the strategies interpreting the proofs

$$\dfrac{\dfrac{\pi}{\vdash \Gamma, A, B}}{\vdash \Gamma, A \,\mathbin{\invamp}\, B}[\mathbin{\invamp}] \qquad \text{and} \qquad \dfrac{\dfrac{\pi}{\vdash \Gamma, A}}{\vdash \Gamma, A \oplus B}[\oplus_L]$$

will contain plays which are either empty or start with a transition $\dagger \longrightarrow \dagger \,\mathbin{\invamp}\, \dagger$ (resp. $\dagger \longrightarrow \oplus_L \dagger$) followed by a play in the strategy interpreting π. The other rules are interpreted in a similar way. To be more precise, since the interpretation of a proof depends on the interpretation of its free variables, the interpretation of a proof will be an uniform family of strategies indexed by the interpretation of the free variables in the formula (as in e.g. [3]) and axioms proving $\vdash A, A^*$ will be interpreted by copy-cat strategies on the game interpreting A. For the lack of space, we will omit details about variables and axioms.

Properties characterizing definable strategies were studied in the case of alternating strategies (where Opponent and Proponent moves should alternate strictly in plays) in [13] and generalized to the non-alternating setting that we are considering here in [14,15]. We recall here the basic properties of definable strategies. One of the interest of these is that they allow one to reason about strategies in a purely local and diagrammatic fashion. It can be shown that every definable strategy σ is

- *positional*: for every three paths $s, t : * \longrightarrow x$ and $u : x \longrightarrow y$, if $s \cdot u \in \sigma$, $s \sim t$ and $t \in \sigma$ then $t \cdot u \in \sigma$. This property essentially means that a strategy is a subgraph of the game: a strategy σ induces a subgraph G_σ of the game which consists of all the positions and transitions contained in at least one play in σ, conversely every play in this subgraph belongs to the strategy when the strategy is positional. In fact, this graph G_σ may be seen itself as an asynchronous graph by equipping it with the tiling relation induced by the underlying game.
- *deterministic*: if the graph G_σ of the strategy contains a transition $n \colon x \longrightarrow y_2$ and a Proponent transition $m : x \longrightarrow y_1$ then it also contains the residual of m along n, this defining a tile of the form (1).

– *receptive*: if σ contains a play $s : * \longrightarrow x$ and there exists an Opponent move $m : x \longrightarrow y$ in the game then the play $s \cdot m : * \longrightarrow y$ is also in σ.
– *total*: if σ contains a play $s : * \longrightarrow x$ and there is no Opponent transition $m : x \longrightarrow y$ in the game then either the position x is terminal (there is no transition with x as source in the game) or there exists a Proponent transition $m : x \longrightarrow y$ such that $s \cdot m$ is also in σ.

2   Focusing in Linear Logic

In linear logic, a proof of the form depicted on the left-hand side of

$$
\cfrac{\cfrac{\cfrac{\pi_1}{\vdash A, B, C}}{\vdash A \,\invamp\, B, C}[\invamp] \quad \cfrac{\pi_2}{\vdash D}}{\vdash A \,\invamp\, B, C \otimes D}[\otimes]
\qquad\qquad
\cfrac{\cfrac{\cfrac{\pi_1}{\vdash A, B, C} \quad \cfrac{\pi_2}{\vdash D}}{\vdash A, B, C \otimes D}[\otimes]}{\vdash A \,\invamp\, B, C \otimes D}[\invamp]
$$

can always be reorganized into the proof depicted on the right-hand side. This proof transformation can be seen as "permuting" the introduction of $\otimes$ after the introduction of $\invamp$ (when looking at proofs bottom-up). From the point of view of the strategies associated to the proofs, the game corresponding to the proven sequent contains

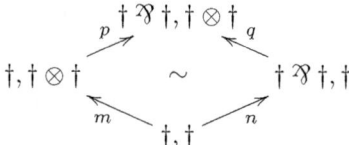

and the transformation corresponds to replacing the path $m \cdot p$ by the path $n \cdot q$ in the strategy associated to the proof. More generally, the introduction rules of two negative connectives can always be permuted, as well as the introduction of two positive connectives, and the introduction rule of a positive connective can always be permuted after the introduction rule of a negative one. Informally, a negative (resp. positive) can always be "done earlier" (resp. "postponed"). We write $\pi \prec \pi'$ when a proof π' can be obtained from a proof π by a series of such permutations of rules.

These permutations of rules are at the heart of Andreoli's work [6] which reveals that if a formula is provable then it can be found using a *focusing* proof search, which satisfies the following discipline: if the sequent contains a negative formula then a negative formula should be decomposed (*negative phase*), otherwise a positive formula should be chosen and decomposed repeatedly until a (necessarily unique) formula is produced (*positive phase*) – this can be formalized using a variant of the usual sequent rules for linear logic. From the point of view of game semantics, this says informally that every strategy can be reorganized into one playing alternatively a "bunch" of Opponent moves and a "bunch" of Proponent moves.

All this suggests that proofs in sequent calculus are too sequential: they contain inessential information about the ordering of rules, and we would like to work with proofs modulo the congruence generated by the $\prec$ relation. Semantically, this can be expressed as follows. A strategy σ is *courteous* when for every tile of the form (1) of the game, such that the path $m \cdot p$ is in (the graph G_σ of) the strategy σ, and either m is a Proponent transition or p is an Opponent transition, the path $n \cdot q$ is also in σ. We write $\tilde{\sigma}$ for the smallest courteous strategy containing σ. Courteous strategies are less sequential than usual strategies: suppose that σ is the strategy interpreting a proof π of a formula A, then a play s is in $\tilde{\sigma}$ if and only if it is a play in the strategy interpreting some proof π' such that $\pi \prec \pi'$.

Strategies which are positional, deterministic, receptive, total, courteous, are closed under residuals of transitions and satisfy some other properties such as the *cube property* (enforcing a variant of the domain-theoretic stability property) are called *ingenuous* and are very well behaved: they form a compact closed category, with games as objects and ingenuous strategies σ on $A^*\|B$ as morphisms $\sigma : A \to B$, which is a denotational model of MLL, which can be refined into a model of MALL by suitably quotienting morphisms. Composition of strategies $\sigma : A \to B$ and $\tau : B \to C$ is defined as usual in game semantics by "parallel composition and hiding": the plays in $\tau \circ \sigma$ are obtained from *interactions* of σ and τ, which are the plays on the game $A\|B\|C$ whose projection on $A^*\|B$ (resp. $B^*\|C$) is in σ (resp. τ) up to polarities of moves, by restricting them to $A^*\|C$. Associativity of the composition is not trivial and relies mainly on the determinism property, which implies that if a play in $\tau \circ \sigma$ comes from two different interactions s and t then there it also comes from a third interaction u which is greater than both wrt the prefix modulo homotopy order.

3   Concurrent Games

We recall here briefly the model of concurrent games [5]. A *concurrent strategy* ς on a complete lattice $(D, \leqslant)$ is a continuous closure operator on this lattice. Recall that a closure operator is a function $\varsigma : D \to D$ which is

1. *increasing*: $\forall x \in D, x \leqslant \varsigma(x)$
2. *idempotent*: $\forall x \in D, \varsigma \circ \varsigma(x) = \varsigma(x)$
3. *monotone*: $\forall x, y \in D, x \leqslant y \Rightarrow \varsigma(x) \leqslant \varsigma(y)$

Such a function is *continuous* when it preserves joins of directed subsets. Informally, an order relation $x \leqslant y$ means that the position y contains more information than x. With this intuition in mind, the first property expresses the fact that playing a strategy increases the amount of information in the game, the second that a strategy gives all the information it can given its knowledge in the current position (so that if it is asked to play again it does not have anything to add), and the third that the more information the strategy has from the current position the more information it has to deliver when playing.

Every such concurrent strategy ς induces a set of fixpoints defined as the set $\text{fix}(\varsigma) = \{ x \in D \mid \varsigma(x) = x \}$. This set is (M) closed under arbitrary meets and (J) closed under joins of directed subsets and conversely, every set $X \subseteq D$ of positions which satisfies these two properties (M) and (J) induces a concurrent strategy $X^\bullet$ defined by $X^\bullet(x) = \bigwedge \{\, y \in X \mid x \leqslant y \,\}$, whose set of fixpoints is precisely X.

Suppose that G is a game. Without loss of generality, we can suppose that G is *simply connected*, meaning that every position is reachable from the initial position $*$ and two plays $s, t : * \longrightarrow x$ with the same target are homotopic. This game induces a partial order on its set of positions, defined by $x \leqslant y$ iff there exists a path $x \longrightarrow y$, which can be completed into a complete lattice D by formally adding a top element $\top$. Now, consider a strategy σ on the game G. A position x of the graph G_σ induced by σ is *halting* when there is no Proponent move $m : x \longrightarrow y$ in σ: in such a position, the strategy is either done or is waiting for its Opponent to play. It can be shown that the set σ° of halting positions of an ingenuous strategy σ satisfies the properties (M) and (J) and thus induces a concurrent strategy $(\sigma^\circ)^\bullet$. Conversely, if for every positions $x, y \in D$ we write $x \leqslant_P y$ when $y \neq \top$ and there exists a path $x \longrightarrow y$ containing only Proponent moves, then every concurrent strategy ς induces a strategy ς^i defined as the set of plays in G whose intermediate positions x satisfy $x \leqslant_P \varsigma(x)$ – and these can be shown to be ingenuous. This thus establishes a precise relation between the two models:

Theorem 1. *The two operations above describe a retraction of the ingenuous strategies on a game G into the concurrent strategies on the domain D induced by the game G.*

Moreover, the concurrent strategies which correspond to ingenuous strategies can be characterized directly. In this sense, concurrent strategies are close to the intuition of focused proofs: given a position x, they play at once many Proponent moves in order to reach the position which is the image of x.

However, the correspondence described above is not functorial: it does not preserve composition. This comes essentially from the fact that the category of ingenuous strategies is compact closed, which means that it has the same strategies on the games interpreting the formulas $A \otimes B$ and $A \,\mathfrak{N}\, B$ (up to the polarity of the first move). For example, if X (resp. Y) is interpreted by the game with one Proponent transition $\dagger \longrightarrow x$ (resp. $\dagger \longrightarrow y$), the interpretations of $X^* \,\mathfrak{N}\, Y^*$ and $X \otimes Y$ are respectively

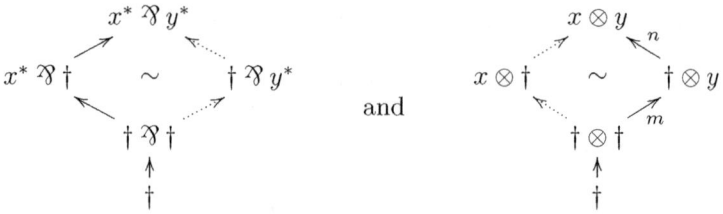

and

Now, consider the strategy $\sigma : X^* \,\mathfrak{N}\, Y^*$ which contains only the prefixes of the bold path $\dagger \longrightarrow (x^* \,\mathfrak{N}\, y^*)$ and the strategy $\tau : X \otimes Y$ which contains only the prefixes of bold path $\dagger \longrightarrow (x \otimes y)$. The fixpoints of the corresponding concurrent games are respectively $\sigma^\circ = \{\, \dagger,\ \dagger\mathfrak{N}\dagger,\ x^*\mathfrak{N}\dagger,\ x^*\mathfrak{N}y^* \,\}$ and $\tau^\circ = \{\, x \otimes y \,\}$. From the point of view of asynchronous strategies, the only reachable positions by both of the strategies in $X \otimes Y$ are † and † $\otimes$ †. However, from the point of view of the associated concurrent strategies, they admit the position $x \otimes y$ as a common position in $X \otimes Y$. From this observation, it is easy to build two strategies $\sigma : A \to B$ and $\tau : B \to C$ such that $((\tau \circ \sigma)^\circ)^\bullet \neq (\tau^\circ)^\bullet \circ (\sigma^\circ)^\bullet$ (we refer the reader to [5] for the definition of the composition of closure operators). In the example above, the strategy τ is the culprit: as mentioned in the introduction, the two strategies on X and Y should be independent in a proof of $X \otimes Y$, whereas here the strategy τ makes the move n depends on the move m. Formally, this dependence expresses the fact that the move m occurs after the move n in every play of the strategy τ. In [14], we have introduced a *scheduling criterion* which dynamically enforces this independence between the components of a tensor: a strategy satisfying this criterion is essentially a strategy such that in a substrategy on a formula of the form $A \otimes B$ no move of A depends on a move of B and vice versa. Every definable strategy satisfies this criterion and moreover,

Theorem 2. *Strategies satisfying the scheduling criterion form a subcategory of the category of ingenuous strategies and the operation $\sigma \mapsto (\sigma^\circ)^\bullet$ extends into a functor from this category to the category of concurrent games.*

This property enables us to recover more precisely the focusing property directly at the level of strategies as follows. Suppose that σ is an ingenuous strategy interpreting a proof π of a sequent $\vdash \Gamma$. Suppose moreover that $s : x \longrightarrow\!\!\!\!\rightarrow y$ is a maximal play in σ. By receptivity and courtesy of the strategy, this play is homotopic in the graph G_σ to the concatenation of a path $s_1 : x \longrightarrow\!\!\!\!\rightarrow x_1$ containing only Opponent moves, where x_1 is a position such that there exists no Opponent transition $m : x_1 \longrightarrow x_1'$, and a path $s_2 : x_1 \longrightarrow\!\!\!\!\rightarrow y$. Similarly, by totality of the strategy, the path s_2 is homotopic to the concatenation of a path $s_2' : x_1 \longrightarrow\!\!\!\!\rightarrow y_1$ containing only Proponent moves, where y_1 is a position which is either terminal or such that there exists an Opponent transition $m : y_1 \longrightarrow y_1'$, and a path $s_2'' : y_1 \longrightarrow\!\!\!\!\rightarrow y$. The path s_2' consists in the partial exploration of positive formulas, one of them being explored until a negative subformula is reached. By courtesy of the strategy, Proponent moves permute in a strategy and we can suppose that s_2' consists only in such a maximal exploration of one of the formulas available at the position x. If at some point a branch of a tensor formula is explored, then by the scheduling criterion it must be able to also explore the other branch of the formula. By repeating this construction on the play s_2'', every play of σ can be transformed into one which alternatively explores all the negative formulas and explores one positive formula until negative formulas are reached. By formalizing further this reasoning, one can therefore show that

Theorem 3. *In every asynchronous strategy interpreting a proof in MALL is included a strategy interpreting a focusing proof of the same sequent.*

A motivation for introducing concurrent games was to solve the well-known *Blass problem* which reveals that the "obvious" game model for the multiplicative fragment of linear logic has a non-associative composition. Abramsky explains in [1] that there are two ways to solve this: either syntactically by considering a focused proof system or semantically by using concurrent games. Thanks to asynchronous games, we understand here the link between the two points of view: every proof in linear logic can be reorganized into a focused one, which is semantically understood as a strategy playing multiple moves of the same polarity at once, and is thus naturally described by a concurrent strategy. In this sense, concurrent strategies can be seen as a semantical generalization of focusing, where the negative connectives are not necessarily all introduced at first during proof-search. It should be noted that some concurrent strategies are less sequential than focused ones, for example the strategies interpreting the multi-focused proofs [7], where the focus can be done on multiple positives formulas in the positive phase of the focused proof-search. Those multi-focused proofs were shown to provide canonical representatives of focused proofs (the interpretation of the canonical multi-focused proof can be recovered by generalizing Theorem 3). We are currently investigating a generalization of this result by finding canonical representatives for concurrent strategies.

The author is much indebted to Paul-André Melliès and Martin Hyland.

References

1. Abramsky, S.: Sequentiality vs. concurrency in games and logic. Mathematical Structures in Computer Science 13(4), 531–565 (2003)
2. Abramsky, S., Ghica, D.R., Murawski, A.S., Ong, L.: Applying game semantics to compositional software modeling and verification. In: Jensen, K., Podelski, A. (eds.) TACAS 2004. LNCS, vol. 2988, pp. 421–435. Springer, Heidelberg (2004)
3. Abramsky, S., Jagadeesan, R.: Games and Full Completeness for Multiplicative Linear Logic. The Journ. of Symb. Logic 59(2), 543–574 (1994)
4. Abramsky, S., Jagadeesan, R., Malacaria, P.: Full abstraction for PCF. Information and Computation 163(2), 409–470 (2000)
5. Abramsky, S., Melliès, P.-A.: Concurrent games and full completeness. LICS 99, 431–442 (1999)
6. Andreoli, J.M.: Logic Programming with Focusing Proofs in Linear Logic. Journal of Logic and Computation 2(3), 297–347 (1992)
7. Chaudhuri, K., Miller, D., Saurin, A.: Canonical Sequent Proofs via Multi-Focusing. In: International Conference on Theoretical Computer Science (2008)
8. Curien, P.L.: Definability and Full Abstraction. ENTCS 172, 301–310 (2007)
9. Ghica, D.R., Murawski, A.S.: Angelic Semantics of Fine-Grained Concurrency. In: Walukiewicz, I. (ed.) FOSSACS 2004. LNCS, vol. 2987, pp. 211–225. Springer, Heidelberg (2004)
10. Hyland, M., Ong, L.: On Full Abstraction for PCF: I, II and III. Information and Computation 163(2), 285–408 (2000)
11. Laird, J.: A game semantics of the asynchronous π-calculus. In: Abadi, M., de Alfaro, L. (eds.) CONCUR 2005. LNCS, vol. 3653, pp. 51–65. Springer, Heidelberg (2005)
12. Melliès, P.-A.: Asynchronous games 2: the true concurrency of innocence. In: Gardner, P., Yoshida, N. (eds.) CONCUR 2004. LNCS, vol. 3170, pp. 448–465. Springer, Heidelberg (2004)

13. Melliès, P.-A.: Asynchronous games 4: a fully complete model of propositional linear logic. In: Proceedings of 20th LICS, pp. 386–395 (2005)
14. Melliès, P.-A., Mimram, S.: Asynchronous Games: Innocence without Alternation. In: Caires, L., Vasconcelos, V.T. (eds.) CONCUR 2007. LNCS, vol. 4703, pp. 395–411. Springer, Heidelberg (2007)
15. Mimram, S.: Sémantique des jeux asynchrones et réécriture 2-dimensionnelle. PhD thesis, PPS, CNRS–Univ. Paris Diderot (2008)
16. Plotkin, G.D.: LCF considered as a programming language. TCS 5(3), 223–255 (1977)
17. Winskel, G., Nielsen, M.: Models for concurrency. In: Handbook of Logic in Computer Science, vol. 3, pp. 1–148. Oxford University Press, Oxford (1995)

A Note on the Least Informative Model of a Theory

Jeff B. Paris and Soroush R. Rad*

School of Mathematics
The University of Manchester
Manchester M13 9PL
jeff.paris@manchester.ac.uk,
soroush.rafiee-rad@postgrad.manchester.ac.uk

Abstract. We consider one possible interpretation of the 'least informative model' of a relational and finite theory and show that it is well defined for a particular class of Π_1 theories. We conjecture that it is always defined for Π_1 theories.

Keywords: Uncertain reasoning, probability logic, inference processes, Polyadic Inductive Logic.

1 Introduction

Let Ψ be a consistent sentence of a first order language L and let $\theta(x_1, x_2, \ldots, x_n)$ be a formula of L. Then given a structure M for L with universe $\{\, a_i \mid i \in \mathbb{N}^+ \,\}$ about which we know only that M is a model of Ψ a natural question one might ask is how likely, or probable, is it that $M \models \theta(a_1, a_2, \ldots, a_n)$? Since we are supposed to know only that M is a model of Ψ the probabilities we assign should not weight M in any direction which would imply that we had some special inside information. In other words these probabilities should in some sense minimize the information we have about M beyond it being a model of Ψ.

In this short note we consider one (limited) approach to answering this question which was originally (see [4]) based on adapting methods of uncertain reasoning developed for propositional probabilistic knowledge bases, in particular the Maximum Entropy Inference Process (see for example [16], [17], [21], [23]), though for the purposes of this paper we will not need to recall that history.

2 Notation

From now on we assume of L that it is a finite relational first order language without functions or constants. We shall also assume that L does not contain equality. Let $\mathcal{T}$ be the set of structures for L with universe $\{\, a_i \mid i \in \mathbb{N}^+ \,\}$ and let

---

* Supported by a MATHLOGAPS Research Studentship, MEST-CT-2004-504029.

F. Ferreira et al. (Eds.): CiE 2010, LNCS 6158, pp. 342–351, 2010.

L_a be L augmented with constants a_i for $i \in \mathbb{N}^+$, which of course are interpreted in the structures $M \in \mathcal{T}$ by the elements a_i of M. Let $\mathcal{T}^{(n)}$ be the (finite) set of structures for L with universe $\{ a_i \mid 1 \leq i \leq n \}$ and let $L_a^{(n)}$ be L augmented with the constants a_i for $1 \leq i \leq n$.

Let Ψ be a consistent sentence in L, as above. Then provided

$$\{ M \in \mathcal{T}^{(n)} \mid M \models \Psi \} \neq \emptyset$$

we can define a probability function[1] w_n on the quantifier free sentences of $SL_a^{(n)}$ by

$$w_n(\theta) = \frac{|\{ M \in \mathcal{T}^{(n)} \mid M \models \theta \wedge \Psi \}|}{|\{ M \in \mathcal{T}^{(n)} \mid M \models \Psi \}|} .$$

Provided the limit is well defined we now set w to be the probability function on the quantifier free sentences of SL_a which is the limit of the w_n, so in this case w satisfies

(P1) $\models \theta \Rightarrow w(\theta) = 1$,
(P2) $\models \neg(\theta \wedge \phi) \Rightarrow w(\theta \vee \phi) = w(\theta) + w(\phi)$,

on the quantifier free θ, ϕ of SL_a. By a theorem of Gaifman, see [8], w has a unique extension to a probability function on SL_a, meaning that it satisfies (P1), (P2) for any sentences θ, ϕ and for $\exists x\, \psi(x) \in SL_a$,

(P3) $w(\exists x\, \psi(x)) = \lim_{n \to \infty} w(\psi(a_1) \vee \psi(a_2) \vee \ldots \vee \psi(a_n))$.

Then provided w is well defined and $w(\Psi) = 1$ we putatively propose $w(\theta)$, for $\theta \in SL_a$, as the probability that $M \models \theta$ for $M \in \mathcal{T}$ an otherwise unknown model of Ψ. Indeed we could think here of w as representing a 'random model' of Ψ, and arguably the 'least informative' such random model. The basis for this assertion is that we may equally think of w_n as a probability functions on the *propositional language* with propositional variables $R(b_1, \ldots, b_k)$ for each predicate R of L and (not necessarily distinct) choice of b_i from the $a_1, \ldots, a_n$, when we identify $\Phi \in SL_a^{(n)}$ with the sentence of the propositional language formed by iteratively replacing each subformula $\forall x\, \Phi'(x)$ of Φ by $\bigwedge_{i=1}^n \Phi'(a_i)$ and analogously for the existential quantifier. Under this identification w_n becomes the maximum entropy, equivalently minimal information content (see [20]) solution of the corresponding propositional constraint corresponding to $w(\Psi) = 1$. In turn we would argue that justifies proposing the same title for any limit of the w_n.[2]

---

[1] Meaning that w_n is a map from the set $SL_a^{(n)}$ of sentences of $L_a^{(n)}$ to $[0,1]$ such that for $\theta, \phi, \exists x\, \psi(x) \in SL_a^{(n)}$,
(P1) $\models \theta \Rightarrow w_n(\theta) = 1$,
(P2) $\models \neg(\theta \wedge \phi) \Rightarrow w_n(\theta \vee \phi) = w_n(\theta) + w_n(\phi)$,
(P3) $w_n(\exists x\, \psi(x)) = w_n(\psi(a_1) \vee \psi(a_2) \vee \ldots \vee \psi(a_n))$.

[2] For a more detailed exposition of this rationale see [4] where this approach is described in the case of L purely unary, not only for a finite theory as here but more generally for a finite, linear, knowledge base.

The first thing to say about this proposal is that the w_n may not be well defined simply because Ψ has no finite models. So suppose from now on that Ψ does have a finite model, say of cardinality n. Then by just making 'clones' of some element in that model we can construct models of any finite cardinality greater than n so all the subsequent w_m will be defined. Still however the limit need not be well defined.

As an obvious example here let L have relations G, R and P of arities 3, 2, 1 respectively, let $x =_G y$ abbreviate

$$\forall u, t \,(G(x, u, t) \leftrightarrow G(y, u, t))$$

and let Ψ_1 be the conjunction of:

$$\forall x, y, z \,(x =_G y \to (R(x, z) \to R(y, z))), \qquad \forall x, y \,(R(x, y) \leftrightarrow R(y, x)),$$

$$\forall x, y, z \,((R(x, y) \wedge R(x, z)) \to (y =_G z)),$$

$$\forall x \exists y \, R(x, y), \qquad \forall x \, \neg R(x, x),$$

and Ψ_2 be the conjunction of:

$$\forall x, y, z \,(x =_G y \to (R(x, z) \to R(y, z))), \qquad \forall x, y \,(R(x, y) \leftrightarrow R(y, x)),$$

$$\forall x, y, z \,((R(x, y) \wedge R(x, z)) \to (y =_G z)), \qquad \forall x, y \,((R(x, x) \wedge R(y, y)) \wedge (x =_G y)),$$

$$\forall x \exists y \, R(x, y), \qquad \exists x \, R(x, x).$$

Then $\models \neg(\Psi_1 \wedge \Psi_2)$ and for n odd the proportion of $M \in T^{(n)}$ satisfying $M \models \Psi_1 \vee \Psi_2$ which also satisfy Ψ_1 tends to zero as $n \to \infty$ whereas for n even it is the proportion which satisfy Ψ_2 which tend to zero. Hence if we take $\Psi = \Psi_1 \vee \Psi_2$ then $w_n(\Psi_1)$ bobs backwards and forwards between being close to 1 and close to 0 as $n \to \infty$ and no overall limit exists. [For fuller details see [18].]

A second problem with this approach is that whilst the w_n may have a limit w this probability function, once extended to all sentences of L_a via Gaifman's Theorem, may no longer give $w(\Psi)$ probability 1. This happens for example in the case when L has a single binary relation R and Ψ is $\exists x \forall y \, R(x, y)$. The informal explanation of why this happens is that the overriding majority of structures in $T^{(n)}$ which model this Ψ will have just one a_i satisfying $\forall y \, R(x, y)$. Thus as $n \to \infty$ the probability that any one a_i satisfies this tends to zero with the result that in the limit none of them have non-zero probability of satisfying it. Again details can be found in [18].

On a more optimistic note however the limit w does exist for (consistent) Ψ when L is purely unary, see for example [1], [2], [4], (and also the developments in [10], [11], [12]). So from now on we shall assume that L has at least one non-unary relation.

3   The Conjecture

A feature of these two failures relevant to this paper is that in the first Ψ is of quantifier complexity Π_2 whilst in the second it is Σ_2.[3] On the other hand it is not difficult to check that if Ψ is Σ_1 then the limit w exists and $w(\Psi) = 1$, see [18]. This leaves open the question of what happens when Ψ is Π_1.

We would conjecture that in the case Ψ is Π_1 and consistent then the limit w always exists and furthermore satisfies $w(\Psi) = 1$.

The main contribution of this paper towards confirming this conjecture is to show that it is true for such Ψ when

$$|\{\, M \in \mathcal{T}^{(n)} \,|\, M \models \Psi \,\}|$$

is polynomially bounded.

There is some evidence that confirming the conjecture in general when we also allow in equality may not be very easy. This comes from results in Graph Theory concerned with what is there called the 'speed of hereditary classes'. A class of graphs $\mathcal{F}$ is said to be *hereditary* if it is closed under isomorphisms and subgraphs, in other words whenever a graph G is in $\mathcal{F}$ and H is the subgraph of G formed by restricting the edges to some subset of the vertices of G then $H \in \mathcal{F}$. Given a hereditary class $\mathcal{F}$ of graphs let f_n be the number of graphs in $\mathcal{F}$ with vertices $\{\, a_i \,|\, 1 \leq i \leq n \,\}$. A prominent question in Graph Theory over the last 15 years (see for example [3], [19]) is what f_n may look like as a function of n. From results obtained to date it would appear that f_n must fall into one of 4 bands. The lowest of these, which is well understood, is when f_n is bounded by a polynomial. After that however the bands are much wider and within them there *seems*, as far as is currently known, to be scope for f_n to behave uncommonly badly.

The reason these results, or lack of, are relevant to our conjecture here is that if L has a single binary relation R then the models in $\mathcal{T}^{(n)}$ of a Π_1 sentence Ψ which implies that R is symmetric are just the graphs with vertices $\{\, a_i \,|\, 1 \leq i \leq n \,\}$ in some hereditary class $\mathcal{F}$. From this then it would seem that understanding the behavior of the w_n may not come so easily.

The reader familiar with the 0-1 laws of Fagin [7] and Glebskiǐ et al [9] and subsequent developments (see for example [5], [6], [13], [24]) concerning the asymptotic frequency of models of Ψ in $\mathcal{T}^{(n)}$ may wonder if everything that we are conjecturing has not already been answered in the course of that body of research. As far as we have been able to discover it seems that it has not. The problem here is that Ψ will only have non-zero asymptotic frequency (assuming as we are that it does not mention any constants) in the rather exceptional circumstance that it is true in a random structure for that language. Thus the obvious idea that the limit of the $w_n(\theta)$ will be the ration of the limiting frequencies of $\Psi \wedge \theta$ and Ψ generally leads to just 0/0 !

---

[3] Note however that this failure is not invariably the case. For example when Ψ is the (necessarily) Π_2 sentence asserting that $R(x, y)$ defines a linear order without end points the limit does exist and gives Ψ probability 1.

4   Slow Sentences

In this section we shall confirm the above conjecture in the case that Ψ is *slow*, meaning that

$$|\{\, M \in \mathcal{T}^{(n)} \mid M \models \Psi \,\}|$$

is polynomially bounded. First however we introduce some notation and give a syntactic characterization of 'slowness'.

Suppose L has relations $R_1, \ldots, R_q$ of arities $h_1, \ldots, h_q$. Let $b_1, b_2, \ldots, b_n$ stand for some distinct choices from the constants a_i (a convention we adopt throughout). Then a *state description* for $b_1, \ldots, b_m$ is a sentence of the form

$$\Theta(b_1, b_2, \ldots, b_m) \;=\; \bigwedge_{s=1}^{q} \bigwedge_{i_1, i_2, \ldots, i_{h_s} \in \{1, \ldots, m\}} \pm R_s(b_{i_1}, b_{i_2}, \ldots, b_{i_{h_s}}), \qquad (1)$$

where $\pm R$ stands for R or $\neg R$ respectively. In other words a state description for $b_1, \ldots, b_m$ determines a structure for L whose universe is the set $\{b_1, \ldots, b_m\}$.

Given such a state description we say that b_i and b_j are *indistinguishable* with respect to $\Theta(b_1, \ldots, b_m)$ if

$$\Theta(b_1, \ldots, b_m) \wedge b_i = b_j$$

is consistent (with the axioms of equality) and in this case write $b_i \sim_\Theta b_j$. Clearly $\sim_\Theta$ is an equivalence relation. Let $\|\Theta(b_1, \ldots, b_m)\|$ be the number of equivalence classes with respect to $\sim_\Theta$.

For $\Theta(b_1, \ldots, b_m)$ a state description and $i_1, \ldots, i_k \in \{1, 2, \ldots, m\}$, set $\Theta[b_{i_1}, \ldots, b_{i_k}]$ to be the (unique) state description for $b_{i_1}, \ldots, b_{i_k}$ consistent with $\Theta(b_1, \ldots, b_m)$.

The following rather technical lemma will be useful in what follows.

Lemma 1. *Let r be at least the largest arity of any relation in L, let $p \geq k \geq r$ and let $\Phi(a_1, \ldots, a_m)$ be a state description (of L) with $\|\Phi(a_1, \ldots, a_m)\| = p$. Then there is some $k \leq s \leq k + r$ and $1 \leq i_1 < i_2 < \ldots < i_s \leq m$ such that $\|\Phi[a_{i_1}, a_{i_2}, \ldots, a_{i_s}]\| = s$.*

Proof. If $p \leq k + r$ taking $a_{i_1}, a_{i_2}, \ldots, a_{i_p}$ to be representatives from the equivalence classes of $\sim_\Phi$ will give $\|\Phi(a_{i_1}, a_{i_2}, \ldots, a_{i_p})\| = p$ so assume $k + r < p$. Suppose we have picked $a_{i_1}, a_{i_2}, \ldots, a_{i_t}$ with $\|\Phi[a_{i_1}, a_{i_2}, \ldots, a_{i_t}]\| = t$. It is enough to show that we can find some $j_1, j_2, \ldots, j_s \leq m$ with $a_{j_1}, a_{j_2}, \ldots, a_{j_s}$ distinct from the $a_{i_1}, \ldots, a_{i_t}$ such that $1 \leq s \leq r$ and

$$\|\Phi[a_{i_1}, a_{i_2}, \ldots, a_{i_t}, a_{j_1}, \ldots, a_{j_s}]\| = t + s.$$

To this end let a_{j_1} be inequivalent to each of the $a_{i_1}, a_{i_2}, \ldots, a_{i_t}$ modulo $\sim_\Phi$. If $\|\Phi[a_{i_1}, a_{i_2}, \ldots, a_{i_t}, a_{j_1}]\| = t + 1$ we are done. Otherwise, according to this state description $\Phi[a_{i_1}, a_{i_2}, \ldots, a_{i_t}, a_{j_1}]$ a_{j_1} and a_{i_g} are indistinguishable for some $1 \leq g \leq t$. Then since they *are* distinguishable modulo $\sim_\Phi$ and r is the largest arity of any relation in L we can find $j_2, \ldots, j_q$ with $q \leq r$ such that a_{j_1}

and a_{i_g} are distinguishable according to $\Phi[a_{i_g}, a_{j_1}, a_{j_2}, a_{j_3}, \ldots, a_{j_q}]$. Indeed we may further assume that none of these $a_{j_2}, a_{j_3}, \ldots, a_{j_q}$ are indistinguishable in $\Phi[a_{i_1}, a_{i_2}, \ldots, a_{i_t}, a_{j_1}, a_{j_2}, \ldots, a_{j_q}]$ from any of these other constants mentioned there, otherwise we could simply remove them. It follows then that $\|\Phi[a_{i_1}, a_{i_2}, \ldots, a_{i_t}, a_{j_1}, a_{j_2}, \ldots, a_{j_q}]\| = t + q$, as required.

In the case $r = 2$ it can be shown, see [22], that we can take $s = k$ in this Lemma. However we cannot hope to have this result for $s = k$ when $r > 2$. For example let L have just the ternary relation R and let $\Phi(a_1, a_2, a_3, a_4, a_5, a_6)$ imply $R(a_i, a_j, a_k)$ just if (i, j, k) is one of $(1, 2, 3)$ or $(4, 5, 6)$ (so it implies $\neg R(a_i, a_j, a_k)$ otherwise). In this case

$$\|\Phi(a_1, a_2, a_3, a_4, a_5, a_6)\| = 6$$

but we cannot find $a_{i_1}, a_{i_2}, a_{i_3}, a_{i_4}, a_{i_5}$ such that $\|\Phi[a_{i_1}, a_{i_2}, a_{i_3}, a_{i_4}, a_{i_5}]\| = 5$.

The next theorem gives a characterization of the slow Π_1 sentences.

Theorem 2. *Let r be at least the largest arity of any relation in L and Ψ a consistent Π_1 sentence of L. Then*

$$|\{ M \in \mathcal{T}^{(n)} \mid M \models \Psi \}| = o(k^n)$$

iff for some state descriptions $\Phi_i(a_1, a_2, \ldots, a_{k+r})$, $i = 1, \ldots, h$, with $\|\Phi_i(a_1, a_2, \ldots, a_{k+r})\| < k$,

$$\Psi \equiv \forall x_1, \ldots, x_{k+r} \bigvee_{j=1}^{h} \Phi_j(x_1, x_2, \ldots, x_{k+r}). \tag{2}$$

Proof. Let the $\Phi_i(a_1, \ldots, a_{k+r})$ for $i = 1, \ldots, h$ list all state descriptions with $\|\Phi_i(a_1, a_2, \ldots, a_{k+r})\| < k$ which are consistent with Ψ. If

$$\Psi \wedge \neg \forall x_1, \ldots, x_{k+r} \bigvee_{j=1}^{h} \Phi_j(x_1, x_2, \ldots, x_{k+r}) \tag{3}$$

was consistent it would have a model M in $\mathcal{T}$ and hence there would be a state description $\Theta(a_{i_1}, a_{i_2}, \ldots, a_{i_{k+r}})$ true in M with $\|\Theta(a_{i_1}, a_{i_2}, \ldots, a_{i_{k+r}})\| \geq k$. Clearly by permuting these constants we may assume a_{i_j} is just a_j for $j = 1, \ldots, k+r$. In this case take n large and let $\Xi(a_1, a_2, \ldots, a_{k+r}, a_{k+r+1}, \ldots, a_n)$ be a state description which implies (equivalently extends) $\Theta(a_1, a_2, \ldots, a_{k+r})$ such that each of the $a_{k+r+1}, \ldots, a_n$ is equivalent according to $\sim_\Xi$ to some a_i with $i \leq k + r$. In other words no new equivalence classes are created in going from $\sim_\Theta$ to $\sim_\Xi$, they just enlarge. Then just as the structure $M_\Theta \in \mathcal{T}^{(k+r)}$ determined by Θ is a model of Ψ so also is $M_\Xi \in \mathcal{T}^{(n)}$. However since $\sim_\Theta$ has at least k equivalence classes such a Ξ can be formed in at least k^{n-k-r} ways, contradicting the given bound.

From (3) then

$$\vdash \Psi \rightarrow \forall x_1, \ldots, x_{k+r} \bigvee_{j=1}^{h} \Phi_j(x_1, x_2, \ldots, x_{k+r}). \tag{4}$$

For the provability of the other direction here suppose on the contrary that

$$\neg\Psi \wedge \forall x_1, \ldots, x_{k+r} \bigvee_{j=1}^{h} \Phi_j(x_1, x_2, \ldots, x_{k+r})$$

was consistent, so had a model $M \in T$. Then since Ψ is Π_1 there is some large m such that the state description $\Phi(a_1, \ldots, a_m)$ determined by M is inconsistent with Ψ. If $\|\Phi(a_1, \ldots, a_m)\| \geq k$ then by Lemma 1 we could find some $i_1, i_2, \ldots, i_s \leq m$ with $s \leq k+r$ such that $\|\Phi[a_{i_1}, \ldots, a_{i_s}]\| \geq k$. Since we can permute the elements of M we may suppose that $i_j = j$ for $j = 1, \ldots, s$. But in that case since M is a model of the second conjunct of (4), $\Phi[a_1, \ldots, a_s, a_{s+1}, \ldots, a_{k+r}]$ would have to be one of the $\Phi_j(a_1, \ldots, a_{k+r})$ which is a contradiction because it has too many equivalence classes.

Hence $\|\Phi(a_1, \ldots, a_m)\| < k$ and without loss of generality we may assume that $a_1, a_2, \ldots, a_{k+r}$ contains representatives of all the equivalence classes of $\sim_\Phi$. In that case $\Phi[a_1, a_2, \ldots, a_{k+r}]$ must again, as above, be one of the $\Phi_j(a_1, \ldots, a_{k+r})$. But then $\Phi[a_1, a_2, \ldots, a_{k+r}]$ must be consistent with Ψ, indeed it determines a model of Ψ in $T^{(k+r)}$, so $\Phi(a_1, a_2, \ldots, a_m)$ will also be consistent with Ψ since it is formed by simply duplicating a_i in $\Phi[a_1, a_2, \ldots, a_{k+r}]$.

Turning now to the other direction of the equivalence stated in the Theorem assume that (2) holds and let the state description $\Phi(a_1, a_2, \ldots, a_m)$ determine a model of Ψ in $T^{(n)}$. Then as above if $\|\Phi(a_1, a_2, \ldots, a_m)\| \geq k$ we could cut this down to a $\Phi[a_{i_1}, \ldots, a_{i_{k+r}}]$ satisfying $\|\Phi[a_{i_1}, \ldots, a_{i_{k+r}}]\| \geq k$ and this would still be consistent with Ψ. But clearly this is not consistent with the right hand side of (2), contradiction.

We conclude that $\|\Phi(a_1, a_2, \ldots, a_m)\| < k$ and in turn that if $a_{i_1}, \ldots, a_{i_{k+r}}$ contain representative from all the equivalence classes of $\sim_\Phi$ then $\Phi[a_{i_1}, \ldots, a_{i_{k+r}}]$ is one of the $\Phi_j(a_{i_1}, \ldots, a_{i_{k+r}})$. Hence $\Phi(a_1, a_2, \ldots, a_m)$ is determined by this $j \in \{1, 2, \ldots, h\}$, the choice of $i_1, \ldots, i_{r+k}$ and the choice of which of the (at most) $k - 1$ equivalence classes contain the remaining a_i for $1 \leq i \leq m$, $i \neq i_1, \ldots, i_{k+r}$, which overall amounts to just $o(n^k)$ choices.

We are now in a position to prove the main result of this paper.

Theorem 3. *Let Ψ be a consistent Π_1 slow sentence of L. Then the limit w of the w_n exists and satisfies $w(\Psi) = 1$.*

Proof. It is clear that if the limit w exists then it satisfies $w(\Psi) = 1$ so we only need to show that the limit exists. Let m be large and let $\Theta(a_1, \ldots, a_m)$ be a state description consistent with Ψ, so this sentence defines a model in $T^{(m)}$. We want to count the number of state descriptions $\Phi(a_1, a_2, \ldots, a_n)$ consistent with Ψ and extending $\Theta(a_1, \ldots, a_m)$. Since Ψ is slow let the $\Phi_j(a_1, \ldots, a_{k+r})$ etc. be as in Theorem 2. Let $a_{i_1}, a_{i_2}, \ldots, a_{i_s}$ be the first (viz-a-viz the indices) elements of the distinct equivalence classes of $\sim_\Phi$, so $\|\Phi[a_{i_1}, a_{i_2}, \ldots, a_{i_s}]\| = s$, which is less than k as in the arguments above. Let r be minimal such that $i_r > m$. Then the sentence $\Phi(a_1, a_2, \ldots, a_n)$ is determined by the formula $\Phi'(a_1, \ldots, a_m, x_r, \ldots, x_s)$ such that

$$\Phi'(a_1, \ldots, a_m, a_{i_r}, \ldots, a_{i_s}) = \Phi[a_1, \ldots, a_m, a_{i_r}, \ldots, a_{i_s}]$$

and the assignment of the remaining $n - m - (s - r - 1)$ constants a_i to the equivalence classes determined by $\Phi[a_{i_1}, \ldots, a_{i_s}]$ in such a way that each a_{i_j} for $j \geq r$ is the least element of it's class. In turn we can now see that $\Phi(a_1, \ldots, a_n)$ is uniquely determined by the formulae $\Phi'(a_1, \ldots, a_m, x_r, \ldots, x_s)$, modulo any permutation of $\{x_r, x_{r+1}, \ldots, x_s\}$, and an assignment of the $a_{m+1}, \ldots, a_n$ to s classes so that each of the last $s - r + 1$ classes receives at least one element (in the case we have in mind then the least such elements being $a_{i_r}, \ldots, a_{i_s}$ respectively). Hence the number of such $\Phi(a_1, a_2, \ldots, a_n)$ is $dc^n + o(c^n)$ for some constant d and $c < k$.

The Theorem now follows by noticing that $|\mathcal{T}^{(n)}|$ is just the sum of these $dc^n + o(c^n)$ over the finite number of possible $\Theta(a_1, \ldots, a_m)$.

Notice that (3) provides the definitive list of Π_1 slow sentences for L. More specifically for L having a single binary relation symbol R let $C(x, y) = R(x, x) \wedge R(y, y) \wedge R(x, y) \wedge R(y, x)$ and let $D(x, y) = R(x, x) \wedge R(y, y) \wedge \neg R(x, y) \wedge \neg R(y, x)$. Then

$$\forall x, y, z \, [((C(x, y) \wedge C(x, z) \wedge C(y, z)) \vee ((C(x, y) \wedge D(x, z) \wedge D(y, z)) \vee$$

$$((D(x, y) \wedge D(x, z) \wedge C(y, z)) \vee ((D(x, y) \wedge C(x, z) \wedge D(y, z)))]$$

is slow with k, c as above equal to 3, 2 respectively. Indeed in this case it can be checked that the $M \in \mathcal{T}^{(n)}$ are just equivalence relations with at most two equivalence classes.

5 Discussion

Jon Williamson in [23] has suggested an alternative approach to the problem of what probability $v(\theta)$ to give to a sentence θ of L_a being true in $M \in \mathcal{T}$ when all we know about M is that it is a model of Ψ. The idea is that given a probability function on SL_a we let v_n be the restriction of v to $SL_a^{(n)}$ and define a partial ordering on such v by

$$v \prec v' \quad \text{iff for all } n \text{ eventually } E(v'_n) < E(v_n)$$

where $E(v_n)$ is the entropy of v_n, that is the sum over the state descriptions Θ of $SL_a^{(n)}$ of $-v_n(\Theta) \log(v_n(\Theta))$. Since entropy is generally accepted as a measure of 'lack of information' (see [20]), one might argue that in our current context of knowing only Ψ one should assign θ probability $v(\theta)$ for that probability function v which satisfies $v(\Psi) = 1$ and is $\prec$-minimal amongst all such functions.

As with the approach discussed in this paper this minimal choice exists (and gives the same answers as our approach) in the cases of unary languages and in the cases of Ψ being Σ_1 but can fail to be defined for Σ_2 and Π_2 Ψ (see

[18]). We would conjecture that the method also succeeds for slow Π_1 sentences, and indeed more generally that both approaches are defined and give the same answers for all Π_1 Ψ. However that is not always the case for higher quantifier complexity, Williamson's method can work even when Ψ has no finite models, for example when Ψ defines a dense linear ordering.

The results given in this paper suffer the obvious weakness that the language and structures do not include equality. Of course we could include in the Π_1 sentence Ψ the (finite) Π_1 axiomatization of the equality axioms appropriate to L and carry on as before. However in that case our structures would be what Mendelson refers to in [14] as 'non-normal' structures rather than structures in which $=$ is interpreted as real equality, so this is not much help. Given the results from Graph Theory (which do allow equality) we would conjecture that Theorem 3 also holds when we properly allow equality into the language.

We finally mention that the method of this paper and Williamson's were originally introduced to address the more general problem of inference from predicate probabilistic knowledge bases as in [4] and [15]. However it would seem that the special case problem considered here is really the obstacle to be overcome in this endeavor.

References

1. Bacchus, F., Grove, A.J., Halpern, J.Y., Koller, D.: Generating new beliefs from old. In: de Mantaras, R.L., Poole, D. (eds.) Proceedings of the Tenth Annual Conference on Uncertainty in Artificial Intelligence, pp. 37–45. Morgan Kaufman, San Fransisco (1994)
2. Bacchus, F., Grove, A.J., Halpern, J.Y., Koller, D.: From statistical knowledge to degrees of belief. Artificial Intelligence 87, 75–143 (1996)
3. Balogh, J., Bollobas, B., Weinreich, D.: The penultimate range of growth for graph properties. European Journal of Combinatorics 22(3), 277–289 (2001)
4. Barnett, O.W., Paris, J.B.: Maximum Entropy inference with qualified knowledge. Logic Journal of the IGPL 16(1), 85–98 (2008)
5. Compton, K.J.: A Logical Approach to Asymptotic Combinatorics I. First Order Properties. Advances in Mathematics 65, 65–96 (1987)
6. Compton, K.J.: 0-1 Laws in Logic and Combinatorics. In: Rival, I. (ed.) Proc. 1987 NATO Adv. Study Inst. on Algorithms and Order. Reidel, Dordrecht (1988)
7. Fagin, R.: Probabilities on Finite Models. Journal of Symbolic Logic 41, 50–58 (1976)
8. Gaifman, H.: Concerning measures in first order calculi. Israel J. of Mathematics 24, 1–18 (1964)
9. Glebskiĭ, Y.V., Kogan, D.I., Liogon'kiĭ, M.I., Talanov, V.A.: Range and Degree of Formulas in the Restricted Predicate Calculus. Cybernetics 5, 142–154 (1972)
10. Grove, A.J., Halpern, J.Y., Koller, D.: Random Worlds and Maximum Entropy. Journal of Artificial Intelligence Research 2, 33–88 (1994)
11. Grove, A.J., Halpern, J.Y., Koller, D.: Asymptotic conditional probabilities: the unary case. SIAM Journal of Computing 25(1), 1–51 (1996)
12. Grove, A.J., Halpern, J.Y., Koller, D.: Asymptotic conditional probabilities: the non-unary case. Journal of Symbolic Logic 61(1), 250–276 (1996)

13. Lynch, F.J.: An Extension of 0-1 Laws. Random Structures and Algorithms 5(1), 155–172 (1994)
14. Mendelson, E.: Introduction to Mathematical Logic, 4th edn. International Thompson Publishing (1997)
15. Paris, J.B., Rad, S.R.: Inference Processes for Quantified Predicate Knowledge. In: Hodges, W., de Queiroz, R. (eds.) Logic, Language, Information and Computation. LNCS (LNAI), vol. 5110, pp. 249–259. Springer, Heidelberg (2008)
16. Paris, J.B., Vencovská, A.: A Note on the Inevitability of Maximum Entropy. International Journal of Approximate Reasoning 4(3), 183–224 (1990)
17. Paris, J.B., Vencovská, A.: Common sense and stochastic independence. In: Corfield, D., Williamson, J. (eds.) Foundations of Bayesianism, pp. 203–240. Kluwer Academic Press, Dordrecht (2001)
18. Rad, S.R.: Inference Processes For Probabilistic First Order Languages. Ph.D. Thesis, University of Manchester (2009), http://www.maths.manchester.ac.uk/~jeff/
19. Scheinerman, E.R., Zito, J.: On the size of hereditary classes of graphs. Journal Combinatorial Theory, Ser. B 61, 16–39 (1994)
20. Shannon, C.E., Weaver, W.: The Mathematical Theory of Communication. University of Illinois Press (1949)
21. Shore, J.E., Johnson, R.W.: Axiomatic Derivation of the Principle of Maximum Entropy and the Principle of Minimum Cross-Entropy. IEEE, Trans. Info. Theory IT-26, 26–37 (1980)
22. Vencovská, A.: Binary Induction and Carnap's Continuum. In: Proceedings of the 7th Workshop on Uncertainty Processing (WUPES), Mikulov (2006), http://www.utia.cas.cz/MTR/wupes06_papers
23. Williamson, J.: Objective Bayesian Probabilistic Logic. Journal of Algorithms in Cognition. Informatics and Logic 63, 167–183 (2008)
24. Woods, A.R.: Counting Finite Models. Journal of Symbolic Logic 62(1), 925–949 (1995)

Three Roots for Leibniz's Contribution to the Computational Conception of Reason

Olga Pombo

Centro de Filosofia das Ciências da Universidade de Lisboa
Faculdade de Ciências da Universidade de Lisboa
opombo@fc.ul.pt

Abstract. The aim of this paper is to show that Leibniz's contribution to what we may call today the computational conception of reason must be inscribed in three different roots: 1) the combinatory tradition coming from the XIII century, 2) XVII century attempts at constructing an artificial universal and philosophical language, and 3) the Hobbesian conception of reason.

Keywords: Leibniz, Llull, Hobbes.

1 Lull and Leibniz

Leibniz first great inspiration is the old *Ars Magna* by Ramón Lull (1232-1315). Lull's central idea is that it would be possible, by the combination of a set of simple terms, to establish all possible propositions and thus to discover all possible statements and demonstrate all possible truths to which human knowledge can aspire.

For the accomplishment of this project, Lull a) establishes a set of categories, organized in six great classes of nine categories each; b) defines the syntactic rules of the combinatorial elements; c) conceives a complex system of technical procedures of automatic application capable of effecting the combination of these elements. His aim is to make it possible to determine, automatically, all the possible subjects of a given attribute as well as the conclusion and the middle term of all incompletely known syllogisms[1]. Thus, starting on the basis of categories of universal application and operating with a system of symbolic notations and combinatorial diagrams, Lull establishes the principles of a synthetic and inventive procedure which, unlike the demonstrative logics of Aristotle, would not be limited to the analysis of truths already known but would make possible ways for the discovery of new truths.

Lull's *Ars Magna*, which exerted a deep influence in the renaissance and modern times[2], is thus one of the most far-reaching and prestigious proposals for the mechanization of logical operations.

---

[1] Among the complex set of signs, tables and figures constructed by Lull, there is a set of material circles, rotating in concentric movement of superposition in order to allow the combination of the symbols marked in their limits, cf. Carreras y Artau (1939: 345-455) e também Couturat «Sur L'Ars Magna de Raymond Lulle et Athanase Kircher» (1981: 541/3).

[2] Namely the encyclopaedic lullism of Johann Heinrich Alsted (1588-1638) or Sebastián Izquierdo (1601-1681), and the major work of Athanasius Kircher, Ars Magna sciendi seu porta Scientiarum, published in 1669, which claims to be the development of Lull's Ars Magna.

F. Ferreira et al. (Eds.): CiE 2010, LNCS 6158, pp. 352–361, 2010.

Leibniz was quite aware of Lull's Ars. In fact, since his first writings, namely *De Arte Combinatória (1660)*, Leibniz criticizes a) the arbitrariness of Lullian signs; b) his set of categories; c) the methodological aspects of Lull's *Ars* which Leibniz considers insufficient and rudimentary. These are decisive critiques. Leibniz cannot accept the alphabet of the Lull' *Ars*. He cannot admit his list of categories which he considers as «vagues», arbitrarily chosen and organized in classes of 9 categories each, a number artificially chosen purely for reasons of symmetry[3]. Further, he does not agree with the mechanical combinatorial procedures proposed by Lull. Instead, Leibniz points to the calculatory processes of mathematical analysis (binary, ternary, quaternary groupings, exchanges, substitutions, equivalencies, etc.). On the basis of the analogy between the analysis of the concepts, its associative structure, and the decomposition of a whole number into prime factors, Leibniz proposes, in the *De Arte*, to assign a whole number to each primitive idea, each complex idea therefore being represented by the product that would constitute its definition.

However, behind these criticisms, Leibniz recognizes the epistemological value of Lull's project on two important points: a) the status of combinatory as the foundation for the *ars inveniendi*; b) the possibility of applying the same logical procedures to the totality of knowledge, including new knowledge. That is to say, Leibniz accepts Lull's idea that the knowledge of the primitive ideas and of the laws of their combination would allow us to find all possible predicates of a given subject and vice versa, that is, not only formulate all the truths already known but also to invent/to discover new truths[4]. By taking mathematics as a model, Leibniz conceives the project of arithmetization of all the mechanisms of thought, of submitting the totality of human intellectual activity to calculatory processes able not only to demonstrate all known propositions but also to discover all new propositions, that is, to reduce rational activity to a calculus. However, in my view, Leibniz's greatest difference with respect to Lull concerns the kind of symbolic system able to operate within the calculus. Anticipating his fundamental methodological experiment - the discovery of the algorithm of the differential calculation as a true *organon* of mathematical research[5] - Leibniz is fully aware of the value of the sign as a not merely representative but also prospective and heuristic tool. But for that, Leibniz claims that the signs must not be, as in Lull, arbitrarily chosen. That is, Leibniz's project is not to be reduced to the constitution of a mere formal language, a system of universal signs that would allow the logical treatment of science and would supply a set of simple, necessary and rigorous symbols to express all actual and possible knowledge. Leibniz does not fall into the illusion – in some way present in Lull's project – according to which the automatic functioning of a set of operative rules can allow the development of science. For Leibniz science cannot be reduced to a merely formal, well-made language. That is why heuristics in Leibniz also pass for the requirement of semantics.

---

[3] Cf. for instance (GP 4: 63).

[4] Cf. *De Arte Combinatoria*, §§ 64/79 (GP 4: 64-8).

[5] As Cassirer stress (1923-29.1: 76), Leibniz would have applied his methodological experience with the algorithm of the differential calculation to the construction of the universal language.

2 XVII Century Projects of Philosophical Language and Leibniz

Let us now come to the XVII century and to the several attempts then made for the construction of an universal philosophical language. The XVII century was a time of an intense research on the intimate nature of human languages. The reasons are many and diverse and I will not analyze them here[6]. Let me just point out that the main question concerns the role which language performs in the process of knowledge: Does language help to promote knowledge? Or, conversely, does it prevent its progress? Is language a disturbing factor or a necessary element for the acquisition of knowledge? A mere means for communicating knowledge or an essential *medium* for constituting knowledge?

Two main positions can be identified: a) a critical position, which attributes to natural languages mere communicative functions and moreover emphasizes their insufficiencies and disturbing effects and b) a positive position which, although recognizing some limits and imperfection in human languages, nevertheless stresses its constitutive character, that is, its decisive role as a necessary cognitive device. Both positions were defended by great thinkers. In the first critical position we find Bacon, Locke, Descartes, Arnauld, Malabranche and, in general, all those who look for the construction of new artificial languages – from the baroque English pasigraphers to the constructors of philosophical languages (from Lodwick to Dalgano, from Seth Ward to Wilkins). In the second positive position, we can only find two names in modern times: Thomas Hobbes and some years later Leibniz.

Now, what I would like to stress is the fact that a) both of these two conflicting positions were relevant for the development of a computational conception of reason; and b) that Leibniz was the only one who fully understood the possibility of making those opposite positions converge in a fully developed conception of linguistic reason.

The critical position comes mainly from the awareness of the inadequateness of natural languages for representing the universe that modern science is providing as well as the logical categories of thought that are the basis for that scientific progress.

Bacon (1561-1626) is the great inspiration. He is the first to design the project of constructing a universal language which would represent, «neither letters nor words, but things and notions» (Bacon, 1605: 6.1.439). The idea was to build a language on the basis of a set of (real) characters representing not the sounds or voices of any natural language but the things or notions. This language could thus be read and understood by all men in all different languages. As Bacon writes: «Any book written in characters of this kind can be read of by each nation in their own language» (Bacon, 1605: 6.1, 439). Bacon also points towards the construction of a philosophical grammar, a set of rules common to all vulgar natural tongues. In his words : «If some one well seen in a great number of tongues, learned as well as vulgar, would haudle the various properties of languages, showing in what points each excelled, in what it failed. For so not only many languages be enriched by mutual exchanges, but the several beauties of each may be combined <...> into a most beautiful image and excellent model of speech itself, for the right expressing of the meanings of the mind» (Bacon, 1605: 6.1.421-422).

The programmatic suggestions of Francis Bacon, presented mostly in *The Advancement of Learning* (1605), are at the root of numerous projects of construction of a universal artificial language which emerged in the XVII century. Their aim is the

---

[6] For further developments on these issues, see Pombo (1987).

construction of a universal communicative tool. These projects, which gained great popularity in the XVII century, are clear practical endeavours. Facing mostly the needs created by the development of international contacts (economical, political, scientific and religious), their aim is the construction of a universal language, unique, perfectly regular, able to substitute Latin, to overcome the imperfections and diversity of natural languages and to allow adequate and full communication between men.

Of the many works undertaken in the XVII century, we must refer those of the English school of Baroque pasygraphs, Cave Beck, *The Universal Character, by which all the Nations in the World may Understand one Anothers Conceptions* (1657), Henry Edmundson, *The Natural Languages in a Vocabulary Contrieved and built upon Analogy* (1658), Edward Somerset, *A Century of the Names and Scantlings of such Inventions as at Present I can call to Mind to have Tried and Perfected* (1663). Above all, must be mentioned the works of the German Johan J. Becher, *Character pro Notitia Linguarum Universali* (1661) as well as of Athanasius Kircher, *Polygraphia Nova et Universalis ex Combinatoria Arte Delecta* (1663). Starting out from the assignment of a number to each word of the Latin dictionary and the organization of dictionaries of diverse languages in which each word was to be followed by the same number as attributed to its Latin equivalent, Becher intended to allow the translation, in any language, of a text written solely with numerical signs. The reader would only need to have access to the key-dictionary organized for their own language. More elaborated, Kircher's project consists in the construction of two polyglot dictionaries (Latin, Italian, French and German), each one organized in view of its future communicative role as source and target[7]. In both cases, we are faced with written systems of codification, totally artificial and arbitrary, of the sort we today would call special and technical languages (graphical semiotic systems limited to domains such as codes of telegraphic or maritime signals) rather than true philosophical languages. However, their authors, directly inspired by Bacon's programmatic proposals, looked for the construction of artificial languages able to avoid the ambiguity and irregularity of natural tongues, thus allowing the development of commerce, of knowledge and of true religion. As Cave Beck writes in the preface to *The Universal Character* (1657), such a universal language would be «a singular means of propagating all sorts of learning and true religion in the world».

Philosophical projects are more ambitious. Beyond merely communicational goals, their aim is to construct a symbolic system able adequately to express thought and its articulations and to supply a rigorous symbolic system capable of translating all the actual and possible knowledge, that is, to fulfil an essentially cognitive role. The idea is that it is possible to reduce all our ideas and notions to a small number of basic concepts, simple or primitive ideas. Universal language would have then to consist in the assigning of a character, not to each of the possible ideas as Bacon argued, but to each of the elementary ideas, and in the definition of a grammar (fixed rules of combination of characters). The inspiration is now cartesian. The primary objective is

---

[7] In the dictionary designed for sending the message, the words are ordered alphabetically and each one corresponds to the number of its page (in Arab numbers). In the dictionary designed for the reception of the message, the synonyms of the five languages are displayed in columns according to the alphabetical order of the Latin words[7]. We meet here a project which supposes a sketch of classification of concepts, that is, a philosophical project. For Leibniz's position with regard to Becher and Kircher, cf. (GP 4: 72), (C: 536) and his *letter to Oldenburg July 1670* (GP 7: 5).

to find the small number of simple ideas to which all other can be reduced and, by the combination of an equally limited number of characters, each one corresponding to one of those simple ideas, to construct the whole system of knowledge. That is why, prior to the choice of the characteristic signs, all philosophical projects begun by making an elaborated logical and semantical classification of concepts.

In a celebrated *Letter to Mersenne, 20 th November 1629* (*AT* 1:76-82), Descartes had in fact stated a very relevant thesis about the possibility of a universal language. If the totality of arithmetic may be constructed upon a small number of axioms, in a similar way, it would have to be possible to exhaustively symbolize the totality of the thoughts on the basis of a limited number of linguistic signs. As he writes : «Je trouve qu'on pourroit ajouter à ceci une invention, tant pour composer les mots primitifs de cette langue que pour leurs caractères en sorte qu'elle pourroit être enseignée en fort peu de temps, et ce par le moyen de l'ordre, c'est-à-dire établissant un ordre entre toutes les pensées qui peuvent entrer en l'esprit humain de même qu'il y en a un naturellement établi entre les nombres; et comme on peut apprendre an un jour à nommer toutes les nombres jusques à l'infini et à les écrire en une langue inconnue, qui sont toutes fois une infinité de mots différents, qu'on pût faire le même de toutes les autres mots nécessaires pour exprimer toutes les autres choses qui tombent en l'esprit des hommes» (AT1: 80-81).

The paradigm is thus mathematics as a symbolic system of universal application and, simultaneously, underlying the order of the world and the structure of the creation. That is to say, mathematics is not only a deductive system grounded on the evidence of its axioms or a rigorous method of demonstration, but a universal language which may represent the world because it is isomorphic with its structure, because the world is, in itself, mathematically ordered and structured. In this sense, it is important to remember that the mathematics which is the basis of modern science is, in all cases, warranted by the Pythagorean conception of the mathematical structure of creation (see the cases of Galileo, Descartes, Kepler, Leibniz or Newton). So, the ideal of a universal philosophical language is an extension and a deepening of mathematical procedures able to operate on the basis of universal categories able to develop and uncover the integral system not only of knowledge but also of world and things.

Of the many philosophical languages proposed in the XVII century, the first seems to have been that of Mersenne who, in a letter to Fabri de Peiresc dated of 1636 or 37[8], claims to have finished the construction of a philosophical language, surely of cartesian inspiration.[9] In England too, we ought to mention names such as the mathematian and astronomer Seth Ward, *Vindiciae Academiarum* (1654). In terms close to those used by Descartes, they both point to the decomposition or analysis of concepts into primitive terms[10]. Also in England, Georges Dalgarno and John Wilkins realized the most relevant and complete philosophical languages. Dalgarno publishes *Ars Signorum, vulgo Character Universalis et Lingua Philosophica* (1661) and Wilkins, bishop of Chester and one of the most reputed scholars of the period, publishes in 1668, with the support

---

[8] Partially reproduced in Adam-Tannery, I , Correspondance, pp. 572-3.

[9] Of which hardly any fragment exists except some propositions in the *Traité de l'Harmonie Universelle* (1636).

[10] «...for all discourses being resolved in sentences, those into words, words signifying either simple notions or being resolvable into simple notions, it is manifest that if all the sorts of simple notions be found out, and have symbols assigned to them, those will be extremely few in respect of the other", Ward (1954: 21).

of Royal Society, *An Essay towards a Real Character and a Philosophical Language, with an Alphabetical Dictionary*. On the basis of a hierarchical organization of concepts in classes and sub-classes, Dalgarno points to a logical symbolic system of letters able to express the normal articulation of thought[11]. Similarly, Wilkins starts out from a classification of concepts which he organizes in a complex system of *Summa Genera, Diferentia and Species*. Those categories, with which Wilkins aims to cover the totality of scientific knowledge, [12] are represented by a *real character*, that is, not by letters as in Dalgarno, but by written characters of ideographic nature[13]

Leibniz knew well all this philosophical projects. He praised above all Wilkins's project. «J'ai consideré avec attention le grand ouvrage du Caractere reel et Langage Philosophique de Mons. Wilkins; je trouve qu'il y a mis une infinité de belles choses, et nous n'avons jamais eu une Table des predicateurs plus accomplie» (*Letter to Burnet, 24th August 1697*, GP 3: 216). However, Leibniz does not avoid criticism. His main concern is the insufficient analysis which, in Leibniz opinion, underlies Dalgarno and Wilkins proposal. As Leibniz writes in the same *Letter:* «J'avois consideré cette matière avant le livre de Mr. Wilkins, quand j'estoit un jeune homme de 19 ans, dans mon petit livre de *Arte Combinatoria*, et mon opinion est que les Caracteres veritablement reels et philosophiques doivent repondre à l'Analyse des pensées» (GP 3: 216). This is in fact the most difficult but also the most fundamental task. Only the analysis of the diverse conceptual contents and of their relationships can constitute the limited set of primitive terms, on the basis of which the philosophical language must be constructed and which can support the choice of the non-arbitrary system of signs for the philosophical language. As Leibniz says: «the name which, in this language, should be attributed to gold will be the key of the knowledge which could be obtained from it."[14]

Leibniz was aware of all of his predecessors and contemporary proposals for an universal language, both international and philosophical. He fully realized both the scientific and the logical character of those projects. He was the one who have articulated more deeply the logical, semantic and heuristic objectives of such an endeavour. Against Bacon, he stressed the need of beginning the construction of universal language by the attribution of a character, not to each one of possible ideas as Bacon defended, but to each of the elementary ideas to which all others could be reduced. Against Descartes who finally condemned the project because, in his view, only the complete (encyclopaedic) analysis of the diverse conceptual contents and of their relationships might allow the finding of the primitive terms on the basis of which the philosophical language could be constructed, [15] Leibniz defended that the two tasks -

---

[11] For a detailed description of Dalgarno's philosophical language, cf. Couturat (1901: 544--548), Rossi (1960: 218 segs.) and Cram (1980).

[12] Wilkins had the collaboration of John Ray, Francis Willoughby and Samuel Pepys.

[13] For a detailed description Wilkins project, cf. Couturat, «*Sur la langue philosophique de Wilkins*» (1901: 548-552).

[14] «Nomen tamen quod in hac lingua imponetur, clavis erit eorum omnium quae de auro humanitus, id est ratione at que ordine scriri possunt, cum ex eo etiam illud appariturum sit, quaenam experimenta de eo cum ratione institui debeant» (GP 7: 13).

[15] As Descartes stated in the refered *Letter to Mersenne, 20 th November 1629* «L'invention de cette langue dépend de la vrai philosophie; car il est impossible autrement de dénombrer toutes les pensées des hommes, et de les mettre par ordre, ni seulement de les distinguer en sorte qu'elles soient claires et simples» (AT1: 82).

characteristica universalis and *encyclopaedia* – could be done in a zig-zag process, not as linear and sequential enterprises but as reciprocal and mutual conditioning endeavours: [16] each time analysis advances and is able to determine a new unit, this unit will only become clear and distinct for though when it is designated by a specific sign. On the other hand, from the attribution of this sign will arise new heuristic virtualities as a result of its insertion into the structural network of all previously constituted signs.

Further, Leibniz was quite conscious that, in order to construct such a well grounded *organon* of rational inquiry, a new conception of symbolism and a new conception of reason should be worked out. The first, which I cannot present here[17], was for the most part constructed against Descartes, in the context of their different understanding of the success of mathematical sciences and of the gnoseological status of its symbolism. The second comes mainly from Hobbes.

3 Hobbes and Leibniz

In *Leviathan*, Hobbes states: "reasoning is nothing but reckoning"[18]. And he adds : "when a man Reasoneth, hee does nothing else but conceive a summe totall, from Addition of parcels; or conceive a Remainder, from Substraction of one summe from another" (L: 110). Further he explains: "those operations are not incident to Numbers onely, but to all manner of things that can be added together and taken one of another" (L: 110), that is, to arithmetic, geometry, logic, and also politics, law, "In summe, in what matter so ever there is place for addition and substaction, there also is place for Reason, and where these have no place, there Reason has nothing at all to do» (L: 110-1). Of course, the aim of extending the same kind of rational procedures to the totality of the domains finds an immediate echo in Leibniz. However, if Hobbes is the indisputable predecessor of the computational linguistic theory of cognition, there is not in Hobbes any methodological solution for the effective realization of such a rational enterprise. And given the poor mathematical capacity of Hobbes, [19] his definition of reason is probably not much more than a empty *dictum*, whose rich and profound consequences Hobbes himself was not fully aware. As Couturat shows (1901: 461), Hobbes would not have penetrated all the meaning of his own formula. By contrast, Leibniz was completely prepared to develop that computational conception of reason in the context of his acute mathematical capacity and his huge interest in combinatorial processes.

---

[16] In fact, in the mole of Leibniz manuscripts, there is a copy of the letter from Descartes to Mersenne in which, by his own hand, Leibniz wrote: "Il est vray que ces caractères pressupposeroient la veritable philosophie (...) Cependant, quoique cette langue depende de la vraye philosophie, elle ne depend pas de sa perfection. C'est à dire, cette langue peut estre établie, quoyque la philosophie ne soit pas parfaite: et à mesure que la science des hommes croistra, cette langue croistra aussi.» (C: 28)

[17] For a presentation of this debate, cf. Pombo (1990).

[18] Hobbes (L: 99).

[19] For an analysis of the polemics between Hobbes and some great mathematicians of his time, like Wallis, Ward, Boyle and Huygens, cf. Breidert (1979).

But there is a further aspect in which Hobbes is an important root for Leibniz computational conception of reason: Hobbes's conception of reason as a linguistic activity, a faculty which only operates on the basis of names.[20] According to Hobbes, only language makes us able to "transfer (*translation*) our mental discourse into verbal discourse" (*L.* p. 100), that is, only language stabilizes the fluid of mental discourse by establishing points of orientation around which representations become fixed and isolated. As Hobbes writes: "it is by the very names that we are able to stabilize the multiple conceptions of one representation" (*H. N.* V, § 4). That is to say, only language provides the symbolic elements upon which the activity of calculus may be realised: "For Reason, in this sense, is nothing but Reckoning (that is Adding and Substacting) of the Consequences of generall *names* agreed upon, for the marking and signifying of our thoughts" (L: 111, our emphasis). That means that without language there will be no reason. Ideas are not images but words.

We are here face to a very important contribution of Hobbes to Leibniz's philosophy of language and to the computational conception of reason in general. If language is the verbal *translatio* of a previous *Mentall Discourse*[21], that imagetic, unstable fluid of thoughts[22] in which language alone may establish points of reference, areas of security, marks, notes, names, therefore language is the sensible support for thought providing the material, signifying conditions required for the development of calculation. Only language permits the substitution of calculation of ideas by calculation of names. Only language preserves the possibility of recollecting previous steps, of revising deductive chains, of progressing, gradually and securely, from one consequence to another. That is why Hobbes writes: "The use and End of reason, is not the finding of the *summe*, and truth of one, or a few consequences, remote from the first definitions, and settled significations of names; but to begin at these; and proceed from one consequence to another» (L: 112).

Hobbes is here giving a significant contribution to Leibniz who will fully adopt this thesis. And indeed, Leibniz recognizes his heritage from Hobbes precisely in this point. As he says: "names are not only signs of my present thoughts for the others but notes of my previous thoughts to myself, as Thomas Hobbes has demonstrated."[23]

From this important thesis Leibniz will work out epistemic (heuristic) consequences which Hobbes never suspected. But for doing this, Leibniz will have to overcome the internal limits of Hobbes philosophy of language. Because Hobbes is attached to an evocative conception of the sign, because he claims that names must always be accompanied by the conceptions they mean, Hobbes will refuse vehemently the hypothesis of a manipulation of signs. Without the evocative fulfilment of signs by the ideas that correspondent to them, *ratio*, says Hobbes, would be reduced to *oratio* (cf. HN, V, § 14), to mechanical speech, similar to what happens with "beggars, when they say their paternoster putting together much words, and in such

---

[20] As Hobbes states: "Children therefore are not endued with Reason at all, till they have attained the use of Speech» (*L*: 116).

[21] «The general use of Speech is to transferre our Mentall Discourse, into verbal» (*Leviathan*: 100).

[22] «By consequence, or Trayne of Thoughts I understand that sucession of one thought to another which is called (...) Mentall Discourse» (*L*: 94).

[23] «Verba enim non tantum signa sunt cogitationis meae praensentis ad alios, sed et notae cogitationes meae praeteritae ad me ipsum, ut demonstravit Thomas Hobbes», (A. VI, 1. 278).

manner, as in their education they have learned from their nurses, from their companions, or from their teachers, having no images or conceptions in their minds answering to the words they speak» (HN, V, § 14).

Now, on the contrary, to wish for constantly thinking the meaning of signs which one handles is, not only *de facto* an impossible activity, as it would be *de jure* an obstruction to rigour and invention which both claim for the abandon of the spirit to the formal mechanisms created by the spirit. That is the main point of the celebrated leibnizian theory of *cogitation caecae*. [24]But that will be one of the greatest discoveries from Leibniz's philosophy of language.

4 Concluding Remarks

Leibniz contribution to the computational conception of reason comes thus from three important roots which Leibniz was able to integrate and simultaneously to overcome. 1) Leibniz accepts Lull's idea of a combinatory however being able to substitute the mechanical combinatorial procedures proposed by Lull by calculatory processes of mathematical analysis (in this point anticipating Boole); 2) in the line of Bacon, Leibniz takes the communicative aims of all projects for universal languages however trying to give them a cognitive foundation in order to make them true philosophical languages. In order to do so, Leibniz agrees to the methodological regime proposed by Descartes for the construction of a philosophical language however refuting the linear relationship Descartes established between philosophical language and encyclopaedia (against the separation which later Frege will establish between those two issues[25]), 3) Leibniz works out the computational conception of reason formulated by Thomas Hobbes however building a new conception of sign which makes possible to explore its heuristic potentialities.

References

1. Bacon, F.: The Advancement of Learning. In: The works of Francis Bacon. Faksimile Neudruck der Ausgabe von Spedding. Ellis und Heath, vol. III, pp. 260–491. Stuttgart Bad/Frommann, London/Cannstatt (1963)
2. Becher, J.: Character, pro notítia linguarum universali, inventum steganographicum hactenus inauditam quo quilibet suam legendo vernaculam diversas imo omnes linguas, unius etiam diet informatione, explicare ac intelligere potest. Francofurti: sumtibus Johanis Wilh. Amonii et Wilhelmi Serlini, typis Johannis Georgii Spörlin (1661)
3. Beck, C.: The Universal Character, by which all the Nations in the World may understand one anothers Conceptions, Reading out on one Common Writing their own Mother Tongues. T. Maxey, for W. Weekly, Longon (1657)

---

[24] Which we have studied in Pombo (1998).

[25] Apart from the similarities between the leibnizian *characteristica universalis* and Frege's ideographic project, the two projects should be fundamentally distinguished from one another by the fact that Frege, leaving to the sciences the labour of defining their own concepts, rejects the close articulation established by Leibniz between characteristic language and encyclopaedia.

4. Breidert, W.: Les mathématiques et la méthode mathématique chez Hobbes. Revue Internationale de Philosophie 129, 415–431 (1979)
5. Carreras y Artau, Tomás/Joaquín, Historia de la filosofia espanõla. Filosofia cristiana de los siglos XII a XV. I. Madrid: Real Academia de Ciências Exactas, Físicas y Naturales (1939)
6. Cassirer, E. (1923-29): La philosophie des formes symboliques. I. Le langage (Trad. Ole Hansen Love, Jean Lacoste). III. La phénoménologie de la connaissance (Trad. Claude Fronty). Paris: Minuit (1972); Orig.: Philosophie der symbolischen Formen. Erster teil. Die Spache (Berlin, 1923). Dritter Teil. Phänomenologie der Erkenntnis (Berlin, 1929)
7. Couturat, L.: La logique de Leibniz d'après des documents inédits. Alcan, Paris (1901)
8. Cram, D.: George Dalgarno on Ars Signorum and Wilkins Essay. In: Koerner (ed.) Progress in Linguistic Historiography, pp. 113–121. John Benjamins, Amsterdam (1980)
9. Dalgarno, G. (1661): Ars Signorum, vulgo character universalis et lingua philophica. In: The Works of George Dalgarno of Aberdeen, pp. 1–82. Constable, Eddinburg (1834)
10. Descartes, R.: Euvres de Descartes. Ed. Charles Adam Paul Tannery. Vrin, Paris (1964 76)
11. Hobbes, Thomas, Leviathan. Ed. by C. B. Macpherson. Penguin, London (1968)
12. Hobbes, T.: Human Nature. Or the Fundamental Elements of Policy. In: Hobbes, T. (ed.) The English Works of Thomas Hobbes of Malmesbury, Ed. by Sir William Moleswort, London, vol. 4 (1939)
13. Leibniz, G.: Wilhelm Leibniz Samtliche Schrifften und Briefe. Akademie der Wissenschaften zu Berlin. Reihe I VI. Reichl, Darmstadt (1923 segs)
14. Leibniz, Opuscules et fragments inédits de Leibniz. Extraits des manuscrits de la Bibliothèque royale de Hannover par Louis Couturat. Alcan, Paris (1903)
15. Leibniz, Die philosophischen Schriften von Gottfried Wilhelm Leibniz. Hrsg v. Carl Immanuel Gerhardt. pp. 1–7. Olms, Hildesheim (1960)
16. Llull, R. (1308): Ars Brevis (tradução, introdução e notas de Armand Llinarès). La éditions du Cerf, Paris (1991)
17. Mersenne, M. (1636): Harmonie universelle contenant la théorie et la pratique de la musique, 1636. Ed. facs. de l'exemplaire conservé à la Bibliothèque des Arts et Métiers et annoté par l'autheur. Introduction par François Lesure. Vols. 1, 3. Editions du Centre National de la Recherche Scientifique, Paris (1963)
18. Pombo, O.: Linguagem e verdade em Hobbes. Filosofia 1, 45–61 (1985)
19. Pombo, O.: Leibniz and the problem of a Universal Language. Nodus Publikationen, Münster (1987)
20. Pombo, O.: The Leibnizian theory of representativity of the sign. In: Niederehe, H.J., Koerner, K. (eds.) History and Historiography of linguistics, II, pp. 447–459. John Benjamins Publishing Company, Amsterdam (1990)
21. Pombo, O.: La Théorie Leibnizienne de la Pensée Aveugle en tant que Perspective sur quelques-unes des Apories Linguistiques de la Modernité. Cahiers Ferdinand Saussure 51, 63–75 (1998)
22. Rossi, P., Universalis, C.: Arti Mnemoniche e Logica Combinatoria da Lullo a Leibniz. Riccardo Ricciardi, Milano (1960)
23. Somerset, E.W. (1663): A Century of the Names and Scantlings of such Inventions, as at Present I can call to Mind to Have Tried and Perfected. Printed by J. Grismond and reprinted by J. Adlard, London (1813)
24. Ward, S.: Vindicae academiarum. Containing Some Briefe Animadversions Upon Mr. Websters Book. Stiled. The Examination of Academies. Together with an Appendix Concerning what Mr. Hobbes, and Mr. Dell have Published on the Argument. Leonard Lichfield for Thomas Robinson, Oxford (1654)
25. Wilkins, J.: An Essay Towards a Real Character, and a Philosophical Language. S. Gellibrand & J. Martyn, London (1668)

Development of a Bacteria Computer: From *in silico* Finite Automata to *in vitro* and *in vivo*

Yasubumi Sakakibara[1,2]

[1] Department of Biosciences and Informatics, Keio University,
3-14-1 Hiyoshi, Kohoku-ku, Yokohama, 223-8522, Japan
yasu@bio.keio.ac.jp
[2] Institute for Bioinformatics Research and Development (BIRD),
Japan Science and Technology Agency (JST)

Abstract. We overview a series of our research on implementing finite automata *in vitro* and *in vivo* in the framework of DNA-based computing [1,2]. First, we employ the length-encoding technique proposed and presented in [3,4] to implement finite automata in test tube. In the length-encoding method, the states and state transition functions of a target finite automaton are effectively encoded into DNA sequences, a computation (accepting) process of finite automata is accomplished by self-assembly of encoded complementary DNA strands, and the acceptance of an input string is determined by the detection of a completely hybridized double-strand DNA. Second, we report our intensive *in vitro* experiments in which we have implemented and executed several finite-state automata in test tube. We have designed and developed practical laboratory protocols which combine several *in vitro* operations such as annealing, ligation, PCR, and streptavidin-biotin bonding to execute *in vitro* finite automata based on the length-encoding technique. We have carried laboratory experiments on various finite automata with 2 up to 6 states for several input strings. Third, we present a novel framework to develop a programmable and autonomous *in vivo* computer using *Escherichia coli (E. coli)*, and implement *in vivo* finite-state automata based on the framework by employing the protein-synthesis mechanism of *E. coli*. We show some successful experiments to run an *in vivo* finite-state automaton on *E. coli*.

1 Introduction

Biological molecules such as DNA, RNA and proteins are natural devices to store information, activate (chemical) functions and communicate between systems (such as cells). Studies in DNA computing utilize these biological devices to make a computer [5]. One of the ultimate goals is to make an autonomous cell-based Turing machine applicable to genetic engineering. Our attempts to make a bacteria-based computer make progress toward this goal.

The finite-state automata (machines) are the most basic computational model in Chomsky hierarchy and are the start point to build universal DNA computers. Several works have attempted to develop finite automata *in vitro*. Benenson et al. [6] have successfully implemented the two state finite automata by the

F. Ferreira et al. (Eds.): CiE 2010, LNCS 6158, pp. 362–371, 2010.

sophisticated use of the restriction enzyme (actually, *Fok*I) which cuts outside of its recognition site in a double-stranded DNA. Soreni et al. [7] have extended the method to 3-symbol and 3-state finite automaton. Yokomori et al. [3] have proposed a theoretical framework using length-encoding technique to implement finite automata on DNA molecules. Theoretically, the length-encoding technique has no limitations to implement finite automata of any larger states.

In our first research work [1], we have attempted to implement and execute finite automata of a larger number of states *in vitro*, and carry intensive laboratory experiments on various finite automata of from 2 states to 6 states for several input strings.

On the other hand, in our next research work [2], we have proposed a method using the protein-synthesis mechanism combined with four-base codon techniques [8] to simulate a computation (accepting) process of finite automata *in vitro* [4] (a codon is normally a triplet of base, and different base triplets encode different amino acids in protein). The proposed method is quite promising and has several advanced features such as the protein-synthesis process is very accurate and overcomes mis-hybridization problem in the self-assembly computation and further offers an autonomous computation. Our aim was to extend this novel principle into a living system, by employing the *in vivo* protein-synthesis mechanism of *Escherichia coli (E. coli)*. (Escherichia coli is a typical bacteria living inside our body, large intestine.) This *in vivo* computation possesses the following two novel features, not found in any previous biomolecular computer. First, an *in vivo* finite automaton is implemented in a living *E. coli* cell; it does not mean that it is executed simply by an incubation at a certain temperature. Second, this automaton increases in number very rapidly according to the bacterial growth; one bacterial cell can multiply to over a million cells overnight. The present study explores the feasibility of *in vivo* computation.

The main feature of our *in vivo* computer based on *E. coli* is that we first encode an input string into one plasmid, encode state-transition functions into the other plasmid, and transform *E. coli* cells with these two plasmids by electroporation. Second, we execute a protein-synthesis process in *E. coli* combined with four-base codon techniques to simulate a computation (accepting) process of finite automata, which has been proposed for *in vitro* translation-based computations in [4]. The successful computations are detected by observing the expressions of a reporter gene linked to mRNA encoding an input data. Therefore, when an encoded finite automaton accepts an encoded input string, the reporter gene, *lacZ*, is expressed and hence we observe a blue color. When the automaton rejects the input string, the reporter gene is not expressed and hence we observe no blue color. Our *in vivo* computer system based on *E. coli* is illustrated in Fig. 1. This approach enables us to develop a programmable *in vivo* computer by simply replacing a plasmid encoding a state-transition function with others. Further, our *in vivo* finite automata are autonomous because the protein-synthesis process is autonomously executed in the living *E. coli* cell.

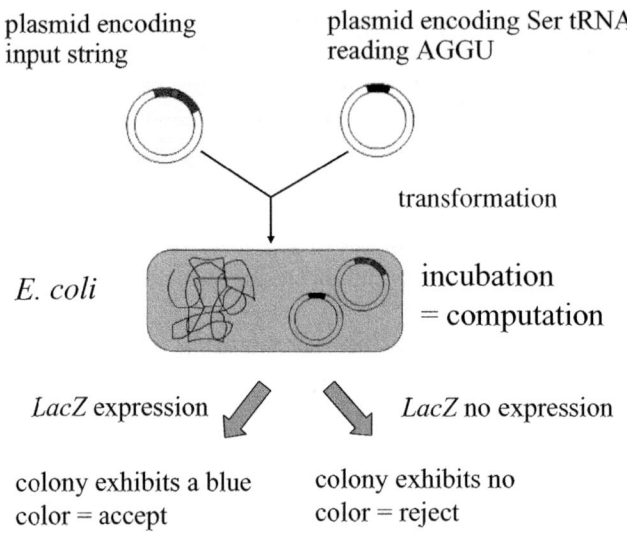

plasmid encoding
input string

plasmid encoding Ser tRNA
reading AGGU

transformation

E. coli

incubation
= computation

LacZ expression

LacZ no expression

colony exhibits a blue
color = accept

colony exhibits no
color = reject

Fig. 1. The framework of our *in vivo* computer system based on *E. coli*

2   Methods

2.1   Length-Encoding Method to Implement Finite-State Automata

Let $M = (Q, \Sigma, \delta, q_0, F)$ be a finite automaton, where Q is a finite set of states numbered from 0 to k, Σ is an alphabet of input symbols, δ is a state-transition function such that $\delta : Q \times \Sigma \longrightarrow Q$, q_0 is the initial state, and F is a set of final states. We adopt the length-encoding technique [3] to encode each state in Q by the length of DNA subsequences.

For the alphabet Σ, we encode each symbol a in Σ into a single-strand DNA subsequence, denoted $e(a)$, of fixed length. For an input string w on Σ, we encode $w = x_1 x_2 \cdots x_m$ into the following single-strand DNA subsequence, denoted $e(w)$:

$$5\text{'-}\ e(x_1) \underbrace{X_1 X_2 \cdots X_k}_{k \text{ times}} e(x_2) \underbrace{X_1 X_2 \cdots X_k}_{k \text{ times}} \cdots e(x_m) \underbrace{X_1 X_2 \cdots X_k}_{k \text{ times}}\ \text{-3'},$$

where X_i is one of four nucleotides A, C, G, T, and the subsequences $X_1 X_2 \cdots X_k$ are used to encode $k + 1$ states of the finite automaton M. For example, when we encode a symbol '1' into a ssDNA subsequence GCGC and a symbol '0' into GGCC, and encode three states into TT, a string "1101" is encoded into the following ssDNA sequence:

$$5\text{'-}\ \overbrace{\text{GCGC}}^{1} \text{TT} \overbrace{\text{GCGC}}^{1} \text{TT} \overbrace{\text{GGCC}}^{0} \text{TT} \overbrace{\text{GCGC}}^{1} \text{TT}\ \text{-3'}$$

In addition, we append two supplementary subsequences at both ends for PCR primers and probes for affinity purifications with magnetic beads which will be used in laboratory protocol:

$$5\text{'-}\ \underbrace{S_1S_2\cdots S_s}_{\text{PCR primer}}\, e(x_1)X_1X_2\cdots X_k\cdots e(x_m)X_1X_2\cdots X_k\ \underbrace{Y_1Y_2\cdots Y_t}_{\text{probe}}\,\underbrace{R_1R_2\cdots R_u}_{\text{PCR primer}}\ \text{-3'}.$$

For a state-transition function from state q_i to state q_j with input symbol $a \in \Sigma$, we encode the state-transition function $\delta(q_i, a) = q_j$ into the following complementary single-strand DNA subsequence:

$$3\text{'-}\ \underbrace{\overline{X}_{i+1}\overline{X}_{i+2}\cdots\overline{X}_k}_{k-i\ \text{times}}\, \overline{e(a)}\, \underbrace{\overline{X}_1\overline{X}_2\cdots\overline{X}_j}_{j\ \text{times}}\ \text{-5'}$$

where $\overline{X}_i$ denotes the complementary nucleotide of X_i, and $\overline{y}$ denotes the complementary sequence of y. Further, we put two more complementary ssDNA sequences for the supplementary subsequences at both ends:

$$3\text{'-}\ \overline{S}_1\overline{S}_2\cdots\overline{S}_s\ \text{-5'}, \qquad 3\text{'-}\ \underbrace{\overline{Y}_1\overline{Y}_2\cdots\overline{Y}_t\overline{R}_1\overline{R}_2\cdots\overline{R}_u}_{\text{biotinylated}}\ \text{-5'},$$

where the second ssDNA is biotinylated for streptavidin-biotin bonding.

Now, we put all those ssDNAs encoding an input string w and encoding state-transition functions and the supplementary subsequences of probes and PCR primers. Then, a computation (accepting) process of the finite automata M is accomplished by self-assembly among those complementary ssDNAs, and the acceptance of an input string w is determined by the detection of a completely hybridized double-strand DNA.

The main idea of length-encoding technique is explained as follows. Two consecutive valid transitions $\delta(h, a_n) = i$ and $\delta(i, a_{n+1}) = j$ are implemented by concatenating two corresponding encoded ssDNAs, that is,

$$3\text{'-}\ \underbrace{AAA\cdots A}_{k-h}\, \overline{e(a_n)}\, \underbrace{AAA\cdots A}_{i}\ \text{-5'},$$

and

$$3\text{'-}\ \underbrace{AAA\cdots A}_{k-i}\, \overline{e(a_{n+1})}\, \underbrace{AAA\cdots A}_{j}\ \text{-5'}$$

together make

$$3\text{'-}\ \underbrace{AAA\cdots A}_{k-h}\, \overline{e(a_n)}\, \underbrace{AAA\cdots A}_{k}\, \overline{e(a_{n+1})}\, \underbrace{AAA\cdots A}_{j}\ \text{-5'}.$$

Thus, the subsequence $\underbrace{AAA\cdots A}_{k}$ plays a role of "joint" between two consecutive state-transitions and it guarantees for the two transitions to be valid in M.

Note that the length-encoding method can be applied to both deterministic and non-deterministic finite automaton.

2.2   Designing Laboratory Protocols to Execute Finite Automata in Test Tube

In order to practically execute the laboratory experiments for the method described in the previous section, we design the following experimental laboratory protocol, which is also illustrated in Fig. 2:

0. **Encoding:** Encode an input string into a long ssDNA, and state-transition functions and supplementary sequences into short pieces of complementary ssDNAs.

1. **Hybridization:** Put all those encoded ssDNAs together into one test tube, and anneal those complementary ssDNAs to be hybridized.

2. **Ligation:** Put DNA "ligase" into the test tube and invoke ligations at temperature of 37 Celsius degree. When two ssDNAs encoding two consecutive valid state-transitions $\delta(h, a_n) = i$ and $\delta(i, a_{n+1}) = j$ are hybridized at adjacent positions on the ssDNA of the input string, these two ssDNAs are ligased and concatenated.

3. **Extraction by affinity purification:** Denature double-stranded DNAs into ssDNAs and extract concatenated ssDNAs containing biotinylated probe subsequence by streptavidin-biotin bonding with magnetic beads.

4. **Amplification by PCR:** Amplify the extracted ssDNAs with PCR primers.

5. **Detection by gel-electrophoresis:** Separate the PCR products by length using gel-electrophoresis and detect a particular band of the full-length. If the full-length band is detected, that means a completely hybridized double-strand DNA is formed, and hence the finite automaton "accepts" the input string. Otherwise, it "rejects" the input string.

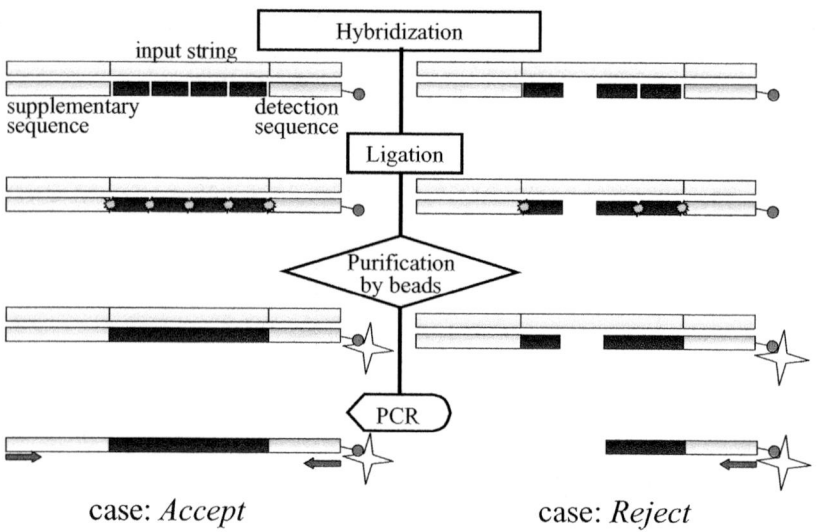

case: *Accept*   case: *Reject*

Fig. 2. The flowchart of laboratory protocol to execute *in vitro* finite automata which consists of five steps: hybridization, ligation, denature and extraction by affinity purification, amplification by PCR, and detection by gel-electrophoresis. The acceptance of the input string by the automata is the left case, and the rejection is the right case.

2.3   Simulating Computation Process of Finite Automata Using Four-Base Codons and Protein-Synthesis Mechanism

Sakakibara and Hohsaka [4] have proposed a method using the protein-synthesis mechanism combined with four-base codon techniques to simulate a computation (accepting) process of finite automata. Our approach to make *in vivo* computer is to execute the proposed method on *E. coli* in order to improve the efficiency of the method and further develop a programmable *in vivo* computer. We describe the proposed method using an example of simple finite automaton, illustrated in Fig. 3, which consists of two states $\{s_0, s_1\}$, defined on one symbol '1', and accepts input strings with even numbers of symbol 1 and rejects input strings with odd numbers of 1s.

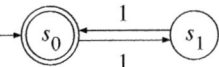

Fig. 3. A simple finite automaton of two states $\{s_0, s_1\}$, defined on one symbol '1', and accepting input strings with even numbers of symbol 1 and rejecting input strings with odd numbers of 1s.

The input symbol '1' is encoded to the four-base subsequence AGGU and an input string is encoded into an mRNA by concatenating AGGU and A alternately and adding AAUAAC at the 3'-end. This one-nucleotide A in between AGGU is used to encode two states $\{s_0, s_1\}$, which is a same technique presented in [3]. For example, a string "111" is encoded into an mRNA:

The four-base anticodon (3')UCCA(5') of tRNA encodes the transition rule $s_0 \xrightarrow{1} s_1$, that is a transition from state s_0 to state s_1 with input symbol 1, and the combination of two three-base anticodons (3')UUC(5') and (3')CAU(5') encodes the rule $s_1 \xrightarrow{1} s_0$. Further, the encoding mRNA is linked to *lacZ*-coding RNA subsequence as a reporter gene for the detection of successful computations. Together with these encodings and tRNAs containing four-base anticodon (3')UCCA(5'), if a given mRNA encodes an input string with odd numbers of symbol 1, an execution of the *in vivo* protein-synthesis system stops at the stop codon, which implies that the finite automaton does not accept the input string, and if a given mRNA encodes even numbers of 1s, the translation goes through the entire mRNA and the detection of acceptance is found by the *blue* signal of *lacZ*. Examples of accepting processes are shown in Fig. 4: (Upper) For an mRNA encoding a string "1111", the translation successfully goes through the entire mRNA and translates the reporter gene of *lacZ* which emits the blue signal. (Lower) For an mRNA encoding a string "111", the translation stops at the stop codon UAA, does not reach the *lacZ* region and produces no blue signal.

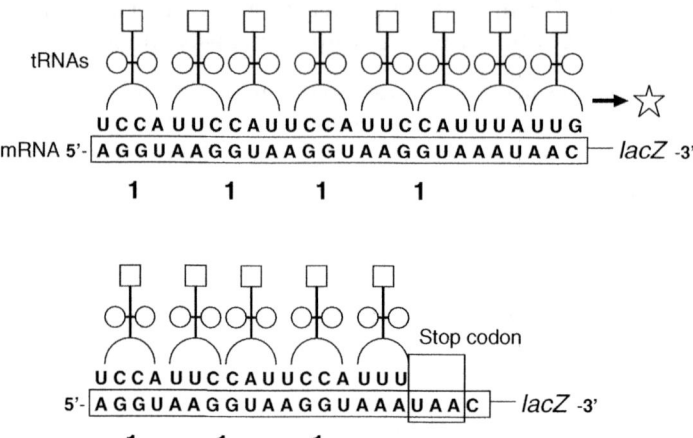

Fig. 4. Examples of accepting processes: (Upper) For an mRNA encoding a string "1111", the translation successfully goes through the mRNA and translates the reporter gene of *lacZ* emitting the blue signal. (Lower) For an mRNA encoding a string "111", the translation stops at the stop codon UAA, does not reach the *lacZ* region and produces no blue signal.

3   In Vitro Experiments

3.1   4-States Automaton with Three Input Strings

Our first experiment consists in the implementation/development of 4-states automaton shown in Fig. 5 (upper left) for the three input strings (a) 1101, (b) 1110, and (c) 1010. This 4-states automaton accepts the language $(1(0 \cup 1)1)^* \cup (1(0 \cup 1)1)^*0$, and hence it accepts 1110 and 1010 and rejects 1101.

The results are shown in Fig. 5 (upper right) in gel-like image. As in the first experiment, the full-length DNA is of 190 bps (mer). Bands at position of 190 mer is detected in lane (b) and lane (c). Hence, our *in vitro* experiments have

Fig. 5. (Left:) A 4-states automaton used for this experiment. (Right:) The results of electrophoresis are displayed in gel-like image. Lane (a) is for the input string 1101, lane (b) for 1110, and lane (c) for 1010. Since the full-length band (190 mer) is detected in lane (b) and (c), we determine the automaton accepts two input strings (b) 1110 and (c) 1010.

successfully detected that the automaton accepts two input string (b) 1110 and
(c) 1010.

3.2   From 2-States to 6-States Automata with One Input String "111111" of Length 6

Our second set of experiments consists in the implementation of the 5 different
automata, having a number of states ranging from 2 to 6, shown in Fig. 6 (upper)
for one input string "111111" of length 6.

The results are shown in Fig. 6 (lower) in gel-like image. For the input string
111111, the full-length DNA is of 240 bps (mer). Bands at position of 240 mer
are detected in lanes (2), (3) and (6) in Fig. 6. Hence, in our *in vitro* experiments,
the automaton (2), (3) and (6) have correctly accepted the input string 111111
and the automaton (4) and (5) have correctly rejected 111111.

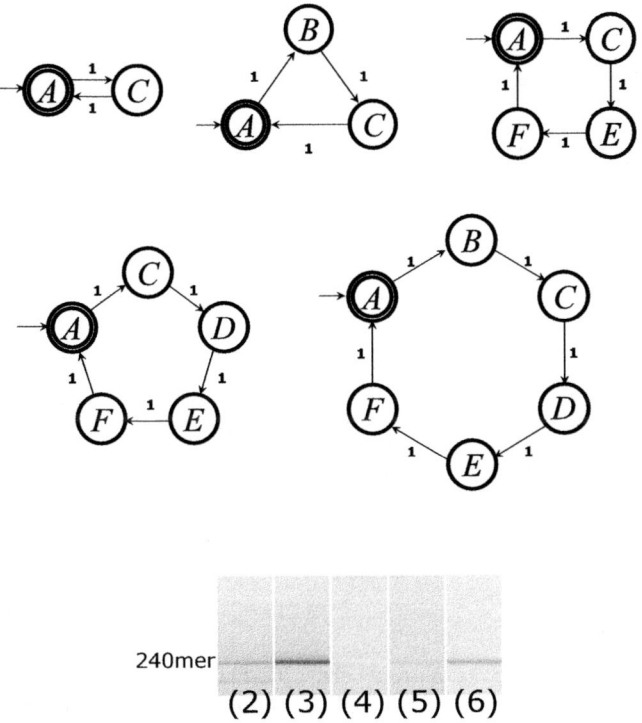

Fig. 6. (Upper:) Five different automata with 2 up to 6 states used for this experiment.
(Lower:) The results of electrophoresis are displayed in gel-like image. Lane (2) is for the
automaton (2), (3) for (3), (4) for (4), (5) for (5), and (6) for (6). Since the full-length
bands (240 mer) are detected in lane (2), (3) and (6), we determine the automata (2),
(3) and (6) accepts the input string 111111.

4   In Vivo Experiments

We have done some laboratory experiments by following the laboratory protocols presented in [2] to execute the finite automaton shown in Fig. 3, which consists of two states $\{s_0, s_1\}$, defined on one symbol '1', and accepts input strings with even numbers of symbol 1 and rejects input strings with odd numbers of 1s.

We tested our method for six input strings, "1", "11", "111", "1111", "11111", and "111111", to see whether the method correctly accepts the input string "11", "1111", "111111", and rejects the strings "1", "111", "11111".

The results are shown in Fig. 7. Blue-colored colonies which indicates the expression of *lacZ* reporter gene have been observed only in the plates for the

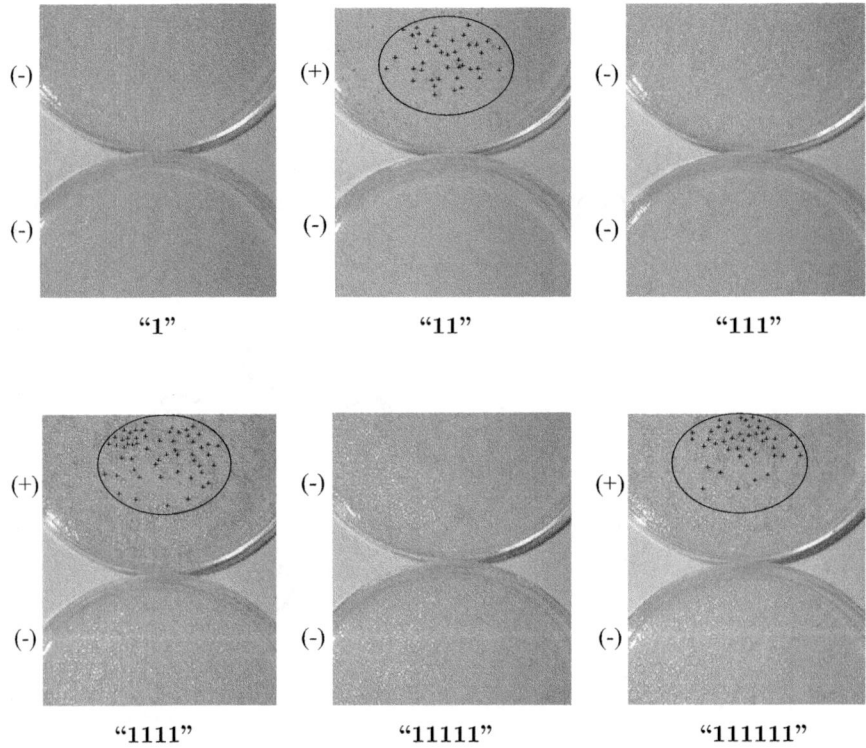

Fig. 7. Computation by the *E. coli* cells with plasmids of the input strings: 1, 11, 111, 1111, 11111, 111111. In each panel, the upper plate (part of a LB plate) shows the result in the presence of the suppressor tRNA with UCCU anticodon in the cell, while the lower plate shows the result of control experiment with no suppressor tRNA expressed. The signs (+) and (-) indicate the theoretical values about the expressions of *lacZ* reporter gene: (+) means that the cultured *E. coli* cells must express *lacZ* theoretically, and (-) means it must not express. Circles indicate the blue-colored colony expressing *lacZ*.

input strings 11, 1111, and 111111. Therefore, our *in vivo* finite automaton has succeeded to correctly compute the six input strings, that is, it correctly accepts the input strings 11, 1111, 111111 of even numbers of symbol '1' and correctly rejects 1, 111, 11111 of odd number of 1s.

References

1. Kuramochi, J., Sakakibara, Y.: Intensive in vitro experiments of implementing and executing finite automata in test tube. In: Proceedings of 11th International Meeting on DNA Based Computers, London, Ontario, pp. 59–67 (2005)
2. Nakagawa, H., Sakamoto, K., Sakakibara, Y.: Development of an in vivo computer based on *Escherichia coli*. In: Proceedings of 11th International Meeting on DNA Based Computers, London, Ontario, pp. 68–77 (2005)
3. Yokomori, T., Sakakibara, Y., Kobayashi, S.: A Magic Pot: Self-assembly computation revisited. In: Brauer, W., Ehrig, H., Karhumäki, J., Salomaa, A. (eds.) Formal and Natural Computing. LNCS, vol. 2300, pp. 418–429. Springer, Heidelberg (2002)
4. Sakakibara, Y., Hohsaka, T.: In vitro translation-based computations. In: Proceedings of 9th International Meeting on DNA Based Computers, Madison, Wisconsin, pp. 175–179 (2003)
5. Păun, G., Rozenberg, G., Salomaa, A.: DNA Computing. Springer, Heidelberg (1998)
6. Benenson, Y., Paz-Ellzur, T., Adar, R., Keinan, E., Livneh, Z., Shapiro, E.: Programmable and autonomous computing machine made of biomolecules. Nature 414, 430–434 (2001)
7. Soreni, M., Yogev, S., Kossoy, E., Shoham, Y., Keinan, E.: Parallel biomolecular computation on surfaces with advanced finite automata. J. Am. Chem. Soc. 127, 3935–3943 (2005)
8. Hohsaka, T., Ashizuka, Y., Taira, H., Murakami, H., Sisido, M.: Incorporation of nonnatural amino acids into proteins by using various four-base codons in an *Escherichia coli* in vitro translation system. Biochemistry 40, 11060–11064 (2001)

The Complexity of Explicit Constructions

Rahul Santhanam

School of Informatics, University of Edinburgh
rsanthan@inf.ed.ac.uk

Abstract. The existence of extremal combinatorial objects, such as Ramsey graphs and expanders, is often shown using the probabilistic method. It is folklore that pseudo-random generators can be used to obtain explicit constructions of these objects, if the test that the object is extremal can be implemented in polynomial time. In this talk, we pose several questions geared towards initiating a structural approach to the relationship between extremal combinatorics and computational complexity. One motivation for such an approach is to understand better why circuit lower bounds are hard. Another is to formalize connections between the two areas, so that progress in one leads automatically to progress in the other.

Keywords: computational complexity, explicit constructions.

1 Explicit Constructions of Extremal Objects

The field of extremal combinatorics deals with objects that are as large or small as possible given certain specified constraints. These objects might be graphs (eg., Ramsey graphs, expanders, extractors), sets or families of sets (eg., error-correcting codes, designs), matrices (eg, rigid matrices),or sequences (eg., universal traversal sequences). The study of such extremal objects is motivated by the fact that they are extremely useful in algorithmic contexts, eg., expanders and extractors are useful in designing efficient data structures and small sorting networks, error-correcting codes are critical to transmission of information in the presence of noise, and universal traversal sequences are useful in derandomizing certain randomized space-bounded algorithms.

In most cases, good bounds on the sizes of extremal objects can be obtained by using the probabilistic method initiated by Paul Erdos - to prove that there is an object of size S satisfying certain specified constraints, prove that a *random* object of this size satisfies the constraints with positive probability. The probabilistic method has been spectacularly successful in obtaining good bounds for various extremal objects [1].

However, the algorithmic applications of extremal objects often require "explicit" constructions of these objects. The standard meaning of "explicit" is that the object should be produced by a deterministic procedure running in polynomial time in the size of the object. The probabilistic method is inherently non-constructive, and so does not yield explicit constructions in this sense.

F. Ferreira et al. (Eds.): CiE 2010, LNCS 6158, pp. 372–375, 2010.
© Springer-Verlag Berlin Heidelberg 2010

Note that there is a naive procedure for deterministically producing an extremal object of size S once we are guaranteed existence. Namely, enumerate all objects of size S, check for each one whether it satisfies the desired contraints, and output the first one which does. However, this procedure takes time exponential in S, even if the check whether the constraints hold can be done in polynomial time.

The question of explicit constructions has been studied intensively for various combinatorial objects. For objects such as expanders, extractors and codes, explicit constructions are known with near-optimal parameters. For other objects such as Ramsey graphs, rigid matrices and universal traversal sequences, explicit constructions are still unknown despite years of effort. In any case, for a long time the question of explicit constructions was typically studied in isolation for each different combinatorial object. Recently, complexity theory and specifically the theory of pseudo-randomness have motivated progress towards a unified theory. In this talk, we raise several questions and initiate directions that promise progress in both the theory of explicit constructions and in complexity theory.

2   The Role of Complexity Theory

The major open problems in complexity theory are about separating complexity classes. Is there a Boolean function in NP which is not in P? Is there a Boolean function in EXP which does not have polynomial-size circuits? A priori, it is not clear what complexity lower bounds have to do with explicit constructions. However, it is a simple observation that virtually all complexity lower bounds we wish to prove hold for a *random Boolean function* - i.e., the probabilistic method applies to these questions. This is a highly "non-explicit" construction; making it explicit corresponds to upper bounding the complexity of a Boolean function for which the lower bound holds, and hence to separating complexity classes. For example, the question of proving that EXP does not have polynomial-size circuits is precisely the question of finding a polynomial-time procedure for outputting the truth table of a function which does not have polynomial-size circuits (here we mean polynomial-time in the size of the truth table, and hence exponential time in the number of input bits). This is an explicit construction question.

The question of constructing hard Boolean functions is not just one explicit construction question among many - it has relevance to all the others. This connection is through the beautiful theory of pseudo-random generators, initiated by Blum,Micali [2] and Yao [8] from cryptographic motivations, and extended by Nisan and Wigderson [4] to a complexity-theoretic setting. Given a class C of statistical tests, a pseudo-random generator with seed $s(n)$ against C is a function G from $s(n)$ bits to n bits such that that no test from C can distinguish the uniform distribution on n bits from the uniform distribution on the range of G. If $s(n)$ is much smaller than n (e.g. $s(n) = O(\log(n))$), this means that the uniform distribution on the range of G "looks" random to C, even though it is in fact very far from random. Thanks to work of Nisan-Wigderson, Impagliazzo-Wigderson [3] and many others, it's known that the existence of pseudo-random generators with seed length $O(\log(n)$ against C which can be computed "efficiently" (i.e., in time polynomial in n) is equivalent to the existence of functions in linear exponential time which do not have sub-exponential size circuits with oracle access to some test in C.

Now, for a given explicit construction question where existence has been shown using the probabilistic method, let C be any class of tests which includes the test A for whether an object is extremal. A random object passes the test A with probability close to 1 - this is what the probabilistic method gives us. Now, if G is a pseudo-random generator with logarithmic seed length against C, then by pseudo-randomness of G, a random element in the range of G satisfies A with probability close to 1; hence at least one element in the range satisfies A. Thus, the explicit construction of a function in linear exponential time with no sub-exponential size C-oracle circuits implies an explicit construction of a polynomial-size set which contains an extremal object for *any* notion of "extremal" which can be tested within C! In the case that the extremality test is in P, we can identify efficiently an object in the set which is indeed extremal. Thus appropriate circuit lower bounds give explicit constructions for any object whose extremality can be tested in P!

These are folklore observations which have certainly played a role in the development of the field of complexity-theoretic pseudo-randomness. Specifically, as surveyed by Trevisan [6] and Vadhan [7], *conditional* constructions of pseudo-random generators from Boolean function hardness are closely tied to explicit constructions of objects such as expanders, extractors and error-correcting codes. However, there are still many avenues to be explored, and the observations in the previous paragraph leave many questions unanswered. Are there plausible complexity-theoretic hypothesis under which explicit constructions follow even when the test for extremality is not in polynomial time? What is the evidence for and against the extremality test being in polynomial time for questions such as whether a graph is Ramsey, and how does this connect with one-way functions in cryptography and the "natural proofs" framework of Razborov and Rudich [5]? Can we define new classes of tests containing natural explicit construction questions for which generic explicit constructions can be shown unconditionally? Is there a reasonable theory of reductions between explicit construction questions, as there is for decision problems, and if so, are there natural "hard" or "complete" questions?

We attempt to formulate these questions in a precise way, and to explore their implications. Underlying our investigation is a fundamental belief - that the connection between complexity lower bounds and explicit constructions , while mysterious, is far from accidental, and that understanding this connection more deeply is key to progress in both areas.

References

1. Alon, N., Spencer, J.H.: The Probabilistic Method, with an appendix on open problems by Paul Erdös. John Wiley & Sons, Chichester (1992)
2. Blum, M., Micali, S.: How to generate cryptographically strong sequence of pseudo-random bits. SIAM Journal on Computing 13, 850–864 (1984)

3. Impagliazzo, R., Wigderson, A.: P = BPP if E requires exponential circuits: Derandomizing the XOR lemma. In: Proceedings of the 29th Annual ACM Symposium on the Theory of Computing, pp. 220–229 (1997)
4. Nisan, N., Wigderson, A.: Hardness vs randomness. Journal of Computer and System Sciences 49(2), 149–167 (1994)
5. Razborov, A., Rudich, S.: Natural proofs. Journal of Computer and System Sciences 55(1), 24–35 (1997)
6. Trevisan, L.: Pseudorandomness and combinatorial constructions. Electronic Colloquium on Computational Complexity (ECCC) 13(13) (2006)
7. Vadhan, S.: The unified theory of pseudorandomness: Guest column. SIGACT News 38(3), 39–54 (2007)
8. Yao, A.: Theory and application of trapdoor functions. In: Proceedings of the 23rd Annual IEEE Symposium on Foundations of Computer Science, pp. 80–91 (1982)

Kolmogorov Complexity Cores

André Souto*

Universidade do Porto and Instituto de Telecomunicações
andresouto@dcc.fc.up.pt

Abstract. We study the relationship between complexity cores of a language and the descriptional complexity of the characteristic sequence of the language based on Kolmogorov complexity.

We prove that a recursive set A has a complexity core if for all constants c, the computational depth (the difference between time-bounded and unbounded Kolmogorov complexities) of the characteristic sequence of A up to length n is larger than c infinitely often. We also show that if a language has a complexity core of exponential density, then it cannot be accepted in average polynomial time, when the strings are distributed according to a time bounded version of the universal distribution.

1 Introduction

A polynomial complexity core of A is the kernel of hard instances, i.e., is the set of strings for which any Turing machine deciding A requires time larger than a polynomial for almost all strings. Lynch [13] showed that any recursive set not in **P** has an infinite polynomial complexity core. Later, Orponen and Schöning [15] showed that, if each algorithm for a language has a non sparse set of "hard" inputs, then the language has a non-sparse proper polynomial complexity core.

Kolmogorov complexity measures the amount of information contained in a string by the size of the smallest program producing it. Almost all strings have nearly maximum Kolmogorov complexity, however one can easily create another string as useful as the first one by using fresh random coins. To capture nonrandom information Antunes *et al* [2] defined a new measure called computational depth, taking the difference between time-bounded Kolmogorov complexity and unbounded Kolmogorov complexity. As a consequence, a deep object has lots of structure but, in the allotted time, we cannot describe that structure.

We study a connection between the computational depth of languages and the existence of complexity cores. We address this issue by studying the computational depth of sets in the computational classes **P**, **EXP** = **DTIME**($2^{poly(n)}$), **FULL-P/log** and **P/poly**. First, we give a characterization of recursive sets not in **P** and thus admitting a polynomial complexity core based on computational depth. Then we prove that if a recursive set is not in **P/poly** then the depth of its characteristic sequence is high infinitey often.

The complexity of a problem is usually measured in terms of the worst case behavior of algorithms. Many algorithms with a bad performance in theory have

---

* The author is partially supported by CSI^2 (PTDC/EIA-CCO/099951/2008) and the grant SFRH/BD/28419/2006 of FCT.

a good performance in practice, since instances requiring a large running time rarely occur. To study this duality, Levin [10] introduced the theory of average case complexity, giving a formal definition of Average Polynomial Time for a language L and a distribution μ. Some languages may remain hard in the worst case but can be solved efficiently in average polynomial time for all reasonable distributions. We show that a Turing machine deciding a proper complexity core of exponential density cannot recognize the language in average polynomial time when a time-bounded version of the universal distribution is used.

2   Preliminaries

We use the binary alphabet $\Sigma = \{0, 1\}$. Elements of Σ^* are called strings. Given a string x, its initial segment of length n is denoted by $x_{[1,n]}$ and its ith bit is represented by $x_{[i,i]}$. $|x|$ denotes the length of x. All languages are subsets of Σ^* and are usually represented by capital letters. Given a set or a language A, χ_A denotes the characteristic sequence of A, i.e., $\chi_{A_{[i,i]}}$ is 1 if and only if the ith string in the length-lexicographic order belongs to A. We denote by $A^{\leq n}$ (resp. $A^{=n}$) the subset of A where the strings considered have length at most n (resp. equal to n). If A is a finite set, $\#A$ denotes the cardinality of A. The model of computation that is used is prefix-free Turing machines. Given a Turing machine M and a string x, $time_M(x)$ is the notation of the running time of M on the input x, i.e., the number of steps that M uses until it stops on the input x.

2.1   Complexity Cores

The concept of a polynomial complexity core was introduced by Lynch [13] and it was developed to capture the intuition of a kernel of hard instances of a language.

Definition 1 (Complexity Core). *Let M be a Turing machine and t a time constructible function. The set of t-hard inputs for M is $H(M, t) = \{x \in \Sigma^* : time_M(x) > t(|x|)\}$. Let $\mathcal{F}$ be a class of functions. A set $C \subset \Sigma^*$ is a $\mathcal{F}$ complexity core for a language $L \subset \Sigma^*$, if given any Turing machine M accepting L and any function p in $\mathcal{F}$, C is almost everywhere contained in $H(M, p)$. A complexity core C for L is proper if $C \subset L$. In particular if $\mathcal{F} = \mathbf{P}$ then the set of hard instances is called a polynomial complexity core.*

Theorem 1 (Lynch). *If A is a recursive set not in $\mathbf{P}$, then there is an infinite recursive set C that is a polynomial complexity core for A.*

In order to study the frequency of intractable problems Meyer and Paterson [14], defined a class of problems for which only a sparse set of the inputs are allowed to run in non-polynomial time.

Definition 2. *Let A be a set. Consider the function $dens_A : \mathbb{N} \to \mathbb{N}$ defined by $dens_A(n) = \#\{x \in A : |x| \leq n\}$. A is called sparse if for some polynomial p, $dens_A(n) \leq p(n)$. A set is co-sparse if its complement is sparse. We say that A has exponential density if, for some $\varepsilon > 0$ and infinitely many n, $dens_A(n) \geq 2^{\varepsilon n}$.*

Definition 3. *The class of almost polynomial time languages is:* **APT** $=$ $\{L(M) : H(M,p)$ *is sparse for some polynomial* $p\}$.

Orponen and Schöning [15] showed an interesting connection of this class with polynomial complexity cores, proving that **APT** identifies exactly the set of recursive languages that have a proper non sparse polynomial complexity core.

Theorem 2. *A recursive set* A *has a non sparse polynomial proper complexity core if and only if* $A \notin$ **APT**.

2.2   Komogorov Complexity and Computational Depth

We refer the reader to the book of Li and Vitányi [12] for a complete study on Kolmogorov complexity.

Definition 4. *Let* U *be a fixed prefix-free universal Turing machine. For any strings* $x, y \in \Sigma^*$, *the prefix-free Kolmogorov Complexity of* x *given* y *is,* $K(x|y) = \min_p\{|p| : U(p,y) = x\}$. *If* t *is a time constructible function, the* t-*time-bounded prefix-free Kolmogorov Complexity of* x *given* y *is* $K^t(x|y) = \min_{p,r}\{|p| + |r| : (\forall i : 1 \le i \le |x|), U^r(p,y,i) = x_i$ *in* $t(|x|)$ *steps*$\}$ *where* r *is given as oracle access with part of the description of the string* x.

The classical definition of t-time-bounded prefix-free Kolmogorov complexity is defined for $t(n) \ge n$, by $K^t(x|y) = \min_p\{|p| : U(p,y) = x$ *in at most* $t(|x|)steps\}$. The definition we use here (which was used, for example, in [2]) is more technical but more general since it allows the use of sublinear time-bounds. When $t(n) \ge n$ the two definitions of time-bounded Kolmogorov complexity are equivalent.

Proposition 1. *For any time constructible function* t, *any* x *and* y, *we have:*

1. $K^t(x|y) \le K^t(x) + O(1)$ *and* $K(x|y) \le K^t(x|y) + O(1)$;
2. $K(x|y) \le K(x) + O(1) \le |x| + O(\log |x|)$;
3. *There are at least* $2^n(1 - 2^{-c})$ *strings* x *of length* n *such that* $K(x) \ge n - c$.
4. *If* A *is a recursive set, then for all* n , $K(\chi_{A_{[1:n]}}|n) \le O(1)$.

Kolmogorov complexity is used to define a universal semi-measure over Σ^*.

Definition 5. *The universal semi-measure* **m** *over* Σ^* *is* $\mathbf{m}(x) = 2^{-K(x)}$.

Imposing restrictions of time on Kolmogorov complexity we can define a time-bounded version of universal semi-measure **m** by $\mathbf{m}^t(x) = 2^{-K^t(x)}$.

Theorem 3 ([12]). *The distribution* $\mathbf{m}^{t'}$ *dominates any distribution which cumulative mass function is* t-*time computable, where* $t'(n) = nt(n)\log(nt(n))$.

The classical Kolmogorov complexity of a string x does not take into account the time effort necessary to produce the string from a description of length $K(x)$. Antunes *et al.* [2] proposed a notion of *Computational Depth* as a measure of nonrandom information in a string (using a similar motivation used by Bennett to define logical depth). Intuitively, strings with high depth have low Kolmogorov complexity (and hence are nonrandom), but a resource-bounded machine cannot identify or describe the structure that allows the compression of these strings. So, in [2] the authors suggest that the difference between two Kolmogorov complexity measures captures the intuitive notion of nonrandom information.

Definition 6. *Let t be a constructible time-bound. For any strings* $x, y \in \Sigma^*$,

$$depth^t(x|y) = K^t(x|y) - K(x|y).$$

A set A *is* k-shallow if for almost all n, $depth^{\log^k}(\chi_{A_{[1:N]}}|n) \leq \log^k n$, *where* $N = 2^{n+1} - 1$. *A is shallow if it is* k-shallow for some k.

2.3   Average Case Complexity

The analysis of efficiency of an algorithm is typically based on the worst case scenario of the running time over all inputs. This analysis is important since this information might be needed to ensure that the algorithm finish on a specific time. In middle eighties, Levin [10], proposed a different measure of complexity based on the average of the performance of the algorithm over all inputs.

Definition 7. *Let* μ *be a probability distribution. A function* $f : \Sigma^* \to \mathbb{N}$ *is polynomial on* μ-average if there is an $\epsilon > 0$ such that $\sum_x \frac{f(x)^\epsilon}{|x|} \mu(x) < \infty$.

In [17] the author proved that a function f is polynomial on μ-average in Levin's sense if and only if there is a polynomial $p : \mathbb{N} \times \mathbb{N} \to \mathbb{R}_0^+$, such that for all $m > 0$, $Pr_\mu[f(x) > p(|x|, m)] \leq \frac{1}{m}$.

Theorem 4 (Antunes *et al.*). *Let* T *and* t *be two constructible time-bounds.* $T(x) \in 2^{O(depth^t(x)+\log|x|)}$ *for all* x *if and only if* T *is polynomial on* $\mathbf{m}^t$-*average.*

The above Theorem has an interesting connection to a result due to Li and Vitányi [11] relating the average case complexity and the worst-case complexity. They showed that when the inputs to any algorithm are distributed according to the universal distribution, the algorithm's average case complexity is of the same order of magnitude as its worst case complexity.

Theorem 5 ([11]). *Let* T *be some constructible time-bound.* $T(x)$ *is bounded by a polynomial in* $|x|$ *for all* x *if and only if* T *is polynomial on* $\mathbf{m}$-*average.*

2.4   The Class FULL-P/log and the Class P/poly

Karp and Lipton [6] initiated the study of advice functions as a tool to provide a connection between uniform models of computation and nonuniform ones. The idea is to give the program for each string length access to a piece of information that "helps" to compute some language and thus incorporating the nonuniformity in the "advice". The two most natural functions for the length of the advice words are polynomials and logarithms on the length of the input size. The class **P/poly** has been widely studied in the literature (see, for example [7,8]).

In this paper we also use a class that derives from **P/log**. It is known that the class **P/log** is not closed under polynomial time Turing reductions (see, for example, [3,9]). To have a more robust complexity class, Ko [9] introduced the class **FULL-P/log** . The robustness derives from the fact that instead of the advices only help to decide strings of a fixed length, they must be used to decide all strings of length smaller or equal to that size. Formally,

Definition 8. *A is in the class* **FULL-P/log** *if:* $(\forall n)(\exists w : |w| \leq c \log n)(\forall x : |x| \leq n) : x \in A \Leftrightarrow \langle x, w \rangle \in B$, *where* $B \in \mathbf{P}$ *and* c *is a constant.*

3   Kolmogorov Complexity and Complexity Cores

All the results of this section could be stated using the time-bounded Kolmogorov complexity instead of computational depth. However as we would like to remove the condition of the sets being recursive and establish a bridge between [1] and complexity cores, we decided to use computational depth.

We assume that all time-bounds t considered in this section are polynomial on the logarithm of the size of the string to be produced. This way when analyzing $\chi_{A_{[1:N]}}$ with $N = 2^{n+1} - 1$ we are only using polynomial time in n, i.e. sub-linear time on N.

Following the work of Juedes and Lutz [4] we start proving an upper bound for the exponential time bounded Kolmogorov complexity of a language A restricted to strings of length n proving the fact that A only admits large complexity cores. The idea of the proof is to consider a set of instances requiring more than polynomial time to be decided. Since that set is a polynomial time complexity core, using the assumption we derive a description in exponential time of $A^{=n}$.

Proposition 2. *Let* A *be a set,* $\varepsilon > 0$ *a constant and* $g : \mathbb{N} \rightarrow [0, \infty[$ *any function. If every polynomial complexity core* C *of* A *satisfies* $dens_C(n) \leq 2^n - g(n)$ *infinitely often then* $K^f(A^{=n}) < 2^n - n^{-\varepsilon}g(n) + O(\log n)$ *infinitely often, where* $f(n) = 2^n \cdot poly(n)$ *for some polynomial poly.*

Proof. Let p be a fixed polynomial and set $k = \lceil 1/\varepsilon \rceil$. Consider $M_1, M_2, ..., $ a standard enumeration of the deterministic Turing machines that decide A. For each m define the sets: $H_m = \{x : time_{M_m}(x) \leq p(|x|)\}$, $B_m = H_m - \Sigma^{\leq m^k}$, $B = \bigcup_m B_m$ and $C = \Sigma^* - B$. It is easy to see that $H_m \cap C = H_m - B \subset H_m - B_m \subset \Sigma^{\leq m^k}$. Thus, $H_m \cap C$ is finite and so C is a polynomial complexity core. Define the set $S = \{n : \#C^{=n} \leq 2^n - g(n)\}$. It follows that $S = \{n : \#B^{=n} \geq g(n)\}$. Hence, for each $n \in S$, $g(n) \leq \#B^{=n} = \# \left(\bigcup_m B_m \right)^{=n} \leq \sum_m \#B_m^{=n} = \sum_{m^k < n} \#(B_m)^{=n} = \sum_{0 < m \leq n^\varepsilon} \#(B_m)^{=n} = \sum_{0 < m < n^\varepsilon} \#(H_m)^{=n}$.

So, there is m such that $0 < m \leq n^\varepsilon$ and $\#(H_m)^{=n} > n^{-\varepsilon}g(n)$. Let $\beta(m)$ be a binary representation of m and $w_1, w_2, ...,$ be the lexicographic ordering of Σ^*. Consider M as the Turing machine implementing the following algorithm:

Let y be a finite string. For $i = 0$ to 2^n do:
1. Simulate $M_m(w_i)$ for at most $p(n)$ steps;
2. If $M_m(w_i) = 1$ or $M_m(w_i) = 0$ then set $z = 0$ or $z = 1$ respectively.

3. Otherwise set $z_i = y_1$ and $y = y_2 y_3 ...$;
Output z.

For all $x \in \Sigma^n$, $time_M(x) \leq 2^n poly(n)$ for some polynomial $poly$. So if, for each $n \in \mathbb{N}$, we choose $m \in \mathbb{N}$ and $y \in \Sigma^*$ such that $0 \leq m < n^\varepsilon$, $\#H_m^{=n} \geq n^{-\varepsilon} g(n)$ and y is the successive $2^n - \#H_m^{=n}$ bits consisting of $\chi_A(w_i)$ with $w_i \in \Sigma^n - H_m$, the output of M is exactly $A^{=m}$. Setting $f(n) = 2^n \cdot poly(n)$, by the assumptions of the Theorem, $K^f(A^{=n}|n) \leq |\langle \beta(m), y \rangle| + O(1) \leq |y| + 2\beta(m) + O(1) \leq 2^n - \#H_m^{=n} + O(\log n) \leq 2^n - n^{-\varepsilon} g(n) + O(\log n)$, infinitely often.

Lemma 1. *For every $\leq_m^p$-hard language H for $\mathbf{E} = \bigcup_{c \in \mathbb{N}} \mathbf{DTIME}(2^{cn})$, there are sets $B, D \in \mathbf{DTIME}(2^n)$ such that $D \cap H = B$ and D has exponential density.*

Proof. Analogous of he proof of Theorem 12 in [4].

Lemma 2. *Any $\mathbf{DTIME}(2^n)$-core of every $\leq_m^p$-hard language H for $\mathbf{E}$ has a exponential dense complement.*

Proof. Let C denote the $\mathbf{DTIME}(2^n)$-core of H and consider B and D as in the previous lemma. Since $D \in \mathbf{DTIME}(2^n)$, by definition of complexity core, $C \cap D$ is finite. As D is exponentially dense it follows that $D - C$ is exponentially dense and thus the complement of C is exponentially dense.

Theorem 6. *For every $\leq_m^p$-hard language H for $\mathbf{E}$, there is a $\varepsilon > 0$ such that $K^f(H^{=n}) \leq 2^n - n^{-\varepsilon} 2^{\varepsilon n} + O(\log n)$ infinitely often, where $f(n) = 2^n \cdot poly(n)$ for some polynomial poly.*

Proof. If H is $\leq_m^p$-hard for $\mathbf{E}$, by last Lemma, there is $\varepsilon > 0$ such that every $\mathbf{DTIME}(2^n)$-core C of H satisfies $dens_C(n) < 2^n - 2^{n \cdot \varepsilon}$. Thus, by Proposition 2, it follows, for $f(n) = 2^n \cdot poly(n)$ for some polynomial $poly$, that $K^f(H^{=n}) \leq 2^n - n^{-\varepsilon} 2^{\varepsilon n} + O(\log n)$.

Theorem 7. *Let t be a fixed polylogarithmic function. A recursive set A is not in $\mathbf{P}$ if and only if for all constant c and for infinitely many n and $N = 2^{n+1} - 1$, $depth^t(\chi_{A_{[1:N]}}|n) > c$.*

Proof. Assume that there is a polylogarithmic function t and a constant c such that, for all n, $depth^t(\chi_{A_{[1:N]}}|n) < c$. Since we have assumed that A is recursive, this means that for all n, $K(\chi_{A_{[1:N]}}|n) < c$. So, if $depth^t(\chi_{A_{[1:N]}}|n) < c$ this implies $K^t(\chi_{A_{[1:N]}}|n) < c$. Consider I the set of all pairs (p, r) such that $|p| + |r| \leq c$ and (p, r) is the least pair in the lexicographic order corresponding to $K^t(\chi_{A_{[1:N]}}|n)$ for some n. Notice that: 1) $\#I \leq 2^c$; 2) If $N_1 < N_2$ and (p_1, r_1) and (p_2, r_2) correspond to $K^t(\chi_{A_{[1:N_1]}}|n_1)$ and $K^t(\chi_{A_{[1:N_2]}}|n_2)$ then, by definition of K^t, $|p_1| + |r_1| \leq |p_2| + |r_2|$. Also, for all $x \in \Sigma^{n-1}$, if $U^{r_2}(p_2, x)$ stops in time $t(N_1)$ then $U^{r_2}(p_2, x) = U^{r_1}(p_1, x) = \chi_A(x)$.

Consider now the following program p: On input x of length n, p runs each pair (p_i, r_i) in lexicographic order for t steps and output the same answer which the largest program in the lexicographic order outputted. From the construction, the running time of p is at most $2^c t(N) + O(1)$, which is polynomial on n. The correctness of the output follows from item 2 above. Thus $A \in \mathbf{P}$.

To prove the converse, assume that $A \in \mathbf{P}$. Thus, there is an algorithm g that, given x, produces $\chi_A(x)$ in polynomial time on the x's length. Since it does not depend on x, the length of g is a constant and then $K^t(\chi_{A_{[1..N]}}|n) \leq O(1)$ and thus $depth^t(\chi_{A_{[1:N]}}|n) < c$.

By Lynch's characterization of complexity cores (Theorem 1) the previous result establishes a connection between recursive sets having a polynomial complexity core and its computational depth.

Corollary 1. *Let t be a fixed polynomial. A recursive set A has a polynomial complexity core if, for all constant $c > 0$, $depth^t(\chi_{A_{[1:N]}}|n) > c$, for infinitely many n and $N = 2^{n+1} - 1$.*

Proposition 3. *Let t be a fixed polylogarithmic function. A is a recursive set not in $\mathbf{P/poly}$ if and only if for all polynomial p, $depth^t(\chi_{A_{[1:N]}}|n) > p(n)$ for infinitely many n and $N = 2^{n+1} - 1$.*

Proof. Assume that there is a polynomial q such that for all n, $depth^t(\chi_{A_{[1:N]}}|n) \leq q(n)$. Since A is recursive this means that $K^t(\chi_{A_{[1:N]}}|n) \leq q(n)$. Thus, there are p and r such that $|p| + |r| \leq q(n)$ and $U^r(p, x) = \chi_A(x)$ in polynomial time for all $x \in \Sigma^{\leq n}$. So, giving p and r as advice we can decide $A^{\leq n}$, hence $A \in \mathbf{P/poly}$.

To prove the converse, assume that A is a recursive set and belongs to $\mathbf{P/poly}$. Then there is a polynomial q such that, given n, there are p and r satisfying $|p| + |r| \leq q(n)$ and $U^r(p, x) = \chi_A(x)$ in polynomial time for all $x \in \Sigma^{\leq n}$. Thus, for all n, $K^t(\chi_{A_{[1:N]}}|n) \leq q(n)$ and then $depth^t(\chi_{A_{[1:N]}}|n) \leq q(n)$.

In [18], the author proved that $\mathbf{APT} \subsetneq \mathbf{P/poly}$. Since all recursive sets not in $\mathbf{APT}$ have a proper complexity core, as a consequence we have:

Corollary 2. *Let t be a fixed polylogarithmic function. A recursive set A has a proper complexity core if for all polynomial p, $depth^t(\chi_{A_{[1:N]}}|n) > p(n)$ for infinitely many n and $N = 2^{n+1} - 1$.*

Notice that if $\mathbf{APT} \subset \mathbf{FULL\text{-}P/log}$ then the next result would give a sharper characterization of sets admitting a proper complexity core.

Proposition 4. *Let t be a fixed polylogarithmic function. A is recursive set not in $\mathbf{FULL\text{-}P/log}$ if and only if there is c such that for infinitely many n, $depth^t(\chi_{A_{[1:N]}}|n) > c \log n$, where $N = 2^{n+1} - 1$.*

We now characterize complexity cores based on Kolmogorov complexity:

Definition 9. *A set A has a Kolmogorov complexity core for the class $\mathcal{C}$ if for all $c \in \mathbb{N}$ and for all $f \in \mathcal{C}$ and for infinitely many n, $depth^{\log^k(f)}(\chi_{A_{[1:N]}}|n) > c$.*

In order to better understand the previous definition we explore it by applying it to the class $\mathbf{EXP}$.

Theorem 8. *Let A be a recursive set. A has a complexity core relative to the class **EXP** if and only if for all sufficiently large functions $f \in$ **EXP** and infinitely many n $depth^{\log^k(f)}(\chi_{A_{[1:N]}}|n) \geq \omega(1)$, where $N = 2^{n+1} - 1$.*

Proof. Assume that for some $f \in$ **EXP**, $depth^{\log^k(f)}(\chi_{A_{[1:N]}}|n) \leq O(1)$ for all n. Then, since A is recursive this implies that $K^f(\chi_{A_{[1:N]}}|n) \leq O(1)$. Thus, among the finite number of programs of constant size, there is one deciding the problem of belonging to A for all $x \in \Sigma^*$ in time f, i.e., there is an algorithm G such that $G(x) = \chi_A(x)$ and $|G| \leq c$, where c is some constant. Thus $A \in$ **EXP** which implies that A does not have a complexity core with respect to the class **EXP**.

On the other hand, if $A \in$ **EXP** then there is an algorithm G such that $G(x) = \chi_A(x)$ for all $x \in \Sigma^*$ in an exponential number of steps. Since G does not depend on x, its length is a constant. So, $depth^{\log^k(f)}(\chi_{A_{[1:N]}}|n) \leq O(1)$, where $N = 2^{n+1} - 1$.

We already proved that if A is recursive and admits a Kolmogorov complexity core for the polynomial time class then it satisfies the condition on Definition 1 and vice versa. In the next result we show how to construct a complexity core.

Proposition 5. *Let $k > 0$ be a fixed integer. If A is a recursive set and for all $c \in \mathbb{N}$ and for infinitely many $n \in \mathbb{N}$, $depth^{\log^k}(\chi_{A_{[1:N]}}|n) > c\log(n)$, where $N = 2^{n+1} - 1$ then A has a polynomial complexity core.*

Proof. We construct a polynomial complexity core inductively. Let n_0 be the first index such that $depth^{\log^k}(\chi_{A_{[1:N]}}|n) > c\log(n)$ for all $n \geq n_0$.

Since A is recursive $K(\chi_{A_{[1:N]}}|n) = O(1)$. So, for all c, $depth^{\log^k}(\chi_{A_{[1:N]}}|n) > c\log(n)$ which is equivalent to say that $K^{\log^k}(\chi_{A_{[1:N]}}|n) > c\log(n)$.

If for all c, $K^{\log^k}(\chi_{A_{[1:N]}}|n) > c\log(n)$ then it means that there are no p and r such that: 1) $|p| + |r| \leq c\log(n)$ and 2) for all $1 \leq i \leq N$, $U^r(p, i) = \chi_{A_{[i,i]}}$ in time $\log^k(N) = poly(n)$.

So, there is at least one i, $1 \leq i \leq N$ such that for all p and r satisfying $|p| + |r| \leq c\log n$ either $U^r(p, i) \neq \chi_{A_{[i,i]}}$ or $U^r(p, i) = \chi_{A_{[i,i]}}$ but $time_p(i) > poly(n)$. Since this is true for all c then there is x of size at most n that cannot be decided by programs of length at most $O(\log n)$ in polynomial time. Such x is an element of the core of A.

To construct the next element in the core we consider the characteristic sequence up to strings of size $n_1 = 2^{2^{n_0+1}}$, i.e., $N = 2^{n_1+1} - 1$. Notice that now we allow programs and oracles up to size $\log n_1 = 2^{n_0+1}$. In particular, we can allow the oracle to be the characteristic sequence of A up to strings of length n_0, and considering the program printing the ith bit of r we conclude that in the allowed time, $poly(n_1)$ we can decide x. But then, by assumption, there is $i_1 \neq i$ with $1 \leq i_1 \leq N_1$ such that no program and oracle of size at most 2^{n_0+1} (logarithmic on the size of the strings we are considering) can decide it in time $poly(n_1)$. The x_1 corresponding to the ith element in the lexicographic order is the next element in the complexity core.

Notice that we can go on forever with this argument and construct an infinite polynomial complexity core. Also, a program that decides A has, in particular, to decide all strings up to a certain size.

4   Complexity Cores and Average Case Complexity

We relate the existence of polynomial complexity cores with distribution $\mathbf{m}$. From the discussion above and from Theorem 5 we can conclude that:

Corollary 3. *Let A be a recursive language such that $A \notin \mathbf{P}$. If M is a Turing machine deciding A, then M is not polynomial on $\mathbf{m}$-average.*

We can go further and prove that in fact A is not $\mathbf{m}^t$ polynomial on average where t is any super polynomial time constructible function.

Theorem 9. *If A is recursive and is not in $\mathbf{P}$ then A is not $\mathbf{m}^t$ polynomial on average where $t(n) = n \cdot t'(n) \log(n \cdot t'(n))$ and t' is any sufficiently large super polynomial time constructible function.*

Proof. If A is recursive and is not in $\mathbf{P}$ then it admits a polynomial complexity core H that is recognizable in time t, where t is any super polynomial time constructible function, as proved in [16]. Let $dens_H(\cdot)$ be the density function of H. Consider the probability distribution over Σ^* defined by $\mu(x) = 1/(dens_H(n) \cdot 2^n)$ if $x \in H^n$ and by 0 otherwise.

By definition of polynomial complexity core we know that, for all polynomial p almost all $x \in H$ satisfies $time_M(x) > p(|x|)$. Thus, there is n_p such that M cannot decide any element of of H of length larger than n_p. Thus, $Pr_\mu[time_M(x) \geq p(|x|)] \geq 1 - \frac{1}{dens_H(n_p)}$. Hence, according to the equivalence of polynomial time average complexity proved in [17], $time_M$ is not μ-polynomial on average. Notice that μ is computable in some super polynomial time t_1. So, by Theorem 3, $\mathbf{m}^{t'}$, where $t'(n) = n \cdot t_1(n) \log(n \cdot t_1(n))$ dominates μ and we conclude that for any sufficiently large super polynomial time constructible function t', A is not polynomial on $\mathbf{m}^{t'}$ average.

If the polynomial complexity core has exponential density we can improve the last result by using the distribution $\mathbf{m}^p$, where p is any polynomial. Since $\mathbf{m}^t \geq \mathbf{m}^p$ where t is some super polynomial time-bound then the result cannot be derive directly from last Theorem.

Proposition 6. *If a set A admits a exponential density polynomial complexity core then A is not $\mathbf{m}^{poly}$ polynomial on average.*

Proof. If A has a polynomial complexity core of exponential density then given any Turing machine M that decides A, there is a $\varepsilon > 0$ such for all polynomial p and sufficiently large n, $\#H^{=n} = \#\{x \in \Sigma^n : time_M(x) > p(n)\} > 2^{\varepsilon n}$.

Let q be a fixed polynomial time-bound and consider the set $D = \{x \in \Sigma^* : depth^q(x) \leq c \log n\}$.

In [2] the authors showed that there are approximately $2^{\varepsilon' n}$ strings of length n such that $depth^{2^n}(x) \geq (1 - \varepsilon')n - c\log n$. Thus $D^{=n} \approx 2^n - 2^{\varepsilon' n}$. So, if $\varepsilon > \varepsilon'$, by a pigeonhole principle, $D^{=n} \cap H^{=n} \neq \emptyset$. For any $x \in D^{=n} \cap H^{=n}$ we know that $O(2^{depth^q(x) + \log|x|})$ is polynomial in the length of x and $time_M(x) > q(|x|)$. Since this works for any polynomial q we conclude by Theorem 4 that $time_M$ is not polynomial time on average with respect to $\mathbf{m}^q$.

Corollary 4. *If* NP *does not have measure* 0 *in* **EXP** *then every* NP-*hard language, with respect to polynomial time many to one reductions, is not in* $\mathbf{m}^{poly}$ *polynomial on average.*

Proof. If NP does not have measure 0 in **EXP** then any NP hard language has a polynomial complexity core of exponential density (Corollary 4.10 of [5]).

References

1. Antunes, L., Fortnow, L., Pinto, A., Souto, A.: Low-depth witnesses are easy to find. In: Proceedings of CCC 2007 (2007)
2. Antunes, L., Fortnow, L., van Melkebeek, D., Vinodchandran, N.: Computational depth: concept and applications. Theor. Comput. Sci. 354(3) (2006)
3. Balcázar, J., Hermo, M., Mayordomo, E.: Characterizations of logarithmic advice complexity classes. In: Proceedings of the WCCASA. NHP (1992)
4. Juedes, D., Lutz, J.: Kolmogorov complexity, complexity cores and the distribution of hardness. In: Kolmogorov Complexity and Computational Complexity. Springer, Heidelberg (1992)
5. Juedes, D., Lutz, J.: The complexity and distribution of hard problems. In: Proceedings of SFCS. IEEE Computer Society, Los Alamitos (1993)
6. Karp, R., Lipton, R.: Some connections between nonuniform and uniform complexity classes. In: Proceedings of STOC. ACM, New York (1980)
7. Karp, R., Lipton, R.: Turing machines that take advice. In: Engeler, E., et al. (eds.) Logic and Algorithmic. L'Enseignement Mathématique (1982)
8. Ko, K.: Some observations on the probabilistic algorithms and np-hard problems. Inf. Process. Lett. 14(1) (1982)
9. Ko, K.: On helping by robust oracle machines. Theor. Comput. Sci. 52(1-2) (1987)
10. Levin, L.: Average case complete problems. SIAM J. Comput. 15(1) (1986)
11. Li, M., Vitányi, P.: Average case complexity under the universal distribution equals worst-case complexity. Inf. Process. Lett. 42(3) (1992)
12. Li, M., Vitányi, P.: An Introduction to Kolmogorov Complexity and Its Applications. Springer Publishing Company, Heidelberg (2008) (incorporated)
13. Lynch, N.: On reducibility to complex or sparse sets. J. ACM 22(3) (1975)
14. Meyer, A., Paterson, M.: With what frequency are apparently intractable problems difficult? MIT technical report TM-126 (1979)
15. Orponen, P., Schöning, U.: The structure of polynomial complexity cores (extended abstract). In: Proceedings of the MFCS. Springer, Heidelberg (1984)
16. Orponen, P., Schöning, U.: The density and complexity of polynomial cores for intractable sets. Inf. Control 70(1) (1986)
17. Schapire, R.: The emerging theory of average case complexity. MIT technical report 431 (1990)
18. Schöning, U.: Complete sets and closeness to complexity classes. Theory of Computing Systems 19(1) (1986)

Every Δ_2^0-Set Is Natural, Up to Turing Equivalence

Dieter Spreen

Lab. for Theoretical Computer Science, Dept. of Mathematics
University of Siegen, 57068 Siegen, Germany
spreen@math.uni-siegen.de

Abstract. The Turing degrees of Δ_2^0 sets as well as the degrees of sets in certain upper segments of the Ershov hierarchy are characterized. In particular, it is shown that, up to Turing equivalence and admissible indexing, the noncomputable Δ_2^0 sets are exactly the decision problems of countable partial orders of tree shape with branches of positive finite length only.

1 Introduction

Call a set of non-negative integers *natural* if it encodes a decision problem of some mathematical structure. In this note we characterize the Turing degrees of Δ_2^0 sets as well as the degrees of sets in certain upper segments of the Ershov hierarchy. Among others, we show that, up to Turing equivalence and admissible indexing, the noncomputable Δ_2^0 sets are exactly the decision problems of countable partial orders of tree shape with branches of positive finite length only.

Partially ordered sets of this kind are special Scott-Ershov domains [9]. These structures have been used with great success in logic and theoretical computer science, e.g. in higher-type computability and computing with exact real numbers [3,6]. Each such domain contains a set of base elements from which all its members can be obtained by taking directed suprema. In applications the base elements can easily be encoded in such a way that the underlying order is semi-decidable. This allows to construct an *admissible* numbering of the constructive domain elements. From each admissible index of a constructive member one can uniformly compute an effective listing of the directed set of base elements lower or equal to the given member. Conversely, there is a procedure which given an effective enumeration of a directed set of base elements computes an admissible index of its least upper bound. Admissible numberings generalize the well-known Gödel numberings for the computable functions. So, they are highly not one-to-one.

In the last years there has been large interest in studying computable structures that have a copy in every degree of a given collection of Turing degrees [2,7,10]. In these cases the structures are supposed to be coded in an one-to-one way so that, up to isomorphism, their universe can be taken as a subset of ω.

The note is organized as follows: Sect. 2 contains basic definitions from computability and numbering theory. Constructive domains are introduced in Sect. 3, where also the above-mentioned results are derived.

F. Ferreira et al. (Eds.): CiE 2010, LNCS 6158, pp. 386–393, 2010.
© Springer-Verlag Berlin Heidelberg 2010

2   Basic Definitions

In what follows, let $\langle\,,\,\rangle\colon \omega^2 \to \omega$ be a computable pairing function. We extend the pairing function in the usual way to an n-tuple encoding. Let $R^{(n)}$ denote the set of all n-ary total computable functions and W_i be the domain of the ith partial computable function φ_i with respect to some Gödel numbering φ. We let $\varphi_i(a)\downarrow_n$ mean that the computation of $\varphi_i(a)$ stops within n steps. In the opposite case we write $\varphi_i(a)\uparrow_n$.

For sets $A, B \subseteq \omega$, we write $A \leq_T B$ if A is Turing reducible to B, and $A \equiv_T B$ in case A is Turing equivalent to B. The cardinality of A is denoted by $|A|$ and c_A is its characteristic function.

Let S be a nonempty set. A *numbering* ν of S is a map $\nu\colon \omega \to T(\text{onto})$. The value of ν at $n \in \omega$ is denoted, interchangeably, by ν_n and $\nu(n)$.

Definition 1. *Let ν and κ be numberings of sets S and S', respectively.*

1. *$\nu \leq \kappa$, read ν is reducible to κ, if there is a function $g \in R^{(1)}$ such that $\nu_m = \kappa_{g(m)}$, for all $m \in \omega$.*
2. *$\nu \leq_1 \kappa$, read ν is one-one reducible to κ, if $\nu \leq \kappa$ and the function g is one-to-one.*
3. *$\nu \equiv_1 \kappa$, read ν is one-one equivalent to κ, if $\nu \leq_1 \kappa$ and $\kappa \leq_1 \nu$.*

Obviously, if $\nu \leq \kappa$ then $S \subseteq S'$.

Definition 2. *A numbering ν of S is said to be*

1. *decidable if $\{\, \langle n, m \rangle \mid n, m \in \omega \wedge \nu_n = \nu_m \,\}$ is recursive*
2. *a cylinder if $\nu \equiv_1 c(\nu)$, where $c(\nu)_{\langle n,m \rangle} = \nu_n$, for $n, m \in \omega$, is the cylindrification of ν.*

For a subset X of S^n, let $\Omega_\nu(X) = \{\, \langle i_1, \ldots, i_n \rangle \mid (\nu_{i_1}, \ldots, \nu_{i_n}) \in X \,\}$. Then X is *completely enumerable*, if $\Omega_\nu(X)$ is computably enumerable. Similarly, X is *completely recursive* if $\Omega_\nu(X)$ is computable.

3   Constructive Domains

Let $\mathcal{Q} = (Q, \sqsubseteq)$ be a partial order with smallest element $\bot$. A nonempty subset S of Q is *directed* if for all $y_1, y_2 \in S$ there is some $u \in S$ with $y_1, y_2 \sqsubseteq u$. $\mathcal{Q}$ is *directed-complete* if every directed subset S of Q has a least upper bound $\bigsqcup S$ in Q.

An element z of a directed-complete partial order $\mathcal{Q}$ is *finite* (or *compact*) if for any directed subset S of Q the relation $z \sqsubseteq \bigsqcup S$ always implies the existence of a $u \in S$ with $z \sqsubseteq u$. Denote the set of all finite members of Q by $\mathcal{K}$. If for any $y \in Q$ the set $\mathcal{K}_y = \{\, z \in \mathcal{K} \mid z \sqsubseteq y \,\}$ is directed and $y = \bigsqcup \mathcal{K}_y$, $\mathcal{Q}$ is said to be *algebraic* or a *domain*. Moreover in this case, we call $\mathcal{Q}$ a *tree domain*, or simply a *tree*, if the diagram of the partial order is a tree.

In this note we will particularly be interested in domains consisting of finite elements only. We refer to such domains as *finitary*. Note that in this case each ascending chain is finite.

A domain Q is said to be *effectively given*, if there is an indexing β of $\mathcal{K}$, called the *canonical indexing*, such that the restriction of the domain order $\sqsubseteq$ to $\mathcal{K}$ is completely enumerable. In this case a member $y \in Q$ is called *computable* if $\mathcal{K}_y$ is completely enumerable. Let Q_c denote the set of all computable elements of Q, then $\mathcal{Q}_c = (Q_c, \sqsubseteq, \beta)$ is called *constructive domain*. We say that $\mathcal{Q}_c$ is a *recursive domain* if $\{\, \langle i, j \rangle \mid \beta_i \sqsubseteq \beta_j \,\}$ is even computable. Obviously, β is decidable in this case.

A numbering η of the computable domain elements is *admissible* if $\{\, \langle i, j \rangle \mid \beta_i \sqsubseteq \eta_j \,\}$ is computably enumerable and there is a function $d \in R^{(1)}$ with $\eta_{d(i)} = \bigsqcup \beta(W_i)$, for all indices $i \in \omega$ such that $\beta(W_i)$ is directed. As is shown in [11] such numberings exist. Moreover, $\beta \leq \eta$. For what follows we always assume that $\mathcal{Q}_c$ is a constructive domain indexed by an admissible numbering η. We will fix a uniform way of approximating the computable domain elements.

Lemma 1 ([8]). *Let β be decidable. Then there is a function* en $\in R^{(1)}$ *such that for all $i \in \omega$, $\mathrm{dom}(\varphi_{\mathrm{en}(i)})$ is an initial segment of ω and the sequence of all $\beta(\varphi_{\mathrm{en}(i)}(c))$ with $c \in \mathrm{dom}(\varphi_{\mathrm{en}(i)})$ is strictly increasing with least upper bound η_i.*

In what follows let $\mathrm{ld}(i)$ be the cardinality of $\mathrm{dom}(\varphi_{\mathrm{en}(i)})$.

In [8] the difficulty of the problem to decide for two computable domain elements x, y whether $x \sqsubseteq y$ was studied in a general topological setting. As was shown, $\Omega_\eta(\sqsubseteq) \in \Pi_2^0$. The more exact localization depends on whether the constructive domain contains nonfinite elements or not.

Proposition 1 ([8]). *If Q_c contains a nonfinite element then $\Omega_\eta(\sqsubseteq)$ is Π_2^0-complete.*

Proposition 2 ([8]). *If Q_c contains only finite elements and the canonical indexing is decidable, then $\Omega_\eta(\sqsubseteq) \in \Delta_2^0$.*

Proof. We have to show that $\Omega_\eta(\sqsubseteq)$ is identified in the limit by some function $g \in R^{(2)}$. In what follows we construct a function g which is different from the one presented in [8] and which will be used again later. Let the function en be as in Lemma 1. Since Q_c is finitary, $\varphi_{\mathrm{en}(i)}$ is a finite function, for all $i \in \omega$. By admissibility of η there is a function $v \in R^{(1)}$ with $W_{v(i)} = \{\, a \mid \beta_a \sqsubseteq \eta_i \,\}$. Define $g, k \in R^{(2)}$ by the following simultaneous recursion:

$$k(\langle i, j \rangle, 0) = 0, \qquad g(\langle i, j \rangle, 0) = 1,$$

$$k(\langle i, j \rangle, s+1) = \begin{cases} k(\langle i, j \rangle, s) + 1 & \text{if } \varphi_{\mathrm{en}(i)}(k(\langle i, j \rangle, s))\!\downarrow_{s+1} \\ & \text{and } g(\langle i, j \rangle, s) = 1, \\ k(\langle i, j \rangle, s) & \text{otherwise,} \end{cases}$$

$$g(\langle i,j\rangle, s+1) = \begin{cases} 0 & \text{if } k(\langle i,j\rangle, s+1) \neq k(\langle i,j\rangle, s), \\ 1 & \text{if } \varphi_{v(j)}(\varphi_{\text{en}(i)}(c))\downarrow_{s+1}, \text{ for all} \\ & c < k(\langle i,j\rangle, s), \text{ and } g(\langle i,j\rangle, s) = 0, \\ g(\langle i,j\rangle, s) & \text{otherwise.} \end{cases}$$

Then it follows for $i, j \in \omega$ that

$$
\begin{aligned}
\eta_i \sqsubseteq \eta_j &\Rightarrow (\forall a \in \text{dom}(\varphi_{\text{en}(i)}))\beta(\varphi_{\text{en}(i)}(a)) \sqsubseteq \eta_j \\
&\Rightarrow (\exists s)(\forall a < \text{ld}(i))\varphi_{\text{en}(i)}(a)\downarrow_s \wedge \varphi_{v(j)}(\varphi_{\text{en}(i)}(a))\downarrow_s \\
&\Rightarrow (\exists s)k(\langle i,j\rangle, s) = \text{ld}(i) \wedge g(\langle i,j\rangle, s) = 1 \\
&\Rightarrow (\exists s)(\forall s' \geq s)g(\langle i,j\rangle, s') = 1 \\
&\Rightarrow \lim_s g(\langle i,j\rangle, s) = 1
\end{aligned}
$$

and

$$
\begin{aligned}
\eta_i \not\sqsubseteq \eta_j &\Rightarrow (\exists a \in \text{dom}(\varphi_{\text{en}(i)}))\beta(\varphi_{\text{en}(i)}(a)) \not\sqsubseteq \eta_j \\
&\Rightarrow (\exists a < \text{ld}(i))[(\exists s)\varphi_{\text{en}(i)}(a)\downarrow_s \wedge (\forall s')\varphi_{v(j)}(\varphi_{\text{en}(i)}(a))\uparrow_{s'}] \\
&\Rightarrow (\exists a < \text{ld}(i))[(\exists s)k(\langle i,j\rangle, s) = a+1 \wedge g(\langle i,j\rangle, s) = 0 \\
&\qquad\qquad \wedge (\forall s')\varphi_{v(j)}(\varphi_{\text{en}(i)}(a))\uparrow_{s'}] \\
&\Rightarrow (\exists s)(\forall s' \geq s)g(\langle i,j\rangle, s') = 0 \\
&\Rightarrow \lim_s g(\langle i,j\rangle, s) = 0. \qquad\qquad\qquad\qquad\qquad \square
\end{aligned}
$$

As a consequence of the above construction we have for all $i \in \omega$ that

$$
\begin{aligned}
\text{ld}(i) &\leq |\{\, s \mid g(\langle i,i\rangle, s) \neq g(\langle i,i\rangle, s+1) \,\}|, \\
|\{\, s \mid g(\langle i,j\rangle, s) &\neq g(\langle i,j\rangle, s+1) \,\}| \leq 2\,\text{ld}(i).
\end{aligned}
\tag{1}
$$

The next result shows that a better localization than in the preceding proposition cannot be obtained without further restrictions. If $\mathcal{Q}_c$ is finitary the order decision problem can be at least as difficult as any Δ_2^0 problem.

Proposition 3 ([8]). *For any $A \in \Delta_2^0$ a finitary recursive tree domain can be be designed such that $A \leq_1 \Omega_\eta(\sqsubseteq)$.*

Note that a finitary tree is exactly one with branches of finite length only. We delineate the construction again as it will be used in the next steps.

Proof. Let $A \in \Delta_2^0$. Then there is a 0-1-valued function $g \in R^{(2)}$ such that $c_A(a) = \lim_s g(a, s)$. The idea is to construct a domain consisting of infinitely many chains which are glued together at their first element such that the length of the ath chain is the number of how often $\lambda s.g(a, s)$ changes its hypothesis enlarged by one, i.e.,

$$1 + |\{\, s \mid g(a, s) \neq g(a, s+1) \,\}|.$$

Let to this end $\perp \notin \omega$ and define

$$Q_A = \{\perp\} \cup \{\langle a, s\rangle \mid [s = 0 \wedge g(a, s) = 1] \vee [s \neq 0 \wedge g(a, s - 1) \neq g(a, s)]\}.$$

Moreover, for $y, z \in Q^A$ let

$$y \sqsubseteq_A z \Leftrightarrow y = \perp \vee (\exists a, s, s' \in \omega) y = \langle a, s\rangle \wedge z = \langle a, s'\rangle \wedge s \leq s'.$$

Then $(Q_A, \sqsubseteq_A)$ is a tree domain. Set

$$\beta^A_{\langle a, s\rangle} = \begin{cases} \perp & \text{if } (\forall s' \leq s)g(a, s') = 0, \\ \langle a, \mu s' \leq s : (\forall s' \leq s'' \leq s)g(a, s'') = g(a, s)\rangle & \text{otherwise.} \end{cases}$$

Obviously, β^A is an indexing of Q_A with respect to which the partial order $\sqsubseteq_A$ is completely recursive. In this partial order every element has only finitely many predecessors and is thus finite as well as computable. It follows that $(Q_A, \sqsubseteq_A, \beta^A)$ is a finitary recursive domain.

Now, let η be an admissible numbering of Q^A, and let $h, k \in R^{(1)}$ with

$$W_{h(a)} = \{\langle a, s\rangle \mid s \in \omega\} \text{ and } W_{k(a)} = \{\langle a, s\rangle \mid g(a, s) = 1\}.$$

Furthermore, let the function $d \in R^{(1)}$ be as in the definition of admissibility. Since $W_{k(a)} \subseteq W_{h(a)}$, we always have that $x_{d(k(a))} \sqsubseteq x_{d(h(a))}$. Let t_a be the smallest step s such that $g(a, s') = g(a, s)$, for all $s' \geq s$. Then it holds that $x_{d(k(a))} = x_{d(h(a))}$ if and only if $g(a, t_a) = 1$. With this we obtain

$$a \in A \Leftrightarrow g(a, t_a) = 1 \Leftrightarrow x_{d(h(a))} = x_{d(k(a))} \Leftrightarrow x_{d(h(a))} \sqsubseteq_A x_{d(k(a))}.$$

As shown in [11], admissible indexings are cylinders. The same is true for the numbering W. Thus, the functions d, h and k can be chosen as one-to-one. It follows that $A \leq_1 \Omega_\eta(\sqsubseteq_A)$. $\qquad\qquad\square$

With respect to the weaker Turing equivalence, $\Omega_\eta(\sqsubseteq_A)$ turns out to be even as difficult as A.

Proposition 4. *For $A \in \Delta_2^0$, $\Omega_\eta(\sqsubseteq_A) \leq_T A$.*

Proof. Let $A \in \Delta_2^0$ and $g \in R^{(2)}$ such that $c_A(a) = \lim_s g(a, s)$. Moreover, be $r \in R^{(1)}$ with $W_{r(i)} = \text{range}(\varphi_i)$, $d \in R^{(1)}$ as in the definition of admissibility, and witness $k \in R^{(1)}$ that $\beta^A \leq \eta$.

For any $a \in \omega$, $\bigsqcup_s \beta^A_{\langle a, s\rangle}$ is the maximal element of the branch given by

$$\{\perp\} \cup \{\langle a, s\rangle \mid [s = 0 \wedge g(a, s) = 1] \vee [s \neq 0 \wedge g(a, s - 1) \neq g(a, s)]\}.$$

Let $h \in R^{(1)}$ with $\varphi_{h(a)}(s) = \langle a, s\rangle$ and $\bar{h} = r \circ h$. Then $W_{\bar{h}(a)} = \{\langle a, s\rangle \mid s \in \omega\}$ and hence

$$\bigsqcup_s \beta^A_{\langle a, s\rangle} = \bigsqcup \beta^A(W_{\bar{h}(a)}) = \eta_{d(\bar{h}(a))}.$$

As follows from the construction of $\mathcal{Q}_A$, $c_A(a) = g(a, \mu s : \beta_{\langle a,s \rangle}^A = \eta_{d(\bar{h}(a))})$, i.e.,

$$c_A(a) = g(a, \mu s : \langle k(\langle a,s \rangle), d(\bar{h}(a)) \rangle \in \Omega_\eta(=)).$$

Note that $\Omega_\eta(=) = \{\, \langle i,j \rangle \mid \langle i,j \rangle, \langle j,i \rangle \in \Omega_\eta(\sqsubseteq_A) \,\}$. Thus we have that $A \leq_T \Omega_\eta(\sqsubseteq_A)$. $\square$

Summarizing what we have seen so far, we obtain the first of our main results.

Theorem 1. *For Turing degrees $\mathbf{a}$ the following three statements are equivalent:*

1. $\mathbf{a} = \deg_T(A)$, *for some set $A \in \Delta_2^0$.*
2. $\mathbf{a} = \deg_T(\Omega_\eta(\sqsubseteq))$, *for some finitary recursive tree.*
3. $\mathbf{a} = \deg_T(\Omega_\eta(\sqsubseteq))$, *for some finitary constructive domain with decidable canonical indexing.*

As is well-known, the class of Δ_2^0 sets is exhausted by the Ershov hierarchy $\{\, \Sigma_\alpha^{-1}, \Pi_\alpha^{-1} \mid \alpha \text{ an ordinal} \,\}$ (also known as the Boolean hierarchy) [4,5,1]. Let K denote the halting problem and $\overline{K}$ its complement. For $m \geq 1$ set

$$Z_m = \bigcup_{i=0}^{m} (K^{2i} \times \overline{K}^{2(m-i)})$$

and let

$$Z_\omega = \{\, \langle n,i \rangle \mid i \in Z_n \wedge n \geq 1 \,\}.$$

Then Z_m and Z_ω, respectively, are m-complete for Π_{2m}^{-1} and Π_ω^{-1}. In certain cases also the order decision problem is m-complete for the corresponding levels of the hierarchy.

Let $\mathcal{Q}$ be finitary and CH be the set of all lengths of increasing chains in $\mathcal{Q}$. In general, this collection will be unbounded. Let us first consider the bounded case. Set $\mathrm{lc} = \max \mathrm{CH}$.

Proposition 5 ([8]). *Let $\mathcal{Q}_c$ be finitary so that CH is bounded and let the canonical indexing be decidable. If $\mathrm{lc} \geq 1$ then $\Omega_\eta(\sqsubseteq)$ is m-complete for $\Pi_{2\,\mathrm{lc}}^{-1}$. Otherwise, $\Omega_\eta(\sqsubseteq)$ is computable.*

Let us now turn to the unbounded case.

Proposition 6 ([8]). *Let $\mathcal{Q}_c$ be finitary so that CH is unbounded and let the canonical indexing be decidable. Then $Z_\omega \leq_m \Omega_\eta(\sqsubseteq)$.*

In the above theorem we have seen that the Turing degrees of the order decision problem of finitary constructive domains that come with a decidable canonical indexing, as well as the Turing degrees of the order decision problem of finitary recursive trees, capture exactly the Δ_2^0 degrees. With the preceding result we are now able to capture exactly the Turing degrees of sets in certain upper segments of Δ_2^0.

Theorem 2. *For* $m \in \omega$ *and Turing degrees* **a** *the following three statements are equivalent:*

1. $\mathbf{a} = \deg_T(A)$, *for some* $A \in \Delta_2^0 \setminus \Pi_m^{-1}$.
2. $\mathbf{a} = \deg_T(\Omega_\eta(\sqsubseteq))$, *for some finitary recursive tree such that either* CH *is unbounded or* CH *is bounded with* $2 \operatorname{lc} > m$.
3. $\mathbf{a} = \deg_T(\Omega_\eta(\sqsubseteq))$, *for some finitary constructive domains with decidable canonical indexing such that either* CH *is unbounded or* CH *is bounded with* $2 \operatorname{lc} > m$.

Proof. $(1 \Rightarrow 2)$ Let $A \in \Delta_2^0 \setminus \Pi_m^{-1}$ and Q_A be as in Proposition 3. Then $A \leq_1 \Omega_\eta(\sqsubseteq_A)$. Assume that CH is bounded with $2 \operatorname{lc} \leq m$. By Proposition 5 it follows that $\Omega_\eta(\sqsubseteq_A) \in \Pi_{2\operatorname{lc}}^{-1} \subseteq \Pi_m^{-1}$, which implies that also $A \in \Pi_m^{-1}$, a contradiction.

The implication $(2 \Rightarrow 3)$ is obvious. For $(3 \Rightarrow 1)$ let Q_c be a finitary constructive domain with decidable canonical indexing. If CH is unbounded, we have that $Z_\omega \leq_m \Omega_\eta(\sqsubseteq)$. Hence $\Omega_\eta(\sqsubseteq) \notin \Pi_m^{-1}$, as otherwise we would have that $Z_\omega \in \Pi_m^{-1}$ and hence that $\Pi_\omega^{-1} = \Pi_m^{-1}$.

If CH is bounded with $2 \operatorname{lc} > m$, $\Omega_\eta(\sqsubseteq)$ is m-complete for $\Pi_{2\operatorname{lc}}^{-1}$, from which we obtain that $\Omega_\eta(\sqsubseteq) \notin \Pi_m^{-1}$. $\square$

Note that for a finitary domain Q, CH is unbounded or bounded with $\operatorname{lc} > 0$, exactly if Q has at least two elements. This leads to the following consequence.

Corollary 1. *For Turing degrees* **a** *the following three statements are equivalent:*

1. **a** *is the Turing degree of a noncomputable set in* Δ_2^0.
2. **a** *is the degree* $\deg_T(\Omega_\eta(\sqsubseteq))$ *of the order decision problem for some finitary recursive tree with branches of positive finite length only.*
3. **a** *is the degree* $\deg_T(\Omega_\eta(\sqsubseteq))$ *of the order decision problem for some finitary constructive domains of at least cardinality two with decidable canonical indexing*

As we have seen in Proposition 6, if Q_c is finitary and CH unbounded then $Z_\omega \leq_m \Omega_\eta(\sqsubseteq)$. In the tree domain Q_A constructed in the proof of Proposition 3 the length of the branches is determined by the number of mind changes of the function identifying the given problem A in the limit. Let us now consider the case that this function is recursively majorized.

Lemma 2. *Let* $C \in \Delta_2^0$ *be identified in the limit by the function* $g \in R^{(2)}$. *Then* $C \leq_m Z_\omega$, *precisely if* $\lambda a.|\{ s \mid g(a, s) \neq g(a, s + 1) \}|$ *is recursively majorized.*

Proof. The "if"-part is obvious and the "only if"-part shown in [8].

Now, let Q_c be a finitary constructive domain with decidable canonical indexing and $g \in R^{(2)}$ be the function constructed in the proof of Proposition 2 which identifies $\Omega_\eta(\sqsubseteq)$ in the limit. By the Inequalities (1) we have that $\lambda a.|\{ s \mid g(a, s) \neq g(a, s + 1) \}|$ is recursively majorized, exactly if ld is.

This gives us the following characterization of the Turing degrees of sets in $\Delta_2^0 \setminus \Pi_\omega^{-1}$.

Theorem 3. *For Turing degrees* **a** *the following three statements are equivalent:*

1. $\mathbf{a} = \deg_T(A)$, *for some set* $A \in \Delta_2^0 \setminus \Pi_\omega^{-1}$.
2. $\mathbf{a} = \deg_T(\Omega_\eta(\sqsubseteq))$, *for some finitary recursive tree such that* CH *is unbounded and* ld *not recursively majorized.*
3. $\mathbf{a} = \deg_T(\Omega_\eta(\sqsubseteq))$, *for some finitary constructive domain with decidable canonical indexing such that* CH *is unbounded and* ld *not recursively majorized.*

Acknowledgement

Thanks are due to Wu Guohua for asking the right questions and providing hints to literature.

References

1. Arslanov, M.M.: The Ershov hierarchy (2007) (manuscript)
2. Miller, R.: The Δ_2^0-spectrum of a linear order. Journal of Symbolic Logic 66, 470–486 (2001)
3. Edalat, A.: Domains for computation in mathematics, physics and exact real arithmetic. Bulletin of Symbolic Logic 3, 401–452 (1997)
4. Ershov, J.L.: On a hierarchy of sets I, Algebra i Logika 7(1), 47–74 (1968) (Russian); English transl., Algebra and Logic 7, 25–43 (1968)
5. Ershov, J.L.: On a hierarchy of sets II, Algebra i Logika 7(4), 15–47 (1968) (Russian); English transl., Algebra and Logic 7, 212–232 (1968)
6. Ershov, J.L.: Model $\mathbb{C}$ of partial continuous functionals. In: Gandy, R., et al. (eds.) Logic Colloquium, vol. 76, pp. 455–467. North-Holland, Amsterdam
7. Slaman, T.: Relative to any nonrecursive set. Proceedings of the American Mathematical Society 126, 2117–2122 (1998)
8. Spreen, D.: On some decision problems in programming. Information and Computation 122, 120–139 (1995); Corrigendum ibid. 148, 241–244 (1999)
9. Stoltenberg-Hansen, V., Lindström, I., Griffor, E.R.: Mathematical Theory of Domains. Cambridge University Press, Cambridge (1994)
10. Wehner, S.: Enumerations, countable structures, and Turing degrees. Proceedings of the American Mathematical Society 126, 2131–2139 (1998)
11. Weihrauch, K., Deil, T.: Berechenbarkeit auf cpo-s, Schriften zur Angewandten Mathematik und Informatik Nr. 63, Rheinisch-Westfälische Technische Hochschule Aachen (1980)

Computable Fields and Weak Truth-Table Reducibility

Rebecca M. Steiner

Graduate Center of the City University of New York
365 Fifth Avenue, New York, NY 10016 USA
rsteiner@gc.cuny.edu

Abstract. For a computable field F, the *splitting set* S_F of F is the set of polynomials with coefficients in F which factor over F, and the *root set* R_F of F is the set of polynomials with coefficients in F which have a root in F.

Results of Frohlich and Shepherdson in [3] imply that for a computable field F, the splitting set S_F and the root set R_F are Turing-equivalent. Much more recently, in [5], Miller showed that for algebraic fields, the root set actually has slightly higher complexity: for algebraic fields F, it is always the case that $S_F \leq_1 R_F$, but there are algebraic fields F where we have $R_F \nleq_1 S_F$.

Here we compare the splitting set and the root set of a computable algebraic field under a different reduction: the weak truth-table reduction. We construct a computable algebraic field for which $R_F \nleq_{wtt} S_F$.

1 Introduction

Given a polynomial $p(X)$ with coefficients in a field F, the questions that are typically asked of $p(X)$ are:

- Does $p(X)$ factor in $F[X]$?
- Does $p(X)$ have a root in F?

It is not always clear which of these two questions is easier to answer. We will address this issue in the case where F is a *computable* field.

Definition 1. A *computable field* F consists of two computable functions f and g from $\omega \times \omega$ into ω, such that ω forms a field under these functions, with f as the addition and g as the multiplication. Sometimes we refer to F as a computable presentation of the isomorphism type of F.

Definition 2. The *splitting set* S_F for a field F is the set of polynomials in $F[X]$ which factor, i.e. split, into two nonconstant factors in $F[X]$. The *root set* R_F of F is the set of polynomials in $F[X]$ which have a root in F.

Both of these sets are computably enumerable; they are defined by existential formulas.

F. Ferreira et al. (Eds.): CiE 2010, LNCS 6158, pp. 394–405, 2010.

Turing reducibility is the most common reducibility used by computability theorists to compare the relative complexity of two sets of natural numbers. Finer reducibilities, such as m-reducibility and 1-reducibility, can further separate the relative complexities of sets that sit inside the same Turing degree. For a computable field F, the set S_F of polynomials in $F[X]$ that factor over F and the set R_F of polynomials in $F[X]$ that have a root in F are actually always Turing-equivalent, which surprises most mathematicians; however, $S_F \leq_1 R_F$ while there are computable fields F for which $R_F \not\leq_1 S_F$.

Another, less-common type of reducibility is the *weak truth-table reducibility*.

Definition 3. For $A, B \subseteq \omega$, A is *weak truth-table reducible to B*, written $A \leq_{wtt} B$, if there exists an oracle Turing functional Φ_e and a total computable function f such that Φ_e^B computes the characteristic function of A and for all $x \in \omega$, the computation $\Phi_e^B(x)$ asks its oracle questions only about the membership in B of elements $\leq f(x)$. In other words, for all x, $\Phi_e^{B \restriction f(x)}(x) \downarrow = \chi_A(x)$.

Weak truth-table (wtt) reducibility is finer than Turing reducibility, but still coarser than 1-reducibility. The purpose of this paper is to find out what happens between S_F and R_F under this intermediate reducibility – are they equivalent, as under Turing reducibility, or is one slightly harder to compute than the other, as under 1-reducibility?

Since $S_F \leq_1 R_F$ for any computable algebraic field F, it follows that $S_F \leq_{wtt} R_F$. But it turns out that we can construct a computable algebraic field in which $R_F \not\leq_{wtt} S_F$. The interesting thing about this is that the lack of wtt-reduction from R_F to S_F is equivalent to a purely Galois-theoretic statement about the splitting set and the root set of a field!

Consider the following fact that will be stated and proved later in Lemma 6:

Let L be a Galois extension of $\mathbb{Q}$. For any fields F_0 and F_1, with $\mathbb{Q} \subsetneq F_0, F_1 \subsetneq L$,

$$(\forall q(X) \in \mathbb{Q}[X]) \, [q \text{ has a root in } F_0 \iff q \text{ has a root in } F_1]$$

if and only if

$$(\exists \sigma \in Gal(L/\mathbb{Q})) \, [\sigma(F_0) = F_1].$$

Also consider the statement $\star$, which is almost the same as above, but with the words "factors over" replacing the words "has a root in."

The reason there cannot be any wtt-reduction from R_F to S_F is that the statement in Lemma 6 is true, while $\star$ is false. In fact, if $\star$ were to be true also, then there we would always have a wtt-reduction from R_F to S_F. This idea will be explored further in sections 3 and 4.

A standard reference for algebraic background is [9], and the canonical reference for computability theoretic background is [8].

2   Algebraic Background

We will need the following lemmas in our construction, so we state them here for reference:

Definition 4. The *elementary symmetric polynomials* in $X_1, \ldots, X_m$ over a field F are the polynomials

$$s_k(X_1, \ldots X_m) = \sum_{1 \leq i_1 < \cdots < i_k \leq m} X_{i_1} X_{i_2} \cdots X_{i_k} \quad (\text{for } 1 \leq k \leq m).$$

The *symmetric polynomials* are the elements of $F[s_1, \ldots, s_m]$ – they are exactly the polynomials in $F[X_1, \ldots, X_m]$ that are invariant under permutations of the variables.

A subfield generated by the elementary symmetric polynomials in some proper nonempty subset of the set of roots of a polynomial $q(X) \in \mathbb{Q}[X]$ is called a *symmetric subfield* in the lattice of subfields of the splitting field of q over $\mathbb{Q}$. We refer the reader to [5] for additional details about these subfields, including the proof of the following lemma.

Lemma 1. *Let $p(X) \in F[X]$ be a polynomial over a field F and let $E \supseteq F$ be a field extension. Let $\overline{E}$ be the algebraic closure of E, and A the set of roots of $p(X)$ in $\overline{E}$. Assume that every root of $p(X)$ has multiplicity 1. Then the following are equivalent:*

- *$p(X)$ is reducible in $E[X]$.*
- *There exists $I = \{x_1, \ldots, x_m\}$ with $\emptyset \subsetneq I \subsetneq A$ such that every symmetric polynomial $h \in F[X_1, \ldots, X_m]$ has $h(x_1, \ldots, x_m) \in E$.*

In other words: p factors over E if and only if E contains one of the symmetric subfields generated by a proper nonempty subset of the set of roots of p.

Definition 5. For field extensions $E \supseteq \mathbb{Q}$ and $F \supseteq \mathbb{Q}$, E and F are *linearly disjoint* if $E \cap F = \mathbb{Q}$.

Lemma 2. *For relatively prime numbers m and n, if ζ_m and ζ_n are primitive m-th and n-th roots of unity, respectively, then $\mathbb{Q}(\zeta_m)$ and $\mathbb{Q}(\zeta_n)$ are linearly disjoint.*

Proof. It is not hard to see that we have $\mathbb{Q} \subseteq \mathbb{Q}(\zeta_m) \cap \mathbb{Q}(\zeta_n)$. For the reverse inclusion, it is sufficient to show that $\mathbb{Q}(\zeta_m) \cap \mathbb{Q}(\zeta_n)$ has degree 1 over $\mathbb{Q}$, and this is shown in [10, Propositions 2.3 & 2.4]. $\qquad\square$

Lemma 3. *The field $\mathbb{Q}(\zeta_{p^2})$, where ζ_{p^2} is a primitive p^2-th root of unity and p is prime with $p \equiv 1 \pmod 4$, has exactly one subfield of index 2, and this subfield contains the unique subfield of $\mathbb{Q}(\zeta_{p^2})$ of degree 2 over $\mathbb{Q}$.*

Proof. Note that $\mathbb{Q}(\zeta_{p^2})$ is a Galois extension of $\mathbb{Q}$, since ζ_{p^2} generates all of its conjugates, and so $\mathbb{Q}(\zeta_{p^2})$ is the splitting field of the polynomial $X^{p^2} - 1$.

First we show that $\mathbb{Q}(\zeta_{p^2})$ has exactly one subfield of index 2. The Galois group G of $\mathbb{Q}(\zeta_{p^2})$ over $\mathbb{Q}$ is the cyclic group of order $p(p-1)$ (see [10, Theorem 2.5]). By the Fundamental Theorem of Galois Theory (see [1, Theorem 14.14]), a subfield of $\mathbb{Q}(\zeta_{p^2})$ of index 2 corresponds to a subgroup of G of order 2. G certainly has a subgroup of order 2 since the order of G is even. Indeed, $G \cong \mathbb{Z}/p(p-1)\mathbb{Z} = \{0, 1, \ldots, p(p-1) - 1\}$ under addition mod $p(p-1)$. So the only possible subgroup of G of order 2 is $\{0, \frac{1}{2}p(p-1)\}$, because the non-identity element must yield identity when doubled. If there is exactly one subgroup of G of order 2, then there is exactly one subfield of $\mathbb{Q}(\zeta_{p^2})$ of index 2. In fact, if ζ is a primitive p^2-th root of unity, then $\mathbb{Q}(\zeta + \zeta^{-1})$ is the subfield of $\mathbb{Q}(\zeta_{p^2})$ such that $[\mathbb{Q}(\zeta_{p^2}) : \mathbb{Q}(\zeta + \zeta^{-1})] = 2$ (see [10, remarks before Proposition 2.15]).

To show that $\mathbb{Q}(\zeta_{p^2})$ has exactly one subfield of degree 2 over $\mathbb{Q}$, the argument is very similar. Again by the Fundamental Theorem of Galois Theory, a subfield of $\mathbb{Q}(\zeta_{p^2})$ of degree 2 over $\mathbb{Q}$ corresponds to a subgroup of G that has index 2 in G, i.e. has order $\frac{1}{2}p(p-1)$. The only possibility for a subgroup of G of order $\frac{1}{2}p(p-1)$ must contain precisely the "even" elements of G, i.e. $\{0, 2, \ldots, p(p-1) - 2\}$ under addition mod $p(p-1)$. If there is exactly one subgroup of G with index 2 in G, then there is exactly one subfield of $\mathbb{Q}(\zeta_{p^2})$ of degree 2 over $\mathbb{Q}$. In fact, $\mathbb{Q}(\sqrt{p})$ is the subfield of $\mathbb{Q}(\zeta_{p^2})$ such that $[\mathbb{Q}(\sqrt{p}) : \mathbb{Q}] = 2$ (see [1, Example 14.5.1]).

Now to explain why the index 2 subfield must contain the degree 2 subfield: The subgroup H of G of order $\frac{1}{2}p(p-1)$ must contain a subgroup of order 2, since $\frac{1}{2}p(p-1)$ is even (recall that we assumed that $p \equiv 1 \pmod 4$), and because G has only one of these, H must contain it. The Fundamental Theorem tells us that H containing the subgroup of order 2 corresponds exactly to the subfield of $\mathbb{Q}(\zeta_{p^2})$ with degree 2 over $\mathbb{Q}$ being contained in the subfield of $\mathbb{Q}(\zeta_{p^2})$ of index 2. And since we have the first containment, we must also have the second one.   □

Corollary 1. *If K is an extension of $\mathbb{Q}$ of finite degree over $\mathbb{Q}$, then K is linearly disjoint from $\mathbb{Q}(\zeta_{p^2})$ for all but finitely many primes $p \equiv 1 \pmod 4$.*

Proof. If $K \cap \mathbb{Q}(\zeta_{p^2}) \supsetneq \mathbb{Q}$ for infinitely many primes p, then since K has only finitely many subfields between itself and $\mathbb{Q}$, infinitely many $\mathbb{Q}(\zeta_{p^2})$ would all contain the same proper extension of $\mathbb{Q}$. But according to Lemma 2, the fields $\mathbb{Q}(\zeta_{p^2})$ are pairwise linearly disjoint.

So K can only intersect finitely many of the extensions $\mathbb{Q}(\zeta_{p^2})$ in a field bigger than $\mathbb{Q}$.   □

The author is grateful to Kenneth Kramer for pointing out the following group-theoretic fact that was crucial to the result of this paper:

Lemma 4. *Suppose L/K is a Galois extension and g is an irreducible polynomial in $L[X]$. Let L_1 be the field extension of K generated by the coefficients of g and let α be a root of g in $\overline{L}$. If $K \subseteq E \subseteq L$, then the minimal polynomial for α over E is*

$$f = \prod \sigma_i(g),$$

where $\{\sigma_i\}$ is the set of distinct embeddings of L_1 into L over E. It follows that

$$\deg f = [L_1 E : E] \deg g.$$

Proof. Let $h = \prod \sigma_i(g)$ and let f be the minimal polynomial for α over E. We will show that $h = f$.

Note first that $h(\alpha) = f(\alpha) = 0$.

We want $\tau(h) = h$ for any $\tau \in \mathrm{Gal}(L/E)$, because then h will be a polynomial over E. But since τ acts by permutation on the cosets, it just rearranges the factors of h, and since polynomial multiplication is commutative, $\tau(h) = h$. Therefore f, the minimal polynomial of α over E, divides h.

Next we show that the $\sigma_i(g)$ are all relatively prime to each other. Suppose that $\sigma_k(g)$ and $\sigma_j(g)$ have a factor in common. Since both of these are irreducible over L, the factor that they have in common must be $\sigma_k(g) = \sigma_j(g)$. In particular, this means that $\sigma_k \sigma_j^{-1}(g) = g$, so $\sigma_k \sigma_j^{-1}$ fixes all coefficients of g, hence fixes L_1, which is generated by the coefficients of g. But $\sigma_k \sigma_j^{-1} \in \mathrm{Gal}(L/E)$, so $\sigma_k \sigma_j^{-1}$ also fixes E. Thus $\sigma_k \sigma_j^{-1}$ is the identity on $L_1 E$, and so $k = j$. So the polynomials $\sigma_i(g)$ are all relatively prime.

Finally, we need irreducibility. We already know that $f|h$, so we just need to show that $h|f$. g is irreducible over L. $\sigma_i(g)$ is likewise irreducible over L for each i. Consider f over L: the minimal polynomial for α over L is g, so $g|f$ over L. Then act by σ_i to get $\sigma_i(g)|\sigma_i(f)$. Since σ_i acts trivially on E, $\sigma_i(f) = f$. So $\sigma_i(g)|f$ for each i. Since the $\sigma_i(g)$ are all relatively prime, their product h must also divide f, and so h and f are actually the same polynomial. $\square$

Corollary 2. *For a prime $p \equiv 1 \pmod 4$, a polynomial $q(X) \in \mathbb{Q}[X]$ splits (i.e. factors nontrivially) in $\mathbb{Q}(\zeta_{p^2})$ if and only if it splits in the unique index 2 subfield E of $\mathbb{Q}(\zeta_{p^2})$.*

Proof. Let $q(X)$ be an irreducible polynomial in $\mathbb{Q}[X]$. Let g be a factor of q with coefficients in $L = \mathbb{Q}(\zeta_{p^2})$ which is irreducible over L. L is Galois over $\mathbb{Q}$, so with $K = \mathbb{Q}$ and with notation as in Lemma 4, the degree of q is $\deg q = [L_1 : \mathbb{Q}] \deg g$. The degree of an irreducible factor h of q in the index 2 subfield E of L, where h is the factor that is a multiple of g, is $\deg h = [L_1 E : E] \deg g$. To have q irreducible over E but split over L, we must have $q = h$ and $[L_1 : \mathbb{Q}] = [L_1 E : E] > 1$. If $[L_1 E : E] > 1$, then $L_1 E = L$ because L is the only proper extension of E within L. Thus $[L_1 : \mathbb{Q}] = [L : E] = 2$. But then by Lemma 3 we must have $L_1 \subseteq E$, which is a contradiction, because since L_1 is generated by the coefficients of g, E would contain these coefficients, and then q would not be irreducible over E. $\square$

The following lemma dates back to a paper [4] written by Kronecker in 1882, so we refer the reader to the more recent account in [2].

Lemma 5. *The splitting set of the field $\mathbb{Q}$ is computable. Furthermore, if L is a c.e. subfield of a computable field K and L has a splitting algorithm, then so does every finite extension of L inside K, and if K is algebraic over L, these splitting algorithms can be found uniformly in the generators of the extension.*

Lemma 6. *Let L be a Galois extension of $\mathbb{Q}$. For any fields F_0 and F_1, with $\mathbb{Q} \subsetneq F_0, F_1 \subsetneq L$,*

$$(\forall q(X) \in \mathbb{Q}[X]) \, [q \text{ has a root in } F_0 \iff q \text{ has a root in } F_1]$$

if and only if

$$(\exists \sigma \in Gal(L/\mathbb{Q})) \, [\sigma(F_0) = F_1].$$

Proof. For the backwards direction, let $\sigma \in Gal(L/\mathbb{Q})$ with $\mathbb{Q} \subsetneq F_0 \subsetneq L$ and $\sigma(F_0) \neq F_0$. Let $F_1 = \sigma(F_0)$. Then, for $q(X) \in \mathbb{Q}[X]$,

$$q \text{ has a root in } F_0 \implies \exists \, c \in F_0 \text{ with } q(c) = 0$$
$$\implies \exists \, c \in F_0 \text{ with } \sigma(q(c)) = \sigma(0)$$
$$\implies \exists \, c \in F_0 \text{ with } \sigma(q(c)) = 0$$
$$\implies \exists \, c \in F_0 \text{ with } q(\sigma(c)) = 0$$

(because any automorphism of L must fix $\mathbb{Q}$ pointwise and therefore must fix the coefficients of q)

$$\implies q \text{ has a root in } F_1.$$

To show that q has a root in F_0 whenever q has a root in F_1, repeat the above argument with σ^{-1} instead of σ.

For the forwards direction, suppose F_0 and F_1 have the same root set with respect to polynomials with rational coefficients, i.e. $R_{F_0} \cap \mathbb{Q}[X] = R_{F_1} \cap \mathbb{Q}[X]$.

Claim: $F_0 \cong F_1$.

According to the Theorem of the Primitive Element (see [9]), every finite algebraic extension of $\mathbb{Q}$ is actually generated by a single element, called a "primitive element." Find a primitive generator r_0 for F_0 and then find the minimal polynomial $p_0(X)$ over $\mathbb{Q}$ for this generator. Then F_0 certainly has a root of p_0, and by assumption, F_1 also has a root of p_0.

If α and β are two roots of the same irreducible polynomial p over $\mathbb{Q}$, then $\mathbb{Q}(\alpha) \cong \mathbb{Q}[X]/(p(X)) \cong \mathbb{Q}(\beta)$. With this fact in mind, we can see that since F_1 also has a root of the polynomial p_0, then F_1 must contain a subfield isomorphic to $\mathbb{Q}(r_0) = F_0$.

Now find a primitive generator r_1 for F_1 and then find the minimal polynomial $p_1(X)$ for this generator. Then F_1 certainly has a root of p_1, and by assumption, F_0 also has a root of p_1.

Then F_0 contains a subfield isomorphic to $\mathbb{Q}(r_1) = F_1$.

Now, if two fields embed into one another, then the composition of those embeddings is an embedding of the first field into itself, and every embedding of an algebraic field into itself must be an automorphism (see [6, Lemma 2.10]). So the reverse embedding is surjective, and so the two fields must actually be isomorphic. Thus $F_0 \cong F_1$ via some isomorphism τ. So our claim is proved.

Any isomorphism of extensions of $\mathbb{Q}$ must fix $\mathbb{Q}$ pointwise, and we know that any isomorphism of two subfields of a Galois extension can be effectively extended to an automorphism of that Galois extension (see [9]). These two facts together tell us that τ can be extended to an automorphism σ of L that fixes $\mathbb{Q}$ pointwise, and thus $\sigma \in Gal(L/\mathbb{Q})$ with $\sigma(F_0) = F_1$. $\qquad\square$

The result in this paper depends almost exclusively on the fact that if we replace the phrase "has a root in" with "factors over" in Lemma 6, the statement is no longer true; it is the forwards direction that fails. Here is a counterexample: Let L be the field $\mathbb{Q}(\sqrt[8]{2}, i)$, which happens to be the splitting field of the polynomial $x^8 - 2$, and so is Galois over $\mathbb{Q}$. Let $F_0 = \mathbb{Q}(\sqrt{2}, i)$ and $F_1 = \mathbb{Q}(\sqrt[4]{2}, i)$. Given any polynomial q with rational coefficients, q factors over F_0 if and only if it factors over F_1; this is a consequence of Lemma 4 (F_0 is a subfield of F_1 of index 2, and F_0 contains all the subfields of F_1 of degree 2 over $\mathbb{Q}$, so apply a Corollary 2-like argument). However, F_1 cannot possibly be the image of F_0 under any element of $\mathrm{Gal}(L/\mathbb{Q})$ because F_0 and F_1 are not isomorphic; F_0 is a proper subfield of F_1, so $[F_0 : \mathbb{Q}] < [F_1 : \mathbb{Q}]$.

In Section 4 we will take a look at how one would go about building a wtt-reduction if the "factors over" analogue of Lemma 6 were to be true.

3   The Construction

Theorem 1. *There exists a computable algebraic field F, with splitting set S_F and root set R_F, for which $R_F \not\leq_{wtt} S_F$.*

We will first construct the computable field F, and then prove that $R_F \not\leq_{wtt} S_F$. We effectively enumerate all the partial computable functions $\varphi_0, \varphi_1, \varphi_2, \dots$ and all the partial computable Turing functionals $\Phi_0, \Phi_1, \Phi_2, \dots$, and we build F to satisfy, for every e and every i, the requirement

> $\mathcal{R}_{e,i}$: If $\Phi_e^{S_F}$ and φ_i are total and $\forall q$ ($\Phi_e^{S_F}(q) \downarrow = 0 \iff q \notin R_F$), then
> $(\exists q \in F[X])(\exists y \geq \varphi_i(q))[\Phi_e^{S_F}(q)$ asks an oracle question about $y]$.

If every requirement is satisfied, then we will have $R_F \not\leq_{wtt} S_F$ because no total computable function can possibly play the role of the bound function on the splitting set oracle.

The construction will happen in stages.

For each e and each i, we will choose a witness polynomial $q_{e,i}(X) \in \mathbb{Q}[X]$, and we will feed it to φ_i and wait, perhaps forever, for $\varphi_i(q_{e,i})$ to halt. If this computation ever does halt, then we will feed $q_{e,i}$ to Φ_e with oracle $S_F \upharpoonright \varphi_i(q_{e,i})$ and wait to see if this second computation halts. If it does, we will ensure that $\Phi_e^{S_F}(q) \downarrow = 0 \iff q \in R_F$.

Given a computable (domain ω) field F, we can list out all the monic polynomials in $F[X]$. We do this by listing out $\omega^{<\omega}(\cong \omega)$ and think of $(a_0, \dots, a_d) \in \omega^{d+1}$ as representing $a_0 + a_1 X + \cdots + a_d X^d + X^{d+1} \in F[X]$.

So we list $F[X] = \{p_0(X), p_1(X), p_2(X), \dots\} \cong \omega$. S_F is a subset of $F[X]$.

If we use the standard map from ω to $\omega^{<\omega}$, then $p_n(X)$ only uses coefficients from $\{0, 1, \dots, n\} \subseteq \omega$, the domain of F. If we ensure that the field elements labeled $\{0, 1, \dots, n\}$ lie within $\mathbb{Q}$, then we will have $\{p_0, p_1, \dots, p_n\} \subseteq \mathbb{Q}[X]$. So, if the elements $0, 1, \dots, \varphi_i(q)$ in the domain of F all lie within $\mathbb{Q}$, then we have $\{p_0, p_1, \dots, p_{\varphi_i(q)}\} \subseteq \mathbb{Q}[X]$. So $S_F \upharpoonright \varphi_i(q)$ only tells us whether elements of $\mathbb{Q}[X]$ split in $F[X]$.

Throughout the construction, we refer to D_s as the set of elements of $\overline{\mathbb{Q}}$ that have been enumerated by the end of stage s, and we refer to F_s as the field that these elements generate. Elements are enumerated into D_s with the goal of eventually enumerating the entire field F_s.

Here is the "basic module" for the construction, i.e. how we satisfy a single requirement $\mathcal{R}_{e,i}$:

Step 0: Initialization.

For the strategy that aims to satisfy the requirement $\mathcal{R}_{e,i}$, choose a prime $p \equiv 1 \pmod 4$, and consider the Galois extension generated by adjoining a primitive p^2-th root of unity ζ_{p^2} to $\mathbb{Q}$, and consider its lattice of subfields over $\mathbb{Q}$. Denote the element ζ_{p^2} by $\beta_{e,i}$ and the element $\zeta_{p^2} + \zeta_{p^2}^{-1}$ by $\alpha_{e,i}$.

We choose, for our "witness" polynomial, the cyclotomic polynomial that has, as its roots, precisely the primitive p^2-th roots of unity. (This is the minimal polynomial of $\beta_{e,i}$ over $\mathbb{Q}$). We do this because we want a polynomial with coefficients in $\mathbb{Q}$ that has a root in $\mathbb{Q}(\beta_{e,i})$ but no root in any proper subfield of $\mathbb{Q}(\beta_{e,i})$. Call this polynomial $q_{e,i}$.

Step 1: Adjoining $\alpha_{e,i}$.

If and when the computation $\varphi_i(q_{e,i})$ halts, say that it does so at stage s_0. Then at stage s_1 with $s_1 = s_0 + 1$, we enumerate the element $\alpha_{e,i}$ into D_{s_1} (and hence eventually all $\mathbb{Q}(\alpha_{e,i})$ into the field F_{s_1}) in such a way that the first $\varphi_{i,s_1}(q_{e,i})$ elements of D_{s_1} are elements of the subfield $\mathbb{Q}$.

Once we've put $\alpha_{e,i}$ into the field, then if and when $\Phi_e^{S_{F_{s_1}} \upharpoonright \varphi_{i,s_1}(q_{e,i})}(q_{e,i})$ halts, say that it does so at stage s_2. There are three possible outcomes of this convergence:

- Case 1: $\Phi_{e,s_2}^{S_{F_{s_2}} \upharpoonright \varphi_{i,s_2}(q_{e,i})}(q_{e,i}) \downarrow = 1$. Then according to the splitting set restricted to the first $\varphi_{i,s_2}(q_{e,i})$ elements of the field, F_{s_2} has a root of $q_{e,i}$, which is clearly false, because we haven't put one in!

- Case 2: $\Phi_{e,s_2}^{S_{F_{s_2}} \upharpoonright \varphi_{i,s_2}(q_{e,i})}(q_{e,i}) \downarrow \notin \{0,1\}$. This case is also an immediate win for us, because convergence to something other than 0 or 1 has no meaning in this context. We treat this case exactly like Case 1.

Step 2: Adjoining $\beta_{e,i}$.

- Case 3: $\Phi_{e,s_2}^{S_{F_{s_2}} \upharpoonright \varphi_{i,s_2}(q_{e,i})}(q_{e,i}) \downarrow = 0$. Here lies the potential difficulty, as the restricted splitting set thinks (correctly) that F_{s_2} has no root of $q_{e,i}$. So at some subsequent stage s_3 with $s_3 = s_2 + 1$, we simply enumerate a root of $q_{e,i}$ (namely $\beta_{e,i}$) into D_{s_3}.

But won't adding this root have possible adverse effects on the splitting set?

Actually, it won't. Recall that the first $\varphi_i(q_{e,i})$ elements of the field lie in $\mathbb{Q}$. Adding an element of $\mathbb{Q}(\beta_{e,i})$ has no effect on the splitting of rational polynomials, because F_{s_1} already contains the field $\mathbb{Q}(\alpha_{e,i})$, and by Corollary 2, we know that $\mathbb{Q}(\beta_{e,i})$ and $\mathbb{Q}(\alpha_{e,i})$ split exactly the same rational polynomials. If no rational polynomial will be caused to split by the addition of this root, then nothing can enter the restricted splitting set, and so the oracle $S_F \upharpoonright \varphi_i(q_{e,i})$

will not change, and thus the computation will still yield convergence to 0 even though we now have a root of $q_{e,i}$ in the field. So our basic module does indeed satisfy the single requirement $\mathcal{R}_{e,i}$.

Dealing with many $\mathcal{R}_{e,i}$ at once is not so easy. It may seem that because of the linear disjointness of $\mathbb{Q}(\zeta_{p^2})$ and $\mathbb{Q}(\zeta_{q^2})$ for primes $p \neq q$ that we end up in a situation where there is no injury between requirements, but unfortunately, this is not the case.

Some requirements will be restricted from putting certain elements of $\overline{\mathbb{Q}}$ into D_s. For example, if a requirement wants to be satisfied by putting a root of a particular polynomial into D_s but this root would cause a polynomial in the oracle for a higher-priority requirement to factor, and thus change the oracle for this higher-priority requirement, we cannot allow this. So we prevent requirements from putting anything into D_s that would cause such a factorization.

For this we use the technique of the finite injury priority construction. We omit the details of the full construction, but here is an idea of what happens:

We say that the convergence of the aforementioned $\varphi_i(q_{e,i})$ constitutes an *injury* to each lower-priority requirement $\mathcal{R}_{c,j}$. Because $\mathcal{R}_{c,j}$ may not put into the field any element that might cause changes in the oracle for a higher-priority requirement, its strategy gets *initialized*, i.e. it must start over again from the beginning by choosing a fresh prime and a new $q_{e,i}$. $\mathcal{R}_{c,j}$ can be injured only finitely many times because injury only happens when a higher-priority requirement goes through "step 1" of the basic module. This can't happen more than $2^n - 1$ times if there are n requirements with higher priority than $\mathcal{R}_{c,j}$. Choosing a new prime each of those finitely many times is a way to guarantee that $\mathcal{R}_{c,j}$ can be satisfied while obeying the restrictions set by higher-priority requirements.

Lemma 7. *For each e and each i, the requirement $\mathcal{R}_{e,i}$ is injured only finitely often and acts only finitely often.*

Proof. We prove both of these with a simultaneous induction. A requirement may not need to act at all, as in the case where there is no convergence of the function $\varphi_i(q_{e,i,s})$. If and when the first requirement $\mathcal{R}_{0,0}$ acts, it injures all other requirements and causes their strategies to be initialized. $\mathcal{R}_{0,0}$ is never injured; its strategy is never initialized, and so it never needs to act again.

The second requirement can therefore be injured at most once, and so there will be a stage after which it is never injured again. If and when this requirement acts after that stage, the requirement's strategy is never initialized, and so it never needs to act again.

The third requirement can be injured at most three times, and so on. By induction, each requirement is injured only finitely often, and if and when it acts after the stage of its final injury, that will be its last action.   □

If $\mathcal{R}_{e,i}$ acts for the last time (if it ever acts at all) at stage s, then its chosen polynomial does not change after stage s. So the polynomial $q_{e,i}$ stabilizes after finitely many stages and remains the same forever after.

Lemma 8. *Suppose the requirement $\mathcal{R}_{e,i}$ has been satisfied and is never injured again. Suppose it was the case that the computation $\varphi_i(q_{e,i})$ did in fact halt, and let t be the stage at which the element $\alpha_{e,i}$ was enumerated into D. Then the above construction guarantees that each of the first $\varphi_{i,t}(q_{e,i,t})$ polynomials in $F_t[X]$ splits over F_t if and only if it splits over F.*

Proof. In the case that $\Phi_e^{S_F \upharpoonright \varphi_i(q_{e,i})}(q_{e,i}) \downarrow = 0$, the element $\beta_{e,i}$ will get enumerated into D, causing the whole field $\mathbb{Q}(\beta_{e,i})$ to go into F, but this will not make any difference to the current splitting set, as a result of Corollary 2. And, regardless of whether $\beta_{e,i}$ went into the field, the construction ensures that no lower-priority requirement will ever be allowed to put anything into the field that would make any of these first $\varphi_i(q_{e,i})$ polynomials split. No polynomial within this bound that hasn't split by stage t will ever split. So none of the first $\varphi_i(q_{e,i})$ polynomials in $F_t[X]$ will ever be caused to split after stage t.   □

Lemma 9. *The requirement $\mathcal{R}_{e,i}$ is satisfied for every $\langle e, i \rangle$.*

Proof. We showed in Lemma 7 that each requirement $\mathcal{R}_{e,i}$ acts only finitely often. Now to show that this last action by $\mathcal{R}_{e,i}$ actually does result in his satisfaction: In the case that $\varphi_i(q_{e,i})$ never halts, φ_i is not a total function, and so the hypothesis of the requirement is false. In the case that $\varphi_i(q_{e,i})$ halts but $\Phi_e^{S_F \upharpoonright \varphi_i(q_{e,i})}(q_{e,i})$ never halts, then $\Phi_e^{S_F \upharpoonright \varphi_i(q_{e,i})}$ is not a total function, and so again the hypothesis of the requirement is false. In the case that both functions converge on $q_{e,i}$, the construction ensures that $\Phi_e^{S_F \upharpoonright \varphi_i(q_{e,i})}(q_{e,i}) \downarrow = 0$ iff $q_{e,i} \in R_F$. We never end up with $\Phi_e^{S_F \upharpoonright \varphi_i(q_{e,i})}(q_{e,i}) \downarrow = 0$ and $q_{e,i} \notin R_F$, because if the convergence to zero happens, we force $q_{e,i}$ into R_F, and this doesn't change the convergence to zero, as explained in the proof of Lemma 8, because no requirement will ever put $q_{e,i}$ into R_F. We also never end up with $\Phi_e^{S_F \upharpoonright \varphi_i(q_{e,i})}(q_{e,i}) \downarrow = 1$ and $q_{e,i} \in R_F$, because we don't put $q_{e,i}$ into R_F unless we have already seen that $\Phi_e^{S_F \upharpoonright \varphi_i(q_{e,i})}(q_{e,i}) \downarrow = 0$. Having the lower-priority requirements respect the restraint set by $E_{e,i}$ ensures this.   □

4   Differentiating between R_F and S_F

As promised at the end of Section 2, we will now consider the (false) analogue to Lemma 6 and describe the wtt-reduction $R_F \leq_{wtt} S_F$ that must always exist if this analogue were true.

Lemma 10. *Let F be a computable algebraic field which is normal over the rationals, and suppose the following property holds:*

$$\text{For any fields } F_0 \text{ and } F_1, \text{ with } \mathbb{Q} \subsetneq F_0, F_1 \subsetneq F,$$

$$(\forall q(X) \in \mathbb{Q}[X]) \, [q \text{ factors over } F_0 \iff q \text{ factors over } F_1]$$

$$\text{if and only if}$$

$$(\exists \sigma \in Gal(F/\mathbb{Q})) \, [\sigma(F_0) = F_1].$$

Then $R_F \leq_{wtt} S_F$.

Proof. In particular, for every pair of subfields F_0 and F_1 of F with $\mathbb{Q} \subseteq F_0 \subsetneq F_1 \subseteq F$, we are assuming that there is a polynomial with rational coefficients that factors over F_1 but not over F_0.

Given a polynomial $q(X) \in F[X]$ that is irreducible over $\mathbb{Q}$, we want to know whether q has a root in F. If we happen to start out with a polynomial reducible in $\mathbb{Q}$, then we would factor it into its irreducible factors in $\mathbb{Q}[X]$, which we can do because $\mathbb{Q}$ has a splitting algorithm, as we saw in Lemma 5, and then continue as follows with each factor.

q must be a factor of some polynomial p with rational coefficients. p has a splitting field over $\mathbb{Q}$. Call this splitting field P.

We note here that to find the roots of q in F, we only need to find all roots of p in F, since once we have the roots of p in F, we can easily check, by substituting each one into q, which of the roots of p are also roots of q.

There are finitely many subfields between $\mathbb{Q}$ and P. For every pair of subfields $E_0 \subsetneq E_1$ of P, we can find a polynomial with rational coefficients that factors over E_1 but not E_0. So we make a list of pairs of these subfields where E_1 is a minimal extension of E_0, and next to each pair, we write down a polynomial in $\mathbb{Q}[X]$ that splits over E_1 but not E_0.

So we have this list M of finitely many polynomials. The claim is that if we have an oracle for the splitting set S_F of F restricted to the smallest initial segment of $F[X]$ that contains M (which is a bound that is computable uniformly in p), we can decide whether p has a root in F.

Let E be a subfield of P generated by a single root of p. Then $E_i = \sigma_i(E)$ are exactly the subfields generated by single roots of p, where $\sigma_i \in \mathrm{Gal}(P/\mathbb{Q})$.

Start with the minimal extensions of $\mathbb{Q}$ in P. For some minimal extension K_1 of $\mathbb{Q}$, if q_1 is the polynomial in M that splits over K_1 but not over $\mathbb{Q}$, then we check whether q_1 splits over F. If so, then there is $\sigma_1 \in \mathrm{Gal}(P/\mathbb{Q})$ for which $\sigma_1(K_1) \subseteq F$. (If not, then move onto another minimal extension of $\mathbb{Q}$.)

Then move to the minimal extensions of K_1. For some minimal extension K_2 of K_1, if q_2 is the polynomial in M that splits over K_2 but not over K_1, then we check whether q_2 splits over F. If so, then there is $\sigma_2 \in \mathrm{Gal}(P/\mathbb{Q})$ such that $\sigma_2 \upharpoonright K_1 = \sigma_1 \upharpoonright K_1$ and $\sigma_2(K_2) \subseteq F$.

We continue this way, until either we reach the splitting field P, or we reach some subfield K of P where there are no minimal extensions of K for which the corresponding polynomial in M splits over F. In either case, suppose $K = \sigma_n(\ldots \sigma_2(\sigma_1(K_1))\ldots)$. If K contains one of the subfields E_i (which we can check because we know exactly which polynomials in M split over E and which don't), then we will know whether F contains a root of p. F contains a root of p iff K contains an E_i.

Now, $K = P \cap F$, and so the roots of p in F are precisely the roots of p in K, and we can effectively find all the roots of p in K, since K is a finite algebraic extension of $\mathbb{Q}$ and thus has a splitting algorithm by Lemma 5. And once we have all the roots of p in F, we check which ones of these are also roots of q. □

References

1. Dummit, D.S., Foote, R.M.: Abstract Algebra, 3rd edn. John Wiley & Sons Inc., Hoboken (2004)
2. Fried, M.D., Jarden, M.: Field Arithmetic. Springer, Berlin (1986)
3. Frohlich, A., Shepherdson, J.C.: Effective procedures in field theory. Phil. Trans. Royal Soc. London, Series A 248, 950, 407–432 (1956)
4. Kronecker, L.: Grundzüge einer arithmetischen Theorie der algebraischen Größen. J. f. Math. 92, 1–122 (1882)
5. Miller, R.G.: Is it harder to factor a polynomial or to find a root? The Transactions of the American Mathematical Society (to appear)
6. Miller, R.G.: d-Computable categoricity for algebraic fields. The Journal of Symbolic Logic (to appear)
7. Rabin, M.: Computable algebra, general theory, and theory of computable fields. Transactions of the American Mathematical Society 95, 341–360 (1960)
8. Soare, R.I.: Recursively Enumerable Sets and Degrees. Springer, New York (1987)
9. van der Waerden, B.L.: Algebra. trans. F. Blum & J.R. Schulenberger I. Springer, New York (1970 hardcover, 2003 softcover)
10. Washington, L.C.: Introduction to Cyclotomic Fields. Springer, New York (1982)

What Is the Problem with Proof Nets for Classical Logic?

Lutz Straßburger

INRIA Saclay – Île-de-France and LIX, École Polytechnique
http://www.lix.polytechnique.fr/~lutz

Abstract. This paper is an informal (and nonexhaustive) overview over some existing notions of proof nets for classical logic, and gives some hints why they might be considered to be unsatisfactory.

1 Introduction

There is a very close and well-understood relationship between proofs in intuitionistic logic, simply typed lambda-terms, and morphisms in Cartesian closed categories. The same relationship can be established for multiplicative linear logic (MLL), where proof nets take the role of the lambda-terms, and star-autonomous categories the role of Cartesian closed categories.

It is certainly desirable to have something similar for classical logic, which can be obtained from intuitionistic logic by adding the law of excluded middle, i.e., $A \vee \bar{A}$, or equivalently, an involutive negation, i.e., $\bar{\bar{A}} = A$. Adding this to a Cartesian closed category $\mathscr{C}$, means adding a contravariant functor $\overline{(-)} \colon \mathscr{C} \to \mathscr{C}$ such that $\bar{\bar{A}} \cong A$ and $\overline{(A \wedge B)} \cong \bar{A} \vee \bar{B}$ where $A \vee B = \bar{A} \Rightarrow B$. However, if we do this we get a collapse: all proofs of the same formula are identified, which leads to a rather boring proof theory. This observation is due to André Joyal, and a proof and discussion can be found in [1,2,3].

Here we will not show the category theoretic proof of the collapse, but will quickly explain the phenomenon in terms of the sequent calculus (the argumentation is due to Yves Lafont [4, Appendix B]). Suppose we have two proofs Π_1 and Π_2 of a formula B in some sequent calculus system. Then we can form, with the help of the rules weakening, contraction, and cut, the following proof of B:

$$\text{cut}\ \dfrac{\text{wk}\ \dfrac{\overset{\displaystyle \Pi_1}{\vdash B}}{\vdash B, A} \qquad \text{wk}\ \dfrac{\overset{\displaystyle \Pi_2}{\vdash B}}{\vdash \bar{A}, B}}{\text{cont}\ \dfrac{\vdash B, B}{\vdash B}} \tag{1}$$

F. Ferreira et al. (Eds.): CiE 2010, LNCS 6158, pp. 406–416, 2010.

If we apply cut elimination to this proof, we get either

$$
\text{cont} \cfrac{\text{wk} \cfrac{\overset{\displaystyle \Pi_1}{\vdash B}}{\vdash B, B}}{\vdash B} \qquad \text{or} \qquad \text{cont} \cfrac{\text{wk} \cfrac{\overset{\displaystyle \Pi_2}{\vdash B}}{\vdash B, B}}{\vdash B} \tag{2}
$$

depending on a nondeterministic choice. On the other hand, if we want the nice relationship between deductive system and category theory, we need a confluent cut elimination, which means that the two proofs in (2) must be the same. Consequently, we have to equate Π_1 and Π_2. Since there was no initial condition on Π_1 and Π_2, we conclude that any two proofs of B must be equal.

The problem with weakening, which could in fact be solved by using the mix-rule

$$
\text{mix} \cfrac{\vdash \Gamma \quad \vdash \Delta}{\vdash \Gamma, \Delta} \quad , \tag{3}
$$

is not the only one. We run into similar problems with the contraction rule. If we try to eliminate the cut from

$$
\text{cut} \cfrac{\text{cont} \cfrac{\overset{\displaystyle \Pi_1}{\vdash \Gamma, A, A}}{\vdash \Gamma, A} \quad \text{cont} \cfrac{\overset{\displaystyle \Pi_2}{\vdash \bar{A}, \bar{A}, \Delta}}{\vdash \bar{A}, \Delta}}{\vdash \Gamma, \Delta} \quad . \tag{4}
$$

we again have to make a nondeterministic choice. In Section 2, we will see a concrete example for this.

There are several possibilities to cope with these problems. Clearly, we have to drop some of the equations that we would like to hold between proofs in classical logic. But which ones should go?

There are now essentially three different approaches, and all three have their advantages and disadvantages.

1. The first says that the axioms of Cartesian closed categories are essential and cannot be dispensed with. Instead, one sacrifices the duality between $\land$ and $\lor$. The motivation for this approach is that a proof system for classical logic can now be seen as an extension of the λ-calculus and the notion of normalization does not change. One has a term calculus for proofs, namely Parigot's $\lambda\mu$-calculus [5] and a denotational semantics [2]. An important aspect is the computational meaning in terms of continuations [6,7]. There is a well explored category theoretical axiomatization [8], and, of course, a theory of proof nets [9], which is based on the proof nets for multiplicative exponential linear logic (MELL).

2. The second approach considers the perfect symmetry between $\land$ and $\lor$ to be an essential facet of Boolean logic, that cannot be dispensed with. Consequently, the axioms of Cartesian closed categories and the close relation to the λ-calculus have to be sacrificed. More precisely, the conjunction $\land$ is no longer a Cartesian product, but merely a tensor-product. Thus, the Cartesian

closed structure is replaced by a star-autonomous structure, as it it known from linear logic. However, the precise category theoretical axiomatization is much less clear than in the first approach (see [10,11,12,3,13]).

3. The third approach keeps the perfect symmetry between ∧ and ∨, as well as the Cartesian product property for ∧. What has to be dropped is the property of being closed, i.e., there is no longer a bijection between the proofs of

$$A \vdash B \Rightarrow C \qquad \text{and} \qquad A \wedge B \vdash C \quad ,$$

which means we lose currying. This approach is studied in [14].

In this paper, we only focus on the second approach. The various notions of proof nets that fall in this setting can be grouped into two different *ideologies*:

Sequent Rule Ideology: A proof net is a graph in which every vertex represents an inference rule application in the corresponding sequent calculus proof, and every edge of the graph stands for a formula appearing in the proof. A sequent calculus proof with conclusion $\vdash A_1, A_2, \ldots, A_n$, written as

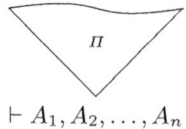

$$\vdash A_1, A_2, \ldots, A_n$$

is translated into a proof net with conclusions $A_1, A_2, \ldots, A_n$, written as

$$\boxed{\pi}$$
$$A_1 \quad A_3 \quad \cdots \quad A_n$$

Flow Graph Ideology: A proof net consists of the formula tree/sequent forest of the conclusion of the proof, together with some additional graph structure capturing the "essence" of the proof (whatever that means).

It should be observed that for multiplicative linear logic (MLL) the two ideologies produce the same notion of proof nets. However, for classical logic the situation is very different.

2   Sequent Calculus Rule Based Proof Nets

The set of *formulas* is defined via

$$\mathscr{F} ::= \mathscr{A} \mid \bar{\mathscr{A}} \mid \mathscr{F} \vee \mathscr{F} \mid \mathscr{F} \wedge \mathscr{F} \tag{5}$$

where $\mathscr{A} = \{a, b, c, \ldots\}$ is a countable set of propositional variables, and $\bar{\mathscr{A}} = \{\bar{a}, \bar{b}, \bar{c}, \ldots\}$ are their duals. In the following, the elements of the set $\mathscr{A} \cup \bar{\mathscr{A}}$ are called *atoms*. The negation ¯ is defined via DeMorgan laws for all formulas. For simplicity, we ignore the units and push negation to the atoms. There are various different sequent calculus systems for classical propositional logic, starting with the one by Gentzen [15]. Figure 1 shows on the top (first and third line) the one we use here (for simplicity, we use a one-sided system). As indicated by Girard [2], and detailed out by Robinson [16], any sequent proof Π can (inductively) be translated into a proof net π, as shown on the bottom (second and forth line) of Figure 1. A proof net is then a graph whose vertices are the rule instances of Π, and whose edges are labeled by the principal formulas of the rules. For simplicity, we consider the outputs to be unordered and ignore the exchange rule. Here is an example of a sequent calculus proof

$$
\begin{array}{c}
\cfrac{
  \cfrac{
    \text{id} \cfrac{}{\vdash \bar{b}, b} \quad
    \text{id} \cfrac{}{\vdash a, \bar{a}}
  }{
    \wedge \cfrac{}{\vdash \bar{b} \wedge a, \bar{a}, b}
  } \quad
  \text{id} \cfrac{}{\vdash \bar{b}, b}
}{
  \wedge \cfrac{}{
    \text{cont} \cfrac{\vdash \bar{b} \wedge a, \bar{a} \wedge \bar{b}, b, b}{\vdash \bar{b} \wedge a, \bar{a} \wedge \bar{b}, b}
  }
}
\quad
\cfrac{
  \text{id} \cfrac{}{\vdash \bar{b}, b} \quad
  \cfrac{
    \text{id} \cfrac{}{\vdash a, \bar{a}} \quad
    \text{id} \cfrac{}{\vdash \bar{b}, b}
  }{
    \wedge \cfrac{}{\vdash \bar{b}, a, \bar{a} \wedge b}
  }
}{
  \wedge \cfrac{}{
    \text{cont} \cfrac{\vdash \bar{b}, \bar{b}, b \wedge a, \bar{a} \wedge b}{\vdash \bar{b}, b \wedge a, \bar{a} \wedge b}
  }
}
}{
\text{cut} \cfrac{}{\vdash \bar{b} \wedge a, \bar{a} \wedge \bar{b}, b \wedge a, \bar{a} \wedge b}
}
\tag{6}
$$

and its translation into a proof net:

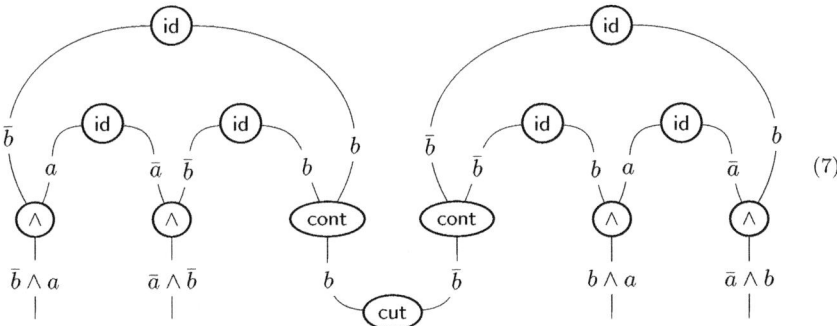

$$\tag{7}$$

The advantage of the sequent-rule-ideology is that all the correctness criteria for MLL proof nets hold unchanged. For example, the switching criterion [17]: A (well-formed) proof net is the translation of a sequent proof if and only if each of its switchings is a connected and acyclic graph, where a switching of a proof net π is a graph that is obtained from π by removing for each $\vee$-node and each cont-node one of the two edges connecting it to its children.

There are however two main disadvantages of the sequent-rule-ideology. The first is that certain proofs are distinguished that should be identified according the rule-permutability-argument. To see a very simple example, consider the following three sequent calculus proofs:

$$
\cfrac{
  \text{wk} \cfrac{\text{id} \cfrac{}{\vdash \bar{a}, a}}{\vdash c, \bar{a}, a} \quad
  \text{id} \cfrac{}{\vdash b, \bar{b}}
}{
  \wedge \cfrac{}{\vdash c, \bar{a}, a \wedge b, \bar{b}}
}
\qquad
\cfrac{
  \cfrac{
    \text{id} \cfrac{}{\vdash \bar{a}, a} \quad
    \text{id} \cfrac{}{\vdash b, \bar{b}}
  }{\wedge \cfrac{}{\vdash \bar{a}, a \wedge b, \bar{b}}}
}{
  \text{wk} \cfrac{}{\vdash c, \bar{a}, a \wedge b, \bar{b}}
}
\qquad
\cfrac{
  \text{id} \cfrac{}{\vdash \bar{a}, a} \quad
  \text{wk} \cfrac{\text{id} \cfrac{}{\vdash b, \bar{b}}}{\vdash c, b, \bar{b}}
}{
  \wedge \cfrac{}{\vdash c, \bar{a}, a \wedge b, \bar{b}}
}
\tag{8}
$$

They differ from each other only via some trivial rule permutation, and should therefore be identified. But they can be translated into five different proof nets.

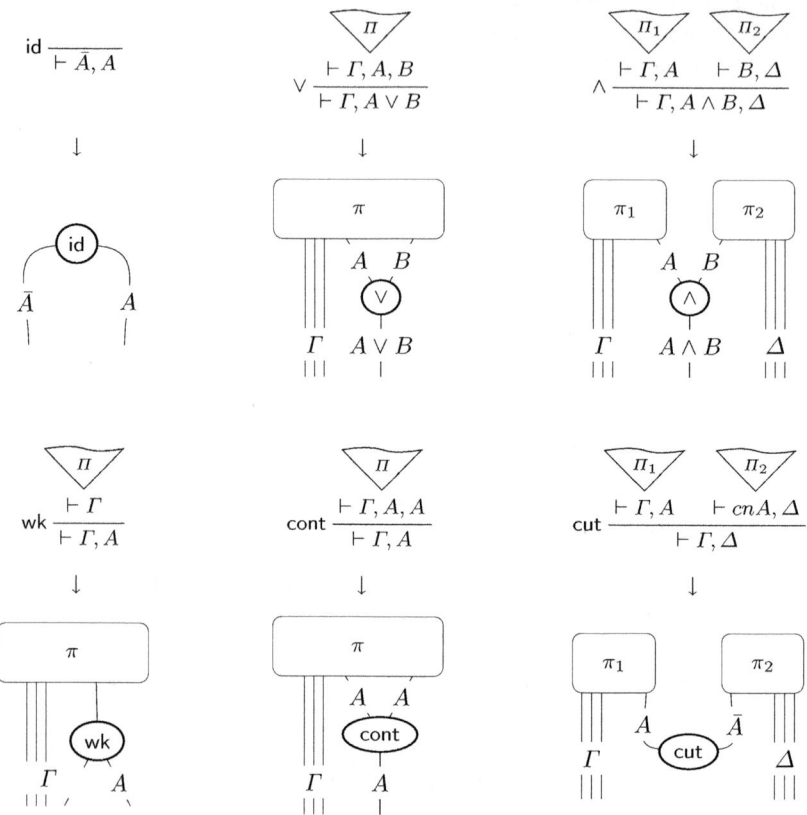

Fig. 1. From sequent calculus to proof nets (sequent rule ideology)

Two of them are shown below:

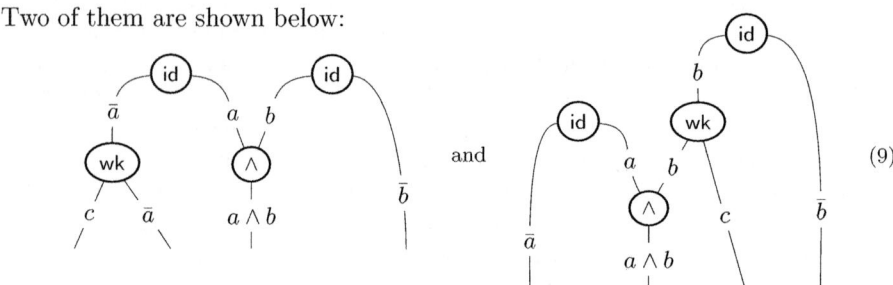

and   (9)

The problem is that there is no canonical choice of where to attach the weakening. A possible solution could be to leave the weakenings unconnected, but this would break the correctness criteria.

The second disadvantage of the sequent-rule-ideology is related to cut elimination. In the introduction we have seen already the problem with weakening. Let us now have a closer look at contraction, when it appears at both sides of a cut, as shown in the example in (6) and (7). For typesetting reasons, let us use

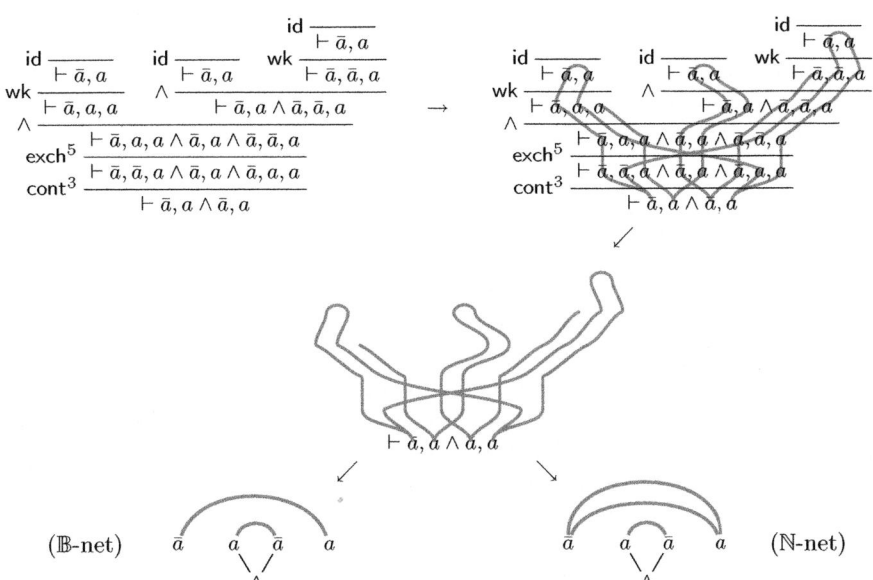

Fig. 2. From sequent calculus to proof nets (flow graph ideology)

the more compact notation:

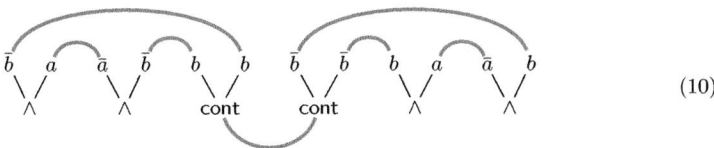

$$\tag{10}$$

We have here an example for the general case in (4). If we want to eliminate the cut from (10), we have to make a nondeterministic choice, which subproof we duplicate. As outcome we get either

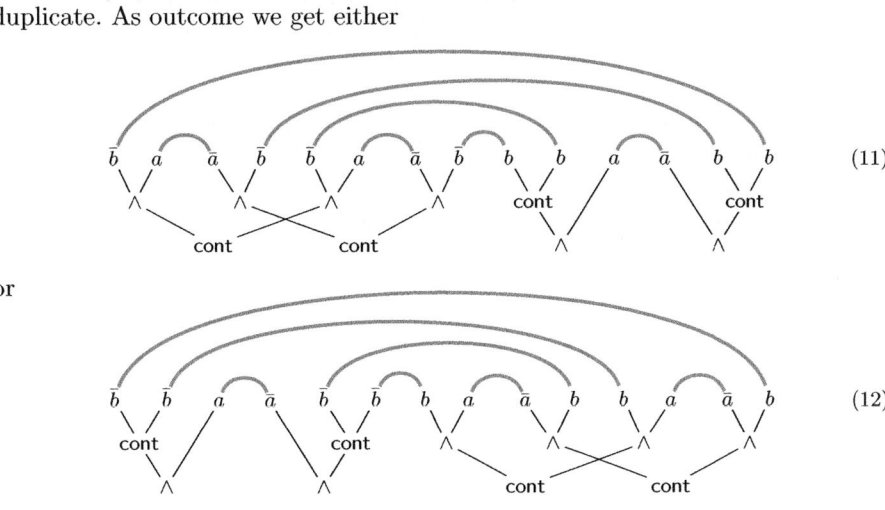

$$\tag{11}$$

or

$$\tag{12}$$

In [2], Girard argues that for this reason it is impossible to have a confluent notion of cut elimination for proof nets for classical logic. Of course, his argumentation is valid only for proof nets following the sequent-rule-ideology.

Thus, for changing the situation with cut elimination, one has to change the ideology.

3   Flow Graph Based Proof Nets

The basic idea is to draw the flow graph [18] of the proof as indicated in Figure 2. The important question is what information should be kept. In [19] there are two proposals. The first takes the sequent forest and adds an edge between a pair of dual atoms if they are connected by a path in the flow graph. This yields the $\mathbb{B}$-nets of [19], and an example is shown on the lower left of Figure 2. The second approach also keeps the number of paths between two atoms. The result is called $\mathbb{N}$-nets in [19]. This is denoted by either labeling the edge between the two atoms by a natural number, or by drawing multiple edges, as shown on the lower right of Figure 2. In both cases it can happen that some atoms have no mate, i.e., live celibate, and that some atoms have more than one mate, i.e., live polygamous. This is the main difference to MLL proof nets, where every single atom lives monogamous.

Cuts are shown as special edges between the roots of the formula trees, as in this example, which is obtained from the sequent proof in (6):

$$\tag{13}$$

The disadvantage of the flow-graph-ideology is that the correctness criteria from linear logic are no longer available. However, for $\mathbb{B}$-nets there is a correctness criterion that is similar to the criterion for matings [20] and matrix proofs [21]: A $\mathbb{B}$-net is the translation of a sequent proof if and only if each of its conjunctive prunings contains at least one axiom link edge, where a conjunctive pruning for π is obtained from π by deleting for each of its $\wedge$-nodes one of the two subtrees including the outgoing axiom link edges.

The main problem with of this criterium is that checking it takes exponential time in the size of the input. This means that checking a given proof is as expensive as finding the proof from scratch.

Furthermore, this criterion does not work for $\mathbb{N}$-nets because it does not take into account how often an axiom link edge is used, and it is an open problem to find some correctness criterion for $\mathbb{N}$-nets.

Let us now look at cut elimination. Reducing cuts on compound formulas is exactly the same as in the previous section:

$$\tag{14}$$

For the cut reduction on atomic cuts, we have to be careful, since the atoms can be connected to many other atoms (or no other atoms). Instead of simply having:

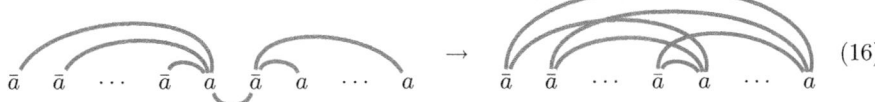
$$\bar{a} \quad a \quad \bar{a} \quad a \qquad \rightarrow \qquad \bar{a} \quad a \tag{15}$$

as in MLL, the reduction looks as follows:

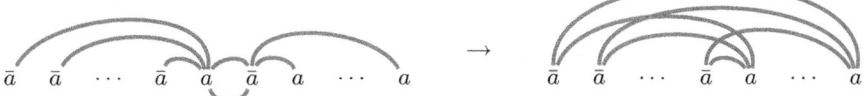

$$\tag{16}$$

If one of the two cut atoms is celibate, no link remains:

$$\tag{17}$$

If the two cut atoms are linked together, then this link is ignored in the reduction (and, of course, removed with the cut):

We certainly have termination of the cut reduction. The interesting observation is that for $\mathbb{B}$-nets, the cut reduction preserves correctness and is confluent.

The natural question that arises now is: How does this confluent cut elimination relate to the non-confluent cut elimination in the sequent calculus?

Let us look again at the two problematic cases (1) and (4). The problem with weakening (1) can easily be solved by using the mix-rule in the sequent calculus:

$$\text{cut} \dfrac{\text{wk} \dfrac{\vdash \Gamma}{\vdash \Gamma, A} \quad \text{wk} \dfrac{\vdash \Delta}{\vdash \bar{A}, \Delta}}{\vdash \Gamma, \Delta} \qquad \rightarrow \qquad \text{mix} \dfrac{\vdash \Gamma \quad \vdash \Delta}{\vdash \Gamma, \Delta}$$

Both subproofs Π_1 and Π_2 are kept in the reduced net, and in $\mathbb{B}$-nets and $\mathbb{N}$-nets it is done in the same way.

For the contraction case (4) the situation is less obvious. Consider again the simple proof net in (13), which corresponds to the sequent calculus proof in (6). If we apply the cut reduction (16), we obtain the following result:

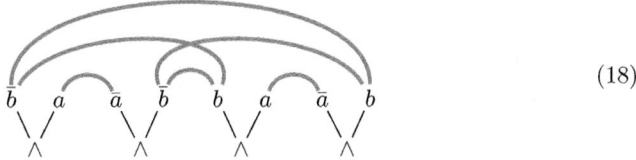

$$\tag{18}$$

which is exactly the $\mathbb{B}$-net obtained from the sequent proofs corresponding to (11) and (12). This correspondence makes crucial use of the fact that we deliberately forget how often an identity link is used in the proof. As $\mathbb{N}$-nets, the proofs in (11) and (12) would be represented by

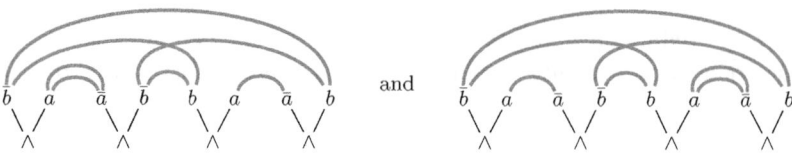

respectively (see [19] for further details). However, although it is not possible to have (18) as N-net of a sequent proof, it can be obtained as a N-net of a proof in the calculus of structures [22], more precisely in system SKS, as presented in [23]:

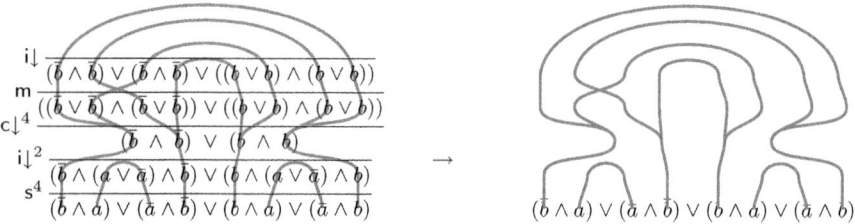

This means that any correctness criterion for N-nets must depend on the chosen deductive system.

The non-confluence of cut-reduction for N-nets has the following reason. When we reduce an atomic cut, we have to multiply the number of edges, and if there are already some links between the remaining pair of atoms, then these links have to be added. For example

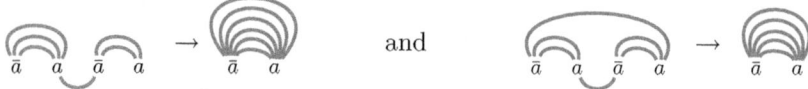

Consider now the following example:

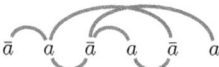

Depending on which cut we reduce first, we get either

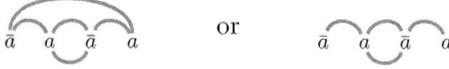

If we reduce the remaining cut, we get

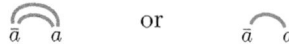

respectively. The solution for circumventing this problem is to reduce atomic cuts only in unproblematic situations like (15) and (17), and leave all atomic cuts like (16) unreduced, as it is done for C-nets in [24]. C-nets are a variant of N-nets that are considered cut-free if they contain only atomic cuts that touch a contraction on both sides. In this way C-nets can also capture the size of a proof, because the reduction (16) is the only one which causes an exponential blow-up of the proof. C-nets can also be used as coherence graphs for SKS-derivations. The same approach is taken by the recently developed atomic flows [25].

References

1. Lambek, J., Scott, P.J.: Introduction to higher order categorical logic. In: Cambridge studies in advanced mathematics, vol. 7. Cambridge University Press, Cambridge (1986)
2. Girard, J.Y.: A new constructive logic: Classical logic. Math. Structures in Comp. Science 1, 255–296 (1991)
3. Straßburger, L.: On the axiomatisation of Boolean categories with and without medial. Theory and Applications of Categories 18(18), 536–601 (2007)
4. Girard, J.Y., Lafont, Y., Taylor, P.: Proofs and Types. In: Cambridge Tracts in Theoretical Computer Science. Cambridge University Press, Cambridge (1989)
5. Parigot, M.: $\lambda\mu$-calculus: An algorithmic interpretation of classical natural deduction. In: Voronkov, A. (ed.) LPAR 1992. LNCS (LNAI), vol. 624, pp. 190–201. Springer, Heidelberg (1992)
6. Thielecke, H.: Categorical Structure of Continuation Passing Style. PhD thesis, University of Edinburgh (1997)
7. Streicher, T., Reus, B.: Classical logic, continuation semantics and abstract machines. J. of Functional Programming 8(6), 543–572 (1998)
8. Selinger, P.: Control categories and duality: on the categorical semantics of the lambda-mu calculus. Math. Structures in Comp. Science 11, 207–260 (2001)
9. Laurent, O.: Polarized proof-nets and $\lambda\mu$-calculus. Theoretical Computer Science 290(1), 161–188 (2003)
10. Führmann, C., Pym, D.: Order-enriched categorical models of the classical sequent calculus. J. of Pure and Applied Algebra (2004) (to appear)
11. Lamarche, F., Straßburger, L.: Constructing free Boolean categories. In: LICS 2005, pp. 209–218 (2005)
12. McKinley, R.: Classical categories and deep inference. In: Structures and Deduction 2005 Satellite Workshop of ICALP 2005 (2005)
13. Lamarche, F.: Exploring the gap between linear and classical logic. Theory and Applications of Categories 18(18), 473–535 (2007)
14. Došen, K., Petrić, Z.: Proof-Theoretical Coherence. KCL Publ., London (2004)
15. Gentzen, G.: Untersuchungen über das logische Schließen. I. Mathematische Zeitschrift 39, 176–210 (1934)
16. Robinson, E.P.: Proof nets for classical logic. Journal of Logic and Computation 13, 777–797 (2003)
17. Danos, V., Regnier, L.: The structure of multiplicatives. Annals of Mathematical Logic 28, 181–203 (1989)
18. Buss, S.R.: The undecidability of k-provability. Annals of Pure and Applied Logic 53, 72–102 (1991)
19. Lamarche, F., Straßburger, L.: Naming proofs in classical propositional logic. In: Urzyczyn, P. (ed.) TLCA 2005. LNCS, vol. 3461, pp. 246–261. Springer, Heidelberg (2005)
20. Andrews, P.B.: Refutations by matings. IEEE Transactions on Computers C-25, 801–807 (1976)
21. Bibel, W.: On matrices with connections. Journal of the ACM 28, 633–645 (1981)
22. Guglielmi, A., Straßburger, L.: Non-commutativity and MELL in the calculus of structures. In: Fribourg, L. (ed.) CSL 2001 and EACSL 2001. LNCS, vol. 2142, pp. 54–68. Springer, Heidelberg (2001)

23. Brünnler, K., Tiu, A.F.: A local system for classical logic. In: Nieuwenhuis, R., Voronkov, A. (eds.) LPAR 2001. LNCS (LNAI), vol. 2250, pp. 347–361. Springer, Heidelberg (2001)
24. Straßburger, L.: From deep inference to proof nets via cut elimination. Journal of Logic and Computation (2009) (To appear)
25. Guglielmi, A., Gundersen, T.: Normalisation control in deep inference via atomic flows. Logical Methods in Computer Science 4(1-9), 1–36 (2008)

Quasi-linear Dialectica Extraction

Trifon Trifonov*

Faculty of Mathematics and Informatics, Sofia University St. Kliment Ohridski
triffon@fmi.uni-sofia.bg

Abstract. Gödel's functional interpretation [1] can be used to extract programs from non-constructive proofs. Though correct by construction, the obtained terms can be computationally inefficient. One reason for slow execution is the re-evaluation of equal subterms due to the use of substitution during the extraction process. In the present paper we define a variant of the interpretation, which avoids subterm repetition and achieves an almost linear bound on the size of extracted programs.

1 Introduction

The idea for extraction of programs from mathematical proofs has received a lot of renewed interest in the recent years. Beside the theoretical investigations related to analysing computational content of proofs, which have yielded new results in various areas of mathematics (see e.g. [4]), there is an ongoing debate whether obtaining correct programs by proof interpretations can be a practical alternative of program verification techniques. One of the reasons for bringing into the spotlight such proof-theoretic techniques, dating back to the middle of the previous century, is the availability of hardware and software, which facilitate experiments with non-trivial operations on large proof objects.

Proofs in constructive logic correspond to functional programs in a natural and direct way (the Curry-Howard isomorphism). On the other hand, methods for extraction from non-constructive proofs need additional computational infrastructure to express the computational meaning hidden in the mathematical argument. In the case of Gödel's functional "Dialectica" interpretation [1], there are two layers of realisers — positive and negative, and if their traditional definitions are implemented directly, unnecessarily inefficient (though correct) programs might be obtained. In the present paper we identify one such shortcoming and show that it can be generally repaired by considering a variant of the Dialectica interpretation, which yields shorter and possibly faster programs.

2 Systems and Notations

We work in a restriction of Heyting Arithmetic with finite types (denoted HA^ω in [8]) to the language of $\rightarrow$ and $\forall$. We refer to the resulting system as Negative Arithmetic (NA^ω).

* The author gratefully acknowledges financial support by the Bulgarian National Science Fund within project DO 02-102/23.04.2009 and by the European Social Fund within project BG 051PO001-3.3.04/28.08.2009.

F. Ferreira et al. (Eds.): CiE 2010, LNCS 6158, pp. 417–426, 2010.

Definition 1. *Types* (ρ, σ), *(object) terms* (s, t) *and formulas* (A, B) *are defined as follows:*

$$\rho, \sigma \quad ::= \quad \mathsf{B} \mid \mathsf{N} \mid \alpha \mid \rho \Rightarrow \sigma \mid \rho \times \sigma$$

$$s, t \quad ::= \quad x^\rho \mid (\lambda_{x^\rho} t^\sigma)^{\rho \Rightarrow \sigma} \mid (s^{\rho \Rightarrow \sigma} t^\rho)^\sigma \mid \langle s^\rho, t^\sigma \rangle^{\rho \times \sigma} \mid (t^{\rho \times \sigma} {}_{\llcorner})^\rho \mid (t^{\rho \times \sigma} {}_{\lrcorner})^\sigma \mid$$
$$\mathsf{tt}^\mathsf{B} \mid \mathsf{ff}^\mathsf{B} \mid \mathsf{0}^\mathsf{N} \mid \mathsf{S}^{\mathsf{N} \Rightarrow \mathsf{N}} \mid \mathcal{C}^{\mathsf{B} \Rightarrow \sigma \Rightarrow \sigma \Rightarrow \sigma} \mid \mathcal{R}^{\mathsf{N} \Rightarrow \sigma \Rightarrow (\mathsf{N} \Rightarrow \sigma \Rightarrow \sigma) \Rightarrow \sigma}$$

$$A, B \quad ::= \quad \mathsf{at}(t^\mathsf{B}) \mid A \rightarrow B \mid \forall_{x^\rho} A$$

The base types of booleans B *and natural numbers* N *are equipped with the usual constructors and structural recursor constants. Here* x *denotes a typed object variable and* α *denotes a type variable. Freely occurring type variables allow for a restricted form of polymorphism. The sets of free variables* $\mathsf{FV}(t)$, $\mathsf{FV}(A)$ *and bound variables* $\mathsf{BV}(t)$, $\mathsf{BV}(A)$ *are defined inductively as usual. Substitution of terms for object variables* $s[x := t]$, $A[x := t]$ *is by default assumed to be capture-free with respect to abstraction and quantification.*

The operational semantics of object terms are given by the usual β-reduction rules and computation rules for the recursor constants:

$$\langle s, t \rangle {}_{\llcorner} \xmapsto{r} s \qquad \mathcal{C} \,\mathsf{tt}\, t_1 \, t_2 \xmapsto{r} t_1 \qquad \mathcal{R} \, 0 \, s \, t \xmapsto{r} s$$
$$\langle s, t \rangle {}_{\lrcorner} \xmapsto{r} t \qquad \mathcal{C} \,\mathsf{ff}\, t_1 \, t_2 \xmapsto{r} t_2 \qquad \mathcal{R} \, (\mathsf{S}n) \, s \, t \xmapsto{r} t \, n \, (\mathcal{R} \, n \, s \, t)$$
$$(\lambda_x s) t \xmapsto{r} s[x := t]$$

We will make use of the following "let" notation for a β-redex:

$$\mathbf{let}\ x := t\ \mathbf{in}\ s \quad := \quad (\lambda_x s) t.$$

We express derivations in a natural deduction system with a similar syntax to that of object terms to stress the Curry-Howard correspondence. Proof terms are typed by their conclusion formulas and are built from *assumption variables*.

Definition 2. *Proof terms* (M, N) *of* NA^ω *are defined as follows:*

$$M, N \quad ::= \quad u^A \mid (\lambda_{u^A} M^B)^{A \rightarrow B} \mid (M^{A \rightarrow B} N^A)^B \mid$$
$$(*) \qquad (\lambda_{x^\rho} M^{A(x)})^{\forall_{x^\rho} A(x)} \mid (M^{\forall_{x^\rho} A(x)} t)^{A(t)} \mid$$
$$\mathsf{AxT} : \mathsf{at}(\mathsf{tt}) \mid \mathsf{Cases}^{A(b)} : \forall_{b^\mathsf{B}} \big(A(\mathsf{tt}) \rightarrow A(\mathsf{ff}) \rightarrow A(b)\big) \mid$$
$$\mathsf{Ind}^{A(n)} : \forall_{n^\mathsf{N}} \big(A(0) \rightarrow \forall_{n^\mathsf{N}} (A(n) \rightarrow A(\mathsf{S}n)) \rightarrow A(n)\big)$$

with the usual variable condition $(*)$ *that the object variable* x *does not occur freely in any of the open assumptions of* M. *The sets of free variables* $\mathsf{FV}(M)$ *and free (open) assumption variables* $\mathsf{FA}(M)$ *as well as capture-free substitutions* $M[x := t]$ *and* $M[u := N]$ *are defined inductively as usual.*

The *truth axiom* AxT defines the logical meaning of $\mathsf{at}(\cdot)$ and allows us to consider any boolean valued function defined in our term system as a decidable predicate.

When we write for example $n = m$, we actually mean at$(\mathrm{Eq}\, n\, m)$, where $\mathrm{Eq}^{\mathrm{N}\Rightarrow\mathrm{N}\Rightarrow\mathrm{B}}$ is a term defining the decidable equality for natural numbers.

In our negative language, defining falsity as $\mathrm{F} := \mathrm{at}(\mathrm{ff})$ already gives us the full power of classical logic. In particular, for a formula A we can prove *Ex falso quodlibet* (efq): $\vdash \mathrm{F} \to A$ and *Stability*: $\vdash ((A \to \mathrm{F}) \to \mathrm{F}) \to A$ by meta induction on A, using AxT and Cases for the base case. We will thus use the abbreviations $\neg A := A \to \mathrm{F}$ and $\tilde{\exists}_{x^\rho} A := \neg\forall_{x^\rho}\neg A$.

The term system in consideration is essentially Gödel's T and the reduction relation $\overset{r}{\mapsto}$ is well known to be strongly normalising and confluent. Thus, instead of insisting that object terms appearing in formulas of proof rules match exactly, we require them only to have the same η-long normal form; this equality will be denoted by $\overset{r}{=}$. Note that we make no such assumption for extracted programs or for proof terms themselves.

Notation. For technical convenience we will use ε for denoting a special *nulltype*. By abuse of notation we also use ε to denote all terms of nulltype. We stipulate that the following simplifications are always carried out implicitly:

$$
\begin{array}{llllll}
\rho \times \varepsilon \rightsquigarrow \rho, & t^{\rho \times \varepsilon}{}_{\mathsf{L}} \rightsquigarrow t, & \langle t, \varepsilon \rangle \rightsquigarrow t, & \rho \Rightarrow \varepsilon \rightsquigarrow \varepsilon, & \lambda_x \varepsilon \rightsquigarrow \varepsilon, & \varepsilon t \rightsquigarrow \varepsilon \\
\varepsilon \times \rho \rightsquigarrow \rho, & t^{\varepsilon \times \rho}{}_{\mathsf{J}} \rightsquigarrow t, & \langle \varepsilon, t \rangle \rightsquigarrow t, & \varepsilon \Rightarrow \rho \rightsquigarrow \rho, & \lambda_x \varepsilon t \rightsquigarrow t, & t \varepsilon \rightsquigarrow t \\
& & & \forall_{x^\varepsilon} A \rightsquigarrow A, & M\varepsilon \rightsquigarrow M
\end{array} \quad (\varepsilon)
$$

3   Dialectica Interpretation

Gödel's functional (Dialectica) interpretation [1] is an embedding of classical logic (expressed in a negative language) in a quantifier-free system employing functionals of higher finite types. In this section we define the interpretation for the system NA^ω, following the approach in [6].

Definition 3. *For a formula A we define the positive ($\tau^+(A)$) and negative ($\tau^-(A)$) computational types by simultaneous induction as follows:*

$$
\begin{array}{ll}
\tau^+(\mathrm{at}(b)) := \tau^-(\mathrm{at}(b)) := \varepsilon & \tau^-(A \to B) := \tau^+(A) \times \tau^-(B) \\
\tau^+(A \to B) := (\tau^+(A) \Rightarrow \tau^+(B)) \times (\tau^+(A) \Rightarrow \tau^-(B) \Rightarrow \tau^-(A)) \\
\tau^+(\forall_{x^\rho} A) := \rho \Rightarrow \tau^+(A) & \tau^-(\forall_{x^\rho} A) := \rho \times \tau^-(A)
\end{array}
$$

Definition 4. *For $r : \tau^+(A)$ and $s : \tau^-(A)$ we define the translation $|A|^r_s$:*

$$
|\mathrm{at}(b)| := \mathrm{at}(b), \qquad |\forall_{x^\rho} A(x)|^r_s := |A(s_{\mathsf{L}})|^{r(s_{\mathsf{L}})}_{s_{\mathsf{J}}}
$$

$$
|A \to B|^r_s := |A|^{s_{\mathsf{L}}}_{r_{\mathsf{J}}(s_{\mathsf{L}})(s_{\mathsf{J}})} \to |B|^{r_{\mathsf{L}}(s_{\mathsf{L}})}_{s_{\mathsf{J}}}
$$

Theorem 1 (Soundness theorem). *Let $\mathcal{P}$ be a proof in NA^ω of the formula A from assumptions $u_i : C_i$. Let $x_{u_i} : \tau^+(C_i)$ be fresh witnessing variables associated uniquely with the assumption variables u_i and let $y_A : \tau^-(A)$ be a fresh challenging variable associated uniquely with the formula A. Then there* are terms $[\![\mathcal{P}]\!]_i^- : \tau^-(C_i)$ and $[\![\mathcal{P}]\!]^+ : \tau^+(A)$, such that $y \notin \mathsf{FV}([\![\mathcal{P}]\!]^+)$ and $|A|_y^{[\![\mathcal{P}]\!]^+}$ is provable from assumptions $|C_i|_{[\![\mathcal{P}]\!]_i^-}^{x_i}$.

The soundness of the interpretation is proved essentially as in [8], but its formulation is adjusted to our natural deduction proof system. We will not present the proof of the soundness theorem here, for a complete proof the reader is referred to [6,7].

4 Recomputation

A noticeable defect of programs extracted by directly applying the Dialectica interpretation is the *repetition of equal subterms*. In the proof of the Soundness theorem the cause can be traced back to an asymmetry in the treatment of elimination rules for the conclusion and for the assumptions. For positive content elimination rules correspond to *application* to a term t, for negative content we *substitute* a challenge variable with a pair of the form $\langle t, y \rangle$. Naturally, the challenge variable could have multiple appearances and this can lead to multiplication of the substituted terms. In case we assume a strict evaluation strategy, we will inevitably have multiple recomputations of the same subterm.[1] The following example demonstrates how recomputation can lead to an exponential slowdown of the extracted program.

Example 1. Let us prove the totality of the function $x, y \mapsto 2^x(x + y)$, i.e., $\forall_{x,y}\tilde{\exists}_z(z = 2^x(x + y))$. A simple constructive proof goes by induction on x, taking $z := y$ for the base case. For the step case our induction hypothesis is $\forall_y\tilde{\exists}_z z = 2^x(x + y)$, so we fix y_0 and look for a $z_0 = 2^{x+1}(x + y_0 + 1)$. We use the induction hypothesis on $y_0 + 1$ to find a z and then take $z_0 := z + z$. Applying modified realisability [5] gives rise to the following program:

$$P := \lambda_x \mathcal{R}\, x\, (\lambda_y y)(\lambda_{x,p,y}\, (\lambda_{x,f} fx)(p(y + 1))(\lambda_z z + z))$$

The **let**-type construction $\lambda_{x,f} fx$ is the computational content of (strong) existence elimination and it is the reason that the number of strict evaluation steps of Pxy is linear in x.

The program extracted by the Dialectica interpretation from a formulation of the same proof in NA^ω would look rather differently:

$$P' := \lambda_x \mathcal{R}\, x\, (\lambda_y y)\, (\lambda_{x,p,y}\, (p(y + 1) + p(y + 1)))\,.$$

Clearly the application $P'xy$ needs an exponential number of steps to reduce because of the two recursive calls.

In the following section we introduce a variant of the Dialectica interpretation, which prevents such kind of recomputation by avoiding syntactic term repetition.

---

[1] This situation is not specific to natural deduction; substitution is also used in [8], for example for the axiom $Q2$, corresponding to $\forall$-elimination.

5 Dialectica Interpretation with Quasi-linear Extraction

The main difference between positive and negative extracted terms is exhibited in the way in which they depend on the negative content of the conclusion. For a proof $M : A$, the *positive* extracted term $[\![M]\!]^+$ is a function, and the types of their parameters correspond to the components of the negative computational type $\tau^-(A)$. On the other hand, the *negative* extracted terms $[\![M]\!]_i^-$ for the assumptions contain freely a challenge variable y_A of type $\tau^-(A)$. Consequently, when the negative content of A needs to be instantiated with a term t, we use an application $[\![M]\!]^+ t$ for the positive program and substitution $[y_A := t]$ in the negative program $[\![M]\!]_i^-$, the latter being the reason for term duplication.

In order to factor out as many common terms as possible, we need to:

1. unify the role of $\tau^-(A)$ in the terms $[\![M]\!]^+$ and $[\![M]\!]_i^-$, and
2. combine $[\![M]\!]^+$ and $[\![M]\!]_i^-$ so that their common subterms appear only once.

Below we will present some notions, which will be used to define a variant of the Dialectica interpretation, for which the size of the extracted terms for will be "almost" linear in the size of the proof.

5.1 Preliminaries

We will use the following notions of size of terms, formulas and proofs:

Definition 5

- $\lceil c \rceil = 1$ *if c is a variable, a constructor or a recursion constant.*
- $\lceil \lambda_x t \rceil = \lceil t_{\mathsf{L}} \rceil = \lceil t_{\mathsf{J}} \rceil = \lceil t \rceil + 1$
- $\lceil st \rceil = \lceil \langle s, t \rangle \rceil = \lceil s \rceil + \lceil t \rceil + 1$
- $\lceil \mathrm{at}(t) \rceil = \lceil t \rceil$
- $\lceil A \to B \rceil = \lceil A \rceil + \lceil B \rceil + 1$
- $\lceil \forall_x A \rceil = \lceil A \rceil + 1$
- $\lceil u^A \rceil = \lceil A \rceil$ *if u is an axiom instance or an assumption variable*
- $\lceil \lambda_{u^A} M \rceil = \lceil \lambda_{x^\rho} M \rceil = \lceil M \rceil + 1$
- $\lceil MN \rceil = \lceil M \rceil + \lceil N \rceil + 1$
- $\lceil Mt \rceil = \lceil M \rceil + \lceil t \rceil + 1$

Definition 6. *For a proof M we define its* maximal sequent length $[\![M]\!]$ *as* $\max_{N \leq M} |\mathrm{FA}(N)|$, *where $N \leq M$ is the subproof relation.*

We will factor out common subterms by using a *definition context* — a term containing a single occurrence of a *hole* $[]$, which can be instantiated with any other term. In order to define this notion, let us reserve a type variable $\circ$ and an object variable of this type $[] : \circ$, which we will call *a hole*.

Definition 7. *A definition context E is a term built by the following rules:*

$$E ::= []^\circ \mid (E^{\rho \Rightarrow \sigma} t^\rho)^\sigma \mid (\lambda_{x^\rho} E^\sigma)^{\rho \Rightarrow \sigma},$$

where t does not contain the type $\circ$. For a definition context E^ρ and term t^σ, we define the term $E[t]$ (t in the context E) as $E[\circ := \sigma][[] := t]$, where, contrary to our usual convention, the free variables of t are allowed to be bound by abstractions in E.

It is easy to see by induction that contexts are closed under composition:

Proposition 1. *Let E_1 and E_2 be definition contexts. Then $E_1[E_2]$ is a definition context.*

Corollary 1. *If E is a definition context, x is a variable and t is a term, then* **let** $x := t$ **in** *E is also a definition context.*

Any definition context E has the type $\vec{\rho} \Rightarrow \circ$ for some list of types $\vec{\rho}$. A key property of the definition contexts is that they correspond to substitutions in a certain way. Indeed, if we have a substitution $\Xi := [\vec{x} := \vec{s}]$, then we can define the context $E := $ **let** $\vec{x} := \vec{s}$ **in** $[]$ and for every term t we will have $E[t] \overset{r}{=} t\Xi$. The reverse correspondence is given by the following property.

Proposition 2. *For every definition context $E : \vec{\rho} \Rightarrow \circ$ and a list of different variables $y_i : \rho_i$ there is a substitution Ξ, such that $(E\vec{y})[t] = t\Xi$ for any term t.*

Proof. Induction on the definition of E.

Case $E = []$. Take Ξ to be the identity substitution.

Case $E = E's$. By induction we have Ξ' such that $(E'y_0\vec{y})[t] = t\Xi'$ for any t. For $\Xi := \Xi'[y_0 := s]$ we obtain the desired property.

Case $E = \lambda_{x^{\rho_1}}E'$. By induction we have Ξ' such that $(E'y_2 \ldots y_n)[t] = t\Xi'$ for any t. Take $\Xi := \Xi'[x := y_1]$. Then we will have $((\lambda_x E')y_1 y_2 \ldots y_n)[t] = ((\lambda_x E'[t])y_1 y_2 \ldots y_n) \overset{r}{=} (E'[t]y_2 \ldots y_n)[x := y_1] = t\Xi$.

Under special circumstances we can permute contexts over application.

Corollary 2. *For every definition context $E : \vec{\rho} \Rightarrow \circ$, variables $y_i : \rho_i$ and terms $t^{\sigma \Rightarrow \tau}, s^\sigma$, such that $\mathsf{FV}(t) \cap \mathsf{BV}(E) = \emptyset$ we have $E[ts] \overset{r}{=} \lambda_{\vec{y}}(tE\vec{y}[s])$.*

Proof. Take the substitution Ξ from Proposition 2. By the variable condition for t we have $t\Xi = t$. Then we obtain

$$E[ts] \overset{r}{=} \lambda_{\vec{y}}E\vec{y}[ts] = \lambda_{\vec{y}}(ts)\Xi = \lambda_{\vec{y}}t(s\Xi) = \lambda_{\vec{y}}t(E\vec{y}[s]).$$

A definition context will hold all common subterms of the extracted computational content, and the hole will contain a tuple of context-dependent terms specifically corresponding to positive and negative computational content. In order to keep terms as small as possible, we will delay hole substitution until the last possible moment. The pairing operation $\langle \cdot, \cdot \rangle$ will allow us to bundle together an arbitrary number of differently typed terms. We will introduce the following notation to easily access components of such tuples:

$$t^\rho \triangleright 0 := t \text{ if } \rho \text{ is not a product}, \quad t^{\rho \times \sigma} \triangleright 0 := t_\mathsf{L}, \quad t^{\rho \times \sigma} \triangleright i+1 := t_\lrcorner \triangleright i.$$

Tuples of terms will usually be of the form $\left\langle [\![M]\!]^+, \ldots, [\![M]\!]_u^-, \ldots \right\rangle$ for some proof M. To make notations more convenient, we will use $t \triangleright u$ for u accessing the respective negative content $[\![M]\!]_u^-$ from the tuple t. In order to unify positive and negative content, we will redefine the computational types of the Dialectica interpretation so that they are normal with respect to the reduction rules below:

$$
\begin{aligned}
\rho \Rightarrow \sigma \Rightarrow \tau \quad &\longmapsto \quad \rho \times \sigma \Rightarrow \tau \\
(\rho \Rightarrow \sigma) \times (\rho \Rightarrow \tau) \quad &\longmapsto \quad \rho \Rightarrow (\sigma \times \tau)
\end{aligned}
\qquad (\longmapsto)
$$

However, this would mean that during extraction we might need to apply a function of type $\rho \times \sigma \Rightarrow \tau$ to a value of type ρ. We will use the following notation for such a partial application:

$$
\begin{aligned}
f^{\rho \Rightarrow \tau} \circ t^\rho &:= f t \\
f^{\rho \times \sigma \Rightarrow \tau} \circ t^\rho &:= \lambda_{x^\sigma} f \langle t, x \rangle, \text{ where } x \text{ is a fresh variable.}
\end{aligned}
$$

Finally, we extend the projection operations $\llcorner$ and $\lrcorner$ to functions as follows:

$$
f^{\rho \Rightarrow \sigma \times \tau} \llcorner := \lambda_{x^\rho} f x \llcorner, \qquad f^{\rho \Rightarrow \sigma \times \tau} \lrcorner := \lambda_{x^\rho} f x \lrcorner.
$$

The following properties can be easily checked.

Proposition 3

1. $\lceil \mathcal{C} \, b \, s \, t \rceil = \lceil b \rceil + \lceil s \rceil + \lceil t \rceil + 4$
2. $\lceil \textbf{let } x := s \textbf{ in } t \rceil = \lceil (\lambda_x t) s \rceil = \lceil s \rceil + \lceil t \rceil + 2$
3. $\lceil E[t] \rceil = \lceil E \rceil + \lceil t \rceil - 1$
4. $\lceil t \triangleright i \rceil \leq \lceil t \rceil + i + 1$
5. $\lceil f^{\rho \times \sigma \Rightarrow \tau} \circ t^\rho \rceil \leq \lceil f \rceil + \lceil t \rceil + 4$
6. $\lceil f^{\rho \Rightarrow \sigma \times \tau} \llcorner \rceil = \lceil f^{\rho \Rightarrow \sigma \times \tau} \lrcorner \rceil = \lceil f \rceil + 4$

5.2   Quasi-linear Extraction

We will revise the definitions of positive and negative computational types, as well as the definition of the Dialectica translation. The purely syntactical changes will aid defining a recomputation-free variant of the interpretation.

Definition 8. *For a formula A we will redefine the positive and negative computational types denoting the new variants as $\sigma^+(A)$ and $\sigma^-(A)$. We will also denote $\sigma^*(A) := \sigma^-(A) \Rightarrow \sigma^+(A)$. We define:*

$$
\begin{aligned}
\sigma^+(\text{at}(b)) &:= \varepsilon, & \sigma^-(\text{at}(b)) &:= \varepsilon, \\
\sigma^+(A \to B) &:= \sigma^+(B) \times \sigma^-(A), & \sigma^-(A \to B) &:= \sigma^*(A) \times \sigma^-(B) \\
\sigma^+(\forall_{x^\rho} A) &:= \sigma^+(A), & \sigma^-(\forall_{x^\rho} B) &:= \rho \times \sigma^-(B)
\end{aligned}
$$

The relation with the original definition of computational types can be established up to the reduction relation $(\longmapsto)$.

Proposition 4. $\tau^-(A) \rightarrowtail^* \sigma^-(A), \qquad \tau^+(A) \rightarrowtail^* \sigma^*(A).$

We also need to adjust the definition of the Dialectica translation accordingly.

Definition 9. *For* $r : \sigma^*(A)$, $s : \sigma^-(A)$ *we define* $|A|_s^r$ *as follows:*

$$|\mathrm{at}(b)| := \mathrm{at}(b), \qquad |\forall_x A|_s^r := |A(s_\llcorner)|_{s_\lrcorner}^{r \circ s_\llcorner}$$

$$|A \to B|_s^r := |A|_{r s_\lrcorner}^{s_\llcorner} \to |B|_{s_\lrcorner}^{(r \circ s_\llcorner)_\llcorner}$$

The soundness theorem for the new variant of the interpretation will follow a pattern similar to Theorem 1 and will be proved by induction on a proof $M : A$ with free assumption variables $u_i : C_i$. On every inductive step we will define:

1. a definition context $[\![M]\!] : \sigma^-(A) \Rightarrow \circ$
2. a context-dependent positive witnessing term $[\![M]\!]^+ : \sigma^+(A)$
3. context-dependent negative witnessing terms $[\![M]\!]_i^- : \sigma^-(C_i)$

The final extracted term will be obtained by putting the context-dependent terms inside the context:

$$\{\![M]\!\} := [\![M]\!][\langle [\![M]\!]^+, \ldots, [\![M]\!]_i^-, \ldots \rangle].$$

We will refer to the separate components put in the context as follows:

$$\{\![M]\!\}^+ := [\![M]\!][[\![M]\!]^+], \qquad \{\![M]\!\}_i^- := [\![M]\!][[\![M]\!]_i^-].$$

Using the $\triangleright$ operation we can gain back the individual parts of $\{\![M]\!\}$:

Proposition 5. $\lambda_y(\{\![M]\!\}y \triangleright 0) \overset{r}{=} \{\![M]\!\}^+, \lambda_y(\{\![M]\!\}y \triangleright i) \overset{r}{=} \{\![M]\!\}_i^-$ *for* $y : \sigma^-(A)$.

Proof. By Proposition 2 we have a substitution Ξ, such that for all terms t we have $([\![M]\!]y)[t] = t\Xi$. Then we have

$$\lambda_y \{\![M]\!\} y \triangleright i = \lambda_y (([\![M]\!]y)[\langle [\![M]\!]^+, \ldots, [\![M]\!]_i^-, \ldots \rangle]) \triangleright i$$

$$\overset{r}{=} \lambda_y (\langle [\![M]\!]^+, \ldots, [\![M]\!]_i^-, \ldots \rangle \Xi) \triangleright i$$

$$= \lambda_y (\langle [\![M]\!]^+, \ldots, [\![M]\!]_i^-, \ldots \rangle \triangleright i) \Xi$$

$$\overset{r}{=} \lambda_y ([\![M]\!]_i^- \Xi) = \lambda_y (\{\![M]\!\}_i^- y) \overset{r}{=} \{\![M]\!\}_i^-.$$

Following the original Dialectica interpretation, for every binary rule instance there needs to be a case distinction on the translation of every assumption shared between the two branches of the proof. It is easy to show that this can be done with a term of linear size.

Lemma 1. *There is a constant* K, *such that for every formula* C *there is a term* $T_C : \sigma^*(C) \Rightarrow \sigma^-(C) \Rightarrow \mathsf{B}$ *such that* $|C|_z^f \leftrightarrow \mathrm{at}(T_C f z)$ *and* $\lceil T_C \rceil \leq K \lceil C \rceil$.

A drawback of the original Dialectica interpretation is that if the assumption variable has n occurrences, a case distinction for the same formula can be repeated $n-1$ times in the extracted term. Thus it is more efficient that for every proof $\mathcal{P}$ we put the extracted terms in a definition context $D : \circ$ containing the definitions of all terms T_C, such that $u : C$ is an assumption variable of M. It is clear that $\lceil D \rceil$ is bounded by the sum of the size of all assumption formulas, which, by our definition, is definitely not greater than $\lceil M \rceil$. In order to keep the presentation simpler, we will not be explicit about the context D; we will just assume that it is the outermost context of the final extracted term and that we have access to variables d_C instantiated with T_C in D.

Theorem 2 (Soundness of quasi-linear extraction). *Let $\mathcal{P} : A$ be a proof from assumptions $u_i : C_i$. Let $x_i : \sigma^*(C_i)$ and $y_A : \sigma^-(A)$ be fresh variables. Then there is a term $\{\!|\mathcal{P}|\!\}$, satisfying the following conditions:*

i. we can prove $|A|_{y_A}^{\{\!|\mathcal{P}|\!\}^+}$ from $|C_i|_{\{\!|\mathcal{P}|\!\}_i^- y_A}^{x_i}$,

ii. $\mathsf{FV}(\{\!|\mathcal{P}|\!\}) \subseteq \mathsf{FV}(\mathcal{P}) \cup \{x_i\}$,

iii. $\lceil \{\!|\mathcal{P}|\!\} \rceil \leq K(\lceil \mathcal{P} \rceil + \lceil\!\lceil \mathcal{P} \rceil\!\rceil^2)$ for a fixed constant K, not depending on $\mathcal{P}$.

Remark 1. Formally, the bound we have obtained in Theorem 2 is not linear, as it depends quadratically on the measure $\lceil\!\lceil \mathcal{P} \rceil\!\rceil$, which in the worst case could be equal to $\lceil \mathcal{P} \rceil$. However, as $\lceil \mathcal{P} \rceil$ increases, $\lceil\!\lceil \mathcal{P} \rceil\!\rceil$ grows much slower, hence we can consider the extracted terms as "almost" linear in the size of the proof.

Remark 2. To the author's knowledge, a linear bound on the Dialectica extracted term was first established by Hernest and Kohlenbach in [2]. One major presentational difference between the current exposition and the results in [2] is the use of a natural deduction system and λ-calculus versus Hilbert-style equational logic with Schönfinkel style combinators Σ and Π. On page 229 the authors comment how a more economical internal representation of terms (keeping pointers to common parts) is crucial for obtaining the linear bound. In our view, an advantage of Theorem 2 is that it obtains concrete λ-terms, which satisfy the given bound regardless of their internal representation.

Remark 3. While the translation $|A|_y^x$ is only a syntactic reformulation of the usual Dialectica, it allows extraction of terms, which are not only smaller, but also reflect the proof structure more precisely, since a cut will correspond to a β-redex, and not to a substitution, as in the original interpretation.

Let us revisit Example 1 from Section 4. Applying directly the results from Theorem 2, we obtain the term[2]

$$P'' := \lambda y_6 \mathbf{let}\ x := y_6\ \mathbf{in}\ \lambda y_7 \mathbf{let}\ f_1 := s\ \mathbf{in}\ f_1 y_7,\ \text{where}$$

$$s := \mathcal{R}\, x\, t_0\, (\lambda x_{,p} \mathbf{let}\ x_p := p\ \mathbf{in}\ t_1),$$

$$t_0 := \lambda y_0 \mathbf{let}\ y := y_0\ \mathbf{in}\ \mathbf{let}\ y_1 := y\ \mathbf{in}\ y_1,$$

$$t_1 := \lambda y_2 \mathbf{let}\ f_0 := (\lambda y_3 \mathbf{let}\ z := y_3\ \mathbf{in}\ \mathbf{let}\ y_4 := z + z\ \mathbf{in}\ y_4)\ \mathbf{in}$$

$$\mathbf{let}\ y := y_2\ \mathbf{in}\ \mathbf{let}\ y_5 := y + 1\ \mathbf{in}\ \mathbf{let}\ z_0 := x_p y_5\ \mathbf{in}\ f_0 z_0.$$

---

[2] Although having pointers would probably simplify such extracted programs, we stress that there is no principal need to complicate the term semantics.

We can simplify the terms above by performing all linear β-reductions, i.e., reductions $(\lambda_x s)t$ in which t is a variable or x has at most one occurrence in s. Thus we obtain

$$t_0 \stackrel{r}{=} \lambda_y y, \qquad t_1 \stackrel{r}{=} \lambda_y(\lambda_z z + z)(x_p(y+1)), \qquad \text{hence}$$

$$P'' \stackrel{r}{=} \lambda_x \mathcal{R}\, x\, (\lambda_y y)\, (\lambda_{x,p,y}(\lambda_z z + z)(p(y+1)))\,.$$

$P''xy$ reduces to $2^x(x+y)$ in a number of steps depending linearly on x, as opposed to its exponentially behaving counterpart obtained in Section 4 by applying the original Dialectica interpretation.

6   Conclusion and Future Work

It is intuitive that the efficiency of the extracted program is tightly related to the source proof; however, for non-constructive proofs this relation has not yet been satisfactorily explored. In our opinion, a proof interpretation should reflect the structure of a mathematical argument as directly as possible in the extracted program, while introducing minimal additional computational complexity. In the present paper we have shown an optimisation of the computational layer inserted by the Dialectica interpretation, namely by preventing a particular kind of unnecessary recomputations. Even though such situations can be avoided by manual proof reformulation, the author believes that a general solution is important for automated extraction. There is certainly room for other improvements, for example, removing redundant computations ([3,7]). A topic for future research will be to find new general optimisations and to analyse how they can be composed with each other.

References

1. Gödel, K.: Über eine bisher noch nicht benützte Erweiterung des finiten Standpunktes. Dialectica 12, 280–287 (1958)
2. Hernest, M.-D., Kohlenbach, U.: A complexity analysis of functional interpretations. Theor. Comput. Sci. 338(1-3), 200–246 (2005)
3. Hernest, M.-D., Trifonov, T.: Light Dialectica revisited. Submitted to APAL (2008), http://www.math.lmu.de/~trifonov/papers/ldrev.pdf .
4. Kohlenbach, U.: Applied Proof Theory: Proof Interpretations and their Use in Mathematics. Springer Monographs in Mathematics. Springer, Heidelberg (2008)
5. Kreisel, G.: Interpretation of analysis by means of constructive functionals of finite types. In: Heyting, A. (ed.) Constructivity in Mathematics, pp. 101–128. North-Holland Publishing Company, Amsterdam (1959)
6. Schwichtenberg, H.: Dialectica interpretation of well-founded induction. Mathematical Logic Quarterly 54(3), 229–239 (2008)
7. Trifonov, T.: Dialectica interpretation with fine computational control. In: CiE 2009, July 19-24. LNCS, vol. 5635, pp. 467–477. Springer, Heidelberg (2009)
8. Troelstra, A.S.: Metamathematical Investigation of Intuitionistic Arithmetic and Analysis. LNM, vol. 344. Springer, Heidelberg (1973)

Computing with Concepts, Computing with Numbers: Llull, Leibniz, and Boole

Sara L. Uckelman*

Institute for Logic, Language, and Computation
S.L.Uckelman@uva.nl

Abstract. We consider two ways to understand "reasoning as computation", one which focuses on the computation of concept symbols and the other on the computation of number symbols. We illustrate these two ways with Llull's *Ars Combinatoria* and Leibniz's attempts to arithmetize language, respectively. We then argue that Boole's development of an algebra of reasoning was in a large part successful due to its ability to marry the two types of computation that are exemplified in Llull's and Leibniz's works.

1 Introduction

The *Oxford English Dictionary* defines 'computation' as

1. a. The action or process of computing, reckoning, or counting; arithmetical or mathematical calculation; an instance of this [1].

As with many dictionary definitions, this gloss is rather vague. There are (at least) two interesting ways in which it may made precise: one focusing on the words 'action, process' and the other on the words 'calculation, reckoning, counting, arithmetical, mathematical'. The first group of words suggests a mechanistic view of computation, whereby computation is the result of a machine running some algorithm or set of instructions. The second group of words, in that they all refer more or less directly to numbers, can be seen as a specification of the first group, by specifying that the actions or processes (mechanisms) involved are numeric or arithmetic. It should be clear that numeric processes and actions are not the only way to do computation; one non-numeric computational process or mechanism involves computation with concepts directly, and not via numerical representation. Thus, the two views of computation that will be considered in this paper are those indicated in the title: computation with concepts and computation with numbers.[1] These two branches or strands of computation are

---

* The author was funded by the project "Dialogical Foundations of Semantics" (DiFoS) in the ESF EuroCoRes programme LogICCC (LogICCC-FP004; DN 231-80-002; CN 2008/08314/GW).

[1] Strictly speaking, we should speak of "computation with concept-*symbols*" and "computation with number-*symbols* (or numerals)", since we are not directly operating on the concepts and numbers themselves. However, we will use the less precise formulation throughout the rest of the paper and trust the reader not to misunderstand our intent.

F. Ferreira et al. (Eds.): CiE 2010, LNCS 6158, pp. 427–437, 2010.

not intended to be exclusive, but rather, as we will see when we look at Boole, as complimentary.

In ordinary use, "computation" has some connotation of mindlessness. Many ordinary users of modern computers would subscribe to this view: computers are something of a black box where the user provides input and receives an output through a computational process which is often invisible, and even when visible, can often be wholly impenetrable (for example, watching LATEX code compile). As we'll see, both Llull's *Ars Combinatoria* and Leibniz's attempts to arithmetize language have this mindlessness component: In so far as these systems are algorithmic, they move the burden of the actual reasoning from the user to the system. This property is illustrated in Boole's algebras by their level of abstraction.

In this paper we survey two different but connected ways that we can understand "reasoning as computation", and illustrate these ways by looking at the works of three figures in the history of logic, Ramon Llull, Gottfried Wilhelm Leibniz, and George Boole. Despite the well-documented inspiration that Leibniz found in Llull's works, Llull's views of the computational side of reasoning contrast with Leibniz's and Boole's development of an algebra of reasoning can be seen as a conceptual synthesis of the two. I argue that the success of Boole's innovation was due in large part to its ability to marry the two types of computation that are exemplified in Llull's and Leibniz's works: computation as rule-based manipulation of concepts and computation as arithmetic calculation.

The plan of the paper is as follows: To illustrate the conceptual view of computation, in the next section we look at a particular aspect of Ramon Llull's system of reasoning developed in the 13th century. Llull, who can rightfully be called a visionary for his ideas concerning the computational side of reasoning, inspired Leibniz in his development of a *calculus universalis* in the 17th century. One of Leibniz's goals along the way to the *calculus universalis* was the arithmetization of language, which serves as our example in Sec. 3 of the numeric or arithmetic view of computation. We then argue in Sec. 4 that these two strands are exemplified together in Boole's development of the algebra of reason, and it is their complementarity that was at least a partial cause of its success.

2 Llull and the Computation of Concepts

Ramon Llull (Catalan; *Raymundus Lullus* or *Lullius*, Latin; *Rámon Lull*, Spanish; *Raymond Lull* or *Lully*, English) was born in 1232 or early 1233 in Palma, the capital of Majorca, to a family that was probably of noble status. In his early years, he served as a courtier in the court of James I and James II of Majorca, which involved extensive travel through Aragon, Catalonia, and Valencia. In 1263 he experienced a religious conversion, and thereafter turned his attentions from secular pursuits such as troubadour poetry lyrics, to theological and philosophical topics. In the 1280s he conceived of a goal of developing a system of argumentation or demonstration which could be used to show the Jew and the Muslim the error of their ways, and the correctness of Christian theology. In addition to knowing Catalan, his native language, he was also conversant in

	Fig. A	Fig. T	Questions & Rules	Subjects	Virtues	Vices
B	goodness	difference	whether?	God	justice	avarice
C	greatness	concordance	what?	angel	prudence	gluttony
D	eternity	contrariety	of what?	heaven	fortitude	pride
E	power	beginning	why?	man	temperance	pride
F	wisdom	middle	how much?	imaginative	faith	accidie
G	will	end	of what kind?	sensitive	hope	envy
H	virtue	majority	when?	vegetative	charity	ire
I	truth	equality	where?	elementative	patience	lying
K	glory	minority	how? and with what?	instrumentative	pity	inconstancy

Fig. 1. The Alphabet of the *Ars brevis* [2, p. 581]

Latin and Arabic, the academic languages of his time. Part of implementing this goal involved missionary travel throughout the Mediterranean. He died during one of these journeys, either in Tunis or on a ship sailing from Tunis back home, sometime between December 1315 and March 1316.[2]

Llull, with his adventurous life, mysticism, and connections in high places, is an exciting figure for computer scientists to read and hear about. Given his adventures (cf. the citations in fn. 2), it's no wonder that computer scientists would want to claim such a celebrity as one of their own. Why they should do so is nicely argued by Sales in [4], who points out that in Llull's work can be found inklings of and first steps towards a number of concepts fundamental in computer science, such as the ideas of a calculus, an 'alphabet of thought', a method, a graph, of logical analysis, heuristics and deduction, generative systems, tableaux, conceptual nets, and diagrams to represent concepts and relations between concepts.[3]

Our interest here, in the context of "reasoning as computation", are just two out of Llull's numerous (263 according to [2, p. 53]) works, the *Ars demonstrativa* (Demonstrative Art, c. 1283–89, hereafter referred to as AD) and the *Ars brevis* (Short Art, 1308, hereafter referred to as AB), the "single most influential work", which builds on and simplifies AD. These works together constitute the 'Art'. In the Art, Llull presents to the reader a mechanism for abstract reasoning with a restricted range of application. The foundation of this mechanism is an alphabet, a system of constants, representing different concepts. The alphabet of AD is two-tiered, with 16 symbols representing basic concepts and then 7 symbols representing what we might call meta-concepts. In AB, the alphabet of AD is simplified to include only 9 symbols, and these symbols have different meanings depending on their usage. Fig. 1 gives the interpretation of the alphabet of AB in different contexts. The full combinatoric power of the Art comes to the fore in AB, where it "became a method for 'finding' all the possible propositions and syllogisms on any given subject and for verifying their truth or falsehood" [2, p. 575]. The allowed combinations of the constant symbols in the alphabet are

---

[2] This is a very compressed biography of Llull; for a more detailed history full of exciting details, see [2, vol. 1, pp. 3–52], which includes extensive excerpts from Llull's autobiography *Via coaetanea*, and [3, ch. 1].

[3] See [5] for further discussion of Llull's status as a 'computer scientist'.

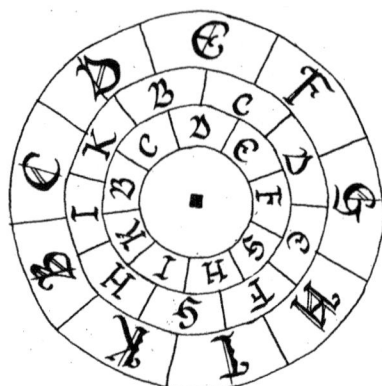

Fig. 2. The Fourth Figure

illustrated by various tables and diagrams[4] of which the most interesting, from the mechanist view of computation, is Figure 4 of the *Ars brevis*, which "has three circles, the outermost of which is fixed and the two inside ones of which are mobile" [2, p. 587]. Figure 4, redrawn from Plate XVIII, is given in Fig. 2. By rotating the moving circles in various ways, one can extract valid syllogisms, where the term on the middle circle is the middle term relating the two extremes, which are located on the outer and inner circles.

The physical nature of the concentric, movable circles of the fourth figure allows us to view this part of Llull's combinatorial system as a crude mechanism for computing new concepts (the output) on the basis of a given set of concepts (the input). As Welch says, "The fourth figure was thus a primitive logical machine" [6, p. 6], and Gardner calls it a "mechanical method" [3, p. 9]. Its primitiveness comes from both the rude nature of its construction, and also the fact that it essentially only handles intersection [7, p. 12]. But the primitiveness of it should not detract from its novelty: It is, so far as is known, the first attempt to provide a mechanistic and 'mindless' (in the sense of the word discussed above) *physical* method of reasoning.[5]

Llull's computational system is based purely on concepts; there is no arithmetic involved.[6] It is for this reason that we have selected Llull as an example of the purely non-arithmetic conception of reasoning as computation. However,

---

[4] The diagrams of the first, second, third, and fourth figures of the *Ars brevis* as found in the Escorial MS are reproduced in [2] between pages 582 and 583; a selection of figures from the Venice MS of AD, along with some interpretational tables by Bonner, are given in the same source between pages 318 and 320.

[5] We recognize that the introduction given here is nowhere near adequate. For further discussion of Llull's system, see [2] and [8].

[6] One should not take this criticism unfairly: Without the notions of the 'intension' and 'extension' of a concept (ideas not developed until the 17th century), it is by no means clear how one would associate numbers with concepts in any useful fashion that would allow the numeric type of computation.

this lack of mathematical foundation is often cited as one of the failures of his system, which is cumbersome and not easily extended (even though, as Zweig notes, "Llull believed his Art could be applied to all fields of knowledge and was therefore a truly universalist system" [9, p. 22]). Llull's mathematical naïvity was one factor which motivated Leibniz's search for a more rigorous and numerically-based system of computation [6, p. 2]. We turn to discuss this in the next section.

Before concluding this section, we note in passing that if we understand "computation" in the broad sense of "(algorithmic) process", then Llull was by no means the first to attempt a mechanistic view of reasoning. John of Salisbury, writing in the middle of the 12th century, tells us that his student, William of Soissons,

> invented a device to revolutionize the old logic by constructing unacceptable conclusions and demolishing the authoritative opinions of the ancients [10, Bk. II, ch. 10, p. 98].[7,8]

According to Kneale and Kneale, some people have thought that William's "machine" was a physical construction, akin to Jevon's logical machine [13, p. 201],[9] but both the Kneales and Martin have argued that "machine" should be understood here in a metaphorical sense, and that it was likely that what William had in mind was a method of argument-construction which, given a contradiction or an impossible statement, would return any other statement [14, p. 565]. Whether William's machine was a concrete object, or merely a procedure for a reasoner to follow, it is an interesting example of a computational method where the user is no longer necessarily the reasoner; rather, it is the "machine" itself which is doing the reasoning.[10]

3   Leibniz and the Computation of Numbers

Leibniz discovered Llull's works at an early age; he discusses Llull in his *Dissertatio de arte combinatoria* (Dissertation on the combinatorial art, 1666), written at the age of 19 [3, p. 3]. As Bonner notes, "the relational nature of Llull's system is fundamental to his idea of an Ars combinatoria" [5, p. 4]. Given Llull's emphasis on binary and ternary relations and his combinatoric approach to reasoning,

---

[7] *Interim Willelmum Suessionensem, qui ad expugnandam, ut aiunt sui, logice uetustatem et consequentias inopinabiles construendas et antiquorum sententias diruendas machinam postmodum fecit* [11, Bk. II, ch. 10, p. 81].

[8] Adamson [12, p. 27] translates John of Salisbury's *machinam* as "method", and the Kneales translate it as "engine" [13, p. 201].

[9] The Kneales do not say who these "some people" are, and I have been unable to find this out myself.

[10] All this talk of machines reasoning sounds anachronistic and, from the point of view of the 12th century, futuristic. Interestingly, from the point of view of contemporary artificial intelligence, while Leibniz believed that it was possible to mechanize (in the physical sense of the word) these algorithmic processes, "he never thought that we might invent a machine which could invent machines" [15, p. 110].

it is no surprise that he was such a fascinating figure to a young Gottfried Wilhelm Leibniz [5,6,16,17, p. 657]. Llull and Leibniz shared much in their guiding philosophy and goals. They both shared the desire to provide a system within which all theological controversies could be definitively solved. (Though they differed in the application of this system: Llull wanted to use it to show the Jew and the Muslims their errors, and the truth of the Christian way, whereas Leibniz intended his to help resolve the splintering of the Catholic church that had started in the previous century.) Both Llull and Leibniz also shared a belief in what Welch calls "conceptual atomism, a belief that the majority of concepts are compounds constructed from a relatively small number of primitives" [6, p. 2] (cf. [4, p. 16] and [17, p. 20]). These primitives could be combined and related in different ways which would allow the reasoner to generate complex information. Llull and Leibniz also both developed combinatorial systems which exploited physical systems of moving wheels [18, p. 274].[11]

However, Leibniz differed from Llull in two ways: As we noted in the previous section, he criticized Llull's mathematical naïvity and wished to ground his systems arithmetically, and he had access to the recently-developed notions of the intension (or comprehension) and extension of a concept. If we wish to speak anachronistically, we could say that, in so far as Llull's concepts could be represented in a physical medium and manually manipulated, he identified concepts with their extensions. Leibniz, on the other hand, discussed both intensional and extensional interpretations of concepts in his arithmetization of language, in the end preferring the intensional approach [20,21]. We offer Leibniz's attempts to arithmetize syllogistic as an illustration of the numeric interpretation of reasoning as computation. Leibniz made several different attempts to arithmetize Aristotelian syllogistic in his pursuit of developing a universal language, of which only the final attempt was successful. Given his belief in logical atomism, it was a natural step for him to associate primitive concepts with their *numeri characteristici*. If this is done properly, then more complex concepts could be reduced to their constituent parts merely by knowing both the rules for the combination of primitives and the mapping associating numbers with primitives, and further, syllogistic statements, which assert relations between complex concepts, could be represented by numeric relations between different numbers.

We consider two of the arithmetizations of syllogistic that Leibniz developed. Recall that a syllogism is a set of three categorical proposition, where a categorical proposition is one of the form "All S are P" or "Some S are P" (or their negations, "Some S are not P" and "No S are P", respectively, but since their

---

[11] Hence it should be clear in the following that we are not trying to argue that Leibniz did not have a mechanistic or purely concept-based view of reasoning; by no means is that the case, as is amply illustrated by his combinatorial theory, e.g., as presented in *Dissertatio de Arte Combinatoria*, which is much closer to the Llullian-style conceptual computation, relying less heavily on numeric support. This system is extensively discussed in [18, ch. 5] and [19, ch. 3]. Since in this paper we are interested in his developments which illustrate the approach of computing with numbers, we do not discuss his developments which fall under the approach of computing with concepts further here.

truth conditions fall straightforwardly out of the truth conditions for the affirmative claims, we will not consider them), where S and P are variables for terms representing primitive or complex concepts. Leibniz focused on the syllogism because he believed that all propositions could be reduced to subject-predicate ones, meaning that all reasoning could be simulated with syllogistic reasoning [19, p. 13]. In the first arithmetization attempt, each primitive concept is associated with a single positive integer; the characteristic number of a complex concept is the multiplication of all the primitive concepts it contains (for example, if 2 is 'animal' and 3 is 'rational', then 6 is 'rational animal'='man').[12] Then, a universal affirmative proposition is true iff the characteristic number of the subject term is divisible by the characteristic number of the predicate term [22, p. 42]. These truth conditions are acceptable if one adopts the constraint that no factor can appear in the characteristic number of a term more than once (cf. [21, p. 3]); that is, he recognized the idempotence of properties. How to treat particular affirmative propositions is less clear; Leibniz first offers the rule that such a proposition is true iff the characteristic number of the predicate is divisible by the characteristic number of the subject, or vice versa [22, p. 43]. However, this has the unfortunate consequence that a particular affirmative proposition "Some S is P" implies either that every S is P or that every P is S, which is not generally true. In a later manuscript, Leibniz revised the truth conditions for particular affirmative statements so that they are true whenever the characteristic number of the subject term is multiplied by another integer, it is then divisible by the characteristic number of the predicate term [22, pp. 58,69]. However, taken at face-value, this rule is also problematic; there is always some integer n such that sn is divisible by p, namely p itself. Thus, this rule implies that "Some S is P" is always true, which is unacceptable.

The second attempt at arithmetization [22, pp. 77–82] that we consider is more sophisticated. We follow the presentation of Marshall:

> Each term is assigned an ordered sequence of two relatively prime numbers, the first positive, the second negative. If each number assigned the predicate term divides the corresponding number of the subject term, the proposition is of the form 'All a is b'. The negation of this form... will be given if one of the two conditions is not met... If two of the non-corresponding numbers (i.e., the positive assigned one term and the negative assigned the other) have a common divisor, the proposition is of the type 'No a is b. If this is not the case..., the proposition is of the form 'Some a is b' [23, pp. 238–39].

This system has a number of points in its favor: It solves the problem of the triviality of the affirmative particular sentences, and it validates all the Aristotelian laws of conversion and valid syllogisms, and the square of opposition. But it comes with its own problems. Consider the following assignment of characteristic numbers to concepts: Let 'pious man' be assigned $\langle 2 \times 5, -3 \rangle$; 'fortunate

---

[12] The association of primitives with numbers which can be uniquely factored out of complex combinations of the primitives should strike the reader as reminiscent of Gödel-numbering.

man' be assigned $\langle 2^3, -11 \rangle$, and 'happy man' be assigned $\langle 5, -1 \rangle$; and the following syllogism:

> All pious men are happy.
> Some pious man is not fortunate.
> Some fortunate man is not happy.

On this assignment, the syllogism is verified (we leave the determination of this to the reader: it is straightforward). However, it should be immediately clear that this is not a valid argument; if we instead use the assignment of $\langle 2^4 \times 7, -3^3 \times 5 \rangle$, $\langle 2^2, -3^9 \rangle$, and $\langle 2, -3^3 \rangle$, for 'pious', 'fortunate', and 'happy' respectively, then the syllogism is not validated. Leibniz concluded that there was an error in his arithmetization, since, while it validates all of the valid syllogisms, it failed to invalidate the invalid ones ([23, p. 240], [22, p. 334]). Note that if he had taken as a rule that "If invalidating instantiations can be produced, the mood is invalid; it is valid if they cannot" [23, p. 241], then Leibniz's system would have been a success (this is the argument of Marshall's article). Not recognizing this, Leibniz expressed dissatisfaction with the system presented above and ultimately gave up his attempt at arithematizing language. Leibniz's error, in his attempts, was in trying to identify the atomistic primitive concepts too closely with numbers, so that he could assign numeric properties to them. In the context of the present paper, we can say that in this endeavor, he went too far on the computation as (arithmetic) calculation interpretation.

4   Boole and the Synthesis

In Llull, we saw an example of a purely mechanistic, non-numeric system of computation, based on concepts. With Leibniz's attempts to arithmetize the syllogistic, we had an example of computation from the other end of the spectrum. Both of these systems have their shortcomings, but both have points in their favor. It was George Boole who took the best of both systems and created a reasoning-system that combines both types of computation, a system which continues today to be widely used and extremely fruitful: Boole's abstract algebras, which lead to the development of Boolean algebras (cf. [29]).

Conceptually, Boole's achievements rest heavily on Leibniz's [22, ch. viii]. As the Kneales note:

> Leibniz realized already in the seventeenth century that there is some resemblance between disjunction and conjunction of concepts on the one hand and addition and multiplication of numbers on the other, but he did not find it easy to formulate the resemblance precisely and then to use it as the basis of a calculus of logic. It was this that George Boole (1815–64) achieved in his *Mathematical Analysis of Logic* [13, p. 404].

Leibniz made "an attempt to improve the presentation of logic ... by the use of algebraic symbolism" [24, p. 158], though he was not as successful in doing as he had hoped [23, p. 241]. However, Boole was not familiar with Leibniz's work until *after* both *The Mathematical Analysis of Logic* and *Laws of Thought* were published [25, § 5].[13] (My thanks to one of the anonymous referees for this reference.) Kneale argues that the immediately effective sources for Boole were Gregory's "On the Real Nature of Symbolical Algebra" and De Morgan's four papers on "The Foundation of Algebra" [24, p. 160] and that:

> From these sources it was possible to collect two important discoveries: (i) that there could be an algebra of entities which were not numbers in any ordinary sense, and (ii) that the laws which hold for types of numbers up to and including complex numbers need not all be retained together in an algebraic system not applicable to such numbers [24, p. 160].

The importance of the *calculus of operations* in the development of English mathematics in the 19th century is discussed in great detail in [26]. Boole's crucial discovery was that "there could be an algebra of entities which were not numbers in any ordinary sense" and that the 'numeric' laws governing these entities need not be arithmetic [13, p. 405]. Boole's elements of study were class-designating concepts; that is, unlike Leibniz, he focused on the extension of concepts, rather than their intensions [27, pp. 4–5], [28, p. 29]. In this way, Boole's algebras can be seen as synthesizing the insights of both Llull, who treated concepts in an extensional fashion (speaking anachronistically) but without exploiting any arithmetic tools, and Leibniz, who saw the utility of associating arithmetic properties with intensions, but went too far the other direction in trying to map arithmetic properties directly onto the properties of intension.

The only question left for us to address in the remaining space is the extent to which Boole's algebras, viewed as a method of computation, have the 'mindlessness' property discussed in the opening section. It is certainly the case that one can do these computations by rote, without having any idea of the meaning (interpretation) of the symbols being manipulated.[14] On the other hand, Boole himself allowed the mindless application of symbolic manipulation only if this was an intermediary step eventually followed by an active final step of reasoning involving expanding the symbols to their meanings [29, pp. 173, 179]. He says:

> It is of most material consequence, whether those symbols are used with a full understanding of their meaning, with a perfect comprehension of that which renders their use lawful, and an ability to expand the abbreviated forms of reasoning which they induce, into their full syllogistic

---

[13] Even after Boole was introduced to Leibniz's work, it is not clear what the extent of his access was; during Boole's lifetime many of Leibniz's works languished unedited, and it was not until Couturat's edition in 1903 [22] that Leibniz's important works became generally available. In particular, it is not known whether Boole knew of the *Non Inelegans Specimen Demonstrandi in Abstractis* (in the Erdmann edition of 1840), which "was the only important piece of Leibniz on mathematical logic then generally available... [Leibniz's] most interesting papers lay still unread in the library at Hanover" [24, p. 150].

[14] This fact is recognized in the first sentence of the introduction to Boole's *Mathematical Analysis of Logic*: "[T]he validity of the processes of analysis does not depend upon the interpretation of the symbols which are employed, but solely upon the laws of their combination" [27, p. 3].

development; or whether they are mere unsuggestive characters, the use
of which is suffered to rest upon authority [27, p. 10].

Thus, we end with a two-faced observation: The first successful attempt at developing a system of computational reasoning was done by someone who would
only admit the mechanistic approach towards deduction as a preliminary step.
Even if we can have computers do our reasoning, it is still up to us to interpret
and apply the results.

References

1. OED: Computation, *n*. In: Oxford English Dictionary. September 2009 edn. OUP (2009), http://dictionary.oed.com/cgi/entry/50045985 (January 5, 2010)
2. Llull, R., Bonner, A. (ed. and trans.): Selected Works of Ramon Llull, 1232–1316, 2 volumes. Princeton University Press, Princeton (1985)
3. Gardner, M.: Logic Machines and Diagrams. McGraw-Hill, New York (1958)
4. Sales, T.: Llull as computer scientist. In: Bertran, M., Rus, T. (eds.) AMAST-ARTS 1997, ARTS 1997, and AMAST-WS 1997. LNCS, vol. 1231, pp. 15–21. Springer, Heidelberg (1997)
5. Bonner, A.: What was Llull up to? In: Bertran, M., Rus, T. (eds.) AMAST-ARTS 1997, ARTS 1997, and AMAST-WS 1997. LNCS, vol. 1231, pp. 1–14. Springer, Heidelberg (1997)
6. Welch, J.R.: Llull, Leibniz, and the logic of discovery (Revised version of "Llull and Leibniz: The Logic of Discovery". Catalan Review 4(1-2), 75–83 (1990), http://spain.slu.edu/faculty_&_staff/directory/files/llull.pdf
7. Styazhkin, N.I.: History of Mathematical Logic from Leibniz to Peano. MIT Press, Cambridge (1969)
8. Bonner, A.: The Art and Logic of Ramon Llull: A User's Guide. Brill (2007)
9. Zweig, J.: Ars Combinatoria: Mythical systems, procedural art, and the computer. Art Journal, 20–29 (Fall 1997)
10. Salisbury, J., McGarry, D.D. (trans. & ed.): The Metalogicon of John of Salisbury. University of California Press, California (1955)
11. Salisbury, J., Webb, C. (ed.): Ioannis Saresberiensis episcopi Carnotensis Metalogicon. Libri IV. E Typographeo Clarendoniano (1929)
12. Adamson, J.W.: A Short History of Education. Cambridge University Press, Cambridge (1919)
13. Kneale, W., Kneale, M.: The Development of Logic. rev. edn., Clarendon (1984)
14. Martin, C.J.: William's machine. J. Phil 83(10), 564–572 (1986)
15. Schrecker, P.: Leibniz and the art of inventing algorisms. J. Hist. Ideas 8(1), 107–116 (1947)
16. Wilson, C.: Leibniz's Metaphysics. Princeton University Press, Princeton (1989)
17. Leibniz, G.W., Loemker, L.E. (eds. & trans.): Philosophical Papers and Letters, 2nd edn. D. Reidel (1969)
18. Maat, J.: Philosophical Languages in the Seventeenth Century: Dalgarno, Wilkins, Leibniz. Kluwer Academic, Dordrecht (2004)
19. Ishiguro, H.: Leibniz's Philosophy of Logic and Language, Duckworth (1972)
20. Rescher, N.: Leibniz's interpretation of his logical calculi. J. Sym. Log. 19(1), 1–13 (1954)

21. Sotirov, V.: Arithmetizations of syllogistic à la Leibniz. J. App. Non-Class. Log. 9(2-3), 387–405 (1999)
22. Couturat, L.: Opuscules at Fragments inédits de Leibniz. F. Alcan (1903)
23. Marshall Jr., D.: Łukasiewicz, Leibniz, and the arithmetization of the syllogism. Notre Dame J. Form. Log. 18, 235–242 (1977)
24. Kneale, W.: Boole and the revival of logic. Mind 57(226), 149–175 (1948)
25. Peckhaus, V.: Leibniz's influence on 19th century logic. In: Zalta, E.N. (ed.) Stanford Encyclopedia of Philosophy (Fall 2009 edn.),
    http://plato.stanford.edu/archives/fall2009/entries/
    leibniz-logic-influence/
26. Koppelman, E.: The calculus of operations and the rise of abstract algebra. Archiv. Hist. Exact Sci. 8, 155–241 (1971)
27. Boole, G.: The Mathematical Analysis of Logic. MacMillan, Basingstoke (1847)
28. Boole, G.: Laws of Thought. Dover Publications, Inc., New York (1854)
29. Hailperin, T.: Boole's algebra isn't boolean algebra. Math. Mag. 54(4), 173–184 (1981)

Inference Concerning Physical Systems

David H. Wolpert

NASA Ames Research Center, MailStop 269-1, Moffett Field, CA 94035-1000, USA
david.h.wolpert@nasa.gov, ti.arc.nasa.gov/people/dhw

Abstract. The question of whether the universe "is" just an information-processing system has been extensively studied in physics. To address this issue, the canonical forms of information processing in physical systems — observation, prediction, control and memory — were analyzed in [24]. Those forms of information processing are all inherently epistemological; they transfer information concerning the universe as a whole into a scientist's mind. Accordingly, [24] formalized the logical relationship that must hold between the state of a scientist's mind and the state of the universe containing the scientist whenever one of those processes is successful. This formalization has close analogs in the analysis of Turing machines. In particular, it can be used to define an "informational analog" of algorithmic information complexity. In addition, this formalization allows us to establish existence and impossibility results concerning observation, prediction, control and memory. The impossibility results establish that Laplace was wrong to claim that even in a classical, non-chaotic universe the future can be unerringly predicted, given sufficient knowledge of the present. Alternatively, the impossibility results can be viewed as a non-quantum mechanical "uncertainty principle". Here I present a novel motivation of the formalization introduced in [24] and extend some of the associated impossibility results.

Keywords: physical inference, dynamical systems.

1 Introduction

The question of whether the universe "is" just a computer, just an information-processing system, has been extensively studied in physics [22, 15, 14, 25, 11, 12, 18, 20, 9, 8, 3, 2, 4, 2, 1, 21]. Typically in that work 'computers' were defined as Turing Machine (TM) or some other members of the Chomsky hierarchy [10]. This has raised the issue of whether members of the Chomsky hierarchy are the appropriate formalization of information processing in physical systems, or whether some other formalization more accurately models such information processing. To help address this issue, the canonical forms of information processing in physical systems — observation, prediction, control and memory — were analyzed in [24]. (See commentary in [5, 6].)

Those forms of information processing are inherently epistemological; they all transfer information concerning the universe as a whole into a scientist's mind . Accordingly, [24] formalized the logical relationship that must hold between the state of a scientist's mind and the state of the universe containing the scientist whenever one of those processes is successful.

F. Ferreira et al. (Eds.): CiE 2010, LNCS 6158, pp. 438–447, 2010.

A crucial property of this formalization is that recognizes the fact that inference devices are embedded in the very physical system (namely the physical universe) about which they are making inferences. This embedding is similar to the self-reference capabilities of TM's. As a result, there is a tight relationship between inference devices and TM's. In particular, there are impossibility results for inference devices that are similar to the Halting theorem for TM's.[1] As an example, one impossibility result shows that Laplace was wrong to claim that even in a classical, non-chaotic universe the future can be unerringly predicted, given sufficient knowledge of the present [13]. The other impossibility result reveals that *if they can be initialized independently of each other*, then two separate inference devices cannot both infer each other.

The impossibility results of inference devices have nothing to do with the precise laws governing the underlying universe; they follow simply from the logical formalization of what it means to have information concerning a universe be transferred into a subsystem of the universe. In particular, those results do not invoke chaotic dynamics as in [18, 20], nor quantum mechanical indeterminism. Similarly, they apply independent of the values of any physical constants (in contrast, for example, to the work in [15]), and more generally apply to every universe in a multiverse. Nor do the results presume limitations on where in the Chomsky hierarchy an inference device lies. So for example they would apply to oracles, if there can be oracles in our universe.

The mathematics of inference devices extends beyond these impossibility results. For example, one can analyze extensions to the basic definition to allow a probability distribution across possible states of the universe. Doing this transforms the impossibility results into limitations on the joint probability of two distinguishable inference devices both correctly perform their prediction / observation / control / memory. In essence, these stochastic versions of the impossibility results comprise a first-principles, non-quantum mechanical "uncertainty principle".

In addition, one can define an inference device analog of algorithmic information complexity. This "inference complexity" obeys analogs of many of the theorems concerning algorithmic information complexity. However instead of quantifying the complexity of a sequence of bits that can be read by a TM, it quantifies the complexity of successfully performing a given task of prediction, observation, control or memory.

The contribution of this paper is new, simpler examples motivating the definition of inference devices and an extension of the impossibility results first presented in [24].

2   Observation, Prediction, Recollection and Control

In [24] is a detailed analysis of the logical relationship connecting the state of scientist's mind and the state of the universe containing the scientist in any instance of the scientist perfectly predicting the future value of a physical variable in the universe. The shared nature of those relationships was then formalized, as an "inference device". That

---

[1] However the mathematics of inference device is in some respects far simpler than that of TM's. In particular, the inference device impossibility results do not involve diagonalization over an infinite set, as in the proof of the Halting theorem. Rather they involve "diagonalization over a two-element set", i.e., Cretan liar's arguments.

formalization relies on a novel definition of semantic information (as opposed to the syntactic information of Shannon information theory).

In this section I present a motivational example for that formalization of perfect prediction without invoking semantic information. Analogous examples hold for observation, control and recollection, but are not presented here for reasons of space.

Example 1. Consider a scientist who claims to be able to predict $S(t_3)$, the value of some physical variable at time t_3, at some time before t_3. If the claim is correct, then for any question of the form "Does $S(t_3) = L$?" where L is a possible value of the variable, the scientist is able to consider that question at some $t_1 < t_3$, and produce an answer at $t_2 < t_3$, where the answer is "yes" if $S(t_3) = L$ and "no" otherwise. So loosely speaking, if the scientist's claim is correct, then by considering the appropriate question at t_1, he can provide any bit of the delta function representation of $S(t_3)$, at t_2.

It may be that the scientist has to use some prediction computer to make these predictions; in that case the definition of the "scientist" should be expanded to include that computer. Similarly, it may be that the scientist has to program that computer appropriately at t_1. In this case, the definition of the scientist's "pondering the appropriate question" should be expanded to include his programming the computer appropriately, in addition to the cognitive event of his pondering that question.

To formalize this, recall a "worldine" is a set of the values of all variable in the universe across all time. Let U refer to a set of worldlines with the following properties:

  i) All $u \in U$ are consistent with the laws of physics;
 ii) The scientist exists throughout the interval $[t_1, t_2]$, and the system S exists at t_3;
iii) At t_1 the scientist considers some question q of the form "Does $S(t_3) = L$?", in light of some information he has concerning the worldline;
 iv) At t_2 the scientist provides the (binary) answer to q he believes to be correct.[2]

The scientist's claim is that for any question q of the form "Does $S(t_2) = L$?", for all u in the set U such that at t_1 the scientist ponders q, the laws of physics imply that the scientist provides the correct answer to q at t_2. (Any prior knowledge that the scientist relies on to make this claim is embodied in the precise choice of the set U.)

The value $S(t_3)$ is a function of the actual worldline of the entire universe, $u \in U$, which I will write as $\Gamma(u)$. Similarly, what the scientist ponders at t_1 is a function of the state of his brain at t_1, and the "information he has concerning the worldline" is a restriction on u. Therefore together, what he ponders at t_1 and the associated information concerning u are a function of u. Write that function as $X(u)$. Finally, the binary answer the scientist provides at t_2 is a function of the state of his brain at t_2, and therefore a function of u. Write that function as $Y(u)$.

So the predicting scientist is just a pair of functions (X, Y), both with the domain U defined above, where Y has the range $\{-1, 1\}$. The scientist can indeed predict $S(t_3)$ if for the space U defined above, for any $\gamma \in \Gamma(U)$, there is some associated X value x such that, no matter what precise worldline $u \in U$ we are in, due to the laws of physics, if $X(u) = x$ then the associated $Y(u)$ equals 1 iff $\Gamma(u) = \gamma$.

---

[2] This means in particular that the scientist does not lie when giving his answer, and also does not believe he was distracted from the question during $[t_1, t_2]$.

In particular, if our device is a computer programmed by a human to predict the value of some physical variable, then $X(u)$ of the computer specifies a particular one of the possible values of that variable (in addition to other information like what simulation to run, the information the scientist has about the current value of the physical variable, etc.). Our hope is that the computer's conclusion bit $Y(u)$ correctly answers whether the variable has that value specified in how the computer is set up.

As demonstrated in [24], in all instances of a scientist successfully doing observation, prediction recollection, or control, there are a pair of functions X, Y related as in Ex. 1. To formalize that relationship and analyze it, we need to introduce some notation. I will take the set of binary numbers $\mathbb{B}$ to equal $\{-1, 1\}$, so that logical negation is indicated by the minus sign. For any function Γ with domain U, I will write the image of U under Γ as $\Gamma(U)$. For any function Γ with domain U that I will consider, I implicitly assume that $\Gamma(U)$ contains at least two distinct elements. For any (potentially infinite) set R, $|R|$ is the cardinality of R. Given a function Γ with domain U, I say that the partition **induced** by Γ is the family of subsets $\{\Gamma^{-1}(\gamma) : \gamma \in \Gamma(U)\}$, and write it as $\overline{\Gamma}$. Intuitively, it is the family of subsets of U each of which consists of all elements having the same image under Γ. I say that two functions A and B over U are **(partition) equivalent** iff $\overline{A} = \overline{B}$. Equivalent functions can differ only in the labels of the elements of their codomains. Functions that are not (partition) equivalent are said to be **inequivalent**. Recall that a partition A over a space U is a **refinement** of a partition B over U iff every $a \in A$ is a subset of some $b \in B$. Similarly, for any $R \subset U$ and function A, "R refines A" (or "A is refined by R") iff R is a subset of some element of $\overline{A}$.

Definition 1. *An* **(inference) device** *over a set U is a pair of functions (X, Y), both with domain U. Y is called the* **conclusion** *function of the device, and is surjective onto $\mathbb{B}$. X is called the* **setup** *function of the device.*

Unless specified otherwise, a device written as "C_i" for any integer i is implicitly presumed to have domain U, with setup function X_i and conclusion function Y_i (and similarly for no subscript). Similarly, unless specified otherwise, expressions like "$\min_{x_i}$" mean $\min_{x_i \in X_i(U)}$, and any function on U implicitly has at least two distinct values.

Given some function Γ with domain U and some $\gamma \in \Gamma(U)$, we are interested in setting up a device so that it is assured of correctly answering whether $\Gamma(u) = \gamma$ for the actual universe u. Loosely speaking, this is formalized with the condition that $Y(u) = 1$ iff $\Gamma(u) = \gamma$ for all u that are consistent with some associated setup value of the device, i.e., such that $X(u) = x$.

Note that this desired condition relating X, Y and Γ can hold even if $X(u) = x$ doesn't fix a unique value for $Y(u)$. Such non-uniqueness is typical when the device is being used for observation. Setting up a device to observe a variable outside of that device restricts the set of possible universes; only those u are allowed that are consistent with the observation device being set up that way to make the desired observation. But typically just setting up an observation device to observe what value a variable has doesn't uniquely fix the value of that variable; hence the possibility of multiple $Y(u)$'s.

To formalize this with minimal notation, I will use the following shorthand:

Definition 2. *Let A be a set having at least two elements. A* **probe** *of A is a mapping from A onto $\mathbb{B}$ that equals 1 for one and only one argument $a \in A$.*

So a probe of A is a function that picks out a single one of A's possible values. As an example, if A is countable, a probe of A is a Kronecker delta function whose second argument is fixed, and with its image value 0 replaced by -1. Note though there is no restriction on the countability of A in Def. 2. I denote the set of all probes of A by $\pi(A)$.

I can now formalize inference:

Definition 3. *A device C (**weakly**) **infers** a function Γ over U iff $\forall f \in \pi(\Gamma)$, $\exists x \in X(U)$ such that $\forall u \in U, X(u) = x \Rightarrow Y(u) = f(\Gamma(u))$.*

Recall our stipulation that all functions over U take on at least two values, and so in particular Γ must. Therefore $\pi(\Gamma)$ is non-empty.

I will write $C > \Gamma$ if C infers Γ. Expanding the shorthand notation, $C > \Gamma$ means that for all $\gamma \in \Gamma(U)$, $\exists x \in X(U)$ with the following property: $\forall u \in U : X(u) = x$, it must be that $Y(u) = f_\gamma(\Gamma(u))$, where $f_\gamma : \Gamma(U) \to \mathbb{B}$ is the probe of Γ's range that equals 1 iff $\Gamma(u) = \gamma$. I say that a device C infers a set of functions if it infers every function in that set. In general inference among devices is non-transitive.

In the sequel I sometimes consider situations involving multiple inference devices, $(X_1, Y_1), (X_2, Y_2), \ldots$, with associated domains $U_1, U_2, \ldots$. For example, I will consider scenarios where scientists try to observe one another. In such situations, when referring to "U", I implicitly mean $\cap_i U_i$, implicitly restrict the domain of all X_i, Y_i to U, and implicitly assume that the codomain of each such restricted Y_i is binary.

As discussed in App. B of [24], the definition of weak inference is very unrestrictive. For example, a device C is 'given credit' for correctly answering probe f_γ if there is *any* $x \in X(U)$ such that $X(u) = x \Rightarrow Y(u) = f(\Gamma(u))$. In particular, C is given credit even if the binary question associated with x is not whether $\Gamma(u) = \gamma$, but some other question. In essence, the device receives credit even if it gets the right answer by accident.

Having such an unrestrictive definition of inference means that the definition applies very broadly. Accordingly, any impossibility results based on that definition apply very broadly. (Adding extra structure to a definition cannot negate impossibility results based on that definition; see App. B. of [24].) Arguably, the definition of weak inference does not fully capture the physical processes of prediction, observation, control or recollection. (In particular, nothing is specified about how the scientist comes to ponder the question, i.e., about how the value $\gamma \in \Gamma(U)$ is mapped to a value x and u is then made to obey $X(u) = x$.) The important point is that the structure presented in these examples is always found in real-world instances of the associated physical processes. Whether there is additional structure that "should" be assumed is not relevant; the structure that is assumed in the examples is sufficient to establish our formal results below.

Nonetheless, one might want to investigate devices that have more structure than inference devices, to more precisely capture the phenomena of observation, prediction, control and memory. Sec. 9 and App. C of [24] contain discussions of some such strengthened types of devices. The relationship among these types of device is analogous to the relationship between the machines comprising the Chomsky hierarchy.

3   Impossibility Results Concerning Inference Devices

The number of functions X over any U and the number of functions Γ over U both equal the number of partitions of U. Accordingly, the set of all inference devices over

U is at least as big as the set of all functions over U. Despite this relative freedom in specifying inference devices compared to specifying functions, there are limitations in how well we can design an inference device to infer functions.

3.1 The First Demon Theorem

The first such limitation is established in [24]:

Proposition 1. *Let $\{\Gamma_i\}$ be a set of functions with domain U and $R \subsetneq U$.*
i) *If $\forall i$, $|\Gamma_i(R)| \geq 2$, then there is a device over U that infers $\{\Gamma_i\}$.*
ii) *For any device C, there is a function that C does not infer that is equivalent to Y.*

The proof of Prop. 1(ii) is very simple; basically just a Cretan liar's paradox. However Prop. 1(ii) has far-ranging implications. For example, when applied to prediction (cf. Ex. 1), Prop. 1(ii) means that Laplace was wrong: even if the universe were a giant clock, he could not reliably predict the universe's future state before it occurs. More precisely, for all $\{\Gamma_i\}$ as in Prop. 1(i), Laplace could build a demon C that can infer $\{\Gamma_i\}$. However Prop. 1(ii) tells us that there must exist a Γ that any such single demon cannot infer. This is true even if all $u \in U$ comprise only Laplace together with a single particle external to him with dynamics given by a readily solved differential equation.[3]

One might suppose that Laplace could circumvent this restriction by simply constructing a different demon to infer that confounding Γ. Continuing in this way, one might suppose, he could construct a set of demons that, among them, could infer any function. However this is not possible. To see this, have the device C in Prop. 1 be the combination of Laplace with all of his demon-constructing tools. In this case Prop. 1(ii) says that Laplace is limited in what kinds of demons he can construct; there is always a Γ such that for some probe of Γ, f_r, there is no state Laplace can be in which results in his constructing a demon whose conclusion bit assuredly equals the value of f_r. For that f_r, whatever demon-building state of mind Laplace has at present, the resultant demon he builds in the future is potentially wrong in its prediction / observation of f_r.

This result is referred to as the "first (Laplace's) demon theorem". Viewed in more modern terms, the first demon theorem means that regardless of noise levels and the dimensions and other characteristics of the underlying attractors of the physical dynamics of various systems, there cannot be a computer running a time-series prediction algorithm that is always correct in its prediction of the future state of such systems.

Note that time does not appear in Ex. 1's model of a prediction system. So in particular in Ex. 1 we could have $t_3 < t_2$ — so that the time when the scientist provides the prediction is *after* the event they are predicting — and the first demon theorem still holds. This proves that any device used for memory (recollection) must be fallible.

---

[3] Similar conclusions have been reached previously [17, 19]. However in addition to concering observation, control, or recollection, that earlier work is quite informal. Furthermore, it unknowingly disputes well-established results in engineering. For example, the claim in [17] that "a prediction concerning the narrator's future ... cannot ... account for the effect of the narrator's learning that prediction" is refuted by adaptive control theory and Bellman's equations. Similarly, those with training in computer science will recognize statements (A3), (A4), and the notion of "structurally identical predictors" in [19] as formally meaningless.

Moreover, consider the variant of Ex. 1 where the scientist programs a computer to do the prediction. In this variant, the program that is input to the prediction computer could even contain the value that we want the scientist to predict. In other words, the program to the computer to explicitly specify the correct answer (!). The first demon theorem would still mean that the conclusion that the scientist using the computer comes to after reading the computer's output cannot be guaranteed to be correct.

This is all true even if the computer / scientist has super-Turing capability, and does not derive from chaotic dynamics, or physical limitations like the speed of light, or the uncertainty principle.[4] Indeed, when applied to an observation apparatus, the first theorem is a sort of non-quantum mechanical "uncertainty principle", establishing that there is no general-purpose, infallible observation device. (See also Prop. 3 below.)

To illustrate the first demon theorem in more detail, consider the scenario where C is a computer making a prediction at time t about the state of the (deterministic, classical) universe at $t' > t$. Let G be the set of all time-t states of the universe in which C's output display is $+1$. The laws of physics can be used to evolve G forward to time t'. Label that set of states of the universe given by evolving G from t to t' as H. Let Γ be the binary-valued question, "does the state of the universe at t' lie outside of H?".

By the first demon theorem, there is no information concerning H that can be programmed into C at some time $t^- < t$ that guarantees that the resultant prediction that C makes at t is a correct answer to that question. This is true no matter what t^- is, i.e., no matter how much time C has to run that program before making its answer at time t. It is also true no matter how much time there is between t' and t. It is even true if the program with which C is initialized explicitly gives the correct answer to the question.

Similar results hold if $t' < t$. In particular, they hold if C is an observation device that we wish to configure at $t^- < t' < t$ so that at time t it correctly completes an observation process saying whether the universe was outside of H at time t'. See [23] for further discussion of these points.

These limitations imposed by the first demon theorem concern whether a device infers an *arbitrary* set of functions. However often we are instead interested in whether a given device infers some specified subset of all functions. Prop. 1(i) addresses that situation. In particular, given our assumption that any function over U must contain at least two values in its range, it immediately implies the following:

Corollary 1. *Consider a set U.*

i) *Let $\{\Gamma_i\}$ be a set of functions with domain U and $R \subset U$. If $\forall i$, $\Gamma_i(U) = \Gamma_i(R)$, then there is a device that infers $\{\Gamma_i\}$.*

ii) *For any function Γ with domain U there is a device that infers Γ.*

Another implication of Prop. 1(i) is the following:

Corollary 2. *Let $C = (X, Y)$ be a device over U where X refines Y. Then $|X(U)| > 2$ iff there is a function that C infers.*

---

[4] In this breadth, the first demon theorem is an extension of the results in [18, 20], which focused on scenarios where the computer cannot have super-Turing ability, chaotic dynamics is allowed, etc., i.e., which rely on assumptions concerning the nature of the physical universe.

3.2   The Second Demon Theorem and Stochastic Inference

The first demon theorem tells us that any inference device C can be thwarted by an associated function. However it does not forbid the possibility of some "independent" second device that can infer that function that thwarts C. To analyze issues of this sort, and more generally to analyze the inference relationships within sets of multiple functions and multiple devices, it is useful to make the following definition:

Definition 4. *Two devices* (X_1, Y_1) *and* (X_2, Y_2) *are* **(setup) distinguishable** *iff* $\forall x_1, x_2$, $\exists u \in U$ *such that* $X_1(u) = x_1, X_2(u) = x_2$.

No device is distinguishable from itself. Distinguishability is non-transitive in general. Setup distinguishability is a stronger condition than having the setup functions not be equivalent. (If two devices are distinguishable then their setup functions are inequivalent, but not necessarily vice-versa.)

Having two devices be distinguishable means that no matter how the first device is set up, it is always possible to set up the second one in an arbitrary fashion; the setting up of the first device does not preclude any options for setting up the second one. Intuitively, if two devices are not distinguishable, then the setup function of one of the devices is partially "controlled" by the setup function of the other one. In such a situation, they are not two fully separate, independent devices.

By choosing the negation probe $f(y \in \mathbb{B}) = -y$ we see that no device can weakly infer itself. In addition, given devices C_1 and C_2, I say $C_1 > C_2$ to mean $C_1 > Y_2$, i.e., C_1 infers C_2 if it can infer what C_2 will conclude.

Proposition 2. *No two distinguishable devices* (X, Y) *and* (X', Y') *can weakly infer each other. If* (X', Y') *can weakly infer* Y, *then* (X, Y) *can infer neither of the two binary-valued functions equivalent to* Y'.

Proof. Let C_1 and C_2 be the two devices. Since Y for any inference device is surjective, $Y_2(U) = \mathbb{B}$, and therefore there are two probes of $Y_2(U)$. Since by hypothesis C_1 weakly infers C_2, using the identity probe $f(y \in \mathbb{B}) = y$ establishes that $\exists\ x_1$ s.t. $X_1(u) = x_1 \Rightarrow Y_1(u) = Y_2$. Similarly, since C_2 weakly infers C_1, using the negation probe $f(y) = -y$ establishes that $\exists\ x_2$ s.t. $X_2(u) = x_2 \Rightarrow Y_2(u) = -Y_1(u)$. Finally, by the hypothesis of setup distinguishability, $\exists\ u^* \in U$ s.t. $X_1(u^*) = x_1, X_2(u^*) = x_2$. Combining, we get the contradiction $Y_1(u^*) = Y_2(u^*) = -Y_1(u^*)$. $\square$

This is the second (Laplace's) demon theorem. Note the distinguishability condition is crucial to it; mutual weak inference can occur between non-distinguishable devices.

The first demon theorem is a worst-case result, relying on a function Γ specially matched to C. It says that a devil's advocate can always create a function that confounds Simon de Laplace. In contrast, the second one is universal, in that it has no role for a devil's advocate. The second demon theorem says that no matter how clever I am in designing a pair of inference devices, so long as they are distinguishable from each another, one of them must thwart the other, providing a function that the other device cannot infer. So in particular, Laplace cannot construct two distinguishable demons that can infer each other. Alternatively, the second demon theorem says that while Laplace might be able predict the future actions of any other person, *if that person is independent of Laplace*, they will not be able to predict the future actions of Laplace. Viewed differently, the second demon theorem establishes that a whole class of functions cannot be inferred by C (namely the conclusion functions of devices that are distinguishable from C and also can infer C). More generally, let $\mathscr{S}$ be a set of devices, all of which are distinguishable from one another. Then the second demon theorem says that there can be at most one device in $\mathscr{C}$ that can infer all other devices in $\mathscr{S}$.

Combining the two demon theorems establishes the following:

Corollary 3. *Consider a pair of devices* $C = (X, Y)$ *and* $C' = (X', Y')$ *that are both distinguishable from one another and whose conclusion functions are inequivalent. Say that* C' *weakly infers* C. *Then there are at least three inequivalent surjective binary functions* Γ *that* C *does not infer.*

Proof. First, since $C' = (X', Y')$ can infer C, we know from Prop. 2 that there are two distinct binary, surjective functions that cannot be inferred by C, both of which are equivalent to Y'. Next, from Prop. 1 we know that C cannot infer some function Γ that is equivalent to Y. Since Y and Y' are not equivalent, Γ is also not equivalent to Y'. So Γ is a third inequivalent surjective binary function that C cannot infer.   □

Corollary 3 is a "conservation of non-inference" law. It means that if any device from a set of distinguishable devices with inequivalent conclusion functions can infer all the others, then each of those others must fail to infer at least three inequivalent functions.

There are several ways to extend the analysis to incorporate probability measure P over U, so that functions over U become random variables. One is as follows.

Definition 5. *Let* $P(u \in U)$ *be a probability measure,* Γ *a function with domain U and finite range, and* $\epsilon \in [-1.0, 1.0]$. *Then we say that a device* (X, Y) *(weakly) infers* Γ **with (covariance) accuracy** ϵ *iff*

$$\frac{\sum_{f \in \pi(\Gamma)} max_x [\mathbb{E}_P(Y f(\Gamma) \mid x)]}{|(\Gamma(U)|} = \epsilon.$$

As an example, if P is nowhere 0 and C weakly infers Γ, then C infers Γ with accuracy 1.0. We also define two device (X_1, Y_1) and (X_2, Y_2) to be **stochastically distinguishable** under P if the X_1 and X_2 are statistically independent under P.

The math of stochastic inference seems to be intrinsically more challenging than the math of deterministic inference; it has only been possible to analyze special cases so far. To present the simplest of these special cases, define H as the four-dimensional hypercube $\{0, 1\}^4$, $k(z)$ as the map taking any $z = (z_1, z_2, z_3, z_4) \in H$ to $z_1 + z_4 - z_2 - z_3$. Also define $m(z)$ as the map taking any $z \in H$ to $(z_2 - z_4)$, and $n(z)$ as the map taking any $z \in H$ to $(z_3 - z_4)$. The following is derived in [24]:

Proposition 3. *Let P be a probability measure over U, and C_1 and C_2 two devices that are stochastically distinguishable, where $X_1(U) = X_2(U) = \mathbb{B}$. Define $P(X_1 = -1) \equiv \alpha$ and $P(X_2 = -1) \equiv \beta$. Say that C_1 infers C_2 with accuracy ϵ_1, while C_2 infers C_2 with accuracy ϵ_2. Then*

$$\epsilon_1 \epsilon_2 \leq max_{z \in H} | \alpha\beta[k(z)]^2 + \alpha k(z)m(z) + \beta k(z)n(z) + m(z)n(z) |.$$

In particular, if $\alpha = \beta = 1/2$, then

$$\epsilon_1 \epsilon_2 \leq \frac{max_{z \in H} \mid (z_1 - z_4)^2 - (z_2 - z_3)^2 \mid}{4}$$
$$= 1/4.$$

Note that unlike the uncertainty principle in quantum mechanics, the bound on the right-hand side of Prop. (3) does not involve a unit of action like Planck's constant. This is because covariance accuracy does not involve a measure of degree of error. Future work involves trying to extend the definition to include such a measure.

References

1. Aaronson, S.: quant-ph/0502072 (2005)
2. Bennett, C.H.: International Journal of Theoretical Physics 21 (1982)
3. Bennett, C.H.: IBM Journal of Research and Development 17, 525–532 (1973)
4. Bennett, C.H.: Emerging Syntheses in Science. In: Pines, D. (ed.), p. 297. Addison Wesley, Reading (1987)
5. Binder, P.: Theories of almost everything. Nature 455, 884–885 (2008)
6. Collins, G.P.: Impossible inferences, pp. 12–21. Scientific American, Singapore (2009)
7. Cover, T., Thomas, J.: Elements of information theory. Wiley-Interscience, New York (1991)
8. Feynman, R.: Foundations of Physics 16, 507 (1986)
9. Fredkin, E., Toffoli, T.: International Journal of Theoretical Physics 21, 219 (1982)
10. Hopcroft, J.E., Ullman, J.D.: Introduction to automata theory, languages and computation. Addison Wesley, Reading (1979)
11. Landauer, R.: IBM Journal of Research and Development 5, 183 (1961)
12. Landauer, R.: Nature 335, 779–784 (1988)
13. Laplace, S.P.: Philosophical essays on probabilities, Dover (1985), Originally in 1825; Translated by Emory, F.L., Truscott, F.W.
14. Lloyd, S.: Nature 406, 1047 (2000)
15. Lloyd, S.: Programming the universe. Random House (2006)
16. Mackay, D.J.C.: Information theory, inference, and learning algorithms. Cambridge University Press, Cambridge (2003)
17. MacKay, D.M.: On the logical indeterminacy of a free choice. Mind, New Series 69(273), 31–40 (1960)
18. Moore, C.: Physical Review Letters 64, 2354–2357 (1990)
19. Popper, K.: The impossibility of self-prediction, The Open Universe: From the Postscript to the Logic of Scientific Discovery, p. 68. Routledge, New York (1988)
20. Pour-El, M., Richards, I.: International Journal of Theoretical Physics 21, 553 (1982)
21. Wolfram, S.: A new kind of science. Wolfram Media, Inc. (2002)
22. Wolpert, D.: Memory systems, computation, and the second law of thermodynamics. International Journal of Theoretical Physics 31, 743–785 (1992); Revised version available from author
23. Wolpert, D.: Computational capabilities of physical systems. Physical Review E 65, 016128 (2001)
24. Wolpert, D.H.: Physical limits of inference. Physica D 237, 1257–1281 (2008), More recent version at http://arxiv.org/abs/0708.1362
25. Zurek, W.: Nature 341, 119 (1984)

Author Index